Return To
ENVIRONMENTAL REFERENCE CENTER
Bechtel Corporation
San Francisco

HANDBOOK OF INDUSTRIAL NOISE CONTROL

Contributors

E. E. ALLEN *Fisher Controls Company, Marshalltown, Iowa*

MALCOLM J. CROCKER *Professor of Mechanical Engineering, Ray W. Herrick Laboratories, School of Mechanical Engineering, Purdue University, West Lafayette, Indiana*

L. L. FAULKNER *Editor-in-Chief, Department of Mechanical Engineering, The Ohio State University, Columbus, Ohio 43017*

J. BARRIE GRAHAM *Director of Research, Buffalo Forge Company, Buffalo, New York*

JAMES F. HAMILTON *Professor of Mechanical Engineering, Ray W. Herrick Laboratories, School of Mechanical Engineering, Purdue University, West Lafayette, Indiana*

DAVID I. G. JONES *B.Sc., Ph.D., Materials Research Engineer, Air Force Materials Laboratory, Wright-Patterson AFB, Ohio 45433*

GEORGE KAMPERMAN *Kamperman and Associates, Downers Grove, Illinois*

S. M. MARCO *Past Chairman, Department of Mechanical Engineering, The Ohio State University, Columbus, Ohio 43017*

ABBOTT A. PUTNAM *Research Engineer, Battelle, Columbus Laboratories, 505 King Ave., Columbus, Ohio*

CHARLES W. RODMAN *Staff Engineer, The Applied Dynamics and Acoustics Section, Battelle Memorial Institute, Columbus Laboratories, Columbus, Ohio*

JAMES E. SHAHAN *Senior Consultant, Kamperman Associates, Downers Grove, Illinois*

WILLIAM SIEKMAN *Technical Counsel, Hixton, Wisconsin*

J. W. SULLIVAN *Contributing Editor, Ray W. Herrick Laboratories, School of Mechanical Engineering, Purdue University, West Lafayette, Indiana*

HANDBOOK OF INDUSTRIAL NOISE CONTROL

Edited by L. L. FAULKNER, Ph.D.
Associate Professor, Department of Mechanical Engineering
The Ohio State University

INDUSTRIAL PRESS INC.
200 Madison Avenue, New York, N.Y. 10016

Library of Congress Cataloging in Publication Data

Main entry under title:
Handbook of industrial noise control.
   Includes bibliographical references and index.
   1. Noise control—Handbooks, manuals, etc.
I. Faulkner, Lynn L., 1941–
TD892.H33   620.2'3   75-41315
ISBN 0-8311-1110-0

HANDBOOK OF INDUSTRIAL NOISE CONTROL

Copyright © 1976 by Industrial Press Inc., New York, N.Y. Printed in the United States of America. All rights reserved. This book, or parts thereof, may not be reproduced in any form without permission of the publishers.

Contents

List of Contributors	ii
Preface	vi
1. Fundamentals of Sound and Vibration *Joseph W. Sullivan and L. L. Faulkner*	1
2. Noise Measurements and Instrumentation *Lynn L. Faulkner*	67
3. Standards for Sound Measurement *Charles W. Rodman*	98
4. Criteria for Machinery Noise *S. M. Marco*	113
5. Acoustic Absorption and Transmission-Loss Materials *William Siekman and Joseph W. Sullivan*	146
6. Damping Materials for Vibration and Sound Control *David I. G. Jones*	207
7. Vibration Control for Noise Reduction *James F. Hamilton and Malcolm J. Crocker*	262
8. Machine Element Noise *James E. Shahan and George Kamperman*	329
9. Fan and Flow System Noise *J. Barrie Graham and L. L. Faulkner*	386
10. Combustion and Furnace Noise *Abbott A. Putnam*	439
11. Fluid Piping System Noise *E. E. Allen*	473
12. Industrial Noise Control Studies *Lynn L. Faulkner*	506
References	556
Appendix	566
Index	578

Preface

This book was written for the industrial noise control engineer who has already completed basic engineering and mathematics courses. The authors have fully presented the fundamentals of noise control applicable for new design or for the retrofit of existing designs.

The field of engineering acoustics and noise control assimilates knowledge and data from many disciplines, and presented here are practical engineering principles that demonstrate how functional and economically feasible noise-reduction solutions can be achieved through proper design methods; i.e., the consideration of machine elements that produce noise; industrial systems such as hydraulic or process air supplies; and the factory space within which a noise source may be located.

The text may be used for study by engineers who have not had specific noise-control training and, to this end, many examples—complete with step-by-step calculations—are presented. This approach is particularly helpful in making the transition from the theory and science of noise control to engineering applications. The assumptions and approach to the solutions of practical noise-control problems are discussed by the authors to provide a means of developing confidence in the reader in applying the text material.

The first chapter is devoted to the fundamentals of sound and vibration appropriate for industrial noise control. The concepts presented are those necessary for understanding the remaining chapters, while presenting analytical concepts along with a physical understanding of the nature of sound propagation and the principles of vibration. Those readers who have not had exposure to noise control either through formal education or past experience, are urged to master the concepts in Chapter 1 before attempting to understand the remainder of the book.

Chapter 2 presents the fundamentals of the measurement of sound necessary for effective noise control. Space does not permit, however, an exhaustive survey of all instrumentation presently available for the analysis of sound and vibration phenomena. Here the intent is to present the essential concepts of sound measurement and analysis; not to review all instruments currently available. Manufacturers and suppliers are the best sources for the latest information and details of noise measurement instrumentation.

Chapter 3 lists the standards and recommended practices of importance in industrial noise measurement.

Chapters 4 through 11 treat specific topics pertinent to industrial noise control. In each chapter the fundamental concepts are illustrated with case studies and examples from engineering practice. Extensive calculations are given in the text. Chapter 12 contains actual case histories from industrial practice, which are intended to illustrate how the material in previous chapters is utilized in

PREFACE

practical solutions to certain noise-control situations. These case histories, though rather lengthy, will afford the reader valuable insight in the use of this book.

Sincere appreciation is expressed to Mr. Graham Garratt, formerly of Industrial Press Inc., for his help and direction during the conception of this book and his guidance in shaping the organization of the material. The invaluable assistance of Mr. Karl Hans Moltrecht, Technical Editor for Industrial Press Inc., is gratefully acknowledged; his patience, thoughtful criticism, and constructive suggestions are appreciated. Professor Joseph K. Sullivan of Purdue University receives special recognition for his review of the manuscript, his constructive comments, and his contributions in Chapters 1 and 5. Gratitude is expressed to all contributing authors, without whose help and effort this book could not have been written.

Lynn L. Faulkner, Editor-in-Chief
The Ohio State University,
Department of Mechanical Engineering.

1

Fundamentals of Sound and Vibration

JOSEPH W. SULLIVAN
and
L. L. FAULKNER

Introduction

The subject of noise control is not an easy one to master. Part of the difficulty lies in its interdisciplinary nature: it borrows generously from the fields of mathematics, dynamics, vibrations, fluid mechanics, thermodynamics, electronics, psychoacoustics, materials science, and, of course, engineering acoustics. Nomenclature and jargon are other obstacles to be overcome and noise-control engineering, perhaps more than any other field, has an abundance of new terms which must be mastered when learning the subject. Also, the seemingly insidious nature of the way sound is generated and propagated often defies intuitional noise-reduction solutions which, in other areas of engineering, are based upon experience. Such solutions are embarrassing at best, and are sometimes catastrophically costly. The cost in materials and man-hours in arriving at more satisfactory solutions demands that "cut-and-try" approaches be avoided.

An understanding of the fundamentals of sound generation and propagation, then, is essential to the engineer responsible for initiating and carrying out noise-control programs. In this beginning chapter, some fundamentals of sound and vibration, pertinent to the subject, will be covered. This will be attempted, more or less in a qualitative manner, because a physical understanding of the material is more important at this stage than a mathematical treatment. The chapter is divided into two major sections, the first is concerned with the generation of sound; the propagation of sound waves through air; how sound is reflected at boundaries; how the ear responds to and is affected by sound; and the basics of controlling unwanted sound, or noise. The second part deals with mechanical vibrations in terms of the basic elements necessary for vibratory motion, what causes such motion, and the basics of reducing the motion as related to controlling noise.

Physical Units

It is very unfortunate that a uniform system of physical units is impossible to use throughout this book. The science and engineering application of acoustics and noise control have evolved in rather widely diversified laboratories and engineering offices; small but important contributions were usually made by individuals or independent groups who used the particular units common to their own branch of science or engineering and to their communities in which the results were applied. Thus, important banks of data, mathematical coefficients and, to some degree, engineering concepts, are now available only in a given set of units, often quite different from each other. While commendable in principle, a unified system of physical units does not, in fact, exist at present in acoustics. As this book is intended to be representative of the current state of the art and science of acoustics, it has been necessary to use those units in which the useful information appears. To do otherwise would create only a greater hardship in applying the information contained in this book. Of course, it is possible to convert these units when required, and conversion tables are included for this purpose in the appendix.

FUNDAMENTALS OF SOUND

Sound, in the physiological sense, is the result of pressure variations in the air which act on the surface of the eardrum. By processes which will be described in the following discussion, the ear converts these pressure variations into electrical signals which are then interpreted by the brain as "sound." The pressure variations associated with sound in air are quite small. For example, an intense sound such as that caused by a pneumatic chipper would have an associated pressure variation of only $\pm.001$ psi (6.894 K/m^2) above and below the local mean atmospheric pressure of approximately 14.7 psi. For normal conversational speech, the pressure variation would be on the order of $\pm.00001$ psi ($.06894$ K/m^2)!

The fundamentals of sound in relation to noise control are the generation of sound by a source and the properties of propagation of this sound from the source to the receiver, which for noise control, is a human being. All noise control techniques are based on the alteration of the sound generation mechanism of a source or the alteration of the sound pressure as it propagates in the atmosphere from the source to the receiver.

PROPAGATION OF SOUND WAVES

In order for sound to reach the ear, it must travel or propagate, some distance through the air, from the point where it originated. Since air has both mass and elasticity, propagation can best be thought of in terms of momentum and elastic restoring forces. It is also very helpful to imagine the air medium sliced into very thin layers of particles; each of these layers being thick compared with the average molecular spacing. Thus, each particle contains a large number of molecules all of which are in random motion such that the average of all the particle velocities is zero. This is the state of the medium when there is no imposed disturbance. Consider now a rigid machinery panel that is moving back and forth (vibrating) from its static equilibrium position, as shown in Fig. 1-1. As the panel moves to the right, the particles of air against the panel must also move to the right. Since the particles have acquired a net velocity they have momentum and thus are capable of applying a force, or pressure, on neighboring particles to the right. Having done this, the neighboring particles are also set in motion which, in turn, transfers their momentum to their neighbors, and so on. Seen here is a disturbance, or wave, traveling to the right. This can be called the "positive pressure" wave, or more properly, the *condensative* wave. While this wave is progressing, the vibrating panel has attained its maximum displacement to the right and has reversed direction so that it is now traveling to the left. Again, the particles of air next to the panel remain in contact with the panel and must faithfully move to the left with momentum in that direction in synchrony with the panel. Momentum again is transferred to the neighboring particles by the imposed motion to the left. This process is repeated with subsequent neighboring particles; each particle moving, in turn, to the left as the disturbance passes through. This can be called the "negative pressure" wave, or,

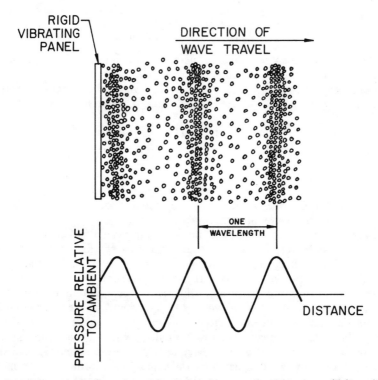

Fig. 1-1. Schematic diagram of particle motion due to a vibrating source which results in a sound wave.

more properly, the *rarefactive* wave. Note that the rarefactive wave travels in the same direction as the condensative wave, away from the source of the disturbance, or to the right as in Fig. 1-1.

The reader should be aware of two different types of motion here: particle motion and wave motion. In the first type each of the particles is moving back and forth about its equilibrium position in oscillatory fashion with momentum transfer. The second type of motion, wave motion, is really a manifestation of the first in that it is a measure of the speed with which *momentum transfer* occurs through the particles. Note that this is the speed at which the disturbance (momentum transfer from particle to particle) propagates through the medium; matter or mass does *not* propagate to give the transfer of energy and the resulting pressure variations associated with sound. Although there is no net mass transfer, the sound wave does have energy by virtue of the pressure variation that occurs. The rate at which this momentum transfer occurs is known as the *speed of sound*, or simply, the *sound speed*. The speed of sound is a function of the density and the elasticity of the medium and thus, for a gaseous medium, it will be strongly dependent on temperature. In air the sound speed is given by

$$c = 49.03\sqrt{460 + T_F} \text{ ft/sec} \qquad (1\text{-}1)$$

or

$$c = 25.05\sqrt{273 + T_C} \text{ meter/sec}$$

where c is the sound speed, in feet or meters per second and T_F or T_C is the ambient temperature in degrees Fahrenheit or Celcius. At a temperature of 70 °F, the sound speed in air is 1129 ft/sec (344 m/sec). In water it is about 4,700 ft/sec (1433 m/sec); in wood, 13,000 ft/sec (3962 m/sec); and in steel it is 15,500 ft/sec (5029 m/sec).

The mention of the words "waves" and "water" invariably conjures up visions of ripples on the surface of water. Such is not the case here! In ripple wave motion, the particle motion is perpendicular (or nearly so) to the surface of the water or to the direction in which the waves are traveling, as witnessed by the "bobbing" of a cork. In acoustic waves, the particle motion is always *parallel* to the wave travel. The former is known as *transverse* wave motion and the latter as *longitudinal* wave motion.

To return now to the example of our vibrating panel, it is assumed that both condensative and rarefactive waves travel at the same speed; that is, at the speed of sound. This will be true for the magnitude of sound pressure fluctuations normally encountered in industry. If the wave motion were to be stopped at some instant in time and the pressure fluctuation were plotted as a function of distance something like that shown in Fig. 1-1 would be seen. Each positive pressure fluctuation is a condensative wave and each negative one a rarefactive wave. The distance between successive similar points on the combined waveform, such as the distance between successive pressure maximum, is called the *wavelength* of the wave, designated by the symbol λ. If now the wave motion were to be viewed through a narrow slit, then the number of successive similar points on the wave which passed by the slit in one second would be called the

Ch. 1   FUNDAMENTALS OF SOUND AND VIBRATION

frequency of the sound wave. For example, if 100 pressure maximums passed by the slit in one second then the frequency would be 100 cycles per second. There is a simple unique relationship between frequency, wavelength, and sound speed and it is given by

$$f = \frac{c}{\lambda} \text{ cycles/sec or Hz}^1 \qquad (1\text{-}2)$$

where:

f = frequency, Hz[1]
c = sound speed, ft/sec or meter/sec
λ = wavelength, feet or meter

This relationship is worth remembering because it is used often in noise-control work, particularly in the form: $\lambda = c/f$. For example, at a frequency of 1000 Hz, the wavelength in air at room temperature is

$$\lambda = \frac{1129}{1000} = 1.13 \simeq 1 \text{ foot at 1000 Hz.} \qquad (1\text{-}3)$$

Thus, at 500 Hz it is roughly 2 feet; at 250 Hz it is 4 feet. At 2000 Hz it is roughly 6 inches; at 4000 Hz it is 3 inches, etc.

The range of audible frequency sound extends from 20 to 20,000 Hz. But for the situations encountered in industry frequencies above 10,000 Hz are seldom of interest; however, there are some instances with ultrasonic equipment when these high frequencies are of concern. At the lower end of the scale, however, we may be concerned with sound (infrasound) below 20 Hz.

The manner in which the sound waves propagate is dependent upon the relationship between the size of the source and the wavelength. In Fig. 1-1, the panel is assumed to be large in relation to the wavelength, in which case a plane wave will be formed; i.e., at any instant the sound pressure throughout a plane, perpendicular to the direction of the wave propagation, is constant. This is called plane wave propagation. If, on the other hand, the size of the panel were small as compared to the wavelength, the wave will spread out in all directions thereby forming a spherical wave; this is called "spherical wave propagation." Since the wavelength has been shown to vary inversely with frequency, the manner of wave propagation, whether planar or spherical, is also a function of the frequency and the size of the source. Thus, for a given size panel, the wave propagation will tend to be spherical, or omnidirectional, at low frequencies and large wavelengths; at high frequencies and short wavelengths it will tend to be planar, or directional.

In practice, plane waves usually occur near a large source, in terms of its physical size; as the distance from this source increases the wave tends to assume a spherical condition, regardless of the size of the source. Plane waves most commonly occur in pipes or ducts, where the wavelength is large compared to the diameter of the pipe or duct cross-section. This condition is given by the following expression, which has been determined experimentally.

---

[1] It is accepted practice, since about 1968, to use Hz (after H. R. Hertz, 1857–1897) to designate cycles per second.

For plane waves to exist in a pipe or a duct:

$$\frac{\lambda}{d} \geq 6$$

where:

λ = wavelength of sound (given by Equation [1-2])
d = diameter of pipe or duct.

Functions Common in Noise Control

In the preceding discussion the mechanism by which an acoustic wave is generated by the motion of a rigid panel was presented. To understand the various types of sound waves which can be generated in this manner, an understanding of the mathematical functions which describe the possible types of motion will now be given. It is not the intent here to pursue a detailed mathematical presentation of functions but rather to give a physical understanding of the types of functions encountered in noise control.

The most frequently encountered functions in noise control are classified as *periodic*, *transient*, or *random* (see Fig. 1-2). A sinusoidal signal (or motion) is a special case of the periodic function which can be described by the simple sinusoidal function $A \sin(2\pi f t)$ where A is the amplitude; f, the frequency in Hz; and t, the time in seconds. A periodic function is one which repeats itself at a definite interval over and over again with definite regularity. Transient functions have definite beginnings and endings; a series of transient functions may be produced by a series of one-time events such as impacts, air blasts, etc. Random signals have neither a definite frequency nor a defined amplitude.

The scope of this text is limited to the treatment of periodic functions which do, however, form a basis for the majority of machinery noise applications. The most commonly used technique by engineers is to identify the frequency content of a periodic signal. This can be accomplished by analytical frequency spectrum techniques; instruments called *spectrum analyzers* automatically perform this operation on electrical signals from transducers. The basis for the analytical analysis, or the analysis performed by spectrum analyzers, is the result of the theory of Fourier[2] analysis which states that any periodic function can be formed from a series of simple sinusoidal functions

$$y(t) = y_{\text{average value}} + \sum_{n=1}^{\infty} A_n \cos(2\pi n f t) + \sum_{n=1}^{\infty} B_n \sin(2\pi n f t) \qquad (1\text{-}4)$$

or in expanded form,

$$y(t) = y_{\text{avg}} + A_1 \cos(2\pi f t) + A_2 \cos(4\pi f t) + \cdots$$
$$+ B_1 \sin(2\pi f t) + B_2 \sin(4\pi f t) + \cdots \qquad (1\text{-}5)$$

where:

y = ordinate, or magnitude of the periodic function
t = time, seconds

---

[2] After J. B. Fourier (1768–1830).

Ch. 1 FUNDAMENTALS OF SOUND AND VIBRATION 7

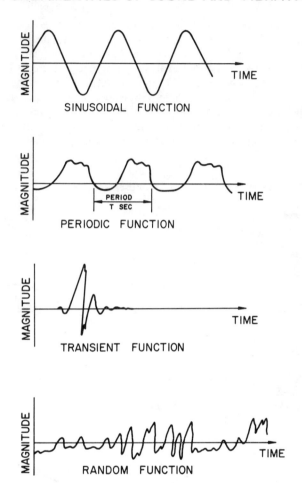

Fig. 1-2. Illustrations of common time functions encountered in noise control.

$y(t)$ = the ordinate, y, as a function of time, or the magnitude of the periodic function with respect to time.

f = frequency of the periodic function, Hz, equal to $1/T$, where T is the period in seconds.

n = integers, 1, 2, 3, . . .

A_n, B_n = coefficients of the sine and cosine terms

Since the frequency of the periodic function is equal to $1/T$ Hz where T is the period in seconds, the frequency, f, may be replaced by $1/T$ in all the above forms.

The concept (which is beyond the scope of this book to prove) is that any periodic function such as force, displacement, velocity, acoustic pressure, etc., can be physically interpreted as an infinite series of sine and cosine terms as illustrated in Fig. 1-3. Note in Equation (1-5) that the frequencies of successive terms are integer multiples of the first or fundamental frequency, f. This concept forms the basis of Fourier analysis (sometimes referred to as harmonic analysis), and is one of the basic concepts for noise analysis in that the frequencies of the so-called higher harmonics are related to some specific periodically occurring event in a machine. In practice the actual sine and cosine functions are not usually obtained, but rather the concept of frequency multiples or harmonics is used to relate specific frequencies to the appropriate sources.

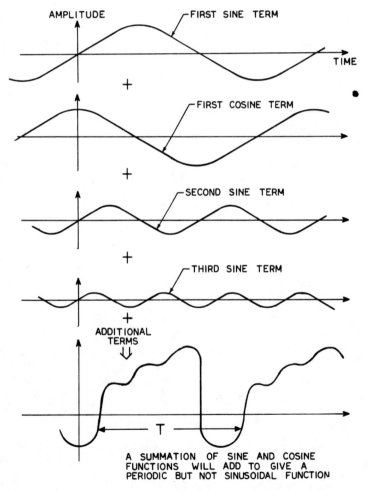

Fig. 1-3. Illustration of the summation of sine and cosine functions which results in a periodic function.

Simple and Complex Sound Waves

The simplest form of sound wave that can be generated is composed of only one frequency and is sinusoidal with time, as shown by the top form of Fig. 1-2. In noise control this is called a *pure tone* because it has only one frequency. Common examples of devices which emit pure tones are tuning forks and organ pipes. It is rare in industry for a process or machine to emit only a single pure tone although there are a number of cases where a single pure tone dominates, such as in furnaces (combustion driven oscillations), material mixers (electromagnetic), and in various electric motors, generators, transformers, or the blade tone of a fan or blower. Acoustic or mechanical resonances may also result in a pure-tone sound. A pure tone can be generated with a rigid vibrating panel by forcing it to move back and forth about an equilibrium position with a displacement given by the trigonometric relation:

$$\xi = A \sin(2\pi ft), \tag{1-6}$$

where:

ξ = displacement from equilibrium, inches
A = maximum displacement from equilibrium, inches
f = frequency of pure tone, Hz
t = time, seconds

One way of generating such *sinusoidal motion* with a maximum displacement A is shown in Fig. 1-4, where a point is rotating on a circle of radius A at a con-

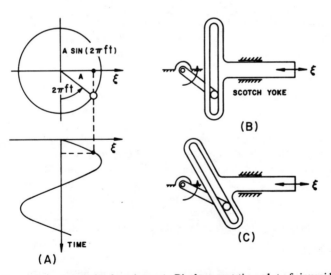

Fig. 1-4. Generation of periodic functions: A. Displacement-time plot of sinusoidal motion generated by scotch-yoke mechanism with perpendicular slot; B. Scotch-yoke mechanism with perpendicular slot; C. Scotch-yoke mechanism with canted slot that gives a periodic but not a sinusoidal motion.

stant angular velocity of f revolutions per second. Starting at the bottom of the circle and rotating counterclockwise, after time, t, the projection of the point on the horizontal axis is given by $A \sin(2\pi ft)$. If now the displacement of this projected point were to be plotted as a function of time, the sinusoidal curve shown in Fig. 1-2 would be obtained. Thus, if the projected point on the horizontal axis were to be attached to the plate, it would be driven with sinusoidal motion and would produce a pure tone. Mechanically, sinusoidal motion can be produced by a scotch-yoke mechanism, and electrically by an electromagnet in which the electric field of the magnet is varied sinusoidally in time. Another name for sinusoidal motion is *simple harmonic motion*.

Let us now consider a more complicated situation. Suppose instead of the straight vertical slide in the scotch-yoke mechanism, a curved or canted slide were substituted, as shown in Fig. 1-4. The motion would still be periodic since it would repeat itself once per revolution. However, the motion would no longer be sinusoidal, but one described by a more complicated trigonometric function. Whatever this function is, the function or the resulting motion can be expressed in the following form, based on the theory of Fourier analysis presented in the preceeding section.

$$\xi = A_1 \sin(2\pi ft) + A_2 \sin(4\pi ft) + A_3 \sin(6\pi ft) + \cdots A_n \sin(2\pi nft) \quad (1\text{-}7)$$

In comparing this equation with Equation (1-6), it is evident that the first term, or *fundamental* waveform, in Equation (1-7) is the same as in Equation (1-6), but that there are *higher harmonics* in Equation (1-7), each one of which has a frequency equal to an integral multiple of the fundamental frequency $2\pi f$. Thus, the nth harmonic of the periodic motion has a frequency equal to $n(2\pi f)$. The amplitudes of the higher harmonics $A_2, A_3, \cdots A_n$, relative to the amplitude of the fundamental or 1st harmonic A_1, will depend on the motion ξ. For example, if the yoke profile of the scotch-yoke mechanism was straight but canted slightly from the vertical, the motion could be described accurately by using only the first few harmonics. On the other hand, if there were a rapid change in the slide profile (such as a "dog-leg") many more harmonics would be needed to describe the motion. Such motion applied to a vibrating plate would, in the first case, produce a sound very much like a pure tone; in the second case, the sound would be quite different. In general, the sharper or more discontinuous the motion producing the sound, the greater would be the number of significant harmonics in the sound wave and the greater would be the frequency range over which they extended. For example, diesel engines are notable for an extremely rapid rise in pressure in the combustion chamber during the combustion process, and the radiated sound from such engines contains hundreds of measurable harmonics extending over the entire audio range.

Random Sound

If a sound wave were to be constructed by closely grouping sine waves of different frequency, but not harmonically related, and further, if each sine wave were to have an amplitude that varied individually with time, a form of *random sound*, as illustrated in Fig. 1-2, would result. Such sound would not have any discernible periodic behavior and the pressure amplitude would have to be deter-

Ch. 1   FUNDAMENTALS OF SOUND AND VIBRATION

mined on a time average. If the time average amplitude of the random sound was approximately constant over a broad range of frequencies, the result would be by definition called a *white noise*.[3] White noise occurs often in nature; e.g., falling water and rustling leaves, and is a by-product of some machinery and machining processes. Of note is the noise of escaping steam or air, and the noise from weaving mills. Random sounds, in general, accompany all machinery and machining processes in combination with single and complex sounds.

Energy in Sound Waves

Since sound is propagated by momentum transfer, there must be energy involved in the process. This is dramatically evident in the destruction caused by shock waves from explosives or sonic booms. On a lesser scale the process is the same: the energy from the vibrating surface (or other source) is transmitted in the propagating wave at the sound speed. At any point in the propagating path, the acoustic energy consists of the sum of the potential and kinetic energies associated with the particle motion at that point. The potential energy arises from the elastic compression of the medium and the kinetic energy from the velocity of the particle mass.

The average energy flowing through a given area perpendicular to the direction of wave propagation is equivalent to the average work on that area. Since work is defined as force times distance, or pressure times area times distance, and since power is defined as the time rate at which work is performed, *sound power* may be defined as follows:

W = Sound Power

= Average of $\left(\dfrac{\text{acoustic pressure} \times \text{acoustic particle displacement}}{\text{time}} \right) \times$ area

= Average of (acoustic pressure $\times$ acoustic particle velocity) $\times$ area

W = Average of $(pu)\, a$

$$W = \overline{pu}\, a \text{ watts} \tag{1-8}$$

where:

p = acoustic pressure, N/m^2, or psf
u = acoustic particle velocity, m/sec, or ft/sec
a = area, m^2, ft^2

The bar over p and u denotes the average of the product.[4] If pressure, p, and particle velocity, u, are constant over the area of interest the average value becomes a time average defined as

$$\overline{pu} = \lim_{T \to \infty} \frac{1}{2T} \int_{-T}^{T} pu\, dt \tag{1-9}$$

---

[3] So called because of the analogy to white light which contains all the visible colors of the spectrum. White noise contains all frequencies.
[4] $\overline{pu}$ or $\langle pu \rangle$ are symbols used in the literature to denote the average of a quantity.

or since it is inconvenient to obtain data for negative time we can say

$$\overline{pu} = \lim_{T \to \infty} \frac{1}{T} \int_0^T pu\, dt \qquad (1\text{-}10)$$

For the special case of sinusoidal functions for p and u, the time average becomes

$$\overline{pu} = \frac{\text{magnitude of } p}{\sqrt{2}} \times \frac{\text{magnitude of } u}{\sqrt{2}} \qquad (1\text{-}11)$$

or

$$\overline{pu} = \frac{|pu|}{2} \qquad (1\text{-}12)$$

The *sound intensity*, an often used measure of sound, is defined as the amount of *sound power* passing through a unit area

I = sound intensity = average of (acoustic pressure $\times$ acoustic particle velocity)

I = average of (pu)

$$I = \overline{pu} \text{ watt/m}^2 \qquad (1\text{-}13)$$

This form for sound intensity is not convenient for computational purposes because particle velocity is not easily measured due to the extremely small magnitudes involved. A form for sound intensity better suited for engineering computations will be presented later in this chapter.

Reflection of Sound Waves

When a sound wave strikes a wall or plane surface, part of the energy is reflected back into the space. This is a common experience and terms like *echo* and *reverberation* are used to describe the phenomenon. However, since sound can easily pass through air it may be interesting to know why the sound is reflected at all. Although a lengthy mathematical treatment is necessary to describe the process in quantitative terms, it is sufficient here to use a mechanical analog familiar to most of us to get at least a qualitative feel for what is occurring.

First, consider the suspended rigid spheres shown in Fig. 1-5; they all are of the same size and mass. If Sphere A were displaced outward and released it would collide with Sphere B with a certain amount of momentum.

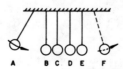

Fig. 1-5. Rigid-sphere analog to show non-reflection of sound in a homogeneous media.

Ch. 1   FUNDAMENTALS OF SOUND AND VIBRATION   13

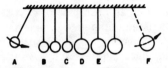

Fig. 1-6. Rigid-sphere analog to illustrate sound reflection between two different media.

Since the masses are identical the principles of conservation of momentum tell us that all of the momentum of Sphere A will be transferred to Sphere B, and A will come to rest. Meanwhile B collides with C and comes to rest, C with D, etc. After a finite length of time F is kicked out; if caught in time it will not fall back against E to start a reflected pulse. It can be observed that there are no reflected pulses in the line of spheres. This is an analog of a progressive wave in a homogeneous medium where the input is a single pulse. Consider now a more complicated analog where spheres of two different masses are used, as shown in Fig. 1-6. Again, if Sphere A is swung back and released, it collides with B and comes to rest. The process continues as before until C impacts the heavier sphere D. Instead of coming to rest, C rebounds causing a reflected pulse in the lighter spheres which eventually kicks A outward; meanwhile, D moves to the right having absorbed some of the momentum from C. Since Spheres D through F have the same mass, impacts will occur smoothly until F is kicked out. If A and F are caught in time no further motion occurs. Note, however, the single reflection at the interface of the different size spheres.

What happens if the motion from the right side is initiated by swinging Sphere F outward and releasing it? The collisions will be smooth until D collides with C. Instead of stopping, Sphere D (being more massive than C) continues a short distance to the left, then swings backward and impacts E, causing a reflected pulse which eventually kicks F outward. The transmitted impulse from C continues to the left until A is kicked out. Again, note the single reflection at the interface of the different size spheres.

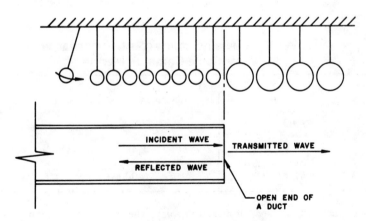

Fig. 1-7. Sound reflection from an open duct.

The reflection of sound from an open pipe or duct can be explained with the help of Fig. 1-7. The situation in simplified form is similar to the spheres of different mass in that all the momentum cannot transfer from the wave inside the duct to a wave outside the duct and the result is a reflected momentum wave at the opening. Although the mathematics needed to completely describe the three waves at the duct opening are beyond this text, the simple concept is adequate to understand reflected waves at discontinuities.

Impedance

In discussing the phenomena of reflection it is helpful to introduce the concept of *impedance*. As the word implies, impedance is a measure of resistance to movement from an applied force or pressure. For the mechanical analog above, think in terms of a *mechanical impedance* which is defined as

$$Z_m = \text{Mechanical Impedance} = \frac{\text{force}}{\text{velocity}} \tag{1-14}$$

Comparing a heavy sphere with a light sphere, it is intuitive that the heavy sphere will have less motion (velocity) for a given applied force than the lighter sphere, and thus, will have a larger mechanical impedance. The process of reflection is often loosely referred to as resulting from an "impedance mismatching."

In wave propagation, *specific acoustic impedance* (at a point) is a convenient ratio which is defined for a single propagating wave as:

$$Z_s = \text{Specific acoustic impedance} = \frac{\text{Acoustic pressure}}{\text{Acoustic particle velocity}} \tag{1-15}$$

For a progressive plane wave or for a spherical wave at a sufficient distance from the source the specific acoustic impedance of the medium is simply ρc. That is, the resistance of the medium (solid, liquid, or gas) to the propagating sound wave is equal to the product of the density of the medium (ρ) and the speed of sound in that medium (c). Another name for this impedance is *characteristic impedance* of the medium. The characteristic impedance for air and other media is listed in Table 1-1.

Table 1-1. Characteristic Acoustic Impedance (ρc) for Some Common Media

Medium	Characteristic Impedance, ρc kilogram/meter^2-sec ($\times 10^6$)
Steel	47.0
Wood (pine)	1.57
Water (fresh)	1.48
Rubber (soft)	1.0
Air (20°C)	0.000415

The product, ρc, in wave propagation, plays the same part as the mass of the spheres in the mechanical analog of reflection.

Reflection and Transmission Coefficients

Now understanding of the mechanical case of reflection is ready to be extended to that of real acoustic media. Consider two acoustic media with characteristic impedances $\rho_1 c_1$ and $\rho_2 c_2$, and where a plane wave from the left (incident wave) is propagating perpendicular to the interface. (See Fig. 1-8.) If $\rho_2 c_2$ is different from $\rho_1 c_1$, then all of the energy from the incident wave cannot be transmitted across the plane interface; any energy remaining must go into a re-

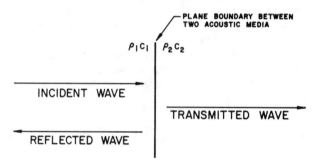

Fig. 1-8. Reflection of sound at a plane interface of two acoustic media.

flected wave, as shown. By considering the conservation of energy and momentum at the interface, it can be shown that

$$A_R = \left(\frac{\rho_2 c_2 - \rho_1 c_1}{\rho_1 c_1 + \rho_2 c_2}\right)^2 A_I \quad \text{or,} \quad A_R = \alpha_R A_I \tag{1-16}$$

and

$$A_T = \frac{4\rho_1 c_1 \rho_2 c_2}{(\rho_1 c_1 + \rho_2 c_2)^2} A_I \quad \text{or,} \quad A_T = \alpha_T A_I \tag{1-17}$$

where:

A_I = amplitude of incident wave
A_T = amplitude of transmitted wave
A_R = amplitude of reflected wave
α_R = acoustic reflection coefficient
α_T = acoustic transmission coefficient

From the equations, we can see that when $\rho_1 c_1 = \rho_2 c_2$, $A_T = 1$, $A_R = 0$. That is, when there is no change in the characteristic impedance of the medium, wave propagation continues undisturbed. When $\rho_2 c_2$ is larger than $\rho_1 c_1$, A_R has the same sign as A_I, that is, the reflected wave is in phase with the incident wave. But when $\rho_2 c_2$ is smaller than $\rho_1 c_1$, the signs are opposite so that they are exactly 180 degrees out of phase. In either case note that a reflected wave occurs at the interface.

In practical engineering situations it is rare that a simple situation such as that described above, occurs. When both incident and reflected acoustic waves are present in a media a term called *acoustic impedance* is used. This is defined as the total acoustic pressure composed of the sum of the incident and reflected pressure wave divided by the sum of the particle velocity in the incident and reflected wave at a point.

$$Z_A = \text{Acoustic impedance} = \frac{\text{Total acoustic pressure}}{\text{Total particle velocity}} \quad (1\text{-}18)$$

The difference between Equations (1-15) and (1-18) is that the *specific acoustic impedance* given by Equation (1-15) is defined for one propagating wave in one direction while the *acoustic impedance* given by Equation (1-18) accounts for waves in opposite directions. In general, the waves traveling in opposite directions will be out of phase with each other and the associated particle velocities will also be out of phase. The typical situation is that the acoustic impedance can be expressed as a complex quantity of the form

$$\text{Acoustic impedance} = r + jx \quad (1\text{-}19)$$

where r is the real part and x the imaginary part of the ratio of acoustic pressure to particle velocity. The symbol j denotes that the term x is out of phase 90° as compared to the r term. This is the usual form for the acoustic impedance at the surface of an acoustic material. For a sound wave incident on an acoustic material with acoustic impedance Z_A, the magnitude of the reflected wave is obtained from Equation (1-16) with $\rho_2 c_2$ replaced by the complex acoustic impedance of the material Z_A and $\rho_1 c_1$ denotes the characteristic impedance of the acoustic media, usually air, in contact with the material.

$$A_R = \left(\frac{Z_A - \rho_1 c_1}{\rho_1 c_1 + Z_A}\right)^2 A_I \quad (1\text{-}20)$$

Note that in Equation (1-20) there is a magnitude and a phase angle associated with the term in brackets such that the reflected wave has a magnitude and a phase angle with respect to the incident wave. Z_A typically is a function of the frequency of the incident wave.

Absorption Coefficients

In most noise-control applications of a sound-wave incident on an acoustic material the precise magnitude and phase angle of the reflected wave form of the acoustic surface is not of interest but the fraction of the acoustic wave that is absorbed or dissipated by the material is of interest. The *acoustic absorption coefficient* is a measure of the fraction of acoustic energy absorbed and subsequently dissipated by an acoustic material. It is related to the characteristic impedance of the acoustic media and the acoustic impedance of the material as given by the relation

$$\alpha = \text{acoustic absorption coefficient} = \left|\frac{4\rho_1 c_1 Z_A}{(\rho_1 c_1 + Z_A)^2}\right| \quad (1\text{-}21)$$

or an equivalent term is given by

$$\alpha = 1 - \left|\frac{Z_A - \rho_1 c_1}{\rho_1 c_1 + Z_A}\right|^2 \quad (1\text{-}22)$$

Ch. 1   FUNDAMENTALS OF SOUND AND VIBRATION   17

The vertical bars denote the magnitude of the quantity since the phase angle is not of concern in this form. In engineering practice the quantity given by Equation (1-22) is not calculated because Z_A is unknown, and is a function of frequency. The acoustic absorption coefficient is measured and reported as a property of the material by the manufacturer. The measurement method is described in Chapter 5 and illustrated by Fig. 5-3.

Standing Waves

Reflection of sound from walls or from a duct termination can result in a *standing wave*. This condition is a result of a pure-tone sound wave reflected between two planes (walls or duct discontinuities) such that the spacing is equal to an exact multiple of the wave length of the pure-tone sound wave. Standing waves can be generated in factory spaces which have reflective surfaces, inside machinery enclosures, in duct systems, and in piping systems.

In order to describe a standing wave, the nature of a traveling pure-tone sound wave must first be considered. Illustrated in Fig. 1-9 is a traveling sinusoidal wave in three successive positions at three successive instants of time. It is evident that the wave has a dependence on the variables of time and distance because at a fixed time, the wave has a space dependence and at a fixed distance, the wave has a time dependence. The functional form which describes a positive

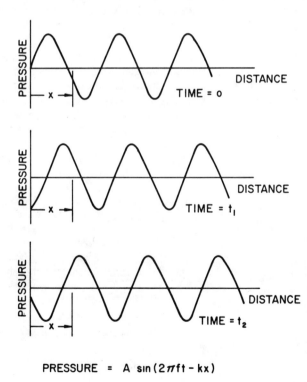

PRESSURE = A sin(2πft − kx)

Fig. 1-9. Schematic of a propagating sound wave.

traveling sinusoidal pressure wave is given by

$$\text{Acoustic Pressure} = p(x, t) = A \sin(2\pi ft - kx) \quad (1\text{-}23)$$

where:

A = peak amplitude, of the pressure, N/m^2
f = frequency, Hz
t = time, sec
k = wave number = $\dfrac{2\pi f}{c}$, 1/meter
c = speed of sound, meter/sec
x = distance from some reference, meter

For a wave propagating in the negative x direction the form is

$$p(x, t) = B \sin(2\pi ft + kx) \quad (1\text{-}24)$$

The wave propagating in the negative x direction may be the result of a reflection. The sum of the right and left traveling waves is given by

$$p(x, t) = A \sin(2\pi ft - kx) - A \sin(2\pi ft + kx) \quad (1\text{-}25)$$

if the reflected wave is equal in magnitude and opposite in sign ($A = -B$) to the incident wave. By trigonometric identities, this equation can be represented as

$$p(x, t) = 2A \sin(2\pi ft) \sin(kx) \quad (1\text{-}26)$$

This is the form of a *standing wave* as illustrated in Fig. 1-10. Since the form given by Equation (1-26) is a spacial function, sin (kx); times a time function, sin ($2\pi ft$); the result is a sinusoidal pressure distribution in space with a time-varying magnitude. The node points in Fig. 1-10 remain fixed in space and for this simple case they always have a zero pressure variation compared to the ambient pressure. At the antinode points the pressure varies with time from a maximum value above the ambient to a minimum value below the ambient. For this idealized case shown, a sound-level meter placed at the node point would register 0 dB for all time, while a sound-level meter placed at the antinode would register

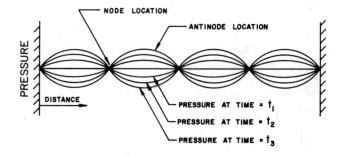

PRESSURE = 2(A+B) sin(2πft) sin(kx)

Fig. 1-10. Geometry of a standing wave in a space.

Ch. 1   FUNDAMENTALS OF SOUND AND VIBRATION

a maximum dB level. In an actual engineering situation the reflected wave will not exactly equal the incident wave and the node points of the standing wave will not be an exact cancellation, but rather will be minimum pressure variation locations. In actual practice the sound-pressure level can vary as much as 20 dB from the node to the antinode locations. Elimination of the standing wave is accomplished by acoustic absorbing material on walls or acoustical material along the propagation path of the wave.

Description of Decibel Scales

The decibel is a mathematical scale similar in use to a logarithmic scale. It is used to describe the intensity or energy level of a physical quantity, compressing the quantity into numbers that are convenient for data presentation. In addition to its application in acoustics, the decibel is used for electrical power, and in certain cases, mechanical energy presentations. Fundamentally, it is ten times the base ten logarithm of the ratio of a power or energy quantity with respect to a reference base of the same physical quantity. The basic equation for expressing this concept is:

$$L_E = 10 \log_{10} \left(\frac{E}{E_{ref}} \right) \text{ dB} \qquad (1\text{-}27)$$

where:

L_E = Energy or power level, in decibels
E = Energy or power quantity of interest, watts
E_{ref} = Reference energy or power quantity of interest, watts

It is evident that the original units of energy or power quantity of interest are cancelled in the ratio. Moreover, energy or power can be described in various ways, such as in terms of velocity, acceleration, pressure, voltage, etc.

In acoustics, the decibel scale is used for acoustic power, acoustic intensity, and acoustic pressure, as expressed by the following equations:

$$L_W = 10 \log_{10} \left(\frac{W}{W_{ref}} \right) \text{ dB} \qquad (1\text{-}28)$$

$$L_I = 10 \log_{10} \left(\frac{I}{I_{ref}} \right) \text{ dB} \qquad (1\text{-}29)$$

$$L_p = 10 \log_{10} \frac{p^2}{p_{ref}^2} = 10 \log_{10} \left(\frac{p}{p_{ref}} \right)^2 \text{ dB} \qquad (1\text{-}30)$$

where:

L_W = Sound-power level, dB
L_I = Sound-intensity level, dB
L_p = Sound-pressure level, dB

It should be noted in Equation (1-30) that energy is proportional to the square of the pressure; therefore, the pressure term must be squared since the decibel is a ratio of the power.

Sound-Pressure Level (L_p)

The measurement of sound pressure is done on a logarithmic (or decibel) scale. There are two compelling reasons for going from a linear to a logarithmic scale. First, the pressure range of interest is exceedingly large; a normal human ear is capable of functioning over a pressure range of about one million to one. Secondly, experiments have shown that the ear responds in a logarithmic manner to the perceived loudness of a sound.

The *sound-pressure level* is defined by:[5]

$$L_p = 10 \log_{10}\left(\frac{p^2}{p_{\text{ref}}^2}\right) \text{ dB}$$

$$= 10 \log_{10}\left(\frac{p}{p_{\text{ref}}}\right)^2 \text{ dB} \qquad (1\text{-}30)$$

where:

L_p = sound-pressure level, in decibels, dB
p = root mean square, rms, of the fluctuating sound pressure, psi, N/m², or pascals
p_{ref} = standard reference value to which all pressures are compared.

The sound-pressure level is proportional to the logarithm (base 10) of a pressure ratio. The pressure p_{ref} is called the *reference pressure* and has a value of 0.00002 dynes/cm² or 0.00002 microbar. The pressure p is the sound pressure that is being measured. The word "level" is used to emphasize that the value of L_p is above some reference level. The above definition for L_p holds whether p is the peak amplitude or the effective or rms average amplitude of the sound wave. However, *all measuring instruments are calibrated to read in terms of rms pressure, so it is generally understood that L_p is an rms level unless otherwise stated*.

The root mean square, rms, of the sound pressure is utilized to obtain an "effective" constant pressure above the local ambient atmospheric pressure. The rms pressure is the equivalent static pressure above the ambient pressure which would represent the same acoustic energy as a fluctuating pressure. Mathematically the root mean square of a series of values is an average (but not the arithmetical average) that is expressed by the following equation:

$$\text{rms} = \sqrt{\frac{Y_A^2 + Y_B^2 + Y_C^2 + \cdots Y_N^2}{N}} \qquad (1\text{-}31)$$

This numerical method for calculating the rms value of a function is illustrated in Fig. 1-11. For this type of calculation, the accuracy is improved as the number of points increases.

An illustration of the root mean square pressure, p_{rms}, for a sinusoidal acoustic pressure variation is given in Fig. 1-12. The sinusoidal or "pure-tone" sound

---

[5] Various symbols are used for sound-pressure level, *SPL* and L_p being the most common in the literature.

Ch. 1 FUNDAMENTALS OF SOUND AND VIBRATION 21

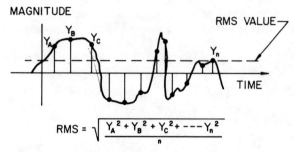

Fig. 1-11. Determination of the root mean square (rms) value of a signal by discrete values.

pressure as a function of time is shown as the top trace of Fig. 1-12. This would be typical of a whistle, squeal, or whine. To obtain the rms value of the pressure-time signal, the signal would be squared, the average of the squared signal taken, and finally, the square root of the average of the squared signal determined. The p_{rms} value represents an equivalent constant value which has the same energy content as the time-varying quantity, thus giving a convenient con-

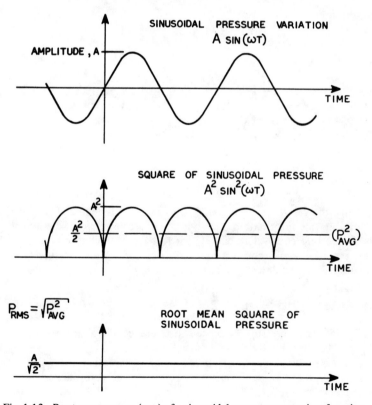

Fig. 1-12. Root mean square (rms) of a sinusoidal pressure versus time function.

stant number value equivalent to the actual time function. For sinusoidal functions, the rms value is equal to the magnitude of the function divided by the square root of two or multiplied by .707.

This is exactly analogous to the effective voltage available from the typical wall outlet. The peak value of electrical voltage is given by

$$\text{Voltage} = 163 \sin(2\pi 60t) \text{ volts}$$

for 60-cycle voltage. The rms value of a sinusoidal function is the amplitude divided by $\sqrt{2}$.

$$V_{rms} = \frac{163}{\sqrt{2}} = 115 \text{ volts (rms)}$$

The rms value is the equivalent constant voltage that provides the same electrical power as the sinusoidal time-varying quantity and this value is referred to as the rms voltage from a common wall outlet. Acoustic pressure is also a time-varying quantity above and below the ambient pressure.

Example: If the *peak* sound pressure, p_{peak}, is 0.03 dyne/cm^2, calculate the sound pressure level. We have from Equation (1-30)

$$L_p = 10 \log_{10} \left(\frac{p}{p_{ref}}\right)^2 \text{ dB}$$

where p is the rms pressure equal to $.707 p_{peak}$ therefore,

$$L_p = 10 \log_{10} \left(\frac{.707 p_{peak}}{p_{ref}}\right)^2 = 10 \log_{10} \left(\frac{.707 \times 0.03}{0.00002}\right)^2$$

$$= 10 \log_{10} \left(\frac{0.02121}{0.00002}\right)^2 = 10 \log_{10} (1060.5)^2$$

$$= 10 \log_{10} (1{,}124{,}660) = 10(6.05)$$

$$= 60.5 \text{ dB (re 0.00002 dyne/cm}^2\text{)}$$

Notice that the reference "re 0.00002 dynes/cm^2" is included in the answer since without this information the sound-pressure level would be meaningless. Although the reference value may appear to have been chosen arbitrarily, it in fact has practical significance: it approximates the lowest pressure at which a normal human ear can detect a pure tone at 1000 Hz.

i.e., $$10 \log_{10} \left(\frac{.00002}{.00002}\right)^2 = 10 \log_{10}(1) = 0 \text{ dB}$$

Thus, the reference pressure is chosen to reference all sound-pressure levels to the threshold of hearing, 0 dB, for a 1000-Hz sound.

Figure 1-13 shows the correlation between p in microbar and L_p re 0.00002 microbar for a range of familiar sound sources. The convenience of the logarithmic scale is obvious.

Ch. 1 FUNDAMENTALS OF SOUND AND VIBRATION 23

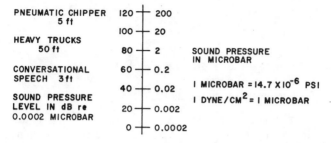

Fig. 1-13. Comparison between sound pressure and sound-pressure level.

Sound-Power Level (L_W)

Although sound power cannot be measured directly, it can, under certain conditions, be calculated based on space averaging of sound-pressure measurements. Conversely, if the sound power radiated from a sound source is known, under certain conditions the sound pressure at a point some distance from the source can be estimated. The link between sound power and sound pressure is the sound intensity, which has been previously defined.

Equation (1-13) gives the relation sound intensity = average of (pressure × acoustic particle velocity); I = average (pu). Also, from Equation (1-15) the specific acoustic impedance is given by

$$Z_s = \frac{\text{acoustic pressure}}{\text{acoustic particle velocity}} = \frac{p}{u} = \rho c$$

or

$$p = \rho c u \quad \text{and} \quad u = \frac{p}{\rho c}$$

for a free progressive sound wave in the positive displacement direction. The acoustic intensity can then be written as

$$I = \overline{pu} = \frac{\overline{pp}}{\rho c} = \frac{\overline{p^2}}{\rho c} \text{ watt/m}^2 \tag{1-32}$$

This form is convenient because the pressure can be measured with a microphone; in previous forms containing the particle velocity, u, the forms were not directly applicable since the measurement of u is extremely difficult.

The *sound power* can then be determined from the *acoustic intensity* by multiplying by the area, a,

$$W = Ia \text{ watts} \tag{1-33}$$

The *sound power level* (L_W) is defined by Equation (1-28).[6]

---

[6] The symbols L_W and PWL are used in the literature to denote sound-power level.

24 FUNDAMENTALS OF SOUND AND VIBRATION Ch. 1

$$L_W = 10 \log_{10} \left(\frac{W}{W_{\text{ref}}}\right) \text{dB} \qquad (1\text{-}28)$$

where:

L_W = sound-power level, dB
W = acoustic power, watts
W_{ref} = reference power chosen as 10^{-12} watts

In some older texts the reference power used was 10^{-13} watts. It is especially important then that the reference-power level be stated when reporting sound-power levels. For example,

$$L_W = 112 \text{ dB re } 10^{-12} \text{ watts}$$

is the proper manner in which to report sound-power levels so that there is no question of the reference quantity used.

To illustrate the fact that the amount of power radiated as sound from typical machinery is small compared to the mechanical power of the machine, consider a 200 hp diesel engine which has a sound-power level of, $L_W = 141$ dB re 10^{-12} watts. The actual sound power radiated can be obtained from Equation (1-28).

$$L_W = 10 \log_{10} \left(\frac{W}{W_{\text{ref}}}\right) \text{dB}$$

$$141 = 10 \log_{10} \left(\frac{W}{W_{\text{ref}}}\right)$$

$$14.1 = \log_{10} \left(\frac{W}{W_{\text{ref}}}\right)$$

$$10^{14.1} = \frac{W}{W_{\text{ref}}}$$

$$1.259 \times 10^{14} = \frac{W}{W_{\text{ref}}}$$

$$W = W_{\text{ref}}(1.259 \times 10^{14}) = 10^{-12}(1.259 \times 10^{14}) \text{ watt}$$

$$W = 125.9 \text{ watt}$$

or

$$W = 125.9(1.341 \times 10^{-3}) \text{ hp} = .17 \text{ hp}$$

Thus, .17 hp is radiated as sound power compared to the 200 hp of mechanical power available. This small amount of power cannot be detected by a dynamometer test and this illustrates the fact that constructing more efficient machinery will not guarantee significant reduction in sound-power levels. To significantly reduce the sound-power level, a reduction of sound power by the ratio of 100 to 1 will result in 20 dB reduction. This cannot be achieved by more efficient machinery in general; usually the basic construction of the machine must be altered to achieve significant noise reduction.

Ch. 1   FUNDAMENTALS OF SOUND AND VIBRATION   25

Equations (1-32) and (1-33) only apply for a sound wave propagating away from the source if the positive distance is measured from the source; if a reflected wave exists and propagates back toward the source, the relation for this wave is:

$$Z_s = \frac{p}{u} = -\rho c \qquad (1\text{-}34)$$

Then

$$u = -\frac{p}{\rho c}$$

and

$$I = -\frac{pp}{\rho c} = -\frac{p^2}{\rho c} \qquad (1\text{-}35\text{A})$$

Thus, if both incident and reflected sound waves are present the total sound intensity is given by

$$I_{total} = I_{incident} + I_{reflected} = \frac{p^2_{incident}}{\rho c} - \frac{p^2_{reflected}}{\rho c} \qquad (1\text{-}35\text{B})$$

In a space or room where incident and reflected sound waves are present, the acoustic intensity in the reverberant sound field (at a distance greater than one wavelength from the sound source) is related to the sound power of the source by the following equation:

$$I = W\left(\frac{4}{R'} + \frac{1}{S_T}\right) \qquad (1\text{-}36)$$

where:

I = Acoustic intensity, watts/meter^2
W = Sound power, watts
R' = Enclosure constant = $S_T \bar{\alpha}/(1 - \bar{\alpha})$, meter^2
S_T = Surface area of enclosure, meter^2
$\bar{\alpha}$ = Average acoustic absorption coefficient

and

$$\bar{\alpha} = \frac{1}{S_T} \sum_{i=1}^{i=N} \alpha_i S_i$$

α_i = Absorption of i'th surface
S_i = Area of i'th surface, meter^2

With a microphone in the sound field, the total pressure is composed of the pressures associated with the incident and reflected waves and it is not possible to separate the energy in the incident wave from that due to the reflected wave. The method used to obtain only the intensity due to the incident wave is to construct a sound room which has no reflected sound waves. This type of acoustic laboratory is called an *anechoic chamber* which is necessary to determine the intensity of the sound wave propagating away from the source. Hence the computed sound power from data taken in such a room is a measure of the sound power of the machine; the variation in intensity with direction is called the *directivity* of the source.

The Addition of Decibels

Decibels are logarithmic quantities, as shown by Equations (1-28), (1-29), and (1-30). Logarithmic quantities cannot be added together in the usual manner. First the antilogs must be found and it is the antilogs that are added together; then the logarithm of the sum of the antilogs is the true sum of the logarithms. While this process seems to be simple enough, special considerations must be given when adding decibels because the acoustic pressure (p) to be used is a square quantity. The summation of root mean square quantities; for example, the rms sound pressure is equal to the summation of their squares. In order to understand the basis of adding decibels, it is worthwhile to follow the derivation of the basic equation for performing this operation. To start with, the equation of two different sound-pressure levels, L_{p1} and L_{p2}, are added together:

$$L_{p1} = 10 \log_{10} \left(\frac{p_1}{p_{ref}}\right)^2 \quad \text{and} \quad L_{p2} = 10 \log_{10} \left(\frac{p_2}{p_{ref}}\right)^2$$

$$L_{p1} + L_{p2} = 10 \log_{10} \left(\frac{p_1}{p_{ref}}\right)^2 + 10 \log_{10} \left(\frac{p_2}{p_{ref}}\right)^2$$

$$L_{p1} + L_{p2} = 10 \log_{10} \left[\left(\frac{p_1}{p_{ref}}\right)^2 + \left(\frac{p_2}{p_{ref}}\right)^2\right] \tag{a}$$

Note that for pressure levels the sum of the pressure ratios *squared* are added together, not the square of the sum of the pressure ratios. This equation can be made more convenient by again writing the equation for the sound-pressure level and rearranging this equation according to the mathematical rules of logartihms.

$$L_p = 10 \log_{10} \left(\frac{p}{p_{ref}}\right)^2 \quad (X = \log_{10} Y)$$

$$\left(\frac{p}{p_{ref}}\right)^2 = \frac{L_p}{10^{10}} \quad (Y = 10^X)$$

This result can be substituted in Equation (a).

$$L_{p1} + L_{p2} = 10 \log_{10} (10^{L_{p1}/10} + 10^{L_{p2}/10}) \tag{b}$$

Thus, the basic equation for the summation of decibels, whether pressure or power, can be written as follows:

$$L_p = 10 \log_{10} (10^{L_{p1}/10} + 10^{L_{p2}/10} + \cdots 10^{L_{pn}/10}) \tag{1-37}$$

$$L_W = 10 \log_{10} (10^{L_{W1}/10} + 10^{L_{W2}/10} + \cdots 10^{L_{Wn}/10}) \tag{1-38}$$

Ch. 1 FUNDAMENTALS OF SOUND AND VIBRATION

As an example of the application of these equations, consider the addition of 90 dB and 94 dB acoustic sound-pressure level.

$$\begin{aligned}
L_p = 90 \text{ dB} + 94 \text{ dB} &= 10 \log_{10} (10^{90/10} + 10^{94/10}) \\
&= 10 \log_{10} (10^9 + 10^{9.4}) \\
&= 10 \log_{10} [10^9 + (10^{.4} \times 10^9)] \\
&= 10 \log_{10} (1 \times 10^9 + 2.51 \times 10^9) \\
&= 10 \log_{10} (3.51 \times 10^9) \\
&= 10 (\log_{10} 3.51 + \log_{10} 10^9) \\
&= 10 (.5453 + 9)
\end{aligned}$$

$$90 \text{ dB} + 94 \text{ dB} = 95.5 \text{ dB}$$

The computation involved in solving this equation can be accomplished without difficulty using a modern engineering calculator. With the calculator the intermediate steps shown above do not need to be performed. Where a calculator is not available, the graph in Fig. 1-14 can be used. To use this graph, the numerical difference between the two decibel values to be added must first be obtained. This difference is shown as the abscissa of the graph in Fig. 1-14; the ordinate is the decibel value that must be added to the largest term to obtain the sum of the decibels. If more than two decibel values are to be added together (using Fig. 1-14), it is necessary to add a pair of values and, as a separate step, add the next value to this sum, etc.

For example, to add 90 dB and 94 dB as in the previous example but this time using Fig. 1-14, first find their difference, which is 4 dB. In Fig. 1-14, the 4 dB value of the abscissa intersects the curve of the 1.5 dB value of the ordinate;

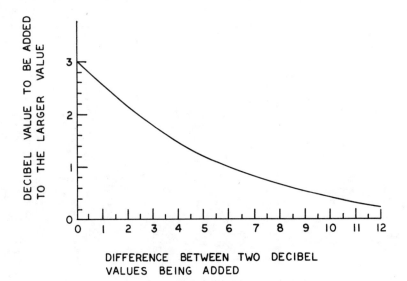

Fig. 1-14. Chart for combining two decibel values.

Expansion Chamber Mufflers

therefore, to obtain the sum 90 dB and 94 dB, 1.5 dB must be added to the larger value, or 94 dB, to obtain the answer, which is 95.5 dB.

From the concepts of sound-wave reflection at discontinuities and the definition of sound-pressure level the principle of operation of an expansion chamber muffler can be presented. Illustrated in Fig. 1-15 is an expansion chamber in a piping system which could be an exhaust system, an air supply system, the suction line for a compressor, etc. In the previous section entitled "Reflection of Sound Waves," on page 18, the concept of a reflected sound wave of a sudden change of cross section was illustrated by the fact that the momentum in the incident wave cannot all be transferred across the discontinuity. This phenomenon occurs whether the cross section expands or contracts as in Fig. 1-15. At both the expansion and the contraction a reflected wave will exist as explained by the momentum concept.

The *transmission loss* of the expansion chamber is defined as the difference in sound-pressure level of the incident sound wave at Section A and the transmitted sound-pressure level at Section B of Fig. 1-15.

$$\text{TL} = \text{transmission loss} = L_p \text{ at } A - L_p \text{ at } B \quad (1\text{-}39)$$

The formulation to compute this difference is given by

$$\text{TL} = \log_{10}\left[1 + .25\left(\frac{S_C}{S_E} - \frac{S_E}{S_C}\right)^2 \sin^2\left(\frac{2\pi L_C}{\lambda}\right)\right] \text{ dB} \quad (1\text{-}40)$$

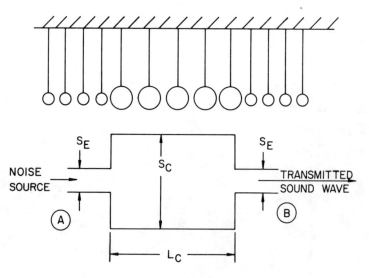

Fig. 1-15. Geometry of an expansion chamber muffler.

Ch. 1   FUNDAMENTALS OF SOUND AND VIBRATION

where:

TL = transmission loss, dB
S_C = cross-sectional area of expansion chamber, ft^2 or meter^2
S_E = cross-sectional area of inlet to expansion chamber, ft^2 or meter^2
L_C = length of expansion chamber, ft or meter
λ = wavelength of sound, ft or meter
$\frac{2\pi L_C}{\lambda}$ = angle, in radians

Equation (1-40) above, is valid for expansion chamber mufflers with cross sections which need not be circular. The equation is limited to geometries such that the largest transverse dimension of the expansion chamber is less than ¾ of the wavelength of sound of interest in the system.

The wavelength of sound is computed from Equation (1-2) in the following form

$$\lambda = \frac{c}{f}$$

where:

c = speed of sound = $49.03\sqrt{460 + T}$, ft/sec
f = frequency of sound, Hz
T = temperature in °F

Equation (1-40) is valid for the cross-sectional dimension less than one wavelength for the frequency of sound of interest. The expansion cross section need not be round for the equation to be valid. Steady flow through the expansion chamber of less than 100 ft/sec will not appreciably alter the transmission loss. Flow velocities at high velocities will generate flow noise by turbulence and flow separation.

Note from Equation (1-40) that when the term $\left(\frac{2\pi L_C}{\lambda}\right)$ is zero, the term in brackets will be 1 and TL will be zero. This occurs for frequencies of sound such that the length of the expansion chamber is equal to $\lambda/2, \lambda, 3\lambda/2, \ldots$ etc. For these frequencies of incident sound, the muffler transmission-loss will be zero. Transmission-loss for various expansion ratios, $S_C/S_E = e$, is illustrated in Fig. 1-16.

The following general rules apply for the design of expansion chamber mufflers:
1. The greater the area ratio S_C/S_E, the greater the transmission loss.
2. Increasing the mean flow-velocity through the muffler up to approximately 100 ft/sec tends to increase transmission loss.
3. In practice, transmission loss of 40 dB or more is not possible with sheet metal construction because of sound transmission through the walls of the muffler. This upper limit would be higher for cast iron or other heavy construction.
4. Large, flat, muffler walls should be avoided because they tend to vibrate and radiate sound.

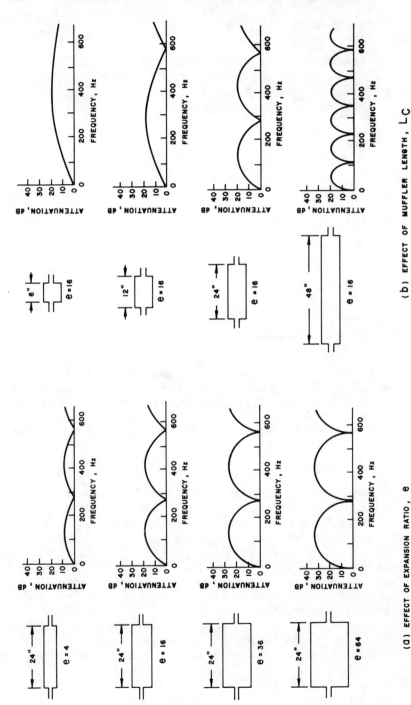

Fig. 1-16. Comparison of muffler attenuation as a function of expansion ratio, e, and length. After Davis, Stokes, Moore, and Stevens.

Ch. 1   FUNDAMENTALS OF SOUND AND VIBRATION   31

5. The length of the expansion chamber controls the frequency at which there is maximum attenuation.

Sound Transmission through Panels

When sound impinges on a plane panel a portion of the sound is transmitted and a portion is reflected back toward the source. The transmission-loss of sound through a panel can be expressed

$$TL = 10 \log_{10} \left[\frac{\text{incident rms sound pressure}}{\text{transmitted rms sound pressure}} \right]^2 \text{dB} \qquad (1\text{-}41)$$

$$TL = 10 \log_{10} \left[\frac{R^2 + (\omega M - k/\omega)^2}{4Z_s^2} \right] \text{dB} \qquad (1\text{-}42)$$

where:

M = mass per unit area, lb/ft^2 or Kg/meter^2
K = stiffness of the panel lb/ft or N/meter
R = damping of the panel lb-sec/ft or N-s/meter
Z_s = characteristic impedance = 87 lb/ft^2-sec for air
ω = frequency of incident sound, rad/sec
A_p = area of the panel, ft^2 or meter^2

The character of this equation is shown graphically in Fig. 1-17 as a function of the incident sound frequency, ω. As can be seen there are four distinct regions: (1) The region below the first natural frequency of the panel where the attenuation depends upon the stiffness of the panel; (2) The resonant range at or near the fundamental frequency of the panel where attenuation is low;

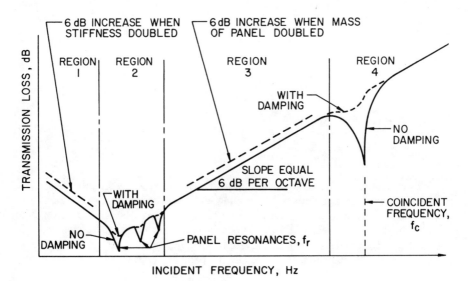

Fig. 1-17. Transmission-loss of acoustic energy through a panel.

(3) The range above the resonance of the panel where the attenuation varies as the mass per unit area of the panel; (4) The region where the wavelength of sound in air coincides with the bending wavelength on the panel. In region (1) the mass and damping of the panel are unimportant and only the stiffness of the panel will alter the attenuation of acoustic energy through the panel. The transmission-loss decreases by 6 dB per octave as frequency increases, until the first resonance frequency of the panel is approached. The attenuation increases by 6 dB in this region, as the stiffness is doubled. For practical construction purposes it is impossible to add enough stiffness to the panel to increase the attenuation to acceptable values. For large panels the first natural frequency will be low, and for most cases the excitation will be larger than this value.

For most panels the boundaries will be something between simple supported and clamped. Thus, the lowest resonant frequency will be between the limits given by the following:

Simply Supported Panel:

$$f_r = 0.454 \, hC_L \left[\frac{1}{L_x^2} + \frac{1}{L_y^2} \right] \text{Hz} \tag{1-43}$$

where:

f_r = resonant frequency (Hz)
h = thickness (ft)
L_x, L_y = dimensions of panel (ft)
C_L = speed of sound in panel material = 17,000 ft/sec for steel, aluminum, glass, and most metals

Clamped Panel:

$$f_r = 0.454 \, \frac{h \Lambda C_L}{L_x^2} \tag{1-44}$$

where:

$$\Lambda^2 = (1.506)^4 \left[1 + \left(\frac{L_x}{L_y}\right)^4 \right] + 2 \left[1.248 \frac{L_x}{L_y} \right]^2 \tag{1-45}$$

L_x = length of panel (ft)
L_y = width of panel (ft)
h = thickness (ft)
C_L = 17,000 ft/sec

Within region (2) the damping of the panel will increase the attenuation somewhat; however, great amounts of damping will never result in large values of attenuation. The mass and stiffness of the panel have little effect in this region. The transition from region (2) to (3) occurs when the resonant frequencies are no longer predominant. This usually occurs when the excitation frequency is 3 to 4 times the fundamental frequency of the panel; i.e., $3f_r$ to $4f_r$.

In region (3) the attenuation varies as the product of the mass per unit area and frequency given by

Ch. 1 FUNDAMENTALS OF SOUND AND VIBRATION

$$TL = 10 \log_{10} \left[\frac{\omega^2 M^2}{4Z_s^2}\right] \text{ dB} \qquad (1\text{-}46)$$

This is the so-called "limp mass law" which means the heavier the panel the greater will be the attenuation for the same incident sound frequency. For every doubling of mass the attenuation increases by 6 dB, independent of the stiffness or damping. At higher frequencies the damping and stiffness again become important and a noticeable "dip" in the transmission-loss occurs which separates regions (3) and (4). The critical frequency which determines where this occurs is called the *coincidence frequency* and is given by

$$f_c = \frac{c^2}{1.8 h C_L} \text{ Hz} \qquad (1\text{-}47)$$

where:

c = speed of sound in air (ft/sec)
h = thickness of panel (ft)
C_L = 17,000 ft/sec for most metals

For most steel, aluminum, iron, and glass this number is

$$f_c = \frac{500}{h \text{ (inches)}} \text{ Hz} \qquad (1\text{-}48)$$

Near this value the transmission-loss can be increased by decreasing the thickness of the panel so that f_c is greater than the frequency of the incident sound. With damping this dip is not as predominant, and with enough damping, the dip may be unimportant.

The usual procedure is to design a panel such that the frequency of incident sound is in the limp mass law region approaching the coincidence frequency. Best results can be obtained for single panels having the greatest amount of mass. It is difficult to design for the region above coincidence for thin panels unless the excitation frequency is very high. For a $1/32$-inch panel coincidence would be 16,000 Hz which is usually above the frequency region of interest.

In general, care must be taken when adding stiffness because the resonant frequency will increase and attenuation will decrease. The resonant frequency of the panel should be well below the excitation frequency for best results. Sometimes addition of stiffeners increases the efficiency of radiation of panels.

The greatest portion of the sound generated by a plate that is vibrating below the coincident frequency will be generated in a region adjacent to the edges. In this case, stiffeners applied from the center of one edge to the center of the opposite edge will, in effect, make two panels having a greater edge area, thereby generating more sound. When the panel is vibrating below the coincident frequency, stiffeners should be placed diagonally from corner to corner in an "X" pattern. If the panel is vibrating above the coincident frequency the sound is generated almost uniformly over its entire area. To reduce the sound generation of the panel in this case, stiffeners can be added in almost any pattern; however, the velocity of the panel vibration must be reduced by the stiffener and great care must be used to make certain that the coincident frequency of the panel is not changed significantly.

Noise Source Characteristics

As we have seen in previous sections, the typical noise source is characterized by some mechanism which imparts mechanical energy to an acoustic medium. Acoustic sources which radiate directly to an acoustic medium can be classified in the abstract such that the basic mechanisms of sound generation can be discussed.

In previous discussions a plane wave generated by a vibrating panel was covered. Experience tells us that most acoustic sources are three-dimensional and radiate energy into three-dimensional enclosures such as rooms, or into three-dimensional space such as the out-of-doors. In general, sound travels from the source in all directions instead of one only.

The acoustic propagational equations can be derived in spherical coordinates with the result as follows:

acoustic pressure,

$$p = \frac{1}{r} f(r - ct) \qquad (1\text{-}49)$$

where f denotes a function of $r - ct$ which represents waves spreading spherically from a source where r is the radial distance from the center of the source and c the speed of sound. Note that the $1/r$ term for pressure has two results.

1. This form does not apply at the location $r = 0$ because all quantities would be infinite. This is no great handicap because all sources have finite size.
2. As distance from the source increases, the magnitude of the physical quantities decreases as the reciprocal of the distance. This does not occur for plane waves.

If the distance to some observer is increased, the difference in sound-pressure level, L_p, computed from p_1 at r_1 to p_2 at r_2 will be,

$$\text{from definition of } L_p = 10 \log_{10} \left(\frac{p}{p_{\text{ref}}}\right)^2$$

$$\text{difference in } L_p = 10 \log_{10} \left(\frac{p_1}{p_{\text{ref}}}\right)^2 - 10 \log_{10} \left(\frac{p_2}{p_{\text{ref}}}\right)^2$$

$$= 10 \log_{10} \left(\frac{p_1}{p_2}\right)^2 = 10 \log_{10} \left(\frac{r_2}{r_1}\right)^2$$

If $r_2 = 2r_1$ the difference in $L_p = 10 \log_{10}(2)^2 = 10(.6) = 6$ dB. Therefore *a 6 dB decrease in L_p is noted for every doubling of distance from a spherical source.*

The total sound power being radiated from a source is given by the following formula:

$$\text{total sound power} = \rho c \overline{V^2} S\sigma \qquad (1\text{-}50)$$

Ch. 1   FUNDAMENTALS OF SOUND AND VIBRATION

where:

ρ = density of the acoustic medium (usually air)
c = speed of sound in the acoustic medium
$\overline{V^2}$ = average of the square of velocity over the surface of the source
S = total surface area of the source
σ = radiation coefficient (indicates how well the source couples to the medium)

Since ρ and c are properties of the acoustic medium, considered to be air in all of the discussions, there is little that can be done to change these quantities. $\overline{V^2}$ is the mean of the square of the velocity over the surface of the source. Note that if the amplitude of the surface vibration can be reduced the radiated sound power will decrease. In fluid type sources this is equivalent to reducing the mean flow velocity or the density fluctuations. The next term in the sound-power equation is the surface area of the source. By reducing the surface area, noise reduction can be achieved, although this might mean smaller mechanical equipment. It is almost impossible to change the surface area without affecting the surface velocity, therefore the two should be considered simultaneously. The radiation efficiency factor, σ, is dependent upon the type of source and how it couples to the air. These, then, are the methods which can be applied to a source of acoustic energy:

1. Reduce the surface velocity
2. Reduce the surface area
3. Reduce the radiation efficiency.

RESPONSE OF THE HUMAN EAR TO SOUND

In noise control it is convenient to systematize the acoustic process into three categories: *source*, *path*, and *receiver*. Thus in the previous section the source of sound was, for example, a vibrating panel and the transmission path was the air medium. In this section the receiver, which is the human ear, will be discussed.

Structure of the Ear

In discussing the ear, it is helpful to consider it as subdivided into three sections: the outer ear, the middle ear, and the inner ear. As a sensor of sound, pressure pulsations in air are gathered and directed by the outer ear, transformed into mechanical motion in the middle ear, and then converted into electrical signals in the inner ear.

The Outer Ear

The outer ear consists of the pinna and the meatus, or auditory canal, as shown in Fig. 1-18, and terminates in the tympanic membrane (eardrum). In man, the pinna is somewhat rudimentary but at high frequencies it does tend to amplify sound so that more sound energy passes into the auditory canal. The canal, about $1\frac{1}{4}$ inches long, directs the sound waves to the eardrum, a stiff, conically shaped diaphragm with flexible boundary attachments. The eardrum is excited by, and vibrates synchronously with, the sound waves.

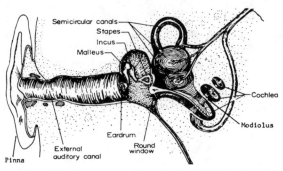

Fig. 1-18. Cross section of the human ear. From *Industrial Noise Manual*, 2nd Ed., American Industrial Hygiene Assn., 1966.

The Middle Ear

The mechanical motion of the eardrum is transmitted through the middle ear by a series of bone linkages known as the malleus, incus, and stapes (commonly known as the hammer, anvil, and stirrup). The middle-ear cavity is filled with air and communicates with the back of the nose through the Eustachian tube. Normally closed, the tube opens when one swallows, allowing the middle-ear cavity to be vented to atmospheric pressure.

Attached directly to the eardrum, the malleus transmits motion to the incus to which it is normally locked, and the incus, in turn, drives the stapes. The stapes is mounted into and sealed to the circumference of the oval "window" by a network of elastic fibers. On the other side of the oval window is the inner-ear cavity which is filled with fluid. The round window, an elastic membrane, terminates the fluid-filled canal of the inner ear.

In function, the stapes acts as a hydraulic piston converting mechanical motion to fluid motion. The motion of the three small bones of the stapes combined with the difference in area between the eardrum and the oval window results in a very effective pressure transformation, or impedance match, between the air in the outer ear and the fluid in the inner ear. This is the principal function of the middle ear.

The Inner Ear

In comparison with the middle ear, the inner ear is exceedingly complex and not yet fully understood. The essence of the middle ear is the cochlea, a snail-shaped bony shell which makes $2^3/_4$ turns around a central hollow passage containing the nerve fibers to the brain. The spiral passageway in the cochlea is filled with fluid and is divided in half by a bony shelf and a membrane called the basilar membrane. About halfway along this spiral, mounted to the basilar membrane, is the all important Organ of Corti. The Corti consists of many thousands of hair cells which are caused to bend under fluid motion. It is the bending of the small hairs which sets off the nerve impulses to the brain. After prolonged stress from excessively high sound levels, the hair cells become fatigued and die. Such damage is irreversible. It is interesting to note that the fluid in the inner

Ch. 1 FUNDAMENTALS OF SOUND AND VIBRATION

ear can also be set in motion through vibrations of the skull. Skull vibrations can be induced by direct transmission from a vibratory source, or by very high sound levels. Under such conditions, external ear protectors are shortcircuited and of limited benefit.

Performance of the Ear

From the discussion of the way the ear works, it is not surprising that the ear does not respond equally well to sound at all frequencies, nor does it respond linearly to changes in the intensity of the sound. In the very young, the range of audible frequencies extends from about 16 Hz to almost 20,000 Hz. The sensitivity of the ear increases from 16 Hz to about 1000 Hz, remains relatively constant from 1000 to 4000 Hz, and then uniformly decreases in sensitivity above 4000 Hz. Thus, the most sensitive range of the human ear is between 1000 and 4000 Hz which, not surprisingly, corresponds to the range of human speech.

Loudness

The pure-tone sensitivity of the ear for a range of sound-pressure levels is shown by the curves in Fig. 1-19. Such curves are known as *Equal Loudness*

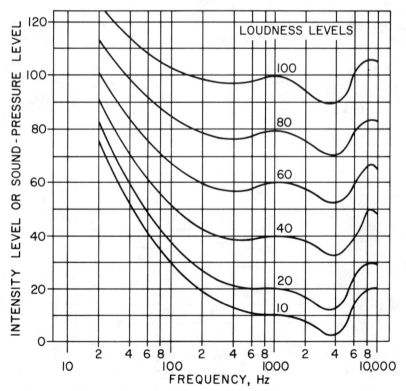

Fig. 1-19. Free field equal-loudness contours of pure tones (Fletcher-Munson curves). Source: ASA Standard Z24.2-1942.

Contours, and are constructed from the average results of hearing tests performed on a large number of people. In the tests, a person is asked to judge whether a tone is as intense, or has the same loudness, as a 1000-Hz tone at a specified sound level. For example, in Fig. 1-19, a 100-Hz tone at 60 dB would be judged to be as loud as a 1000-Hz tone at 40 dB. The *phon* is the unit of loudness level. By definition, the phon is numerically equal to the sound-pressure level at 1000 Hz. Thus, in our example the equal loudness contour which passes through the 40 dB level at 1000 Hz is called the 40 phon curve, and at 100 Hz a 60 dB pure tone has a loudness level of 40 phon.

It is interesting to note from Fig. 1-19 that as the sound level of the 1000-Hz pure tone increases, the loudness contours become more flat. When sound-pressure levels reach 100 to 120 dB, one describes the sound as being uncomfortable; and when they approach 130 dB, one experiences a tingling feeling. Above 130 dB one begins to experience actual physical pain. At these levels, the tolerance threshold does not greatly depend on frequency and one obtains much the same sensation whether the frequencies are high or low.

The response of the human ear to sound is approximated by the A-weighted network, also called the A-weighted scale, on conventional sound-level meters. The A-weighted analyzer (meter) provides a single decibel value that is the weighted summation of the sound-pressure level over the entire audible frequency range; it approximates the 40 phon contour in Fig. 1-8. Table 2-3, in Chapter 2, provides the actual A-weighted responses for various frequencies and in Chapter 12 example calculations are provided to illustrate how the A-weighted valve can be calculated from a known frequency spectrum of the sound field (See Case Studies 3 and 4). The A-weighted scale has been adopted in the Occupational Safety and Health Act of 1970 (OSHA) and in subsequent revisions for

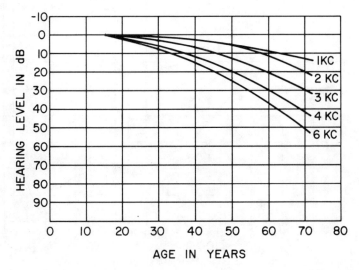

Fig. 1-20. Hearing level as a function of age (non-noise occupation). From *Industrial Noise Manual*, 2nd Ed., American Industrial Hygiene Assn., 1966.

Ch. 1 FUNDAMENTALS OF SOUND AND VIBRATION

all hearing conservation measurements related to occupational noise exposure. Two similar scales, the B-weighted and C-weighted scales, are used; however, they are not indicative of hearing conservation measurements as adopted by OSHA.

Presbycusis

As a person ages, the sensitivity of his hearing changes. The person suffers a hearing loss which begins at the age of about 20 years and becomes progressively greater. The pattern is shown in Fig. 1-20. This phenomenon, called *presbycusis*, shows that significant loss in hearing in the frequencies above 3000 Hz can occur during the working life of the person. It is to be emphasized that these curves are average results and are based on tests with people who were in non-noise occupations. Thus, presbycusis is a naturally occurring loss in hearing, over and above that which might be suffered in a noisy occupation.

BASIC FUNDAMENTALS OF NOISE CONTROL

In the solution of noise problems, it is very convenient to view the system in terms of a source, a path, and a receiver. In block diagram this would appear as shown in Fig. 1-21. The source is the point or points in space where the noise

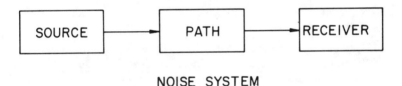

NOISE SYSTEM

Fig. 1-21. Schematic of a noise system.

apparently originates, and the path is the line or lines (rays) in air along which the noise can be considered to propagate to the receiver, or the ear. More complicated situations would, of course, have multiple sources, paths, and receivers with multiple interaction between them, yet the complexity is one of degree only and the approaches to solution are the same.

Acoustic sources were described as fluctuating forces acting on a fluid medium. The origin of the fluctuating forces could be surface motion of solid objects or fluctuating fluid properties such as turbulent flow. A schematic diagram showing the relation of mechanical energy and sound transmission is shown in Fig. 1-22. Energy from time-dependent forces flows along a direct path or by one or more indirect paths. The objective is to reduce the noise at the receiver and this may be accomplished by modifications to either the source, the path, or the receiver, or to all three. The design engineer has the task of specifying the mechanical construction of the system to limit the levels of noise to acceptable values. In accomplishing this, computations or measurements must be made at some point in the system to determine the magnitude and frequency of the time-varying quantities and their proper engineering techniques must be applied to reduce them.

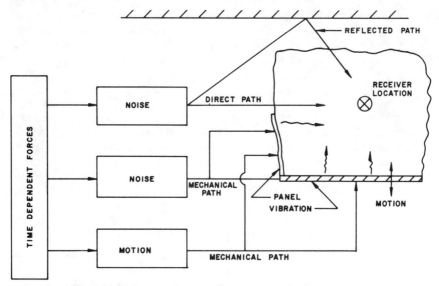

Fig. 1-22. Schematic diagram of noise and motion transmission.

Source Control

Generally, modifications at the source are considered to be the best solution. The noise-producing energy is concentrated at the source, and if the cause can be sufficiently reduced or eliminated there need be no worry about relatively cumbersome and often more expensive path and receiver treatments. There are many instances, however, where source modifications are not practicable or even possible, and the reader will become more aware of this fact as he proceeds into the appropriate chapters of this book and begins designing for and controlling noise in his particular area.

For the present example, in Fig. 1-22, source control may be achieved by reducing 1. the area of the panel, and/or 2. the amplitude of the vibration of the panel. In (1), a panel becomes an efficient radiator of sound when its dimensions begin to approach one wavelength of the sound in air. Thus, if the offending noise is such that the smallest wavelength (highest frequency) of the noise is larger than the smallest permissible panel size, then reductions in noise could be expected by going to a smaller panel. Reducing the amplitude of the vibration (2), is always beneficial; an explanation of the fundamentals will be given in the last section of this chapter.

Noise Control in the Transmission Path

Controlling the noise along its path involves some kind of modification to the space enclosing both source and receiver. For the direct path, a barrier constructed of impervious materials is necessary to reflect the noise back toward the source. For indirect paths, it is necessary to use an absorptive material on the reflecting surfaces to absorb the sound energy. The fundamental differences in these two concepts and the materials used in implementing them cannot be overemphasized, and will be dealt with, in detail, in the following chapters; most particularly in Chapter 5, "Acoustic Absorption and Transmission-Loss Materials." There exists a parallel difference in vibration control regarding the concepts and materials used in *vibration isolation* and *vibration damping* for noise reduction; these, too, will be detailed in the following chapters, particularly in Chapters 6 and 7.

Some of the usual techniques for control of sound in the transmission path are:
1. Construction of enclosures (panels and transmission characteristics)
2. Mufflers
3. Vibration isolation
4. Break mechanical paths (vibration absorption material added between metal parts)
5. Lengthening of transmission paths
6. Absorption of acoustical energy with acoustic treatments (acoustic tile, porous material, etc.) in the acoustic path
7. Constructing heavy air-tight enclosures.

In some applications more than one of the techniques must be applied. For example, in the case of extremely noisy equipment it may be necessary to apply all of these techniques.

Noise Control at the Receiver

The usual receiver of sound is the human ear and, unfortunately, there is nothing that can be done to alter its response to noise. As a result, if all other efforts have failed to lower the noise levels at locations humans must occupy, there are only a few recourses remaining. If noise levels are excessive one solution is to remove humans from the area and use remote control devices to operate machinery.

Another method of receiver control is the use of an operator booth with viewing windows. An operator booth is a three- or four-sided barrier or enclosure in which the operator stays when the operation of his machine requires monitoring only. This is a means for reducing the noise exposure time of the operator without need for wearing personal protectors.

Controlling noise directly at the ear by use of ear plugs or ear muffs can be very effective in most industrial environments; in some, it may well be the only practical means. Nevertheless, there are potential problems in their effective use, thus this method of control has been relegated by legislative bodies to interim use only, should other methods be possible, or to a means of last resort if no other methods are possible. The reasons for objection to the use of ear protectors are that: 1. Warning signals may be inaudible; 2. For hygienic reasons, in some environments; and 3. Because of the difficulties in enforcement of their use. When ear protectors must be used an educational program should also be instituted to make sure that people will use the protection devices.

To summarize this section, the diagram in Fig. 1-23 illustrates the nature of acoustic problems showing the sources, coupling to the acoustic mediums, transmission, and reception of sound. For each of these mechanisms general noise control methods are listed.

42 FUNDAMENTALS OF SOUND AND VIBRATION Ch. 1

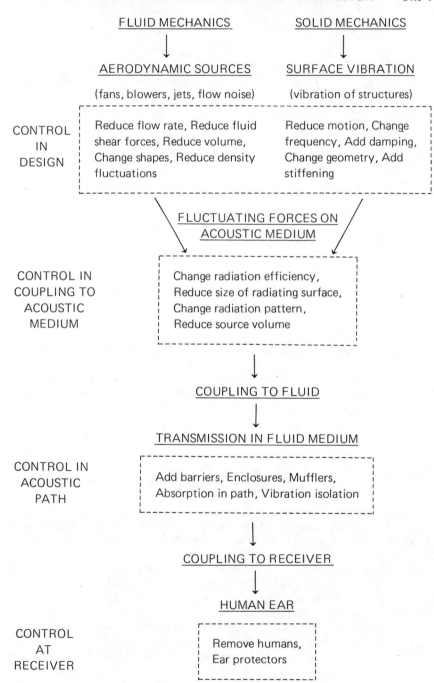

Fig. 1-23. Summary of some noise-control methods.

Ch. 1   FUNDAMENTALS OF SOUND AND VIBRATION   43

FUNDAMENTALS OF VIBRATION

In the practice of controlling noise in industry, one almost invariably encounters a form of mechanical vibration that plays a major role in the generation and transmission of noise. For example, in punch presses, a major radiator of noise is the die plate of the press which is caused to vibrate by the large forces arising in the punching operation. In hydraulic and pneumatic systems, transmitters and radiators of noise include vibrating pump housings, mounting structures, sheet metal enclosures, and piping systems. These are but a few examples.

In this section the basic elements of vibratory motion and the various forces which initiate and sustain such motion will be described and explained. Detailed accounts of controlling vibratory motion by damping and isolation are to be found in Chapters 6 and 7, and a thorough description of vibration in machine elements can be found in Chapter 8.

PHYSICAL DESCRIPTION OF VIBRATORY MOTION

Undamped Vibration

Mechanical vibration is a to-and-fro motion of a body about an equilibrium point in space. The point itself can be moving or stationary. This concept has been alluded to numerous times and illustrated in previous sections, particularly in Fig. 1-1. What must yet be touched upon is the mechanism by which the motion or force of vibration can continue to build up to very high amplitudes exceeding many times that of the input motion or force. The mechanism is illustrated most simply by a child on a swing. If the swing is given a slight push at the right time in each cycle it will move out farther and farther with each push. If the motion does not become excessive, all the ingredients of a simple vibratory system are present: (1) a *mass*, and (2) a *restoring force*. The mass is predominantly that of the child on the swing and the restoring force comes from that component of gravity perpendicular to the swing's support chain.

The swing, or pendulum, represents the simplest of all vibratory systems, having only one identifiable mass and restoring force. Such an arrangement is called a *one-degree-of-freedom* system and is illustrated schematically in Fig. 1-24 in both translational and rotational forms. In the translational form, an oscillating force F (lb) is driving a mass m (lb), which is restrained elastically by a spring K (lb/in). In the rotational form an oscillating torque T (in-lb) is driving a mass moment of inertia I (lb-in^2), which is restrained by a torsional

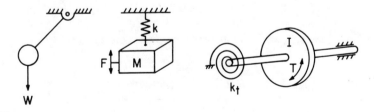

Fig. 1-24. Simple vibratory systems.

spring K_t (in-lb/radian). Note that the force (or torque) could also drive the system from the other end; that is, through the opposite end of the spring.

Energy of Free Vibration

If the applied forces are removed from either system in Fig. 1-24, and the system is given an initial disturbance and then set free, it will seek its own frequency of vibration, called the *natural frequency*. In such a state, the kinetic energy of the mass is always out of phase with the potential energy of the spring such that when one is maximum the other is minimum; the sum of the two, however, remains a constant throughout each cycle of oscillation, as shown in Fig. 1-25.

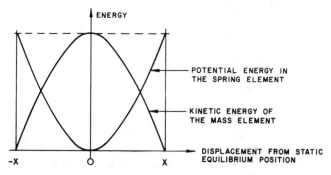

Fig. 1-25. Energy diagram for a simple conservative oscillatory system.

Natural Frequency

The natural frequency for the single-degree system can be calculated by equating the maximum potential energy to the maximum kinetic energy. Thus,

Max. potential energy = Max. kinetic energy

$$\tfrac{1}{2} KX^2 = \tfrac{1}{2} MV^2 \tag{1-51}$$

where X and V are the displacement and velocity amplitude, respectively, in the sinusoidal motion given by

$$x = X \sin(2\pi ft) \tag{1-52}$$

and

$$v = \frac{dx}{dt} = X(2\pi f) \cos 2\pi ft = V \cos 2\pi ft \tag{1-53}$$

Thus $V = 2\pi fx$ which we then substitute into Equation (1-51) to obtain

$$\tfrac{1}{2} KX^2 = \tfrac{1}{2} M 4\pi^2 f^2 X^2 \tag{1-54}$$

or, upon solving for the frequency, f, the result is that

$$f = \frac{1}{2\pi} \sqrt{\frac{K}{M}} \text{ Hz} \tag{1-55}$$

Ch. 1 FUNDAMENTALS OF SOUND AND VIBRATION

The same thing can be done for the rotational system, beginning with

$$\text{Max. } E_p = \text{Max. } E_k$$

$$\tfrac{1}{2} K_t \theta^2 = \tfrac{1}{2} I \omega^2 \tag{1-56}$$

and ending with

$$f = \frac{1}{2\pi} \sqrt{\frac{K_t}{I}} \text{ Hz} \tag{1-57}$$

In either case note that the natural frequency can be raised or lowered by increasing or decreasing the spring stiffness to mass ratio.

Equation of Motion

The equation of motion for the one-degree-of-freedom system can be obtained from the schematic diagrams in Fig. 1-26. The spring symbol represents

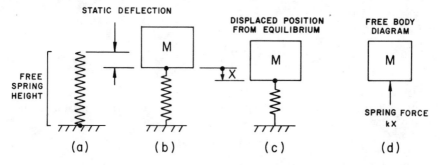

Fig. 1-26. Geometry and free-body diagrams for a spring-mass oscillatory system.

the total spring constant under a machine, and M denotes the mass. In Fig. 1-26a is shown the free length of the spring without the mass. Shown in Fig. 1-26b is the *static deflection* of the spring with the mass resting on the spring. From the spring rate of the spring, K (lb/in.), the static deflection can be determined

$$\text{static deflection} = \Delta_{\text{static}} = \frac{Mg}{K} \tag{1-58}$$

with:

Δ_{static} = static deflection, in.
Mg = weight of mass, lb
g = acceleration of gravity, 32.2 ft/sec^2 = 386 in./sec^2
M = mass
K = spring rate

A general displacement from the static equilibrium position is illustrated in Fig. 1-26c where x is the displacement measured from the static position. The free-body diagram for the mass is given in Fig. 1-26d, with the spring force shown acting upward, equal to Kx. Applying Newton's Second Law to the mass,

$$\text{Change in linear momentum} = \text{sum of applied forces} \quad (1\text{-}59)$$

$$\frac{d}{dt}(MV) = \sum F$$

$$\frac{d}{dt}\left(M\frac{dx}{dt}\right) = \sum F$$

In the simple system of Fig. 1-24 the mass, M, is constant, therefore

$$M\frac{d^2x}{dt^2} = \sum F \quad (1\text{-}60)$$

Since the only applied force to the mass in this system is the spring force, Kx:

$$M\frac{d^2x}{dt^2} = -Kx$$

or

$$M\frac{d^2x}{dt^2} + Kx = 0 \quad (1\text{-}61)$$

which is the equation of motion for free vibration (no applied forcing function). The solution to this differential equation gives the motion of the mass as a function of time:

$$x(t) = X \sin(\sqrt{K/M}\ t) \quad (1\text{-}62)$$

where X is the *amplitude* of the motion and the frequency of the motion is given by $\sqrt{K/M}$ rad/sec. The motion described by Equation (1-62) is called *simple harmonic* motion. Note that for this case of free vibration the amplitude of the motion, X, depends upon the initial displacement of the system.

Forced Undamped Motion

When a vibration is started and sustained by shocks or impulses which are spaced at long intervals of time compared with the period (reciprocal) of the natural frequency, then the vibration occurs mainly at one natural frequency or at several natural frequencies. The gong of a bell, for instance, is one example of vibration at a natural frequency. If an excitation force is applied then the vibration will occur at the frequencies of the excitation force. For example, if the excitation is a sinusoidal force, the vibration will occur at the frequency of the excitation. If it is complex, the vibration will occur at those harmonic frequencies given by Equation (1-5). Finally, if it is a random force then the vibration will be at numerous frequencies. The amplitude of the vibration, however, will not be the same at all frequencies. Firstly, the amplitude will be proportional to the force at that frequency, and secondly, the amplitude will depend on how easily the vibration system responds at that frequency. For a single-

Ch. 1　　FUNDAMENTALS OF SOUND AND VIBRATION　　47

degree-of-freedom system excited by a sinusoidal forcing function, the equation of motion is

$$M\frac{d^2x}{dt^2} + Kx = F \sin (2\pi f_i t) \tag{1-63}$$

where f_i is the forcing frequency. The amplitude of the forced motion is equal to

$$\text{amplitude of } x = \left| \frac{F/K}{1 - \frac{f_i}{f_n}} \right| \tag{1-64}$$

The vertical bars denote the absolute value of the function. f_i is the forcing frequency and f_n is the natural frequency of the system.

The force transmitted to the foundation is the force in the spring which is given by the spring rate times the deflection of the mass, upon the *initial displacement* of the mass before release.

From Equation (1-55) the natural frequency can be formulated in terms of the static deflection as follows:

$$f_n = \frac{1}{2\pi} \sqrt{K/M} \text{ Hz} \tag{1-55}$$

From Equation (1-58),

$$K = \frac{Mg}{\Delta_{static}}$$

therefore

$$f_n = \frac{1}{2\pi} \sqrt{\frac{Mg}{M\Delta_{static}}} = \frac{1}{2\pi} \sqrt{\frac{g}{\Delta_{static}}} \text{ Hz}$$

for Δ_{static}, given in inches, the form becomes,

$$f_n = \frac{1}{2\pi} \sqrt{\frac{386}{\Delta_{static} \text{ (inches)}}} \text{ Hz} \tag{1-65}$$

The natural frequency can thus be related to the static deflection of the mass (machine) on the isolators (springs). Usually, this is the method used when specifying an isolation system, to give the static deflection of the springs.

The force transmitted to the foundation through the spring is equal to the displacement of the mass times the spring rate, as follows:

$$\text{transmitted force} = Kx = \left| \frac{F}{1 - \frac{f_i}{f_n}} \right| \tag{1-66}$$

48 FUNDAMENTALS OF SOUND AND VIBRATION   Ch. 1

The amplitude of the transmitted force divided by the amplitude of the applied force is called the *transmissibility* given by:

$$\left|\frac{F_{transmitted}}{F_{applied}}\right| = \text{transmissibility} = \left|\frac{1}{1 - \frac{f_i}{f_n}}\right| \quad (1\text{-}67)$$

Forced Motion with Damping

If a damping element is connected in parallel with the spring element, as in Fig. 1-26, and a sinusoidal forcing function is applied to the mass, the equation becomes,

$$M\frac{d^2x}{dt^2} + R\frac{dx}{dt} + Kx = F\sin(2\pi f_i t)$$

where R is the damper coefficient lb/(in.-sec). For steady-state amplitude of the mass is

$$\text{amplitude of } x = \frac{F/K}{\sqrt{\left[1 - \left(\frac{f_i}{f_n}\right)^2\right]^2 + \left(\frac{R 2\pi f_i}{K}\right)^2}} \quad (1\text{-}68)$$

and the transmitted force is now the force in the spring plus the force in the damper

$$\text{transmitted force} = Kx + R\frac{dx}{dt} = \frac{F\sqrt{1 + \left(\frac{R 2\pi f_i}{K}\right)^2}}{\sqrt{\left[1 - \left(\frac{f_i}{f_n}\right)^2\right]^2 + \left(\frac{R 2\pi f_i}{K}\right)^2}} \quad (1\text{-}69)$$

With damping, the transmissibility is

$$\text{transmissibility} = \frac{\sqrt{1 + \left(\frac{R 2\pi f_i}{K}\right)^2}}{\sqrt{\left[1 - \left(\frac{f_i}{f_n}\right)^2\right]^2 + \left(\frac{R 2\pi f_i}{K}\right)^2}} \quad (1\text{-}70)$$

For a damped system the frequency of oscillation occurs at the *damped natural frequency*, f_d, given by

$$f_d = f_n \sqrt{1 - \frac{R^2}{4KM}} \text{ Hz} \quad (1\text{-}71)$$

Note that damping lowers the natural frequency of the system.

A graph of transmissibility is given by Fig. 1-27 for a constant magnitude applied force while varying the frequency of the applied force. When the frequency of the applied force is less than the natural frequency, the force transmitted to the foundation is equal to, or greater than, the applied force. When the frequency of the applied force is equal (or near) the natural frequency, the

Ch. 1   FUNDAMENTALS OF SOUND AND VIBRATION   49

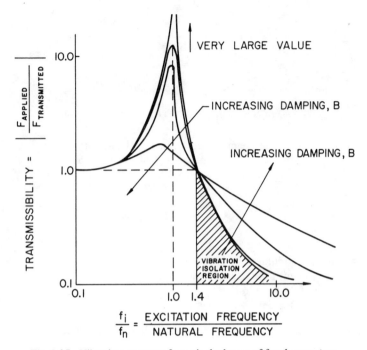

Fig. 1-27. Vibration response for a single-degree-of-freedom system.

force transmitted to the foundation is very large. This is called a *resonant vibration condition*.

When the excitation frequency exceeds the natural frequency by a factor of 1.4, the response amplitude actually becomes less than the excitation amplitude. This is the region of *vibration isolation* for the forced vibration condition.

Note in Fig. 1-27 that when the forcing frequency is greater than 1.4 times the natural frequency, the addition of damping *increases* the transmitted force to the foundation. Thus it can be concluded that, in the isolation region, damping is not desired in an isolation system. For the condition that the forcing frequency is less than 1.4 times the natural frequency, the addition of damping *decreases* the transmitted force. In this region damping is desirable in the isolation system.

Vibration control is based upon designing the system such that the ratio of exciting frequency to natural frequency is in the vibration isolation region. In general, this requires the natural frequency given by $f_n = \frac{1}{2\pi}\sqrt{K/M}$ to be low compared to the applied frequency. This can be achieved in one of two ways, by using a low value of K or by increasing the mass of the machine by adding a base. The details of vibration control are described in detail in Chapter 7.

Equivalent Springs

Quite often a vibratory system that appears complicated can be simplified and analyzed as a single-degree-of-freedom system. Take, for example, a mass cen-

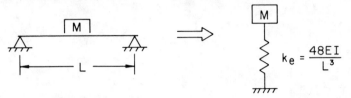

Fig. 1-28. Equivalent spring rate of a simply supported beam.

trally supported on a free beam as shown in Fig. 1-28. From any strength-of-materials handbook we know that the deflection of the beam due to the mass loading is

$$X = \frac{Fl^3}{48\,EI} \tag{1-72}$$

where:

E = Young's modulus, in psi
I = area moment of inertia of beam, in inches
l = length of beam, in inches
$F = Mg$, the force due to the mass.

The equivalent spring rate then, is the force divided by the deflection,

$$K_e = \frac{F}{X} = \frac{48\,EI}{l^3} \tag{1-73}$$

Springs may be combined in several different ways: in parallel, in series, or in a combination of these. The equivalent spring for each combination is shown in Fig. 1-29.

Effective Mass

As with springs, a complicated mass system can often be simplified or approximated in terms of a single, *effective mass*. Take, for example, the rocking-arm arrangement shown in Fig. 1-30. The effective mass of the system can be found readily by calculating the kinetic energy. This kinetic energy equation is

$$\begin{aligned}
T &= \frac{1}{2} J\dot{\theta}^2 + \frac{1}{2} M(a\dot{\theta})^2 \\
&= \frac{1}{2}\left(\frac{J}{a^2} + M\right)(a\dot{\theta})^2 \\
&= \frac{1}{2} M_e v^2
\end{aligned} \tag{1-74}$$

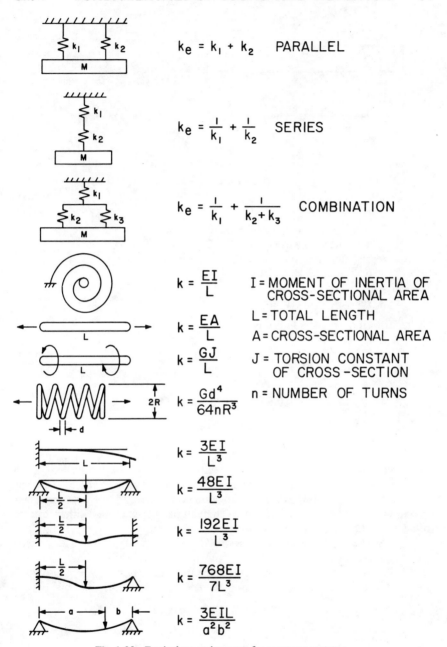

Fig. 1-29. Equivalent spring rates for common systems.

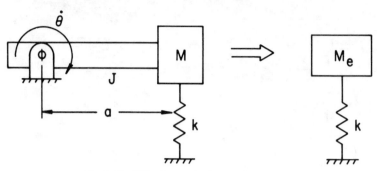

Fig. 1-30. Effective mass of a rotating system.

where:

J = polar mass moment of inertia of the arm
M = mass attached to the arm
a = distance from pivot to spring attachment
$\dot{\theta}$ = peak rotational velocity
M_e = effective mass
v = velocity at the spring attachment.

Hence, the effective mass in this case is:

$$M_e = \frac{J}{a^2} + M \qquad (1\text{-}75)$$

CAUSES OF VIBRATION

Mechanical vibrations are caused by forces which vary in time. By Newton's Law of Motion a time-varying force produces a time-varying change in momentum or velocity of the mass to which the force is applied. These forces may be considered then, to be the input or excitation to the vibration system. There are many different ways in which time-varying forces arise in the operation of machinery. Fortunately, these forces can be categorized for discussion purposes and identified in operation by a characteristic "signature," either in terms of a motion-time plot or a motion-frequency spectrum.

Rotational and Translational Unbalanced Forces: A rotary unbalance occurs when the axis of rotation does not coincide with one of the principal axes of inertia of the rotating body. The unbalance force is predominately sinusoidal in time but it may contain higher harmonics. The fundamental is given by

$$f = \frac{n}{60}$$

where n is the rotational speed in rpm.

Translatory unbalanced forces may occur whenever the velocity, that is, when either speed or direction of a machine member changes. If these forces are not cancelled by equal forces in the opposite directions, then unbalance forces will

Ch. 1   FUNDAMENTALS OF SOUND AND VIBRATION   53

be present. Since the magnitude and direction of such forces depends on the rate of change of velocity, or acceleration of the particular member, the resulting vibration will, in general, be quite complex but still periodic. In a slider crank mechanism, for example, the second harmonic of the rotational frequency is the predominant force component, whereas in a scotch-yoke mechanism it is the first harmonic. In the latter case the slide moves sinusoidally, whereas in the former, it is non-sinusoidal because of the angularity of the connecting rod.

Rotary and translatory forces are transmitted into the machine frame through the support bearings. If the forces are not cancelled by either balancing or countermoving elements, the frame, itself, must move in direct opposition to these forces such that the center of gravity of the machine remains stationary. This is the manner by which internal unbalanced forces result in external motion of the machine.

Forces from Rolling and Sliding Elements

Machine elements such as rolling element bearings, gears, friction brakes, and clutches produce vibratory forces by a rolling and/or sliding action. Rolling action between a ball and a race, or between two gear teeth produces elastic deformations and stress waves which occur periodically every time a roller passes a given point, or when two gear teeth mesh. Bearing-race and gear-teeth deformations introduce additional vibration mechanisms.

Frictional processes cause vibration-producing forces in at least two ways: First, random shearing forces occur between mating surfaces at relative speeds. The second process, known as "stick-slip," is much more complicated and usually depends on the geometry, the surface finish, and the frictional properties of the two mating surfaces. Most often it involves some force feedback between structures supporting the mating surfaces which sustains and amplifies the stick-slip. The screech of chalk against a blackboard is a simple case of stick-slip at work.

Electrical, Hydraulic, and Aerodynamic Forces

Vibratory forces in rotating electrical machinery originate in the air gap between stator and rotor. The magnetic flux oscillates at line frequency, but the magnetic force is proportional to the square of the flux. Therefore, the principal vibratory forces occur not at line frequency but at twice line frequency. This is the frequency of the hum one normally hears from electric motors. The periodic force in the air gap has components in both tangential and radial directions. In the tangential direction, this results in a periodic variation in motor torque, and, in the radial direction, a periodic bending of the rotor shaft and stator housing.

In hydraulic systems, the cyclic compression and release of fluid is a major source of vibration. In piston and vane type pumps, a slug of fluid is drawn into a chamber, compressed, and then suddenly released into the discharge port. The periodic compression and release of fluid produces forces at harmonics of the fundamental discharge frequency. These forces can be particularly large if there is a sizeable difference in the compression and discharge port pressures at the instant the port is opened. Under such conditions, a compressional wave of very high amplitude is propagated in the discharge pipe.

Aerodynamically induced sources of mechanical vibration generally occur when a turbulent flow interacts with the wall of a pipe or duct carrying the flow. This most often occurs when high velocity flow is forced through rapid changes in duct geometry as in orifices, valves, etc.

BASIC FUNDAMENTALS OF VIBRATION CONTROL

As mentioned previously, controlling noise at the source is the most desirable approach. In this context, the word "source" refers to the apparent origin or radiator of the noise. Most often this is some vibrating panel, housing, or other structural or nonstructural member of the machine which is responsive to vibration excitation and which is coupled efficiently to the vibrational energy source within the machine. From this description we can imagine a simple vibration system to consist of a vibration source, a vibration path, and a radiator surface with, possibly, a vibration feedback, as shown in Fig. 1-22, in those cases where self-excited vibrations occur. Examples of the latter are brake squeal and machine-tool chatter. More complicated systems would have multiple sources, paths, and surfaces.

In the simplest case, the radiation source and the vibration source are so closely coupled that they appear to be one and the same. In more extreme cases the vibrational path may be so long and contorted, that identifying the radiator with the vibrational source may be difficult. For example, vibrations fed into pipes, ducts, and the structure of a building can be transmitted great distances before coupling to an efficient radiator such as the wall of a room. The methods of vibration control, therefore, center on modifications to the vibration source, the vibration path, and the radiation surface.

Vibration Source Control

Modifications to the vibration source can be numerous and, of course, depend on the particular process. Rotational unbalance can be corrected by dynamic balancing, and unbalanced forces in translating members can be reduced by minimizing the peak accelerations and the number of times in a cycle that the member must undergo a change in velocity. Improvements in surface finish on rolling and sliding parts will reduce forces caused by roughness and waviness. Impact forces at link pivots can be reduced by reducing the running clearances.

Structural Stiffening

Structural stiffening involves the addition of gussets, ribs, etc., to a structure in order to raise its natural frequencies above the range of the excitation frequencies. In practice this is often difficult to do because stiffeners add mass back into the structure which tends to counter the added stiffness. Remember that natural frequencies are proportional to the ratio of stiffness to mass.

Detuning and Decoupling

Detuning a structure consists of modifications to its component parts so that coincidences of resonance frequencies are avoided. This may involve increasing either stiffness or mass, rearrangement of attachment points, etc. *Decoupling*, by contrast, involves a more rigid attachment of structural components together

Ch. 1 FUNDAMENTALS OF SOUND AND VIBRATION

to reduce the number of degrees-of-freedom and hence, the number of resonance frequencies.

Vibration Isolation and Absorption

The use of resilient supports between a machine and its foundation or between any two components of a machine to reduce the transmissibility is called *vibration isolation*. The objective in isolation is to choose the spring rate of the support such that the natural frequency of the machine on its support is sufficiently low compared to the frequency of excitation. Figure 1-27 shows that if the excitation frequency f_i is constant, the transmissibility is reduced when the natural frequency is reduced. The usual way to reduce the natural frequency is to reduce the spring rate which results in a larger static deflection. Limits on static deflection and stability of the machine usually dictate how soft the mount can be, however, thus restricting the amount of isolation that can be obtained.

Vibration absorption is a technique somewhat the reverse of decoupling, in which an additional mass and spring are added to, say, transform a one-degree-of-freedom system into a system having two degrees of freedom. The technique is useful when a machine is operating unavoidably at or about a resonance frequency. The addition of the second mass-spring unit, in effect, shifts the natural frequencies away from the resonance point. The term is, unfortunately, a misnomer since the vibration is not "absorbed" in the sense of an energy transformation.

Vibration Damping

Vibration damping is an absorption technique in which vibrational energy is transformed into heat. The most common method is to apply a layer of a viscoelastic material to the surface of a vibrating part to reduce the amplitude of vibration at resonance. It is most effective on thin panels of large area; such panels usually being primary noise radiators. In some instances, the internal damping of the material used in machine construction can effectively reduce resonant vibrations; e.g., gray cast iron is used on machine-tool frames because of its vibration damping characteristics. New composite materials can sometimes be used as structural elements to reduce vibration. Where severe vibrations are apt to occur, special consideration should be given to the material properties used in construction. The use of damping materials for vibration control will be considered later, in Chapter 6.

SUGGESTED MACHINERY NOISE CONTROL METHODS

The techniques for noise control, described previously in this chapter, were presented generally and specific machinery applications were not given. The most desirable noise-control technique is to alter the source by redesigning the mechanical elements to reduce forces or by altering the sound-radiating surfaces. The noise radiation characteristics of basic machine elements are described in Chapter 8, "Machine Element Noise," with guidelines presented for preferred designs for minimizing noise.

Based on case histories of industrial noise control, recommended machinery noise treatments, excluding modification of the source, are listed in Table 1-2. Implementation methods for the various recommended noise treatments given in the table are described in detail in later chapters. The use of sound absorbing materials and sound barriers for controlling undesirable noise is presented in Chapter 5. Control of vibration and subsequently the reduction of noise by the use of vibration damping materials is described in Chapter 6: Damping Materials

Table 1-2. Recommended Noise-Control Techniques other than Modifying the Source.*

Equipment Type	Recommended Noise Treatment			
	Sound Absorption	Sound Barrier	Vibration Damping	Vibration Isolators
Screw Machines		X		X
Punch Presses	X	X	X	X
Cold Formers		X		X
Tumblers		X		X
Grinders	X	X		X
Boring Machines		X		X
Lathes				X
Broaching Machines				X
Hobbers				X
Welders		X	X	X
Eyelet Machines	X	X		X
Riveting Machines	X	X	X	X
Slitters		X		X
Saws		X	X	X
Jointers		X		X
Planers		X		X
Grinders	X	X		X
Routers		X		X
Chutes, Bins, Hoppers & Conveyors		X	X	
Vibratory Feeders		X		X
Ductwork (High Pressure & Material Handling)		X	X	
Ductwork (Low Pressure)			X	X
Fans & Blowers		X		X
Pumps		X		X
Valves, Piping		X		X
Precision Equipment				X
HVAC Equipment		X		X
Compressors		X		X
Transformers				X
Generators	X	X		X
Printing Machines	X	X		X
Treatment for Equipment Housings	X	X	X	

*Courtesy of Consolidated Kinetics Corporation.

Ch. 1   FUNDAMENTALS OF SOUND AND VIBRATION   57

For Vibration And Sound Control. The use of vibration isolators for noise reduction is presented in Chapter 7: Vibration Control For Noise Reduction. Detailed case studies utilizing several noise reduction methods are presented in Chapter 12: Case Studies.

GLOSSARY

Acoustic Terms

Absorption Loss—That part of the transmission loss due to the dissipation or conversion of sound energy into other forms of energy (e.g. heat), either within the medium or within an acoustic material upon reflection.

Acoustic Radiation Pressure—A pressure exerted upon a surface exposed to an acoustic wave.

Acoustic Resistance—The real component of the acoustic impedance.

Ambient Noise—All-encompassing noise associated with a given environment, usually a composite of sounds from many sources, near and far.

Attenuation—The reduction in sound power as sound propagates through a duct or enclosure. It is expressed in decibels as the ratio of the sound power at a point upstream, to the sound power at a point downstream, in the direction of the sound propagation. For a duct, attenuation is expressed in units of decibels per foot (dB/ft) indicating the average reduction of sound-power level for each foot of duct length.

Audible Range (of Frequency) (Audio-Frequency Range)—The frequency range 16 Hz to 20,000 Hz (20 kHz). *Note*: This is conventionally taken to be the normal frequency range of human hearing.

Audiometer—An instrument for measuring the threshold, or sensitivity, of hearing.

Audiometry—The measurement of hearing.

Band Pressure Level—The band pressure level of a sound for a specified frequency band is the sound-pressure level for the sound contained within the restricted band. The reference pressure must be stated. The band may be specified by its lower and upper cut-off frequencies, or by its geometric center frequency and bandwidth. The width of the band may be indicated by a prefactory modifier; e.g., octave-band (sound-pressure) level, half-octave-band level, third-octave-band level, 50-Hz-band level.

Beats—Periodic variations in sound pressure that result from combining two pure-tone waves of different frequencies f_1 and f_2. *Example*: Two sound sources radiate pure tones at 250 and 245 Hz, respectively. The beat frequency would be $f_1 - f_2 = 250 - 245 = 5$ Hz.

Broad-Band Noise—Noise whose energy is distributed over a broad range of frequency (generally speaking, more than one octave).

Continuous Noise—On-going noise whose intensity remains at a measurable level (which may vary) without interruption over an indefinite period or a specified period of time.

Continuous Spectrum—The spectrum of a wave the frequency components of which are continuously distributed over a frequency region.

Cycle—The complete sequence of values of a periodic quantity that occur during a period.

Deafness—100 percent impairment of hearing associated with an otological condition. *Note*: This is defined for medical and cognate purposes in terms of the hearing threshold level for speech or the average hearing threshold level for pure tones of 500, 1000, and 2000 Hz in excess of 92 dB.

Decibel—The decibel is a mathematical scale that is similar to a logarithmic scale used to describe the level of a physical quantity compared to a reference; it compresses the quantity into numbers that are convenient for data presentation. Mathematically, it is ten times the logarithm (base 10) of the ratio of a power or energy quantity with respect to a reference base of the same physical quantity.

Diffuse Field—A field opposite to a free field in that multiple reflections from boundaries produce a field where the sound pressure is uniform. In practice, a diffuse field can be approached by having all hard surfaces in a room.

Echo—A wave that has been reflected or otherwise returned with sufficient magnitude and delay to be detected as a wave distinct from that directly transmitted.

Effective Sound Pressure (Root-Mean-Square Sound Pressure)—The effective sound pressure at a point is the root-mean-square value of the instantaneous sound pressures over a time interval at the point under consideration. In the case of periodic sound pressures, the interval must be an integral number of periods or an interval that is long compared to a period.

Efficiency—The efficiency of a device with respect to a physical quantity which may be stored, transferred, or transformed by the device is the ratio of the useful output of the quantity to its total input.

Environmental Noise—By Sec. 3 (Ref. 11) of the Noise Control Act of 1972, the term "environmental noise" means the intensity, duration, and character of sounds from all sources.

Equivalent Sound Level—The level of a constant sound which, in a given situation and time period, has the same sound energy as does a time-varying sound. Technically, it is the level of the time-weighted, mean square, A-weighted sound pressure. The time interval over which the measurement is taken should always be specified.

Free Field—A sound field where there are no boundaries present to reflect sound waves. In practice, it is a sound field in which the effects of the boundaries are negligible over the region of interest. A free field can be approached if walls, floor, and ceiling of a room are covered with absorptive materials.

Frequency—The frequency of a function periodic in time is the reciprocal of the period. The unit is the cycle per unit time and must be specified. The unit cycle per second is commonly called Hertz (Hz).

Hearing Level—The difference in sound-pressure level between the threshold sound for a person (or the median value or the average for a group) and the reference sound-pressure level defining the standard audiometric threshold. *Note*: The term is now commonly used to mean hearing threshold level (qv). Units: decibels.

Hearing Loss—Impairment of auditory sensitivity: an elevation of a hearing threshold level.

Hearing Threshold Level—The amount by which the threshold of hearing for an ear (or the average for a group) exceeds the standard audiometric reference zero. Units: decibels.

Ch. 1   FUNDAMENTALS OF SOUND AND VIBRATION   59

Impulse Noise (Impulsive Noise)—Noise of short duration (typically, less than one second) especially of high intensity, abrupt onset, and rapid decay, and often rapidly changing spectral composition. *Note*: Impulse noise is characteristically associated with such sources as explosions, impacts, the discharge of firearms, the passage of supersonic aircraft (sonic boom), and many industrial processes.

Infrasonic—Having a frequency below the audible range for man (customarily deemed to cut off at 16 Hz).

Infrasonic Frequency—A frequency below the audio-frequency range.

Insertion Loss—The difference in sound-pressure level at a point after a modification has been made in either the source or the transmission path between the point and the source. It is expressed in decibels as the ratio of sound pressure before the modification to the sound pressure after completing the modification. Since it is quite possible for the sound pressure to increase with the modification, the insertion loss can have a negative value.

Intermittent Noise—Fluctuating noise whose level falls one or more times to low or unmeasurable values during an exposure.

Level—In acoustics, the level of a quantity is the logarithm of the ratio of that quantity to a reference quantity of the same kind. The base of the logarithm, the reference quantity, and the kind of level must be specified.

Microbar—A unit of pressure commonly used in acoustics. One microbar is equal to 1 dyne per square centimeter.

Microphone—An electroacoustic transducer that responds to sound waves and delivers essentially equivalent electric waves.

Noise—1. Any undesired sound; and, by extension, any unwanted disturbance within a useful frequency band, such as undesired electric waves in a transmission channel or device. 2. An erratic, intermittent, or statistically random oscillation. Since the definitions of noise are not mutually exclusive, it is usually necessary to depend upon context for the distinction.

Noise Exposure—The cumulative acoustic stimulation reaching the ear or the person over a specified period of time (e.g., a work shift, a day, a working life, or a lifetime).

Noise Hazard (Hazardous Noise)—Acoustic stimulation of the ear which is likely to produce noise-induced permanent threshold shift in some of a population.

Noise-Induced Permanent Threshold Shift (NIPTS)—Permanent threshold shift caused by noise exposure, corrected for the effect of aging (presbycusis).

Noise-Induced Temporary Threshold Shift (NITTS)—Temporary threshold shift caused by noise exposure.

Noise Level—1. The type of level of noise which must be indicated by further modifier or context. The physical quantity measured (e.g., voltage), the reference quantity, the instrument used, and the bandwidth or other weighing characteristic must be indicated. 2. For airborne sound, unless specified to the contrary, noise level is the weighted sound-pressure level called sound level; the weighting must be indicated.

Non-Voluntary Exposure to Environmental Noise—The exposure of an individual to sound which 1. The individual cannot avoid, or 2. The sound serves no useful purpose (e.g., the exposure to traffic noise or to noise from a lawn mower).

Occupational Exposure to Environmental Noise—The noise exposure of an individual defined under P.L. 91-596, Occupational Safety and Health Act of 1970.

Octave—The octave is a term that originated in the field of music meaning a doubling of the frequency.

Octave Band—A range of frequencies such that the highest frequency in the band is double the lowest frequency. Also, for excessive octave bands, the center frequency of the higher band is two times the frequency of the next lower center-frequency band. In acoustics standard center-frequency bands are specified.

Octave-Band Center Frequency—The octave-band center frequency is a specific frequency used to define a particular octave band.

Oscillation—The variation, usually with time, of the magnitude of a quantity with respect to a specified reference when the magnitude is alternately greater and smaller than the reference.

Otologically Normal—Enjoying normal health and freedom from all clinical manifestations and history of ear disease or injury; and having a patent (wax-free) external auditory meatus.

Particle Velocity—The velocity of a given infinitesimal part of the medium, with reference to the medium as a whole, due to a sound wave.

Peak Level—The maximum instantaneous level that occurs during a specified time interval. In acoustics, peak sound-pressure level is to be understood, unless some other kind of level is specified.

Peak Sound Pressure—The absolute maximum value (magnitude) of the instantaneous sound pressure occurring in a specified period of time.

Peak-to-Peak Value—The peak-to-peak value of an oscillating quantity is the algebraic difference between the extremes of the quantity.

Period—The smallest increment of the independent variable for which the periodic function repeats itself.

Periodic Quantity—An oscillating quantity whose values recur for certain increments of the independent variable.

Pink Noise—Noise which has constant energy per octave bandwidth; i.e., each octave band contains the same amount of sound energy such that the sound energy is inversely proportional to the frequency.

Pistonphone—A small chamber with a reciprocating piston that permits the establishment of a known sound pressure in the chamber.

Pitch—A qualitative term used to describe the frequency makeup of a sound. Thus we say that a low-frequency sound has a "low pitch." Although pitch depends primarily upon the frequency, it also depends upon the sound pressure and the waveform of the stimulus.

Plane Sound Wave—At a point in an elastic medium, this is a transverse wave in which the displacements at all times lie in a fixed plane which is parallel to the direction of propagation.

Power Level—Power level, in decibels, is 10 times the logarithm to the base of 10 of the ratio of a given power to a reference power.

Presbyacusia (Presbycusis)—Hearing loss, chiefly involving the higher audiometric frequencies above 3000 Hz, ascribed to advancing age.

Random Noise—An oscillation whose instantaneous magnitude or frequency is not specified for any given instant of time.

Ch. 1 FUNDAMENTALS OF SOUND AND VIBRATION

Reverberation—The persistence of sound in a room or between any two reflective surfaces after the sound source has stopped. It is directly related to the amount of sound absorption; the greater the absorption the smaller the reverberation.

Risk—That percentage of a population whose hearing level, as a result of a given influence, exceeds the specified value, minus that percentage whose hearing level would have exceeded the specified value in the absence of that influence, other factors remaining the same. *Note*: The influence may be noise, age, disease, or a combination of factors.

Sound—1. Sound is an oscillation in pressure, stress, particle displacement, particle velocity, etc., in a medium with internal forces (e.g., elastic, viscous), or the superimposition of such propagated oscillations. 2. An auditory sensation evoked by the oscillation described above.

Sound Absorption—The conversion of sound energy into heat in passing through a space or on striking a boundary. The degree of absorption depends on the particular materials, objects, and constructions used in the space.

Sound Analyzer—A device for measuring the band-pressure level or pressure-spectrum level of a sound as a function of frequency.

Sound Energy—The sound energy of a given part of a medium is the total energy in this part of the medium minus the energy which would exist in the same part of the medium with no sound waves present.

Sound Exposure Level—The level of sound accumulated over a given time interval or event. Technically, the sound exposure level is the level of the time-integrated, mean-square A-weighted sound for a stated time interval or event with a reference time of one second.

Sound Field—A region containing sound waves.

Sound Level—A weighted sound-pressure level obtained by the use of metering characteristics and the weightings A, B, or C as specified in the American National Standard Specification for Sound-Level Meters, ANSI S1.4-1971. The weighting employed must be stated.

Sound-Level Meter—An instrument for the measurement of noise and sound levels.

Sound Pressure—The sound pressure at a point is the total instantaneous pressure at that point in the presence of a sound wave minus the static pressure at that point.

Sound-Pressure Level—The sound-pressure level, in decibels, of a sound is 20 times the logarithm to the base 10 of the ratio of the pressure of this sound to the reference-pressure. The reference pressure must be stated. The following reference pressure is in common use for measurements concerned with hearing and with sound in air and liquids: 2×10^{-5} N/m². Unless otherwise explicitly stated, it is to be understood that the sound pressure is the effective (rms) sound pressure.

Spectrum—1. The spectrum of a function of time is a description of its resolution into components, each of different frequency and (usually) of different amplitude and phase. 2. Spectrum is also used to signify a continuous range of components, usually wide in extent, within which waves have some specified common characteristics; e.g., "audio-frequency spectrum."

Speech Discrimination—The ability to distinguish and understand speech signals.

Spherical Wave—A wave in which the wave fronts are concentric spheres.

Standing Wave—When periodic progressive waves reflect from pairs of parallel hard surfaces, an interference pattern will be set up causing the sound wave to "stand" in space. Standing waves will occur whenever the distances between walls or between a floor and ceiling are an integer multiple of one-half the wavelength of the sound. The severity of the standing wave depends on the hardness of the reflecting surface and it is not uncommon for the sound pressure to vary as much as 20 dB between a *node* (minimum amplitude of standing wave) and an *antinode* (maximum amplitude of standing wave).

Static Pressure—The static pressure at a point is that which would exist at that point in the absence of sound waves.

Temporary Threshold Shift (TTS)—That component of threshold shift which shows a progressive reduction with the passage of time after the apparent cause has been removed.

Threshold of Hearing (Audibility)—The minimum effective sound-pressure level of an acoustic signal capable of exciting the sensation of hearing in a specified proportion of trials in prescribed conditions of listening.

Transmission Loss—The reduction in sound power across a partition or wall. It is expressed in decibels as the ratio of sound power incident on the wall to the sound power transmitted through the wall.

Ultrasonic—Having a frequency above the audible range for man (conventionally deemed to cut off at 20,000 Hz).

Wavelength—The wavelength of a periodic wave in an isotropic medium is the perpendicular distance between two wave fronts in which the displacements have a difference in phase of one complete period.

White Noise—Noise whose spectrum density (or spectrum level) is substantially independent of frequency over a specified range. *Note*: White noise need not be random.

Vibration Terms

Acceleration—A vector quantity denoting the time rate-of-change of velocity.

Angular Frequency (Circular Frequency)—The frequency of a periodic quantity, in radians per unit time, is the frequency in cycles per second multiplied by 2π. The usual symbol is α.

Antiresonance—For a system in forced oscillation, antiresonance exists at a point when any change, however small, in the frequency of excitation causes an increase in the response at this point.

Applied Shock—Any excitation that, if applied to a system, would produce mechanical shock. The excitation may be either a force applied to the system or a motion of its support.

Complex Excitation—An excitation is complex if to every real excitation that is a simple harmonic function of the time, there is combined an imaginary excitation having the same amplitude and frequency but differing in phase by one-quarter of a cycle.

Complex Response—The steady-state response in a linear system to a complex excitation.

Continuous System (Distributed System)—A system that is considered to have an infinite number of possible independent displacements. Its configuration is specified by a function of a continuous spatial variable or variables in contrast to a discrete or lumped parameter system which requires only a finite number of coordinates to specify its configuration.

Coulomb Damping (Dry Friction Damping)—The dissipation of energy that occurs when a particle in a vibrating system is resisted by a force whose magnitude is a constant independent of displacement and velocity, and whose direction is opposite to the direction of the velocity of the particle.

Critical Damping—The minimum viscous damping that will allow a displaced system to return to its initial position without oscillation.

Critical Speed—The critical speed of a rotating shaft is the natural frequency of that shaft in lateral motion. When the rotational speed equals the critical speed, resonance occurs. Generally, a shaft system will have many critical speeds.

Cycle—The complete sequence of values of a periodic quantity that occurs during a period. The cycle is equal to one over the period in seconds and the unit is cycles per second (Hz).

Damped Natural Frequency—The frequency of free vibration of a damped linear system. This will be different from the undamped natural frequency.

Damping Ratio—The ratio of actual damping coefficient to the critical damping coefficient.

Displacement—A vector quantity that specifies the change of position of a body or particle and is usually measured from the mean position or position of rest. In general, it can be represented by a rotation vector, a translation vector, or both.

Dynamic Vibration Absorber. (Tuned Damper)—A device for reducing vibration of a primary system by the transfer of energy to an auxiliary resonant system which is tuned to the frequency of the vibration. The force exerted by the auxiliary system is opposite in phase to the force acting on the primary system.

Equivalent System—A system that may be substituted for another for the purpose of analysis, e.g., an equivalent spring or an equivalent mass.

Excitation—An external force (or other input) applied to a system that causes it to respond in some way.

Forced Oscillation (Forced Vibration)—An oscillation that is imposed by an excitation. If the excitation is periodic and continuing, the oscillation is steady-state.

Foundation—The foundation of a structure supports the gravity load of a mechanical system. It may be fixed in space, or it may undergo a motion that provides excitation for the supported system.

Free Oscillation—The oscillation that occurs in a system in the absence of a forcing function, or forces oscillation or it may occur after a forcing input has been removed from the system.

g—The acceleration produced by the force of gravity, which varies with the latitude and elevation of the observation point. It is denoted by the symbol *g*. By international agreement, the value 980.665 cm/sec^2 = 9.80655 meters/sec^2; 386.087 in./sec^2 = 32.1739 ft/sec^2 has been chosen as the standard acceleration due to gravity.

Harmonic (Harmonic Frequency)—A sinusoidal quantity having a frequency that is an integral multiple of the frequency of a periodic quantity to which it is related.

Impact—A single collision of one mass in motion with a second mass which may be either in motion or at rest.

Impulse—The product of a force and the time during which the force is applied.

Isolation—A reduction in the capacity of a system to respond to an excitation attained by the use of a resilient support. In steady-state forced vibration, isolation is expressed quantitatively as the complement of transmissibility.

Jerk—A vector quantity that specifies the time rate of change of the acceleration; jerk is the third derivative of the displacement with respect to time.

Mechanical Impedance—The ratio of force to velocity during simple harmonic motion.

Mechanical Shock—Mechanical shock occurs when the position of a system is significantly changed in a relatively short time in a nonperiodic manner. It is characterized by suddenness and large displacements, and develops significant internal forces in the system.

Mode of Vibration—A characteristic pattern assumed by a system undergoing vibration in which the motion of every particle is simple harmonic motion with the same frequency. Two or more modes may exist concurrently in a multiple-degree-of-freedom system.

Multiple-Degree-of-Freedom-System—A system which has oscillatory motion in more than one plane or about more than one axis. A system will have at least as many natural frequencies as it has degrees of freedom. In many machinery systems, the motion is primarily in one plane or about a single axis so that it can be analyzed as a one-degree system.

Natural Frequency—The frequency of free oscillation of a system. For a multiple-degree-of-freedom system, the natural frequencies are the frequencies of the normal modes of vibration.

Nonlinear Damping—Damping due to a damping force that is not proportional to velocity.

Oscillation—The variation of the magnitude of a quantity with respect to a specified reference when the magnitude is alternately greater and smaller than the reference.

Pulse Rise Time—The interval of time required for the leading edge of a pulse to rise from some specified small fraction to some specified larger fraction of the maximum value.

Q (Quality Factor)—A measure of the sharpness of resonance or frequency selectivity of a resonant vibratory system having a single degree of freedom.

Resonance—Resonance occurs when the frequency of the exciting force corresponds to a natural frequency in a mechanism or structure. In such a condition, large oscillation can result for only a small exciting force.

Resonance Frequency—The frequency at which a resonance occurs.

Response—The motion (or other output) resulting from an excitation.

Self-Induced (Self-Excited) Vibration—The vibration of a mechanical system, produced and sustained by system motion such as "wheel shimmy."

Shock Absorber—A device for the dissipation of energy to modify the response of a mechanical system to applied excitation.

Shock Isolator (Shock Mount)—A resilient support that tends to isolate a system from applied excitation.

Shock Pulse—A disturbance characterized by a rise and decay of acceleration from a constant value in a short period of time. Shock pulses are normally displayed graphically as curves of acceleration as a function of time.

Shock Spectrum—A plot of the maximum acceleration experienced by a single-degree-of-freedom system as a function of its own natural frequency in response to an applied shock.

Simple Harmonic Quantity—A periodic quantity that is a sinusoidal function of the independent variable. It can be described mathematically as follows: $y = A \cdot \sin(2\pi f t + \phi)$.

Simple Harmonic Motion—Motion where displacement is a sinusoidal function of time.

Single-Degree-of-Freedom System—A system for which only one coordinate is required to define completely the configuration of the system at any instant.

Snubber—A device used to increase the stiffness of an elastic system whenever the displacement becomes larger than a specified amount.

Steady-State Vibration—A vibration in which the amplitude of response continues in a periodic manner.

Stiffness—The ratio of change of force (or torque) to the corresponding change in translational (or rotational) displacement of an elastic element.

Subharmonic Response—The periodic response of a mechanical system exhibiting the characteristic of resonance at a frequency that is a submultiple of the frequency of the periodic excitation.

Transient Vibration—A vibration which is not periodic and which is sustained only for a short length of time. Such vibrations occur from shock or impact forces.

Transmissibility—The transmissibility of a vibratory system is a nondimensional ratio formed by dividing the amplitude of response by the amplitude of the excitation at a given frequency. The amplitude may be either a displacement, velocity, acceleration, or force. The transmissibility is a measure of how well a system responds.

Undamped Natural Frequency—The frequency of free vibration resulting only from elastic and inertial forces of the system.

Velocity—A vector quantity that specifies the time rate-of-change of displacement with respect to a reference frame.

Velocity Shock—Mechanical shock resulting from a non-oscillatory change in velocity of an entire system.

Vibration Isolation—The use of resilient supports to reduce the capacity of a system to respond to external excitation. The ideal is to make the supports sufficiently soft so that the natural frequency is well below the excitation frequencies.

Vibration Isolator—A support that tends to isolate a system from an excitation.

Vibration Meter (Vibrometer)—An instrument for the measurement of displacement, velocity, or acceleration of a vibrating body.

Viscous Damping—The dissipation of energy that occurs when a vibrating system is resisted by a force that has a magnitude proportional to the magnitude of the velocity in a direction opposite to the direction of the system.

2

Noise Measurements and Instrumentation

LYNN L. FAULKNER

Introduction

All noise control programs require measurements for assessment. Instrumentation may range from a single, hand-held sound-level meter to complex laboratory analysis and processing systems. The purpose here is to consider the more frequently used sound measurement instrumentation for industrial uses and to discuss the information to be obtained from such measurements. Knowledge of the techniques and facilities for noise measurement is of general use to differentiate between machinery noise measurements which can be made in a factory space and measurements which must be made in special acoustic laboratories. Methods and information obtainable from factory measurements will be discussed.

Purpose of Noise Measurements

For machinery noise situations there are a number of reasons to make measurements. The most emphasized measurement purpose is to determine whether noise levels are of sufficient magnitude to be a potential hearing-loss hazard to exposed individuals. This situation became of primary interest after the Walsh-Healey Act of 1969, followed by OSHA[1] regulations in 1970. In situations where the noise level is not severe enough to cause hearing damage, there are cases where annoyance and speech interference are of concern. These conditions are discussed in Chapter 4, dealing with criteria for machinery noise.

Noise surveys are required on a continual basis to determine whether the acoustic environment has changed due to tooling changes, process change, new equipment, etc. In most cases it is not sufficient to specify a sound-pressure level for individual new equipment because the combined sound levels of several

---

[1] Occupational Safety and Health Act of 1970, Federal Register, Volume 36, No. 105, May 29, 1971. See Chapter 4, page 121.

machines could exceed the desired sound level in an area although individually they were acceptable. For this reason measurements are made to determine the sound power of a particular piece of equipment.

Measurements are necessary in order to assess the source and character of the noise produced by a particular machine when attempting to control the noise radiated. A measurement program is a necessary part of any noise-control program; when such information is obtained and correctly interpreted, optimum solutions can be prescribed. Acoustic measuring instrumentation is an essential part of the noise-control program and proper measurements must be taken and the measured values analyzed correctly. In order to obtain the proper measurements it is necessary to have an understanding of the type of instrumentation that is available and what measurements they provide, as well as their limitations.

Quantities to be Measured

The physical quantity to be measured is the fluctuating pressure which propagates through the atmosphere as a sound wave. Sound pressures are small fluctuations in pressure above and below the ambient pressure in the atmosphere within which the acoustic wave propagates. The fluctuating magnitude is on the order of 2.94×10^{-9} psi for the threshold of hearing of a person with normal hearing. To avoid the use of a small pressure in terms of pounds per square inch or Newtons per square meter to characterize a sound, the acoustic sound pressure level was defined in Chapter 1 as:

$$L_p = \text{sound-pressure level} = 10 \log_{10} \left(\frac{p}{p_{\text{ref}}}\right)^2 \text{ decibels} \qquad (1\text{-}30)$$

where:

L_p = sound-pressure level in decibels, dB
p = root mean square, rms, of the fluctuating sound pressure, psi or N/m^2 or pascals
p_{ref} = standard reference value to which all pressures are compared.

The first problem encountered is the measurement of the time-varying acoustic pressure. The human ear does not perceive slowly varying static pressure as sound; to illustrate this, consider the variation in static pressure experienced in riding an elevator. The magnitude of the static pressure variation is of the order of 0.1 psi (6.9×10^3 N/m^2) for a 200-ft change in elevation. If this slowly varying pressure were converted to sound-pressure level, as given by Equation (1-30), it would be 170 decibels! This is considerably greater than many common sound-pressure levels, yet the ear does not recognize this change as sound because the change in pressure is too slow. Fluctuating pressure must be in the frequency range of 20 to 20,000 Hz to be classified as audible sound for the human ear.[2] The range of dynamic pressure of interest in noise control is from 2.94×10^{-9} psi (2×10^{-5} N/m^2), which is approximately the threshold of hearing for human beings for a 1000-Hz tone, to 29.4 psi (2×10^5 N/m^2), the sound pressure of large rocket engines. We are thus faced with dynamic pressure mea-

---

[2] Hz is the standard symbol for cycles per second.

Ch. 2 NOISE MEASUREMENTS AND INSTRUMENTATION

surements of pressures which may vary from 2.94×10^{-9} psi (2×10^{-5} N/m^2) to 29.4 psi (2×10^5 N/m^2) a tremendous range. For this reason the decibel scale is used to "compress" the range of magnitudes to a manageable scale as described on page 19 of Chapter 1.

As in any logarithmic quantity, p_{ref} is arbitrarily chosen. The convenient value for this reference pressure is 2.94×10^{-9} psi (2×10^{-5} N/m^2), such that 0 decibels corresponds to the threshold of hearing for a 1000-Hz tone; i.e., $\log_{10}(1) = 0$.[3] Typical sound-pressure levels and corresponding pressures are given in Table 2-1.

A linear frequency scale (100 Hz, 200 Hz, 300 Hz, etc.) is frequently inconvenient for the sound frequencies ranging from 20 to 20,000 Hz. The usual method is to separate this range into ten unequal segments called "octave

Table 2-1. Acoustic Noise Spectrum

Region of Noise	L_p,* decibels	Sound Pressure (rms)			Typical Situation
		N/m^2	Atmospheres	lb/in.^{2}	
Physical Damage	200	2×10^5	2.03	2.94×10	200 yd from Guided Missile at Launch
	180	2×10^4	2.03×10^{-1}	2.94	Instant Deafness
	160	2×10^2	2.03×10^{-2}	2.94×10^{-1}	Threshold of Physical Damage to Tissue
Hearing Damage, Pain	140	2×10^2	2.03×10^{-3}	2.94×10^{-2}	Jet Aircraft Launch, Threshold of Pain
	120	2×10	2.03×10^{-4}	2.94×10^{-3}	Thunder, Gunfire
	100	2	2.03×10^{-5}	2.94×10^{-4}	Heavy Machinery Factory
Annoyance Region	80	2×10^{-1}	2.03×10^{-6}	2.94×10^{-5}	Noisy Factory
	60	2×10^{-2}	2.03×10^{-7}	2.94×10^{-6}	Light Mfg. Factory
	40	2×10^{-3}	2.03×10^{-8}	2.94×10^{-7}	Conversation, Average House
	20	2×10^{-4}	2.03×10^{-9}	2.94×10^{-8}	Whisper, Rustling Leaves
	0	2×10^{-5}	2.03×10^{-10}	2.94×10^{-9}	Threshold of Hearing

$^*L_p = 10 \log_{10} \left(\dfrac{p}{p_{ref}}\right)^2$ dB

where:

p_{ref} = reference pressure = 2×10^{-5} N/m^2
p = root mean-square pressure above ambient

---

[3] Expressed in cgs units, 1 microbar = 1 dyne/cm^2 = 1.45×10^{-5} psi. Therefore, p_{ref} = 0.0002 microbar which is 2×10^{-5} N/m^2 or 2×10^{-5} pascals.

70 NOISE MEASUREMENTS AND INSTRUMENTATION   Ch. 2

bands." In essence, an octave refers to a doubling in frequency; e.g., 63 Hz, 125 Hz, 250 Hz, etc. For convenience in scientific measurement of sound it has been found that certain center frequencies extend over the range of audible sound, from low to high, and that these center frequencies are arranged according to the octave scheme. The center frequencies themselves are extended over a band, or range of frequencies; thus, we have an octave which is defined by its center frequency. For each octave band the frequency of the upper limit is twice that of the lower band limit, as shown in Table 2-2. Moreover, the center frequencies of the octave bands have been standardized for scientific measurements so that each succeeding center frequency is two times the frequency of the previous band, as given in Table 2-2. There are portable sound-measuring instruments that will measure sound-pressure levels with the frequency range of each octave band and for each one-third octave band; the actual meter reading is the rms value of the combined pressures of all of the frequency components within the particular octave band, converted to sound-pressure level in decibels, as given by Equation (1-30).

Each octave band is sometimes separated into three ranges, referred to as "one-third-octave bands." These ranges also have been standardized for scientific measurements as shown in Table 2-2. The one-third-octave bands are sometimes useful in examining the frequency content of the broader octave band. For example, the ball pass frequency of a ball bearing rotating at a constant speed will be discrete and can be more easily identified by the use of the one-third-octave band because it has a narrower range of frequencies. Thus, the one-third-octave-band frequency is sometimes useful in identifying sound sources in machinery.

In some instances it is helpful to measure sound-pressure level in frequency bands narrower than are given by the one-third-octave-band limits. There are instruments available that provide measurements having bandwidths of one Hertz, or even less. The instruments for measuring sound-pressure levels for the different frequency bandwidths are treated under the Electronic Instrumentation section in this chapter.

Sound Power.—Sound power is the total acoustic power, in watts, that a given source radiates (see page 23, Chapter 1). The sound pressure which exists is the result of this power radiation. Again, in dealing with a large range of magnitudes, the decibel is used here as well, to give a reasonable scale. The result is the sound-power level:[4]

$$\text{sound-power level} = L_W = 10 \log_{10}\left(\frac{W}{W_{ref}}\right) \text{(decibels, dB)} \qquad (1\text{-}28)$$

where:

W = radiated power in watts
W_{ref} = reference power = 10^{-12} watts

(Note: In some previous texts W_{ref} of 10^{-13} watts was used before it was internationally agreed to use 10^{-12} watts. This confusion makes it necessary always

---

[4] Some references give *PWL* as the symbol for sound-power level.

Ch. 2 NOISE MEASUREMENTS AND INSTRUMENTATION

Table 2-2. Frequency Limits for Octave Bands and ⅓-Octave Bands

Frequency, Hz					
Octave Bands			⅓-Octave Bands		
Lower Band Limit, Hz	Center Frequency, Hz	Upper Band Limit, Hz	Lower Band Limit, Hz	Center Frequency, Hz	Upper Band Limit, Hz
11	16	22	14.1	16	17.8
			17.8	20	22.4
			22.4	25	28.2
22	31.5	44	28.2	31.5	35.5
			35.5	40	44.7
			44.7	50	56.2
44	63	88	56.2	63	70.8
			70.8	80	89.1
			89.1	100	112
88	125	177	112	125	141
			141	160	178
			178	200	224
177	250	355	224	250	282
			282	315	355
			355	400	447
355	500	710	447	500	562
			562	630	708
			708	800	891
710	1,000	1,420	891	1,000	1,122
			1,122	1,250	1,413
			1,413	1,600	1,778
1,420	2,000	2,840	1,778	2,000	2,239
			2,239	2,500	2,818
			2,818	3,150	3,548
2,840	4,000	5,680	3,548	4,000	4,467
			4,467	5,000	5,623
			5,623	6,300	7,079
5,680	8,000	11,360	7,079	8,000	8,913
			8,913	10,000	11,220
			11,200	12,500	14,130
11,360	16,000	22,720	14,130	16,000	17,780
			17,780	20,000	22,390

to state the reference level when reporting sound-power levels. For example,

$$L_W = 120 \text{ dB re } 10^{-12} \text{ watt}$$

to clearly indicate that the reference used was 10^{-12} watts.)

For an acoustic source in a non-reflective environment (see the following section, "Measuring Environment"), there is a simple relation between sound-power level and sound-pressure level if some conditions are met. The source

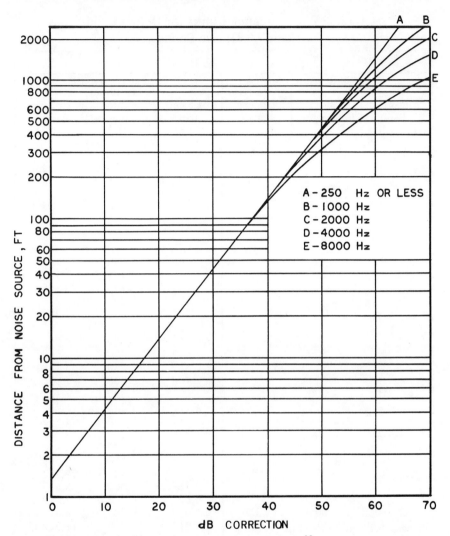

Fig. 2-1. Correction in dB to convert power level, L_W, 10^{-12} watt reference power, to sound-pressure level, L_p, 2×10^{-5} N/m² reference pressure. $L_p = L_W$-correction. The plot accounts for a source mounted on a floor (hemispherical radiation) and absorption of air at 70 °F.

must be such that sound radiation is uniform in all directions; this is referred to as a "spherical" source or a "hemispherical" source if located on a reflecting surface such as a floor. For this case,

$$L_W = L_p + 20 \log(r) + 0.5 \text{ dB } (r, \text{ in feet}) \quad (2\text{-}1a)$$

or

$$L_W = L_p + 20 \log(r) + 8 \text{ dB } (r, \text{ in meters}) \quad (2\text{-}1b)$$

Ch. 2 NOISE MEASUREMENTS AND INSTRUMENTATION

where:

L_p = sound-pressure level, dB, 2×10^{-5} N/m² reference
L_W = sound-power level, dB, 10^{-12} watt reference.

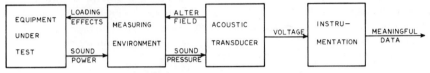

Fig. 2-2. Diagram of instrumentation system.

A plot relating L_W and L_p is given in Fig. 2-1. The relation given above is good for estimating purposes only, for installed machinery. For a very precise determination of sound-power levels, refer to ANSI S1.2-1962, "Standard Method for the Physical Measurement of Sound." For a precise determination of sound pressure in a space from given sound-power levels, see Chapter 4.

The use of the sound-power level is the desirable method to specify or rate machinery because the sound-pressure level can then be computed for any desired environment. The specification of the sound-pressure level is not recommended since this value is a function of the environment into which the noise-producing device will be placed.

The usual method is for equipment suppliers to furnish sound-power levels in octave bands from which the sound-pressure level, with the equipment installed in a particular environment, can be computed in octave bands.

The Summation of Sound-Pressure and Sound-Power Levels

On occasion it is necessary to combine two measured sound-pressure or sound-power levels; e.g., when the effect of two independent sources of sound must be known. Since these measurements are presented in the decibel scale, they must be added logarithmically, as explained in detail in Chapter 1.

Measurement System

It has been previously established that noise control is, by necessity, measurement dependent. A fundamental concept of concern in acoustic measurements, which engineers very often minimize, is the fact that the process of measurement inevitably alters the quantity being measured. In taking acoustic measurements there are at least three aspects of the measuring system that have a direct effect on resulting data. These areas are shown in block diagram form in Fig. 2-2. The dependence of the measuring environment, acoustic transducer, and the electronic instrumentation must be understood in order to extract meaningful data from a particular noise source. These three areas will be discussed in some detail in the following sections.

MEASURING ENVIRONMENT

The space within which a machine is located has a definite effect on the acoustic field to be measured. Acoustic energy is propagated as pressure waves.

74 NOISE MEASUREMENTS AND INSTRUMENTATION Ch. 2

When solid objects are present, these waves are reflected and can propagate back toward the noise source. When this occurs, acoustic measurements can be very much in error because the acoustic field will be a superposition of direct and reflected quantities. For this reason special laboratory rooms are constructed to eliminate reflections, or in some cases, the reflecting properties are utilized in making measurements. The result is that several types of environments are used for laboratory sound measurements, each having special properties.

A. *Anechoic Chamber.*—This essentially is an echoless room which has sound-absorbing materials on all walls, ceiling, and floor as shown in Fig. 2-3. All

Fig. 2-3. Photograph of an anechoic chamber for laboratory sound measurements. A facility of this type provides a free field for acoustic sources where reflections do not occur.

Ch. 2 NOISE MEASUREMENTS AND INSTRUMENTATION

sound energy produced by a piece of machinery propagates radially from the source and is absorbed at the walls. Since no energy reflects back toward the source, no interference pattern results to complicate the measurements. A reflection-free space of this type is necessary to determine the directional properties of the sound radiated from some object. As an illustration, consider the radiation pattern given by Fig. 2-4. We might stipulate for this example that this radiation pattern represents one frequency component from the complete sound spectrum. By taking measurements at various locations a pattern similar to that shown in the figure may have been found. The perimeter shown represents a contour of equal sound-pressure level. Some guesses could be made as to where this particular component of noise is originating. It appears to be somewhat symmetric around the machine.

Radiation patterns of this type can only be made in some type of anechoic environment. This type of facility is also used when narrow-band frequency analysis is to be made. The anechoic room is then essential for acoustic research in that information is obtained here that cannot be obtained in a general acoustic environment. The information usually obtained is as follows:

1. acoustic radiation patterns
2. narrow-band frequency data
3. phase relations of acoustic energy from various source locations.

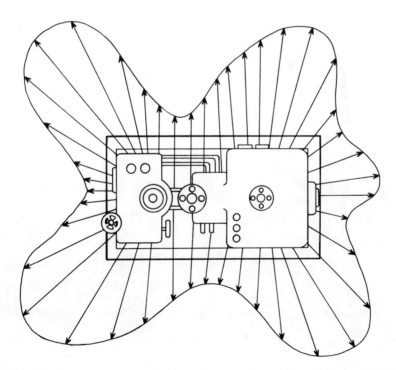

Fig. 2-4. Illustration of a sound radiation pattern around a machine. Patterns of this type are obtained in a free field for a given frequency component of sound.

B. *Reverberant Chamber.*—An environment of this type is just the opposite of the anechoic chamber. Theoretically, in a reverberant room, all acoustic energy is reflected back toward the source by hard reflective walls as shown in Fig. 2-5. Essentially, after a large number of reflections, there is a nearly homogeneous distribution of acoustic energy throughout the room. Sometimes large moving vanes are placed in the room to help insure a homogeneous distribution of sound energy. It is also advisable to have the microphone on a moving boom in order to take average sound-level measurements over some spatial distribution.

The purpose of the reverberant chamber then, is to obtain time and spatial average measurements of the total acoustic energy radiated from some piece of machinery. This room provides a standard environment from which overall noise reductions can be evaluated and comparisons can be made from one unit to another. Sound power can be obtained in a facility of this type. Effects of the sources are all averaged, and for this reason the directionality, the phase relationships, and the exact locations of the various sound sources are not obtained.

The location of the noise source in the room can affect the measurement of the total sound power radiated. Accepted practice is to locate the source in one corner of the room but not closer than $1/2$ wavelength from the walls or floor.

Fig. 2-5. Photograph of a reverberant room used to make measurements of acoustic pressure relatable to the total sound power radiated from a device.

Ch. 2 NOISE MEASUREMENTS AND INSTRUMENTATION

The wavelength being computed from:

$$\lambda = \text{wavelength} = \frac{1130}{\text{frequency (Hz)}} \text{ (ft)} = \frac{344}{\text{frequency (Hz)}} \text{ (m)} \qquad (1\text{-}2)$$

for the lowest frequency of interest. Usually, 100 Hz is the lowest frequency of concern, and for this case the wavelength would be 11.3 feet and the noise source should be no closer than 5.65 feet from walls or floor. Equipment which in normal operation sits on a hard floor can be mounted on the floor of the reverberant room.

C. *Semi-Anechoic Chamber.*—Many times it is not convenient to build an anechoic chamber for acoustic testing. For large, heavy equipment it is essential to have adequate support. In these cases a semi-anechoic test room is utilized. The floor is usually concrete and the other sides and ceiling are covered with sound-absorbing material as shown in Fig. 2-6.

Special considerations must be taken in the utilization of an environment of this type. Even if the noise source has no directional properties, the sound field

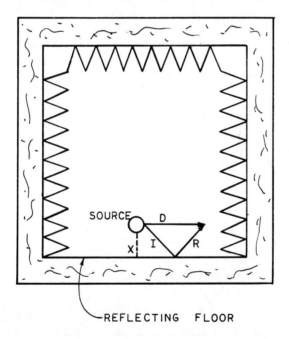

I = INCIDENT WAVE
R = REFLECTED WAVE
D = DIRECT WAVE
H = HEIGHT ABOVE FLOOR

Fig. 2-6. Illustration of a semi-anechoic chamber. The walls and ceiling are covered with sound-absorbing wedges. The floor is concrete.

78 NOISE MEASUREMENTS AND INSTRUMENTATION Ch. 2

may have directivity due to the reflections from the floor. Cancellation will occur whenever the length of the direct sound path and the reflected sound path differ by $1/2$ of a wavelength.

Cancellation when:

$$I + R - D = \frac{n\lambda}{2} \quad \text{(See Fig. 2-6)}$$

where:

$I + R$ = reflected sound-path length
D = direct sound-path length

$$\lambda = \text{wavelength} = \frac{1130}{\text{frequency (Hz)}} \text{ (ft)} = \frac{344}{\text{frequency (Hz)}} \text{ (m)} \quad (1\text{-}2)$$

n = integer = 1, 2, 3 . . . .

This is shown schematically in Fig. 2-7. Reinforcement of sound pressure is represented by regions where the sets of lines intersect (dark areas) and cancellation of sound pressure is represented by regions where the lines are about equally spaced (light areas). The sound field will be a superposition of the direct and reflected radiation consisting of light and dark areas. Close to the reflecting plane the regions always reinforce each other and measurements close to the reflecting plane will be 3 dB higher. The cancellation-reinforcement effect is also dependent upon the distance from the floor to the source, x, and the wavelength of sound, λ.

The following general statements can be made regarding noise sources above reflecting planes:

1. Near the reflecting plane the sound-pressure level is 3 dB greater than the direct-path sound-pressure levels.

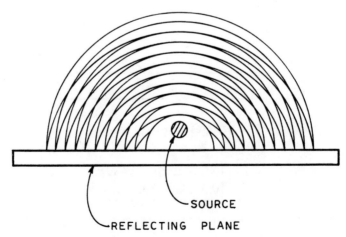

Fig. 2-7. Interference pattern due to direct and reflected acoustic waves from a source above a reflecting plane.

Ch. 2 NOISE MEASUREMENTS AND INSTRUMENTATION

2. If the wavelength is greater than 10 times the distance between source and reflecting plane ($x/\lambda < 0.1$), there is no significant directivity due to reflection. This occurs for low frequencies below 1000 Hz.
3. For medium frequencies such that $x/\lambda = 0.2$ to 1.0, the directivity is pronounced.
4. For high-frequency random sound (no pure tones), the radiated field is nearly uniform except near the reflecting plane.

The location of the noise source, in relation to the reflecting plane, affects the sound-power radiated. This is the result of energy reflected back toward the source, with some phase difference. The following results occur depending on the distance, x, of the source from reflecting planes:

1. If the source is close to a reflecting plane (x less than 0.1 wavelength), the power output of the source is doubled, 3 dB greater.
2. Cancellation occurs when the source is $1/4$ to $1/2$ wavelength away from a reflecting plane and the source output is reduced approximately 1 dB.
3. If the distance from the source to the reflecting plane is greater than one wavelength, there is a negligible effect on radiated sound power.

D. *Out-of-Doors.*—Many times acoustic measurements for machinery can be made outside on a paved area such as a parking lot. The characteristics of the outdoor environment are very similar to those of a semi-anechoic chamber if some conditions are met:

1. Test area should be free of large reflecting surfaces such as signboards, buildings, or hillsides within 100 feet from the machine under test or the microphone.
2. Sound absorption coefficient of the reflecting surface should not be greater than 0.2.
3. Background noise levels should be at least 10 dB below noise level to be measured.
4. Wind noise level should be 10 dB below noise level to be measured. Special microphone covers can be used to reduce wind effects on the microphone; however, the 10 dB differential should be observed with the cover in place.
5. Reflectivity pattern from the ground should be considered as indicated in the previous section, and measurements close to the ground should be avoided.

E. *Installed Machinery.*—The conformance to the OSHA noise-exposure regulation requires measurements in the actual space where the machine and worker are located. This "as installed" or "in situ" condition is the primary measurement situation for industrial noise. Special facilities, as described earlier, are for laboratory analysis involving accurate measurements for detailed acoustic investigation of source characteristics. The "as installed" situation has many undesirable factors such as high background-sound levels, reflecting surfaces, fluctuating noise levels, and intrusive sounds from various sources. All of these conditions affect the accuracy of measured sound-pressure level, L_p. These conditions may exist at a worker location where sound measurements are to be made to conform to OSHA; in this instance the level at the worker location is of interest and the measurement must be made there. For general noise surveys some precautions are necessary to insure greatest accuracy:

Near Reflecting Surfaces.—Near reflecting surfaces such as floors, walls, partitions, cabinets, adjacent machinery, etc., the sound-pressure level will be higher due to pressure doubling which results in a 3-dB increase in L_p close to the surface compared to a location some distance from the surface.

Fluctuating Sound Levels.—In most industrial situations the sound level is not steady with time but rather, it varies with time. The result is a continual needle movement of a sound-level meter. As was previously determined in the section on adding logarithmic quantities, the combined sound-pressure level is not a direct combination of the decibel values but a value corresponding to the summation of the squares of the rms pressure ratios. The mean sound-pressure level, in decibels, is not the mean value of the fluctuating meter readings.

An approximation used in practice to average the fluctuations of the meter is to note the high and low meter indications; if the total fluctuation is 6 dB or less, the average of the two decibel readings is approximately the average of the decibel readings. If the total fluctuation is greater than 6 dB, the average is taken as 3 dB less than the larger value given by the meter.

This result is due to the nature of addition of logarithmic quantities as described earlier. The "fast" and "slow" meter response positions on a typical sound-level meter are designed to minimize some of the needle fluctuations. The two positions on the sound-level meter determine the time response of the instrument; the slow position provides a longer time-average response, thereby reducing the needle fluctuations for rapidly varying pressure levels. The slow response is required by OSHA for all steady-state (non-impulsive) sound-pressure measurements. For most other situations the meter response should also be in the "slow" position to obtain some averaging of the signal with respect to time, although some meter fluctuation may exist in the "slow" response position. A method for evaluating the accumulated noise exposure is given in the chapter on "Criteria," accounting for fluctuating values.

Background Sound Levels.—Many times it is not possible to shut down assembly or production areas to measure sound levels from one particular machine. Measurements could be made at other than normal working hours, at break time, or between shifts. The background level at a given location should be determined by turning off the particular machine in question, if possible, and making sound-pressure level measurements. When the test machine is then operated, the measurements should again be taken at the same location as before. From the chart in Fig. 2-8 it is possible to determine the contribution of the test machine as compared to the background. Note from the chart that if the background is at least 10 dB lower than the sound-pressure level of the test machine when operated, the background contribution can essentially be neglected. If possible, the background should be 10 dB lower than the acoustic signal we are attempting to measure so that the background noise can be neglected.

As an example, consider a case where a machine is operating in a noisy factory area and a portable octave-band analyzer is used to measure the sound-pressure

Ch. 2 NOISE MEASUREMENTS AND INSTRUMENTATION

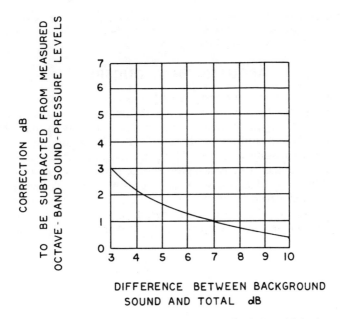

Fig. 2-8. Chart for correction of measured sound-pressure level, L_p, with background levels.

level. In the 1000-Hz octave band, the sound level with machine operating in the noisy environment was found to be 92 dB; when the machine was not operating the 1000-Hz reading was 85 dB. Thus, the difference in readings is 7 dB. Using Fig. 2-8, 1 dB must be subtracted from the total reading (92 dB) to obtain the actual contribution of the machine, or 91 dB. All corrections should be rounded off to the nearest whole decibel. If the difference is 10 dB or more, no correction should be made. This procedure is normally done for each octave-band center frequency.

Air Flow.—In some instances air flow exists from cooling fans, man coolers, blow-off processes, etc. A microphone is a pressure-measuring device; therefore, it will respond to the pressures in an air flow as well as to acoustic pressures. Care must be taken to keep the microphone out of air flows while taking measurements. In the case of man-cooler systems or ventilating fans in the vicinity of workers, a windscreen may be used over the microphone. Contact instrumentation suppliers for information on windscreens for particular microphones and the correction factors to be applied when using these devices.

Effect of Human Observer.—Sound can also be reflected from human observers as well as from walls and machinery. When taking measurements all bystanders should be kept well away from the microphone. The levels can also be influenced by the person reading the sound-level meter. As much as plus or minus 5 dB error can result due to the observer holding the meter too close to his body. The recommended practice for hand-held meters is to hold the meter with extended arms with the hands well away from the microphone. For accurate measurements it is recommended that the microphone be mounted on a tripod with an extension cable to the read-out meter to allow the observer to be at least 10 feet from the microphone.

ACOUSTIC TRANSDUCERS

The second element in the measuring system is the acoustic transducer, typically, a microphone. Of primary concern is the frequency response, sensitivity, and directivity of the microphone; however, another property must be considered before these quantities are discussed. This is the fact that the microphone distorts the acoustic pressure field that is being measured. The reason this distortion occurs is that the physical properties of the microphone are much different from the air in which the sound is propagating. In fact, the microphone may be considered a rigid body. Sound waves impinging on the microphone, as in Fig. 2-9, result in reflections which depend on the frequency, direction of sound propagation, and the microphone size and shape. In this illustration, p_0

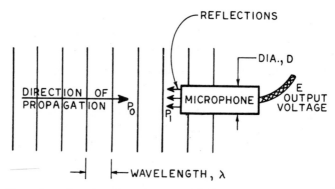

Fig. 2-9. Illustration of a microphone in a sound field. p_0 is the pressure in the acoustic field; p_1 is the reflected pressure at the microphone.

is the pressure of the propagating sound wave and p_1 is the pressure of the sound wave reflected from the microphone. If the wavelength of sound is large compared to the microphone geometry (this occurs for low frequencies), the effect of reflections is negligible. When the wavelength of sound is small compared to the microphone dimension (high frequencies), the microphone acts as a rigid boundary and the pressure at the surface becomes twice that of the free-field incident sound pressure.

The reflection phenomenon is directly related to the wavelength of sound and the size of the microphone. The wavelength of sound in air, λ, in inches, is $13,000/f$, where f is the frequency, in Hz. When λ becomes the same magnitude as the microphone diameter, p_1 becomes the same order as p_0, and the pressure acting on the microphone diaphragm is not that of the free-field sound pressure. For a 1-inch-diameter microphone we would not expect good free-field measurements above 13,000 Hz without some electronic compensation.

Ch. 2 NOISE MEASUREMENTS AND INSTRUMENTATION

The response of the microphone is defined as the output voltage, E, divided by either p_0 or p_1. Microphones are therefore denoted as "free-field" or "pressure" microphones. The frequency response of a typical 1-inch-diameter microphone is illustrated in Fig. 2-10. Below 1000 Hz the response curves are identical because $p_1 = p_0$. Above 1000 Hz, p_1 is greater than p_0 because of reflection waves at the surface of the microphone.

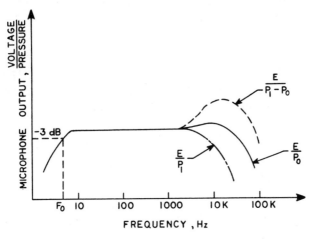

Fig. 2-10. Typical frequency response of a microphone.

Because a microphone is a dynamic device, it will not give accurate measurements at zero frequency. The low frequency limit, f_0, denoted as the "cut-off" frequency, is the point where the response is 3 dB below actual incident sound pressure. Smaller diameter microphones have larger dynamic ranges but in general are not as sensitive because of reduced diaphragm area. Manufacturers' literature should be consulted for particular information regarding use of specific microphones.

ELECTRONIC INSTRUMENTATION

The purpose here is not to explore the electronics of acoustic analyzers in detail, but to discuss the basic differences in the information obtained from the following types of acoustic analyzers:

A. A-, B-, and C-weighted analyzers
B. Octave-band analyzer
C. $1/3$-octave-band analyzer
D. Constant percentage narrow-band analyzer
E. Constant bandwidth narrow-band analyzer.

It is essential that the engineer understand the manner in which each of these devices processes an acoustic signal and he must be able to discern when each should be used. Each has its own usefulness and limitations and of most concern is the information which can be obtained from them. The best method to illustrate the difference between these analyzers is to show the output of each device to the same acoustic signal.

The first example is given in Fig. 2-11 where a noise spectrum is presented for some mechanical device that has definite periodicity in its noise output. The frequency scale is linear and the vertical scale is in decibels. Notice that the 2-Hz *constant bandwidth analyzer* gives a very detailed representation of the sound as a function of frequency. It indicates that a series of single peaks, usually called pure tones, are spaced with some regularity. The base level is approximately 10 dB.

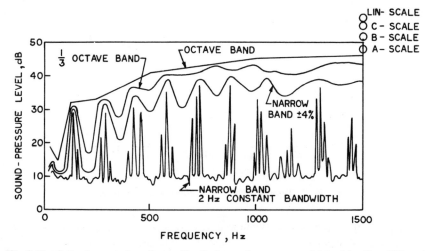

Fig. 2-11. Output comparison for an identical sound-pressure as obtained by four different types of sound analyzers, namely: (1). Octave-band analyzers; (2). 1/3-Octave-band analyzer; (3). ±4% Narrow-band analyzer; and (4). 2 Hz constant bandwidth narrow-band analyzer. The comparative readings of this sound field, as obtained on a portable sound-level meter, are shown in the upper-right-hand corner; i.e., Linear Scale (57 dBLin); C-weighted Scale (55dBC); B-weighted Scale (52dBB); and A-weighted Scale (48dBA).

A constant percentage *narrow-band analyzer* does not give nearly the detail for the same noise input, and in fact, does not give an accurate level for any of the peaks in the spectrum. The reason for this is that all of the peaks within the ±4 percent bandwidth are added together logarithmically. For a center frequency of 1000 Hz the actual bandwidth is 80 Hz, from 960 to 1040 Hz, and all components within this range are combined to give the output of the analyzer. If there were two equal peaks anywhere within this band, the output would be 3 dB higher than either of the peaks. Note also that the mean value continues to increase because of the increasing bandwidth, with an increase in center frequency. The actual details of the sound signal cannot be determined from the output of a constant percentage narrow-band analyzer.

The $1/3$-*octave-band analyzer* gives higher output levels for the same noise signal because it has a wider frequency band; therefore, more components of the

Ch. 2 NOISE MEASUREMENTS AND INSTRUMENTATION

noise are added together. For example, at the 1000-Hz center frequency ⅓-octave-band the actual bandwidth is from 895 to 1128 Hz. For the octave-band analyzer the band spans from 707 to 1414 Hz for a 1000-Hz center frequency.

The A-, B-, and C-weighted analyzer provides a single decibel value that is the weighted summation of the sound-pressure level of the entire audible frequency range. The purpose of this analyzer (sometimes called A-, B-, and C-scale meters) is to provide a single number (decibels) that can be used to assess the

Table 2-3. Relative Response of Sound-Level Meters for Various Weighting Networks

⅓-Octave-Band Center Frequency, Hz	A-weighting Response, dB	B-weighting Response, dB	C-weighting Response, dB
10	-70.4	-38.2	-14.3
12.5	-63.4	-33.2	-11.2
*16	-56.7	-28.5	- 8.5
20	-50.5	-24.2	- 6.2
25	-44.7	-20.4	- 4.4
*31.5	-39.4	-17.1	- 3.0
40	-34.6	-14.2	- 2.0
50	-30.2	-11.6	- 1.3
*63	-26.2	- 9.3	- 0.8
80	-22.5	- 7.4	- 0.5
100	-19.1	- 5.6	- 0.3
*125	-16.1	- 4.2	- 0.2
160	-13.4	- 3.0	- 0.1
200	-10.9	- 2.0	0
*250	- 8.6	- 1.3	0
315	- 6.6	- 0.8	0
400	- 4.8	- 0.5	0
*500	- 3.2	- 0.3	0
630	- 1.9	- 0.1	0
800	- 0.8	0	0
*1000	0	0	0
1250	+ 0.6	0	0
1600	+ 1.0	0	- 0.1
*2000	+ 1.2	- 0.1	- 0.2
2500	+ 1.3	- 0.2	- 0.3
3150	+ 1.2	- 0.4	- 0.5
*4000	+ 1.0	- 0.7	- 0.8
5000	+ 0.5	- 1.2	- 1.3
6300	- 0.1	- 1.9	- 2.0
*8000	- 1.1	- 2.9	- 3.0
10000	- 2.5	- 4.3	- 4.4
12500	- 4.3	- 6.1	- 6.2
*16000	- 6.6	- 8.4	- 8.5
20000	- 9.3	-11.1	-11.2

*Octave-Band Center Frequencies

86 NOISE MEASUREMENTS AND INSTRUMENTATION Ch. 2

sound in the audible frequency range. As shown in the response columns in Table 2-3, the A-scale, or A-weighting response, has greatest sound attenuation below 1000 Hertz, followed by the B-scale, or B-weighting response, with the C-scale having the least. The A-scale is an approximation of the frequency response of the human ear to sound; for this reason OSHA requires the use of the A-scale for occupational noise exposure measurements. The B- and C-scales were developed earlier and are not widely used at present. However, a method of determining whether the measured noise has a high or low frequency content is to measure the noise on all three scales. Then, if the levels on all three scales are approximately equal, the sound predominates at frequencies above 600 Hertz. If the B- and C-scale readings are greater than the A-scale reading, the noise predominates in the range below 600 Hertz.

Some meters may, in addition, have a linear scale. This scale measures a true summation of the sound-pressure level in the entire audible range without any weighting factor. This is not indicative to the response that the human ear has to sound. In some instances this scale has applications in laboratory work. For outdoor sound measurements related to aircraft noise the D- and E-scales are used; they are, however, not used for other industrial applications.

A second example is given in Fig. 2-12 for a different noise spectrum. In this case the frequency scale is logarithmic. The source of this noise is not primarily pure tones as in the previous example, as noted by the change in base level of the constant bandwidth spectrum.

These two examples indicate clearly that only a constant bandwidth narrow-band analyzer can give an accurate record of the noise as a function of frequency. It is possible to make corrections to the output of the other analyzers

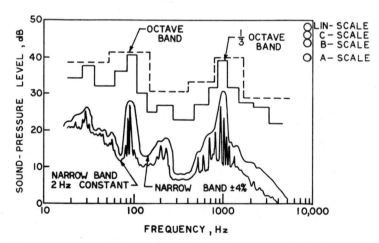

Fig. 2-12. Output comparison for an identical sound-pressure as obtained by four different types of sound analyzers, namely: (1). Octave-band analyzer; (2). 1/3-Octave-band analyzer; (3). ±4% narrow-band analyzer; and (4). 2 Hz constant bandwidth narrow-band analyzer. The comparative readings of this sound field, as obtained on a portable sound-level meter, are shown in the upper-right-hand corner; i.e., Linear Scale (48dBLin); C-weighted Scale (46dBC); B-weighted Scale (44dBB); and A-weighted Scale (39dBA).

Ch. 2 NOISE MEASUREMENTS AND INSTRUMENTATION

if something is known of the actual noise spectrum; i.e., if it is composed of discrete peaks or broad bands. There are some advantages to each of the analyzers described. The narrow-band analysis requires considerable time to obtain a spectrum with conventional analyzers, although with the advent of narrow-band, real-time analyzers this information can be obtained very quickly. There is also a considerable cost difference between narrow-band and full or $\frac{1}{3}$-octave analyzers. For applied research it is essential to have both narrow- and broad-band analyzers to make complete assignments of a given noise source. The narrow-band analysis is commonly used to determine the exact frequency content of the source in order to relate to mechanical operation of the machine. The A-scale analysis is then useful in determining the magnitude of noise reduction in some semblance to the manner the ear would respond and is the scale utilized for industrial noise legislation.

Taking Measurements

The actual process of taking acoustic measurements should be preceded by some consideration of the purpose for which the data is to be taken. Careful preplanning of noise measurements is an important part of obtaining useful information. The science of acoustics and noise control has not yet developed to the point where a step-by-step procedure can be given to obtain standard measurements from which a complete analysis can be made. Each noise-control problem becomes a special case that should be approached carefully. In acoustics it is easy to record noise spectra, but it is quite a different thing to interpret the data. In fact, one can easily obtain much more data than can be analyzed. The point here, is to avoid taking data just for the sake of making measurements.

In noise control the human element is an important part of the system in the selection of a measurement system, procedures, and the analysis of the resulting data. Incorrect measurements or analysis results in completely erroneous conclusions. The engineer becomes a vital part of an acoustic measurement system by considering the following: For what purpose are the noise measurements to be made? What analysis equipment should be used? What measurements should be made? Where should the microphone be located? Load conditions? What other related information will be needed such as speed, forces, pressures, gear ratios, temperatures, etc.? Obviously, not all of these questions can be answered, but some basic guidelines can be given.

What is the Purpose?

This is the first question that should be considered before making measurements because the answer will determine to a large extent what instrumentation will be used. In a great majority of cases, noise measurements are made because someone complains of a machine being "too noisy." This promptly motivates someone to hastily make some measurements, confident that the answer to the problem lies in obtaining a sound-level reading of the objectionable noise and immediately determining where to add damping material or some acoustic tile. The solution to most noise-control problems is rarely, if ever, this simple. The point to be made is that nearly all noise measurements are initiated after some-

88 NOISE MEASUREMENTS AND INSTRUMENTATION Ch. 2

one subjectively assesses a noise to be "loud." The next step should be to determine the desired outcome. The problem can probably be placed into one of the following categories:

A. Noise legislation
B. Product testing
C. Applied research.

The particular aspects of these areas will be considered individually.

A. *Noise Legislation.*—In this particular category the purpose is to determine the liability associated with some machine or process. The measurements to be made usually consist of a magnitude measurement only. The A-weighted noise level, denoted dB (A), is usually the information which is desired to establish the disposition of a piece of machinery. In this particular case the measurement is to be made at a location where the human operator or a bystander will be located. The instrumentation needed is a portable, A-weighted noise analyzer. Two portable meters of this type are shown in Fig. 2-13. Standard sources to calibrate sound-level meters are shown in Fig. 2-14.

At the present time the majority of noise legislation concerns this magnitude measurement only; however, in the future this may change and some frequency information may also be required. Some industry specifications already require the noise levels to be given in octave, or $1/3$-octave, bands.

A. *Courtesy of General Radio Co.*
B. *Courtesy of Brüel & Kjaer*

Fig. 2-13. Typical portable sound-level meters for acoustic surveys.

Ch. 2 NOISE MEASUREMENTS AND INSTRUMENTATION 89

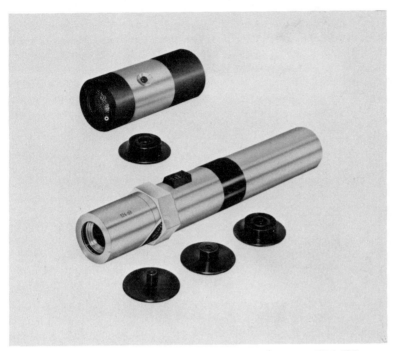

Courtesy of B & K Instruments, Inc.

Fig. 2-14. Standard sources for calibration of sound-level meters.

The criterion for hearing-loss exposure is related to an integrated level-time relation as described on page 122. There are electronic devices which will determine the noise exposure for time-varying sound levels by integrating the level and time to produce an exposure equivalent to the OSHA exposure as explained later on, in Chapter 4. Such a device is shown in Fig. 2-15.

B. *Product Testing.*—The information required in this case may be the magnitude and frequency. The purpose is to compare one device with another or to compare to some standard, and to accept or reject the device on the basis of this comparison. Many times the complete frequency spectrum need not be considered; only those frequencies which are known to be troublesome for this particular machine need be investigated. The procedure can become sophisticated, and sometimes a diagnosis of defects can be made from the noise spectrum. Measuring environment and microphone locations must be selected so that the particular frequency component of interest can be measured. Instrumentation in this case becomes a special consideration for each application. One-third octave or octave-band analyzers are typically used. An analyzer which gives $\frac{1}{10}$-octave-band or $\frac{1}{3}$-octave-band analysis of a signal is shown in Fig. 2-16. A portable octave-band analyzer is shown in Fig. 2-17.

C. *Applied Research.*—The desired outcome of this activity is the reduction of the noise output and the information required is considerably greater. The

90 NOISE MEASUREMENTS AND INSTRUMENTATION Ch. 2

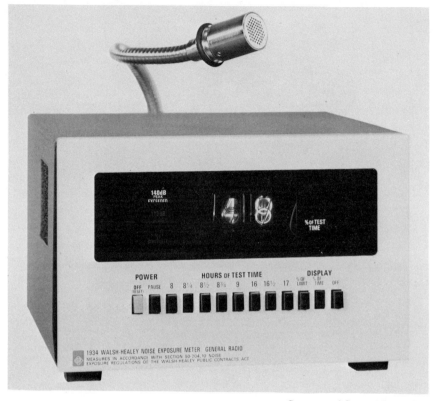

Courtesy of General Radio Co.

Fig. 2-15. Integrating noise exposure monitor for evaluation of exposure for time-varying sound fields.

usual information required of the noise source is magnitude, frequency, directionality, phase relations, and the rate of the noise build-up. Usually this information must be obtained in an acoustic laboratory. The magnitude and frequency information is usually obtained from a narrow-band analyzer; the environment is anechoic, if possible, in order to determine the directional properties of the source. Measurements can be made on installed machinery if precautions on page 97 are taken. Phase information is frequently useful to determine the exact location of a particular source and the transmission paths. Phase information is obtained from phase meters, correlators, and quo-quad analyzers. In some problems the rate of build-up of a particular noise is important. For this type of analysis the intent is to obtain transient information concerning a particular component in the acoustic spectrum. Often the damping in the structure, temperature effects, feedback mechanisms, etc., cause a particular noise component to be a quantity which builds up slowly or waxes and wanes with properties of the machine. The noise-control solution may require the prevention of the build-up of the intensity of the noise. The instrumentation required

Ch. 2 NOISE MEASUREMENTS AND INSTRUMENTATION 91

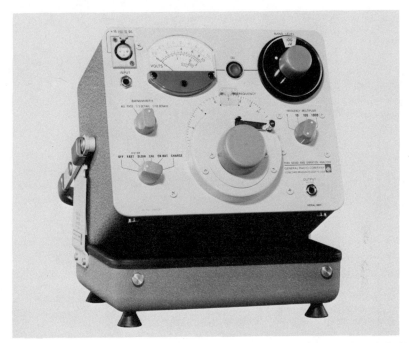

Courtesy of General Radio Co.

Fig. 2-16. Analyzer which provides $1/10$-octave or $1/3$-octave-band analysis of a signal.

for applied research is a function of the complexity of the problem and the information which must be obtained. Required instrumentation varies with sophistication and particular laboratory needs; these are beyond the intent and scope of this discussion.

Microphone Location

Now that the purpose of making noise measurements has been examined, some of the details involved in making the actual measurements will be discussed. The geometric location of the microphone with respect to the noise source and to the geometry of the environment surrounding it is an important factor to be considered. Knowledge of the sound field surrounding a noise source and the properties of reflected and standing waves give indication to where the microphone should be located for valid measurements.

A. *Near and Far Fields.*—Since a microphone is a pressure-sensing device, it will respond to fluctuating pressures if they are acoustic pressures or are only pressure fluctuations due to a flowing fluid. For example, in the turbulent wake of a fan there are pressure fluctuations due to the turbulent fluid motion. Some will contend that this turbulent flow is sound, but it is not the purpose here to debate this point. It is true, however, that these turbulent fluctuations do not propagate as sound far from the wake region. It must then be concluded, if the intent is to measure the sound propagated from a fan, that the microphone

92 NOISE MEASUREMENTS AND INSTRUMENTATION Ch. 2

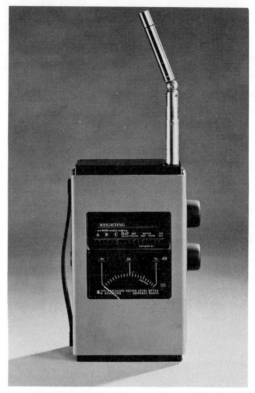

Courtesy of General Radio Co.

Fig. 2-17. Portable sound-level meter and octave-band analyzer.

should not be located in the wake or in the inlet flow regions. The same would apply near the intake or exhaust of an engine where the mean flow is high with turbulence superimposed. In such situations the microphone should be positioned adjacent to the fluid-flow region.

A similar condition exists in the so-called acoustic near-field region which is the acoustic field near an acoustic source; for example, close to a vibrating source. The acoustic near-field describes the actual flow of air that is in motion due to the surface vibrating back and forth. There will be pressure fluctuations from this movement of air which do not propagate as sound. This region is shown, in Fig. 2-18, near the surface of a vibrating source. To avoid measurement of the hydrodynamic pressure fluctuations in this region, the usual practice is to place the microphone at least one acoustic wavelength from the surface determined by the lowest frequency of interest. Again, the acoustic wavelength at room temperature is given by:

$$\text{wavelength} = \lambda = \frac{1130}{\text{frequency (Hz)}} \text{(ft)} = \frac{344}{\text{frequency (Hz)}} \text{(m)} \qquad (1\text{-}2)$$

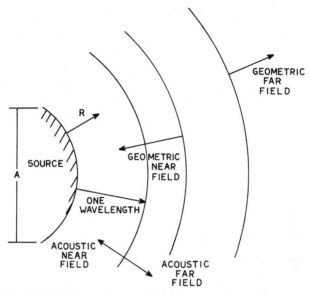

Fig. 2-18. Regions of the acoustic field surrounding a noise source.

If the lowest frequency of interest is 100 Hz, this rule-of-thumb would place the microphone no closer than 11.3 feet. The region beyond the distance of one wavelength is known as the acoustic far-field.

The next region shown in Fig. 2-18 is the geometric near-field which is characterized by the fact that pressure fluctuations in this region are acoustic, but the geometry of the source determines the directivity in the sound field. In other words, the dimension of the source, A, is of the same magnitude as the distance from the source. The sound field is dependent upon the shape of the source.

The geometric far-field is characterized by the condition that the geometry of the source is not distinguishable by measurements in this region. All sources of the same strength and directivity appear identical in this region. This region begins at a distance, R, equal to 3 or 4 times the largest dimension of the source, A. Depending on the actual dimensions of the source, A, and the wavelength, λ, of interest, the acoustic near-field may extend beyond the geometric near-field.

Measurements should not be made in the acoustic near-field if true sound pressures are to be recorded; however, *if an operator is located in this region, measurements must be taken at the operator location and the near- and far-field concept must be ignored.* Within the geometric near-field, acoustic conditions exist but the microphone location directly determines the data obtained. In the geometric far-field, information will be obtained which indicates how acoustic energy will propagate far from the source.

B. *Reverberant Fields.*—The sound field within an enclosure or in a building with reflecting surfaces, can have a definite spatial distribution known as standing wave patterns. A sound field of this type is characterized by sound-pressure levels which are a function of the location of the microphone inside the enclo-

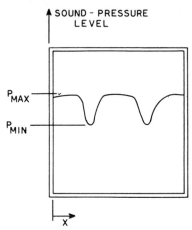

Fig. 2-19. Illustration of a standing acoustic wave pattern in a reverberant space.

sure. This condition is the result of the superposition of sound waves reflected from the sides of the enclosure or walls of a room. An illustration of this type of field is shown in Fig. 2-19. The sound-pressure level represents the output of a microphone as it is moved within the space.

C. *Free-Field Locations.*—In the free field where reflections and standing waves are negligible, the microphone should be located at positions which are indicative of the sound component to be measured. Usually, one selects standard microphone locations based upon the directivity of the particular source of interest. Certain positions may be selected to investigate selected sources such as intake, exhaust, fan, mechanical components, etc. These positions result from a sound survey made about the entire machine. For research purposes it may be acceptable to place the microphone in the near field of one particular source to eliminate contributions from other sources. *For assessing worker noise exposure, the readings must be taken at the work stations regardless of near- or far-field effects.*

The ILG Reference Sound Source

The ILG Reference Sound Source is an instrument that provides a standard sound-power level. It consists of a calibrated, constant-speed fan that generates a standard sound-power level which is independent of the space in which it is located. The sound-power generated by this source, in octave bands or one-third-octave bands, is provided in Table 2-4. It is used whenever a known sound-power-level output is required in an acoustic space to determine the acoustic properties of the space. Whenever sound-pressure levels in a factory space must be calculated, it is necessary to know the acoustic properties of that space. In many instances the acoustic properties in the factory space can be estimated; however, with the ILG source these properties can be much more accurately determined and subsequent sound-pressure-level calculations made with greater assurance. The ILG source is also used to calibrate reverberation rooms, semi-anechoic chambers, and anechoic chambers in acoustic laboratories.

Table 2-4. Sound-Power Levels of the ILG Reference Sound Source

$1/3$-Octave-Band Center Frequency, Hz	Sound-Power Level, dB re 10^{-12} Watts	Octave-Band Center Frequency, Hz	Sound Power Level, dB re 10^{-12} Watts
50	78		
63	77	63	82
80	76		
100	76		
125	76	125	81
160	76		
200	76		
250	76	250	81
315	76		
400	76		
500	76	500	81
630	76		
800	76		
1000	76	1000	81
1250	76		
1600	76		
2000	76	2000	81
2500	76		
3150	75		
4000	75	4000	79
5000	74		
6300	74		
8000	73	8000	78
10,000	71		

Measurement of Impulsive or Impact Sound

Measurement of impulsive sound requires different instrumentation and methods than are used for steady-state or continuous sound. Impulse or impact sound-level meters are available that store the peak sound-pressure level which is then read on the meter scale. OSHA requirements for the peak hold measurements are as follows: The rise time shall be not more than 50 microseconds and the decay rate for the peak hold feature shall be less than .05 decibels per second. The unit of measurement shall be: decibels peak sound-pressure level re 20 micropascals.

In some instances it is necessary to know the duration and the rise time of the impact pulse, and the repetition rate of the impacts in addition to the peak value. A method for obtaining this information is shown in Fig. 2-20. In making this measurement, the microphone must be oriented perpendicularly to the direction of the sound pulse propagation. The amplified signal from the microphone is then displayed on the oscilloscope. The signal can then be recorded with an oscilloscope camera or held on the screen of a storage oscillo-

96 NOISE MEASUREMENTS AND INSTRUMENTATION Ch. 2

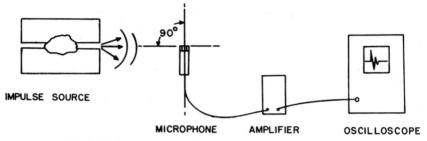

Fig. 2-20. Instrumentation for measurement of impulsive sound.

scope. The vertical scale of the oscilloscope should be calibrated with a standard sound-level calibrator for microphones, shown in Fig. 2-15. For impulsive sound measurement, a small-diameter ($\frac{1}{8}$-inch typical) microphone is recommended to obtain a desirable microphone response.

Development Testing

Throughout this chapter the subject of actual interpretation of data and the diagnosis of noise sources has been carefully avoided. There is some art and intuition involved in noise control, and there seems to be no technical developments on the horizon which will eliminate the dedicated engineer who enjoys mental exercise and solving riddles. Instrumentation has made life much easier for the engineer; he can ask more intelligent questions and can obtain more information, but in the final analysis, he must make some conclusions as to the ailment of the noisy machine.

The noise-control engineer then, is in the same position as a medical doctor. He has impressive instrumentation with elaborate examining rooms within which he can take the temperature of the machine, count its pulse and respiration rate, make electrocardiographs, use trace materials in its system, etc. He can ask the machine to cough and then listen to its resonating cavities for some sign of a symptom. All the sophisticated instruments at his disposal are of little use in the majority of noise problems if he does not ask the correct questions of the machine; therefore he must know some of the physics of acoustics. There are some symptoms which are relatively easy to recognize and the man on the street can make the diagnosis. There are other problems, however, which require training to identify, and there are some problems which the engineer can identify but he goes to a specialist for the answer to the problem. The better a given "patient" is known the easier it becomes to recognize recurring illnesses and potentially troublesome areas, and often it takes time to become familiar enough with a machine to detect the acoustic sources of trouble.

The strategy of the noise-control engineer must be efficient if he is to locate one single source out of several hundred components. He should categorize the problem into groups so that possible sources can be eliminated systematically. He should consider the various systems in the machine as a doctor considers the systems of this patients. The problem can be broken down to the respiratory system, circulatory system, nervous system, etc. This may not always be possible in great detail, but it certainly is more logical than a trial-and-error pro-

Ch. 2 NOISE MEASUREMENTS AND INSTRUMENTATION 97

cedure. Regardless of the plan of attack engineers will seldom find clear-cut solutions. The usual questions are of a relative nature such as: How much of the noise is produced by a given component? How can the energy be shifted to another frequency range? Which source appears to lead to the greatest payoff? Can the noise be muffled or must the source be altered?

With the pressures being applied by legislation and customer acceptance, the investment must be made in sound measurement and analysis equipment. Equally important is the need for an investment in education in the use of instrumentation and the nature of acoustics. Most of all, the engineer needs to have a good physical "feel" for the particular machine he is analyzing in order to apply the optimum techniques.

3

Standards for Noise Control

CHARLES W. RODMAN

The Role of Standards

Standards play an important role in the field of noise control. They provide a common nomenclature and eliminate ambiguity in materials specifications. For example, the development, manufacture, and use of materials for noise control may encompass several disciplines, ranging from chemical engineering to architecture. The diversity of the disciplines involved could easily lead to some lack of understanding unless standard terminology, relative to the acoustic properties of material, is employed. The use of standardized methods of testing materials and reporting the results makes possible the comparison of their relative values which, when properly employed, will achieve the best possible selection of materials for a particular application. In the area of industrial noise control, standard methods of specifying the sound-generation characteristics of machinery are especially important to the plant designer who applies these values to predict the effect on the noise environment in combining several different types of machines to achieve a specific process design. For the environmental engineer, standardized methods of measuring and quantifying noise are essential to predicting the environmental impact of adding a noise source to an existing situation. Standards are frequently cited as technical background in building codes or land-use zoning codes, as well as in other regulatory documents.

The objective of this chapter then, is to point out the importance of standards and to guide the reader toward standards that fall within his area of interest. For obvious reasons, no attempt is made here to reproduce any of the standards, however, a large listing of available standards concerning acoustics and noise is included and the addresses of the organizations where the standards may be obtained will be found at the end of this chapter.

Ch. 3 STANDARDS FOR NOISE CONTROL

Types of Standards

A standard is defined as "an acknowledged measure of comparison for quantitative or qualitative value."[1] Bearing this in mind and examining various standards, the following types of standards can be identified:

1. Definitions of terms and units of measure
2. Specifications for materials or methods
3. Methods of testing
4. Classification according to specific characteristics.

Definitions. In the context of standards, a definition is an explanation of the meaning of a term as applied to the area of usage. By including definitions as a part of standards the possibility of misuse or misunderstanding of the information obtained through the use of a specification or method is minimized. By formalizing the language of technology, standard definitions discourage the use of jargon which is often a source of misunderstanding.

Specifications. A specification, as defined by the American Society for Testing and Materials, is ". . . a precise statement of a set of requirements to be satisfied by a material, product, system, or service . . . ." "When appropriate," the definition states, "the specification shall indicate . . . the procedure by which it can be determined whether the requirements have been satisfied . . . ." The specification establishes acceptable limits for the numerical values of properties, composition, or performance.

Methods of Testing. A "test-method" standard spells out, in detail, the method to be used in determining a set of characteristics and the units and accuracy of quantities, but does not set limits for the quantities. It is this omission of limits which sets the test method apart from a specification.

Classification. A classification standard provides a systematic means of arranging or dividing materials, products, systems, or services into groups based on similar characteristics. For example, ASTM E-413-73 is a standard defining the procedure for classifying acoustic structures by sound transmission class (STC). The standard also includes instructions for determining the STC from laboratory data.

Recommended Practice. Although similar to a test-method standard, the Recommended Practice is intended as an information type document that may or may not be referenced by a test method or specifications.

Standards Writing Organizations. In the United States the best known organizations, which have as their sole purpose the preparation and publication of standards, are the American National Standards Institute (ANSI) and the American Society for Testing and Materials (ASTM). These two organizations generate standards in a wide variety of subject areas. There are also a number of organizations which write standards related to specific subjects. Examples of these are the American Society of Heating, Refrigerating and Airconditioning Engineers (ASHRAE), which includes standards for sound measurement procedures related

---

[1] *American Heritage Dictionary* (Boston: Houghton Mifflin, Co.), 1971.

to air moving and conditioning equipment in its publications; and the Society for Automotive Engineers (SAE) which sponsors standards for measurement of sound related to various types of transportation, including aircraft noise. There are also a number of industry groups that write standards related specifically to the product or service supplied by the group's members. An example is the American Gear Manufacturers Association (AGMA), which includes a standard for gear noise in its list of standards pertaining to the design, manufacture, inspection, and application of gears.

International organizations which prepare standards related to noise include the International Organization for Standardization (ISO) and the International Electrotechnical Commission (IEC).

ISO standards cover many fields. The international standards developed by ISO are the result of agreement between member bodies. The United States' member of ISO is ANSI, where these standards may be obtained directly.

ISO and IEC standards are often adopted as ANSI standards. In some cases, for example, an ANSI standard may consist of an existing ISO standard, rewritten to reflect the need of domestic users.

IEC standards of importance to noise control are those relating to specifications for microphones and noise-measuring instruments.

The United States Bureau of Standards under the sponsorship of the United States Environmental Protection Agency (EPA) has prepared a document, "Fundamentals of Noise: Measurement Rating Schemes, and Standards,"[2] which outlines many of the standards relating to noise and lists a number of standards writing organizations. An expanded version of this list of organizations and their addresses is included at the end of this chapter.

Standards Writing Procedures

The procedures followed by typical standards writing organizations are spelled out in documents variously known as regulations, procedures, operations guides, or bylaws. These documents specify the makeup of the technical committees; the requirements for balance of membership with respect to producers, users, or general interest classification; methods of election and duties of officers; procedures for obtaining approval or consensus, including provision for dealing with negative or dissenting opinions; and methods of review or renewal of standards. Generally, the chairman of a technical committee cannot be a "producer" classification representative if the committee deals with standards related to products, and the membership of the committee must be balanced by classification. While some technical organizations hold open meetings, others require technical committee meetings to be closed to nonmembers.

A standard usually originates as a draft document which, after approval by the originating working group, is submitted to ascending echelons of the organization for approval. The mechanics for resolving negative votes or opinions, and for submitting the standard out of committee, vary considerably from one organization to another. In many cases, the standard is submitted to the voting

---

[2] "National Technical Information Document 300.15," available from Supt. of Documents, Washington, D.C. 20502.

Ch. 3 STANDARDS FOR NOISE CONTROL

membership of the main body of the organization for approval. In some cases the standard is also submitted to a "Standards Committee" and an editorial review board before publication.

Standards Related to Acoustics

A number of standards have been selected as being of particular value to the noise control engineer. For convenience, these have been categorized functionally as follows:

1. Characterization of Sound
2. Community Noise
3. Definitions
4. Acoustic Test Facilities
5. Hearing Protection
6. Sound Measurement Instrumentation
7. Measurement of Acoustic Properties of Materials
8. Measurement of Machinery Noise
9. Measurement of Transportation Noise
10. Measurement of Performance of Acoustic Systems.

Because some of the standards selected cover more than one area, Table 3-1 is provided as a guide to the standards.

Examination of Table 3-1 shows that for the most part the standards prepared by the American National Standards Institute, International Organization for Standardization, and International Electrotechnical Commission are universally applicable to all fields of noise measurement. These standards provide an excellent basis on which to establish a noise measurement facility. Although at present ANSI does not have a standard for community noise measurement, the S.1 Committee of the Acoustical Society of America is working on the preparation of such a standard, but for the most part, the ANSI standards are adequate, providing they are supplemented with those standards which specifically cover the area of interest. The following paragraphs are intended to assist in making an appropriate selection of standards.

Selection of Standards

The titles of many of the standards listed in Table 3-1 are explanatory of the content of the standards; however, the following may be helpful in selecting specific standards for use as reference in machinery noise measurement and control.

Instruments and Facilities. Standard procedures for measuring sound of specific systems or devices usually require the use of instruments meeting the requirements of IEC Publication 179, or ANSI *Recommended Practice* Sl.4-1971, which cover sound level meters. American National Standard ANSI, S1.2-1962 (R1971), *American Standard for the Physical Measurement of Sound*, requires as a minimum, the use of a standard sound-level meter and a standard octave-band filter set and is intended to serve as a basis for test codes. This latter standard includes procedures for measuring and reporting sound pressure levels on

(Continued on page 108)

Table 3-1. Guide to Standards Pertaining to Noise Measurement and Control

Organization*	Standard Designation	Title	Characterization of Sound	Community Noise	Definitions	Facilities	Hearing Protection	Instrumentation	Material Properties	Measurement, Machinery	Measurement, Transportation	System Performance
ANSI	ASA S1.1-1960 (Rev. 1971)	American Standard Acoustical Terminology			X							
ANSI	ASA S1.2-1962 (Rev. 1971)	American Standard Method for the Physical Measurement of Sound**		X		X				X	X	
ANSI	ANSI S1.4-1971	American National Standard Specification for Sound Level Meters						X				
ANSI	USAS S1.6-1967 (Rev. 1971)	USA Standard Preferred Frequencies and Band Numbers for Acoustical Measurements			X			X				
ANSI	ANSI S1.8-1969	American National Standard Preferred Reference Quantities for Acoustical Levels			X			X				
ANSI	ASA S1.10-1966 (Rev. 1971)	American Standard Method for the Calibration of Microphones						X				
ANSI	ASA S1.11-1966 (Rev. 1971)	American Standard Specification for Octave, Half-Octave, and Third-Octave Band Filter Sets						X				
ANSI	USAS S1.12-1967 (Rev. 1972)	USA Standard Specifications for Laboratory Standard Microphones				X		X		X		
ANSI	ANSI S1.13-1971	American National Standard Methods for the Measurement of Sound Pressure Levels***		X	X	X	X					
ANSI	ASA S3.1-1960 (Rev. 1971)	American Standard Criteria for Background Noise in Audiometer Rooms	X	X								
ANSI	USAS S3.4-1968 (Rev. 1972)	USA Standard Procedure for the Computation of Loudness of Noise			X							
ANSI	ANSI S3.6-1969	American National Standard Specifications for Audiometers					X	X				

Ch. 3 STANDARDS FOR NOISE CONTROL

ANSI	ANSI S5.1-1971	American National Standard Test Code for the Measurement of Sound from Pneumatic Equipment		X		X		X	
ANSI	ASA Y10.11-1953 (Rev. 1959)	American Standard Letter Symbols for Acoustics	X						
ANSI	ASA S3.19-1974	American Standard Method for the Measurement of the Real-Ear Attenuation of Ear Protectors at Threshold			X	X			
ASTM	C 384-58 (1972)	Standard Method of Test for Impedance and Absorption of Acoustical Materials by the Tube Method		X		X	X		
ASTM	C 423-66 (1972)	Test for Sound Absorption of Acoustical Materials in Reverberation Rooms		X			X	X	
ASTM	C 522-73	Test for Airflow Resistance of Acoustical Materials					X		
ASTM	C 634-73	Definition of Terms Relating to Acoustical Tests of Building Constructions and Materials	X						X
ASTM	C 635-69	Specification for Metal Suspension Systems for Acoustical Tile and Lay-in Panel Ceilings							X
ASTM	C 636-69	Recommended Practice for Installation of Metal Ceiling Suspension Systems for Acoustical Tile and Lay-in Panels							X
ASTM	C 643-73	Painting Ceiling Materials for Acoustical Absorption Tests		X		X	X		X
ASTM	E 90-70	Recommended Practice for Measurement of Airborne Sound Transmission Loss of Building Partitions		X		X	X		X
ASTM	E 336-71	Recommended Practice for Measurement of Airborne Sound Insulation in Buildings							X

*See Key at end of table.
**Partially revised by ANSI S1.13 (see above) and S1.21, *Methods for the Determination of Sound Power Levels of Small Sources in Reverberation Rooms.*
***Partial revision of S1.2-1962.

Table 3-1 (Cont'd). Guide to Standards Pertaining to Noise Measurement and Control

Organization*	Standard Designation	Title	Characterization of Sound	Community Noise	Definitions	Facilities	Hearing Protection	Instrumentation	Material Properties	Measurement, Machinery	Measurement, Transportation	System Performance
ASTM	E 413-73	Classification for Determination of Sound Transmission Class				X			X			X
ASTM	E 477-73	Testing Duct Liner Materials and Prefabricated Silencers for Acoustical and Airflow Performance				X		X	X			X
ASTM	E 492-73T	Laboratory Measurement of Impact Sound Transmission Through Floor-Ceiling Assemblies Using the Tapping Machine				X			X			X
ASTM	E 497-73T	Recommended Practice for Installation of Fixed Partitions of the Light Frame Type for the Purpose of Conserving Their Sound Insulation Efficiency										X
SAE	Recommended Practice J184 (1970)	Qualifying a Sound Data Acquisition System				X		X				
SAE	Recommended Practice J919 (1966)	Measurement of Sound Level at Operator Station	X			X		X		X	X	
SAE	Recommended Practice J919a (Rev. 1971)	Sound Level Measurements at the Operator Station for Agricultural and Construction Equipment	X			X		X		X	X	
SAE	Standard J952B (1969)	Sound Levels for Engine Powered Equipment	X			X		X		X		

STANDARDS FOR NOISE CONTROL

Org	Number	Title								
SAE	ARP 866	Standard Values of Absorption as a Function of Temperature and Humidity	X							
SAE	AIR 817	A Technique for Narrow Band Analysis of a Transient		X	X					
IEEE	85 (1965)	Test Procedure for Airborne Noise Measurements on Rotating Electric Machinery						X		X
IEEE	151 (1965)	Standard Definitions of Terms for Audio and Electroacoustics			X					
ASHRAE	36-62	Measurement of Sound Power Radiated from Heating, Refrigerating, and Airconditioning Equipment		X		X	X	X	X	
ASHRAE	36A-63	Method of Determining Sound Power Levels of Room Air Conditioners**		X		X	X	X	X	
ASHRAE	36B-63	Method of Testing for Rating the Acoustic Performance of Air Control and Terminal Devices and Similar Equipment		X		X	X	X	X	
ARI	270 (1967)	Standard for the Sound Rating of Outdoor Unitary Equipment		X			X	X	X	
ARI	275 (1969)	Standard for Application of Sound Rated Outdoor Unitary Equipment		X			X	X	X	
ARI	443 (1970)	Standard for Sound Rating of Room Fan-Coil Air Conditioners		X			X	X	X	
AMCA	300-67	Test Code for Sound Rating		X	X		X	XX	XX	XX
ADC	AD-63	Measurement of Room-to-Room Sound Transmission Through Plenum Systems		X	X		X	XX	XX	XX
CAGI	Test Code (1969)	Cagi-Pneurop Test Code for Measurement of Sound from Pneumatic Equipment		X	X		X	X	X	X
AGMA	293.03 (1968)	Specification for Measurement of Sound on High Speed ... Gear Units**		X	X		X	X	X	X
NMTBA	13	Noise Measurement Techniques		XXX	XX		XXX	XXX	XX	XX
AFBMA		Rolling Bearing Vibration and Noise		XXX	XX		XXX	XXX	XX	XX
Military	MIL-S-3151a and Notice 1	Sound Level Measuring and Analyzing Equipment		XXX	XX		XXX	XXX	XX	XX

*See Key at end of table.
**Partial title.

Table 3-1 (Cont'd). Guide to Standards Pertaining to Noise Measurement and Control

Organization*	Standard Designation	Title	Characterization of Sound	Community Noise	Definitions	Facilities	Hearing Protection	Instrumentation	Material Properties	Measurement, Machinery	Measurement, Transportation	System Performance
ISO	Recommendation R31 Part VII	Quantities and Units of Acoustics			X							
ISO	R131	Expression of the Physical and Subjective Magnitudes of Sound or Noise	X		X							
ISO	Recommendation R140 (1960)	Field and Laboratory Measurements of Airborne and Impact Sound Transmission				X						
ISO	Recommendation R266 (1962)	Preferred Frequencies for Acoustical Measurements	X					X				
ISO	Recommendation R354 (1963)	Measurement of Absorption Coefficients in a Reverberation Room				X		X	X			
ISO	Recommendation R362 (1964)	Measurement of Noise Emitted by Vehicles	X					X			X	
ISO	Recommendation R495 (1966)	General Requirements for the Preparation of Test Codes for Measuring the Noise Emitted by Machines	X			X		X		X		X
ISO	Recommendation R1680 (1970)	Test Code for the Measurement of the Airborne Noise Emitted by Rotating Electrical Machinery	X			X		X		X	X	X
ISO	Recommendation R1996 (1971)	Assessment of Noise with Respect to Community Response**	X	X				X		X		
ISO	Recommendation R1999 (1971)	Acoustics–Assessment of Occupational Noise Exposure for Hearing Conservation Purposes	X				X	X		X		

IEC	Recommendation Publication 50 (08) (1960)	International Electrotechnical Vocabulary, Electroacoustics		X	
IEC	Recommendation Publication 123 (1961)	Recommendations for Sound Level Meters***			X
IEC	Recommendation Publication 225 (1966)	Octave, Half-Octave, and Third-Octave Band Filters	X	X	

ADC-Air Diffusion Council
CAGI-Compressed Air and Gas Institute
AGMA-American Gear Manufacturers Association
IEC-International Electrotechnical Commission
NMTBA-National Machine Tool Builders' Association
AFBMA-Anti-Friction Bearing Manufacturers Association
IOS-International Organization for Standardization

*(For complete addresses see end of chapter.)
ANSI-American National Standards Institute
ASTM-American Society for Testing and Materials
SAE-Society of Automotive Engineers
IEEE-Institute of Electrical and Electronics Engineers
ASHRAE-American Society of Heating, Refrigerating and Airconditioning Engineers
ARI-Air-Conditioning and Refrigeration Institute
AMCA-Air Moving and Conditioning Association
**Partial title.
***Precision Sound Level Meters are covered by IEC Publication 179 (1965).

(*Continued from page 101*)

nonimpulsive sound sources operating in air. ANSI S1.13-1971, *Methods for Measurement of Sound Pressure Levels*, gives procedures for making sound pressure measurements under a number of conditions not covered by S1.4.

An excellent standard for qualifying a sound measurement system is SAE Recommended Practice, J184. By applying the techniques outlined in J184, a measurement system can be checked to determine how well it meets the requirements established for an equivalent sound level meter which meets the requirements of IEC Publication 127 and ANSI S1.4-1971. Other standards which are useful to anyone who is setting up a sound measurement laboratory are:

1. ANSI S1.6-1967, *Preferred Frequencies* (Rev. 1971)
2. ANSI S1.8-1969, *Preferred Reference Quantities*
3. ANSI S1.10-1966, *Calibration of Microphones* (Rev. 1971)
4. ANSI S1.11-1966, *Filter Sets* (Rev. 1971)
5. ANSI S1.12-1967, *Laboratory Standard Microphones* (Rev. 1972)
6. ANSI S1.13-1971, *Measurement of Sound Pressure Levels*
7. SAE ARP 866, *Standard Values of Absorption*
8. SAE AIR 817, *Transient Analysis*
9. MIL MIL-S-3151a, *Measuring and Analyzing Equipment*
10. IEC Publication 123, *Sound Level Meters*
11. IEC Publication 225, *Filters*.

Machinery Noise Measurement. The specific standard which will cover the measurement of interest will vary from one measurement to another. One or more of the standards in the following list should apply in most cases.

Pneumatic Equipment: ANSI S5.1-1971 and Cagi-Pneurop Test Code—These standards explain the techniques necessary for accurate measurement and reporting of noise from air compressors of all types, air-operated tools, and air-operated machinery. The Cagi-Pneurop code contains numerous drawings of equipment, showing locations for making measurements. Also shown are sample data sheets.

Electrical Equipment: IEEE 85 and ISO R1680—Both of these standards are intended for use with respect to methods of measuring the noise from rotating electrical equipment, i.e., motors, generators, synchronous condensers, alternators, etc. The ANSI standard S1.13-1971, paragraph 5.7, should be consulted before making measurements in the vicinity of electrical machinery. Electrostatic fields can influence the sound-measuring equipment. S1.13 discusses this and other problems, and gives some precautions which the investigator should follow.

Engine-powered Equipment (including vehicles): SAE J952B (1969) and ISO R362 (1964)—Between these two standards, noise from most engine-powered equipment is covered. An exception is aircraft noise. A number of other SAE standards cover aircraft noise, and truck and bus noises. The SAE standard listed, covers engine-powered tools for nonhighway, nonfactory application. The ISO standard is for motor vehicles. Its provisions include acceleration conditions, transmission noise, etc.

Heating, Ventilating and Air Conditioning (HVAC): ASHRAE 36-62; ASHRAE 36A-63; ASHRAE 36B-63; ASTM E 477-73; ARI 270; ARI 275; ARI 443; AMCA 300-67; and ADC AD63—These standards cover various aspects of HVAC. The titles of the standards (see Table 3-1) provide enough information to permit selection of the standard required for the task at hand.

Mechanical Power Transmission Equipment: AGMA 293.03 (gear noise); NMTBA Noise Measurement Techniques (machine tools); and AFBMA 13 (bearings).

Occupational Safety (hearing conservation): ANSI S3.6; ANSI Z24.22; SAE J919 *Level at Operator Station*; SAE J919a *Agricultural and Construction Equipment*; SAE J952B *Engine-powered Equipment*; ISO R140 *Field and Laboratory Measurements*; ISO R1999 *Assessment of Noise Exposure*; and IEC Publication 123—The first two standards listed are among many having to do with audiometry (measurement of hearing ability). The others, except IEC 123 and ISO R140, deal with specific worker noise exposure problems. IEC 123 discusses sound level meters, including Type 3, which is often used for plant noise surveys.

Architectural Acoustics. Frequently the noise control engineer finds that the only means available to reduce noise levels is to treat the structure in which the source or listener (or both) may be located. The standards prepared by the ASTM Committee E-33 on Environmental Acoustics (see Table 3-1) provide information on the use of various noise control techniques as part of the background portion of the standards. The meanings of the material specifications and the numerical units used to quantify the characteristics are explained. All of the ASTM standards on acoustics may be found in Part 14 (Part 18, 1974, and later) of the Book of Standards.

Community Noise. Community noise is as old as communities, but the interest in it is relatively new. The noise-control engineer may sometimes be called on to solve a factory noise problem because it is producing adverse impact on the community noise environment. ISO Recommendation R1996 was prepared to deal with this subject. Although it does not fall in the category of a standard, a very useful reference is *Noise Assessment Guidelines: Technical Background*,[3] prepared for the U.S. Department of Housing and Urban Development (HUD), by Bolt, Beranek, and Newman. In this document, the authors present numerous criteria for community noise levels and explain the measures which have been devised to describe community noise.

Where to Obtain Standards. The ANSI and ASTM standards are available in many libraries, both public and institutional. However, the copies on file may

---

[3] *HUD Report No. TE/NA 172*, available from Supt. of Documents, Washington, D.C. 20502.

not contain the latest revision. Standards can be ordered directly from the organization which publishes them. The cost for individual standards, when they are available as such, is normally less than $10.00. The ASTM standards are available only as a bound volume (Part 14) which costs in the neighborhood of $25.00. Up-to-date lists of standards, together with instructions for ordering, are available by writing or phoning the appropriate organization at the addresses shown below. (ISO Standards and IEC publications may be obtained through ANSI.)

Air Conditioning and Refrigeration Institute
1815 No. Fort Myer Drive
Arlington, Va. 22209

Air Diffusion Council
435 North Michigan Avenue
Chicago, Ill. 60611
(312) 527-5494

Air Moving and Conditioning Association
30 W. University Drive
Arlington Heights, Ill. 60004
(312) 394-0150

American Boat and Yacht Council
15 East 26th Street
New York, N.Y. 10010
(212) 532-9688

American Gear Manufacturers Association
1330 Massachusetts Avenue, N.W.
Washington, D.C. 20005
(202) 783-3621

American National Standards Institute
1430 Broadway
New York, N.Y. 10018
(212) 868-1220

American Society for Testing and Materials
1916 Race Street
Philadelphia, Pa. 19103
(215) 569-4200

American Society of Heating, Refrigerating and Airconditioning Engineers
United Engineering Center
345 East 47th Street
New York, N.Y. 10017
(212) 752-6800 x360

Anti-Friction Bearing Manufacturers Association
60 East 42nd Street
New York, N.Y. 10017
(212) 687-5028

Ch. 3 STANDARDS FOR NOISE CONTROL

Association of Home Appliance Manufacturers
20 No. Wacker Drive
Chicago, Ill. 60606
(312) 236-2921

California Redwood Association
617 Montgomery Street
San Francisco, Calif. 94111
(415) 392-7880

Compressed Air and Gas Institute
122 East 42nd Street
New York, N.Y. 10017
(212) 697-2668

Electronic Industries Association (Formerly Radio Manufacturers Assn.)
2001 Eye Street N.W.
Washington, D.C. 20006
(202) 659-2200

Factory Mutual Engineering Division
184 High Street
Boston, Mass. 02110

Federal Specifications
Specification Sales (3FRDS)
Building 197, Washington Navy Yard
General Services Administration
Washington, D.C. 20407

Hearing Aid Industry Conference, Inc.
1001 Connecticut Ave., N.W.
Washington, D.C. 20036
(202) 833-1412

Institute of Electrical and Electronics Engineers
345 East 47th Street
New York, N.Y. 10017
(212) 752-6800

International Electrotechnical Commission
1, rue de Varembe
Geneva 20, Switzerland (See ANSI, Sales Representation in U.S.A.)

International Organization for Standardization
1, rue de Varembe
Geneva 20, Switzerland (See ANSI, Sales Representation in U.S.A.)

National Electrical Manufacturers Association
155 East 44th Street
New York, N.Y. 10017
(212) 682-1500

National Machine Tool Builders' Association
7901 Westpark Drive
McLean, Va. 22101
(703) 893-2900

National School Supply and Equipment Association
Folding Partition Subsection
1500 Wilson Blvd.
Arlington, Va. 22209

Power Saw Manufacturers Association
734 15th Street, N.W.
Washington, D.C. 20005
(202) 737-6510

Radio Manufacturers Association (See: [E1A] Electronic Industries Association)

Society of Automotive Engineers
2 Pennsylvania Plaza
New York, N.Y. 10001
(212) 594-5700

4

Criteria for Machinery Noise

S. M. MARCO

Noise is generated by many engineering devices or systems during operation, and like other mechanical characteristics and features such as power output, speed, weight, and operating life, etc., it can be controlled to some degree at the design or development stage. To exercise such control the design engineer must have available quantified noise criteria. Such criteria will also serve as a guide to designers, manufacturers, governmental representatives, and consumers in deciding whether or not the noise performance of a particular machine or system will be acceptable for a specific purpose. Noise criteria, to be useful, must be based upon a considerable body of accumulated empirical information; no simple, straightforward statement will completely cover all aspects of noise performance for any specific purpose. Those authoritative bodies that promulgate noise criteria customarily must also make some arbitrary assumptions and compromises based on information arising from various scientific, engineering, medical, legal, sociological, and economic factors. Much of this type of information has been obtained by planned research and testing as well as from published reports of investigations that already have been made for other purposes. In general, the greater the amount of pertinent data accumulated, the more reliable and useful the noise criteria will be.

In addition to specifying numerical values in any noise criteria it is necessary to specify the types of measurements required, the instruments to be used, the conditions under which measurements are to be made, and the method of comparing the measured values with the specified values. Since reliable and useful criteria are so difficult to develop—as they are actually a measure of the effect of noise on people—there is a real scarcity of satisfactory criteria. Indeed, in some cases there are no useful criteria at all. One example is the lack of a satisfactory criterion to judge the damage-risk to hearing, of impact or impulsive noise. Because this situation is widespread, many investigations are being carried out by researchers in the fields of engineering, physiology, psychology, otolaryngology, speech, etc., to increase data that can be made available for use in formulating some useful criteria.

NOISE CRITERIA CLASSIFICATION

Noise criteria may be classified in several ways, although they may have the same basic objective; that is, to serve as guides to engineers, governmental regulating agencies, manufacturers, and consumers in deciding whether or not the noise performance of a certain machine or system will be acceptable with respect to one or more specific purposes for which it is intended.

Noise criteria are applied for three specific purposes, and one useful method is to classify them in accordance with the special purpose they are to serve:

1. One type of noise criterion is used to control the risk of physiological damage to humans who may be heavily exposed to noise. At the present time the only noise-induced physiological damage that has been sufficiently studied to develop quantified criteria is that related to hearing damage. While it is suspected that other types of noise-induced physiological damage may occur, a sufficiently well-established body of knowledge to develop quantified criteria has not yet been formulated.

2. Another type of noise criterion is used to establish the degree of interference with speech communication.

3. A third type of noise criterion is that used to establish whether a particular noise environment will produce psychological disturbances in people exposed to it. There are, at present, several such criteria in use.

Noise criteria also may be classified in accordance with the organizations or interest groups that promulgate them and who take some responsibility for seeing that they are applied. There are four general categories:

1. Legal bodies including federal, state, and local governments have devised noise regulations, enforceable by the proper authorities who may impose legal sanctions against those who do not comply. These regulations include criteria for the control of hearing damage and those for the control of psychological disturbances. A few local building ordinances also include regulations for the control of interference with speech communications.

2. Professional societies' noise recommendations frequently are adopted as standards for its members. Some typical examples are noise criteria recommended by the American Society for Heating, Refrigerating and Air-conditioning Engineers (ASHRAE), for the determination of acceptable noise environments in occupied spaces, which may be produced by the mechanical equipment in a building; noise criteria recommended by the Society of Automotive Engineers (SAE), for the determination of acceptable noise environments in the passenger compartments of automobiles; and many other organizations with similar noise criteria standards.

While these noise criteria may not be legally enforceable they may, on occasion, serve as guides to the "best prevailing practice," in the settlement of civil suits that may be based on claims of damage produced by excessive noise.

3. Manufacturers' associations, in many cases, recommend noise criteria which are applied to certain specific products. For example, the National Electrical

Manufacturers' Association (NEMA), the American Association of Power Mower Manufacturers, and other such organizations have agreed upon some noise criteria for specific products manufactured by their members. These criteria, as with the noise criteria promulgated by professional societies, are not legal requirements. However, those manufacturers who fail to meet their Association's noise criteria may find themselves at a competitive disadvantage in the sale of their products.

4. Some individual manufacturers have developed their own noise criteria, to be applied in the design, development, and use of their specific product, based upon some type of customer satisfaction determination. These are generally psychological-disturbance-type criteria and the final test of their noise acceptability may be based upon the subjective evaluation of human juries selected for that purpose.

A listing of organizations that have established noise criteria in the form of standards or recommended practices, is given in Chapter 3, "Standards for Sound Measurement." The portions of Standards which provide legal or recommended levels for human exposure are usually considered the "design" criteria.

RELATIONSHIP OF NOISE CRITERIA TO SOUND SPECIFICATIONS

Noise criteria generally are stated in terms of acceptable sound levels on some weighted scale, octave-band sound-pressure levels, or a derived number called the *loudness*,[1] computed from octave-band sound-pressure levels. Of most importance in any noise criterion is the character of the noise at the ears of the occupants of the acoustic space concerned. Thus, any noise criterion must express the acceptable combinations of frequencies and sound-pressure levels in some manner. The sound-pressure levels at various frequencies, at any point in an acoustic environment, are dependent upon all of the acoustic characteristics of the space.

The architect, or engineer, in preparing the specifications for a building or other structure, may suggest that the noise criteria be met in any particular space, based upon the use for which the space is designed; however, in specifying any acoustic treatment for meeting the stated noise criteria, it is necessary for him to have the following facts:

1. The octave based sound-power levels and locations of all sources of sound within the space;

2. The sound absorption and radiation properties of the proposed enclosing surfaces of the space and those of the surfaces of any object located within the space;

3. The octave-band sound-pressure spectrum of all sound transmitted into the space from sources outside it; and

4. The probable location within the space of the receiver or receivers for whom the criteria are intended.

---

[1] "Loudness" is defined on page 126.

With this information it is possible to calculate the estimated octave-band sound-pressure levels at any receiver location and compare them with the requirements of any particular noise criterion. If these calculations show that the noise criterion requirements will be exceeded, in most cases it will be possible with the stated information, to determine what noise reductions will be most effective and what type of treatment is most likely to be effective.

When purchasing mechanical equipment or systems, it is useless to provide specifications to the manufacturer in terms of sound-pressure levels of any kind, since the sound-pressure levels that will result in a particular acoustic environment are dependent upon the acoustic environment as well as upon the operation of the machine or system. What *is* necessary is the determination of the octave-band sound-power levels of the machine which, when operating in the proposed acoustic environment, will result in sound-pressure levels that will not exceed those of the applicable noise criterion. The purchaser may then include in the purchase specifications, the octave-band sound-power levels which are not to be exceeded. With such specifications it will be possible for the supplier to determine if his product will meet the sound-power level specifications and, if it does not, he may be able to determine what other noise control measures must be taken to bring his product into conformance.

Thus, it is obvious that noise criteria are always given either in terms of octave-band sound-pressure levels, weighted sound-pressure levels, or some sound-level measure which is computed from octave-band sound-pressure levels, while noise specifications for machines or mechanical systems must necessarily be given in terms of octave-band sound-power levels which may be used, together with data pertaining to the acoustic environment, in calculating the sound-pressure levels that can be compared with the particular sound criterion to be applied.

In converting sound-power levels, which are a part of the noise performance specifications of a machine, to the sound-pressure levels which will exist at a point or points of interest, the particular technique to be used depends almost entirely on the acoustic characteristics which exist in the environment that is going to be modified by the operation of the machine or system being considered. Steady-state sound-pressure levels at specific points in a reverberant space—as affected by a particular source—are related to the sound-power of the source. This is a directivity factor that accounts for the nonsymmetrical distribution of the sound intensity emanating from the source, the distance from the source to the specific point of interest, the areas of all the surfaces enclosing the space and of any objects enclosed in the space. It also is a factor to account for the acoustic absorption characteristics of the enclosing surfaces and the surfaces of objects enclosed in the space.

The acoustic energy produced by most mechanical machinery varies with frequency, and the acoustic energy dissipated by the materials on the surfaces of a space is dependent upon the frequency of sound. To incorporate these frequency variations, sound-pressure levels are calculated in octave bands, which experience has shown to be an acceptable method. For each octave-band computation the following are required:

1. The sound-power level, L_W, of all sources
2. The acoustic absorption coefficient, α_{sab}, for all materials in the space
3. The acoustic absorption constant, m, for air.

Ch. 4 CRITERIA FOR MACHINERY NOISE

A particular octave band is identified by its center frequency as indicated by the subscript notation. L_{pi} denotes the sound-*pressure* level in the octave band with center frequency, i. L_{Wi} denotes the sound-*power* level of a source in the octave band with center frequency, i. The standardized octave-band center frequencies are: 31.5, 63, 125, 250, 500, 1000, 2000, 8000, and 16,000 Hz. A discussion of octave bands is given in Chapter 2, "Measurement of Sound." Equation (4-1), given below, is very useful for the computation of the sound-pressure level in any frequency band at a particular point in a specific acoustic space produced by a particular acoustic source.

Sound-pressure Level (Decibels) in a Frequency Band for the Space

Sound-power Level of Equipment in a Frequency Band

Correction for Environment Characteristics in a Frequency Band

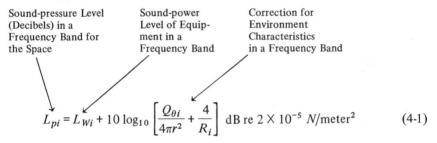

$$L_{pi} = L_{Wi} + 10 \log_{10} \left[\frac{Q_{\theta i}}{4\pi r^2} + \frac{4}{R_i} \right] \text{ dB re } 2 \times 10^{-5} \text{ N/meter}^2 \quad (4\text{-}1)$$

where:

L_{pi} = the sound-pressure level, in the frequency band having the center frequency f_i Hz, generated at the particular point being analyzed by the specific sound source of interest,[2] dB re 2×10^{-5} N/meter squared.

L_{Wi} = the sound-power level, in the frequency band having the center frequency f_i Hz, of the sound source of interest,[3] dB re 10^{-12} watt.

$Q_{\theta i}$ = the directivity factor (given in Fig. 4-1) of the sound source, in the frequency band having the center frequency f_i Hz, and in the direction θ from the source to the particular point for which the sound pressure level is being computed, dimensionless.

r = the distance from the sound source of interest to the particular point for which the sound-pressure is being computed, meters.

R_i = room constant [3] for the frequency band having a center frequency f_i, Hz, square meters.

Equation (4-1) is a slight modification of Equation (9-13) of Reference [9]. (For a complete discussion see pp. 219-243 of this Reference.)

The directivity factor, $Q_{\theta i}$ (see Fig. 4-1), is the ratio of the intensity of the acoustic energy of the actual source, in the direction θ and at a distance d, to the intensity measured at the same distance d, for a nondirectional source having the same sound-power level and the same frequency, as shown in Fig. 4-2. It is necessary to assume that both intensities are determined in an anechoic space.[4]

---

[2] The term dB re 2×10^{-5} N/meter^2 is frequently encountered; it means decibels referred to a reference pressure of 2×10^{-5} newtons per meter squared.
[3] The term dB re 10^{-12} watt means decibels referred to a reference power of 10^{-12} watt (power).
[4] *Anechoic*, Greek word for echoless. An anechoic space—a bounded region, or room, within which no sound waves are reflected from the surfaces.

118 CRITERIA FOR MACHINERY NOISE Ch. 4

UNIFORM RADIATION OVER
1/4 OF SPHERE, Q=4 TWO
REFLECTING SURFACES

UNIFORM HEMISPHERICAL
RADIATION, Q=2 SINGLE
REFLECTING SURFACE

UNIFORM SPHERICAL
RADIATION, Q=1 NO
REFLECTING SURFACES

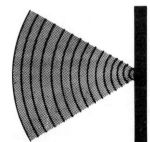

NONUNIFORM RADIATION,
Q DEPENDS ON DIRECTION

Fig. 4-1. Directivity factors for various acoustic radiation conditions.

Directivity factors may be determined when sound-power levels are being determined.

The room constant R_i, is determined from the expression:

$$R_i = S_1(\alpha_{sab})_{i,1} + S_2(\alpha_{sab})_{i,2} + S_2(\alpha_{sab})_{i,3} \cdots 4m_i V \quad (4\text{-}1\text{a})$$

$$R_i = \sum_{j=1}^{n} S_j(\alpha_{sab})_{i,j} + 4m_i V \text{ meter}^2 \quad (4\text{-}1\text{b})$$

where S_j is the area of the jth sound absorbing surface in the enclosure, meter^2; i denotes the ith absorption coefficient for the frequency band having a center frequency, f_i; $(\alpha_{sab})_{i,j}$ is the Sabin sound absorption coefficient of the jth surface for the acoustic energy in the frequency band having a center frequency f_i, Hz, dimensionless. m_i is the air absorption constant for acoustic energy in the frequency band having a center frequency f_i, Hz, and units meter^{-1} (see Tables 4-7 and 4-8). V is the volume of the room for which the sound-pressure level, L_{pi}, at the point of interest, is to be computed, in cubic meters.

Values of $(\alpha_{sab})_i$ for many materials may be obtained from Reference [14]. Where such information is not available in Reference [14] it is suggested that the material manufacturer might be able to supply the information. If this is

Ch. 4 CRITERIA FOR MACHINERY NOISE 119

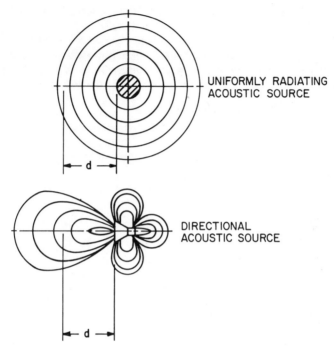

Fig. 4-2. Illustrations of a uniformly radiating and a directional acoustic source.

not available, it is possible for $(\alpha_{sab})_i$ to be measured, or as a last resort, estimated from the data for a similar material for which $(\alpha_{sab})_i$ are available. Values of the absorption coefficient for some materials are given in Chapter 5.

Equation (4-1) may be used for computing the band sound-pressure levels that will be produced by the introduction of a sound source for which the band sound-power levels are known. Conversely, the equation may also be used for computing the band sound-power levels which a sound source must not exceed if the band sound-pressure levels (which it also must not exceed) are known. The band sound-pressure levels may be determined from the band sound-pressure levels of the noise criterion which is acceptable and the existing sound-pressure levels at the point of interest. Examples of these methods of computation will be illustrated after the various, commonly used noise criteria are discussed in some detail.

SPECIFIC NOISE CRITERIA

In the first section, Noise Criteria Classification, it was suggested that three general purposes are served by the use of noise criteria for which data are presently available. These are: 1. Criteria for the prevention of permanent noise-induced hearing damage; 2. Criteria for minimizing noise interference with speech communication; and 3. Criteria for minimizing noise-induced psy-

CRITERIA FOR MACHINERY NOISE Ch. 4

chological disturbance or annoyance. Each of these criteria now will be discussed in enough detail so that the reader will have some basic understanding of the validity of the data used, the source of the data, and the application of the data in each instance. However, space does not permit exhaustive discussions of all these matters and if more background information is desired the reader should investigate the appropriate references given at the end of this chapter.

Damage-Risk Criteria for Hearing

For many years there have been attempts to develop quantified noise criteria that could be used effectively to protect against premature, permanent, auditory threshold-shifts in people exposed to high sound-pressure levels over long periods of time. Early in this century it was established (see Reference [1]) that injury to the receptor organ of the ear resulted from excessive exposure to high levels of noise. Subsequent observations have led to the conclusion that the loss of hearing ability depends, in a complicated way, on the extent of such injury. The mechanism by which excessive exposure to noise damages the auditory receptor is still not well understood. It is known, however, that both *rapid structural damage*, which may be caused by loud sound impulses such as those associated with explosions, and *gradual structural degeneration* associated with long-time exposure to steady noise such as that generated by various types of machines and mechanical systems, will lead to hearing loss. It is also known that while some hearing loss may be diminished after relatively short exposure to high-level noise, a hearing loss which is caused by complete degeneration of any of the primary auditory receptor cells can never be recovered since these cells, once destroyed, cannot be regenerated.

A number of observations may be made regarding data on which damage-risk criteria for hearing loss have been based. These observations have an important bearing on an understanding of the validity of using such criteria. While *n*oise-*i*nduced hearing loss or *p*ermanent *t*hreshold *s*hift (NIPTS) accumulates over many years, it is well established [2] that, beginning at age 30, most individuals experience some normal-hearing threshold shift that is not attributable to excessive noise, and this increases with age. This hearing loss, called *presbycusis*, is separated from that hearing loss attributable to noise, by the simple process of subtraction. Thus, all data used in developing damage-risk criteria for hearing loss are corrected for presbycusis.

The safe limits of permissible noise exposure (Damage-Risk Criteria for hearing loss) have been chosen to protect exposed groups of workers against noise-induced loss of hearing for everyday speech related, by agreement, to the arithmetic average of the hearing losses for pure tones at 500, 1000, and 2000 Hz. Noise-induced loss of hearing for everyday speech for an individual is based on an average loss of 25 dB at 500, 1000, and 2000 Hz above the loss due to presbycusis. In other words, an average hearing theshold of 25 dB above the effects of presbyscusis, would be considered to be a zero noise-induced hearing loss. Similarly, an average total loss of 40 dB, above the presbyscus effect, for the 500, 1000, and 2000 Hz frequencies would be considered to be a 15 dB loss attributable to noise exposure.

Although noise-induced permanent threshold shift has been noted in some

Ch. 4 CRITERIA FOR MACHINERY NOISE

individuals exposed for long periods of time to sound-pressure levels below 80 dBA, it has been noted [3] that exposure to sound-pressure levels below 80 dBA are not statistically significant with respect to damage-risk criteria for hearing. The amount of noise-induced hearing loss and the percentage of exposed individuals suffering such hearing loss *both* increase as the sound-pressure level and the exposure time increase. Based on a large background of data, most of which is summarized in the reports of the National Research Council, Committee on Hearing, Bioacoustics and Biomechanics, (CHABA), [4], a set of hearing-damage risk criteria, which are assumed to protect 90 percent of the people exposed daily to either continuous or intermittent noise levels of 90 dBA, or more, over long periods of time, were adopted by the Occupational Safety and Health Administration. The details of these criteria were first adopted as the Walsh-Healey Public Contracts Act [5], and later incorporated in the Rules and Regulations, Title 29—Labor, of Chapter XVII—Occupational Safety and Health Administration (OSHA), as Part 1910.95 of the Occupational Safety and Health Standards to cover most workers in industry. Since these criteria are, or soon will be, applicable to all workers in manufacturing construction and other areas of employment it is desirable to reproduce the criteria which have come to be known as the OSHA standards [6], as they are published in the Federal Register of May 29, 1971, and as revised October 24, 1974.

OSHA Noise Criteria

The OSHA criteria on occupational noise exposure are given in §1910.95, Subpart G-Occupational Health and Environmental Control, Part 1910—Occupational Safety and Health Standards, Chapter XVII—Occupational Safety and Health Administration, Department of Labor, Title 29—Labor, United States Code, in the following wording [6]:

§1910.95 Occupational noise exposure.
(a) Protection against the effects of noise exposure shall be provided when the sound levels exceed those shown in Table 4-1 when measured on the A scale of a standard sound level meter at slow response. When noise levels are determined by octave band analysis, the equivalent A-weighted sound level may be determined from Fig. 4-3.

(b) (1) When employees are subjected to sound levels exceeding those listed in Table 4-1, feasible administration or engineering controls shall be utilized. If such controls fail to reduce sound levels within the levels of Table 4-1, personal protection equipment shall be provided and used to reduce sound levels within the levels of the table.

(2) If the variations in noise level involve maxima at intervals of one-half second or less, it is to be considered continuous.

(3) To be in compliance with OSHA regulations a hearing conservation program must be established and maintained for employees who receive a daily dose, D, equal to or exceeding 0.5. A noise dose of 0.5 is equivalent to an eight-hour time weighted exposure of 85 dB(A) on the slow meter response. Further details are available in the Federal Register (39 FR 37773), October 24, 1974, or from the Department of Labor.

Table 4-1. Permissible Exposure Limits for Steady-State Noise[1, 2]

Sound Level dB(A) (Slow response)	T, Time Permitted	Sound Level dB(A) (Slow response)	T, Time Permitted
85	16 hr 00 min	101	1 hr 44 min
86	13 56	102	1 31
87	12 08	103	1 19
88	10 34	104	1 09
89	9 11	105	1 00
90	8 00	106	0 52
91	6 58	107	0 46
92	6 04	108	0 40
93	5 17	109	0 34
94	4 36	110	0 30
95	4 00	111	0 26
96	3 29	112	0 23
97	3 02	113	0 20
98	2 50	114	0 17
99	2 15	115	0 15
100	2 00		

[1] Where Table 4-1 does not reflect actual exposure times and levels, the permissible exposure to continuous noise at a single level shall not exceed the time "T" (in hours) computed from the formula:

$$T = \frac{16}{2^{[.2(L-85)]}}$$

Where L is the workplace sound level measured in dB(A) on the slow meter scale of a standard sound-level meter.

[2] When the daily noise exposure is composed of two or more periods of different levels, their combined effect should be considered rather than the individual effect of each. The combined levels may not exceed a daily noise dose, D, of unity (1) where D is computed by the formula:

$$D = \frac{C_1}{T_1} + \frac{C_2}{T_2} + \cdots + \frac{C_n}{T_n}$$

Where, $C_1, C_2, \cdots C_n$ indicate the total duration of exposure (in hours) at a given steady-state noise level; and, $T_1, T_2 \cdots T_n$ are the noise exposure limits (in hours) for the respective levels given in the table or computed from the equation in Footnote 1. Exposures to continuous noise shall not exceed 115 dB(A) regardless of any value computed by the formula for the daily noise dose, D, or by the equation in Footnote 1, above.

Some Clarification of OSHA Criteria Applications

The OSHA criteria cover practically all workers in the United States employed by all employers whose business affects interstate commerce except the United States Government or any state or political subdivisions of a state, mining, railroads or those employers under regulation of the Atomic Energy Commission. United States Government workers are covered by special regulations of the various governmental agencies that exercise statutory authority to prescribe or enforce standards or regulations affecting occupational safety or health. State

Ch. 4 CRITERIA FOR MACHINERY NOISE 123

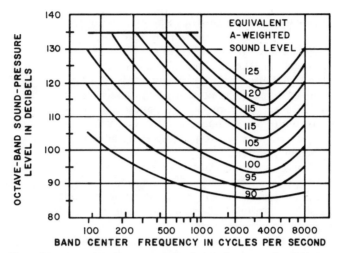

Fig. 4-3. Chart for conversion from octave-band sound-pressure levels to an equivalent A-weighted level. Octave-band levels are plotted on the figure and the equivalent A-weighted level is the value of the contour with the highest penetration of any of the octave-band values.

employees are covered by the OSHA criteria if that State's plan for occupational safety and health is approved by the U.S. Secretary of Labor.

The OSHA noise criteria, presented previously, are enforceable by the U.S. Secretary of Labor or his designated agents. This includes the power to make noise surveys, or request the employer to have them made, for the purpose of determining if any spaces occupied by workers are out of compliance with the criteria as set forth in Part 1910.95 of the Occupational Safety and Health Standards. If such surveys show that employees are subjected to environments exceeding the permissible noise exposure of Table 4-1 (including the footnote), the employer will be required to attempt engineering controls, suggested by engineering studies, to bring the situation into compliance. If no feasible engineering controls can be found to reduce the noise levels below 90 dBA then the employer is required to institute feasible administrative controls. These controls could modify the daily work schedule of each worker in such a way as to divide his exposure time among noisy operations and less noisy operations so that the workers' *noise exposure rating*, $(\Sigma\, C_n/T_n$, as defined in the footnote of Table 4-1 which is given in the OSHA regulation.)[5] would have a value less than unity.

It is further required that if engineering controls are found not to be feasible, in bringing noise exposure into compliance with the permissible values, then the employer must install a personal protection program which shall include either

---

[5] This form of evaluating the acceptable cumulative effect of daily exposure to the various combinations of sound levels, duration, and repetition was derived from the report of the CHABA cited in Reference [4], and the term *Noise Exposure Rating* was first proposed by the Physical Agents Committee of the American Conference of Government Industrial Hygienists [7]. This concept was adopted in the Walsh-Healey Act of 1971.

earplugs or earmuffs and audiometric measurements for each worker excessively exposed. These measurements shall be made at the beginning of the program and at appropriate intervals thereafter, to assess the effectiveness of the personal protection program. The continuing audiometric measurement program is required also if administrative control, without personal protection equipment, is resorted to in bringing the noise exposure levels into compliance with the regulations. In any event, bringing the noise exposure of workers into compliance by the use of administrative controls or personal protection equipment does not relieve the employer from the necessity of continuing to seek effective feasible engineering controls.

Impulse or impact noise is defined as a sound with a rise time of not more than 35 milliseconds to peak intensity and a duration of not more than 500 milliseconds to the time when the level is 20 dB below the peak. If impulses occur at intervals of less than one-half second they shall be considered as continuous sound. Exposure to impulse or impact noise shall not exceed a peak sound pressure level of 140 dB. Exposure to impulses of 140 dB shall not exceed 100 such impulses per day. For each decrease of 10 dB in the peak sound pressure level of the impulse the number of impulses to which employees are exposed may be increased a factor of 10. If for any reason these requirements are exceeded a hearing conservation program must be established.

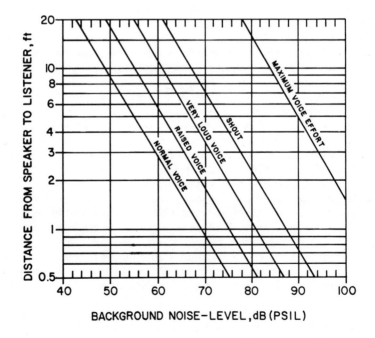

Fig. 4-4. Relation of required voice effort for communication in terms of PSIL and distance between speaker and listener. [10].

Speech-Interference Criteria

Interference with speech communications is caused by a phenomenon of sound which is commonly known as *masking*. In the masking process sound of a particular frequency and sound-pressure level is rendered indistinguishable by another sound of the same frequency but a somewhat higher sound-pressure level originating from another source. The noise levels that cause interference with speech communication are well known. An excellent review of the relationship among the variables of distance from speaker to listener, voice levels, and single number or noise spectra sound-pressure levels, is presented in Reference [10]. The most common speech-interference criterion is the one called the *preferred frequency speech-interference level (PSIL)*, which is computed by taking the simple average of the background noise, measured for the octave bands having center frequencies of 500, 1000, and 2000 Hz. Thus,

$$\text{PSIL} = \tfrac{1}{3} [L_{p_{500}} + L_{p_{1000}} + L_{p_{2000}}], \quad \text{dB re } 2 \times 10^{-5} N/\text{meter squared} \quad (4\text{-}2)$$

The relationship among PSIL, distance between speaker and listener, and voice level is shown in Fig. 4-4. To illustrate its use let us assume that the measured octave-band sound-pressure-level spectrum in a room has a distribution of acoustic energy as given in Table 4-2:

Table 4-2. Example of Measured Octave-Band Sound-Pressure Levels

Octave-Band Center Frequency, Hz	63	125	250	500	1000	2000	4000	8000
Sound-Pressure Level, decibels, dB, reference pressure 2×10^{-5} N/meter^2	76	70	70	62	60	58	58	55

The octave-band sound-pressure levels at the 500, 1000, and 2000 Hz center frequencies were found to be 62, 60, and 58 dB re 2×10^{-5} N/meter^2; then from Equation (4-2) the speech interference level is $\tfrac{1}{3}$ (62 + 60 + 58) = 60 dB PSIL. An examination of Fig. 4-4 shows that a normal voice level in face-to-face speech communication between two individuals 3 feet (1 meter) apart would provide satisfactory speech intelligibility.[6] If the two individuals were 6 feet apart it would be necessary to use raised voice levels. It may be observed that good speech intelligibility would result from the use of normal voice level in face-to-face communications for two individuals 6 feet apart if the background PSIL is reduced to 53 dB. For computation of the preferred speech-interference level it is necessary to have an octave-band spectrum of the background noise levels. In some situations only the A-weighted sound-pressure level for a space is

---

[6] Measurements which were utilized in arriving at the values shown in Fig. 4-4 are given in Reference [11].

known either by measurement or by estimation from some sound criterion which is to be specified, in which case computation of a value for the PSIL is not possible. It has been found [11], from much of the work done in measuring speech interference, that a fairly good approximation is given by the relationship:

$$\text{PSIL} = L_p(A) - 7 \quad \text{dB re } 2 \times 10^{-5} \, N/\text{meter squared} \quad (4\text{-}3)$$

That is to say, the dBA equivalent, for purposes of speech-interference determinations, is approximately 7 dB higher than the effective value of the PSIL.

It should be noted that speech interference criteria may be implied as a part of another type of criterion which is primarily designed to describe the acceptable maximum band sound-pressure levels for general comfort or freedom from annoyance. Such criteria, which are described in the next section, are frequently designed to take into account the fact that speech interference, in addition to the fact that it obscures needed information transfer, is also associated with some annoyance factors.

Annoyance Type Noise Criteria

The noise criteria that have been discussed up to this point are capable of being described in terms of objective measurements. The annoyance type criteria, by contrast, are based upon subjective answers to questions asked regarding some quality of the noise being rated. To measure the effect of noise exposure on hearing or to measure the intelligibility of speech communication in the presence of noise, while difficult, may be accomplished well enough so that the results may be used with a considerable degree of confidence. To make measurements by means of which predictions can be made as to whether or not a noise will annoy people, and how they will react if annoyed, is extremely difficult, and conclusions drawn from such measurements are subject to very great variability.

To be annoyed, irritated, distracted, or disturbed by sound is a commonplace experience. There are several possible reasons for the annoyance or disturbance caused by the noise. The annoyance caused by the noise of a neighbor's power mower may be produced because it interrupts a conversation, because it awakens one from sleep, or because it has some property which is irritating to the particular individual. There are a number of different reasons for annoyance and it would seem to be almost impossible to determine useful relationships between the physical characteristics of a noise and the amount of annoyance or irritation caused by it. It has been possible, however, to relate some of the physical characteristics of noise to its acceptability, or unacceptability, in particular situations, through field and laboratory studies in which the subjects are asked to judge such subjective properties of the noise as *loudness, noisiness,* or *unacceptability*. After many years of such studies there have evolved several techniques which have proved useful in predicting the characteristics of noise which will be acceptable or unacceptable in a variety of situations, and will rank ordering noises for their subjective effect.

Loudness

Loudness is an attribute of auditory experience. It is generally a term used to attempt to compare a complex property of sound that is probably related to a

Ch. 4 CRITERIA FOR MACHINERY NOISE

number of its physical properties which, when combined in certain ways, produce an auditory response related to acceptability. The term (loudness) was used in some very early experimental work [15] that was related to defining the curves known as *equal-loudness contours* [16] shown in Fig. 4-5.[7]

The levels shown by the curves are known as "phons," and were obtained by experiments with human subjects who were asked to compare pure tone sounds of different frequencies. The curves are equal-loudness levels versus frequency as the human ear would respond to pure tones. The term *phon* is used to designate the levels of the curves. Forty phons is used to describe the curve which passes through the 40 dB sound-pressure-level value at a frequency of 1000 Hz; 60 phons is used to describe the curve which passes through the 60 dB sound-pressure-level value at 1000 Hz, and so on, for the other curves. A *phon* is defined [17] as a numerical designation of the strength of a sound which is proportional to the subjective magnitude as estimated by normal observers: 1 *sone* is defined as the loudness of a sound whose loudness level is 40 phons.

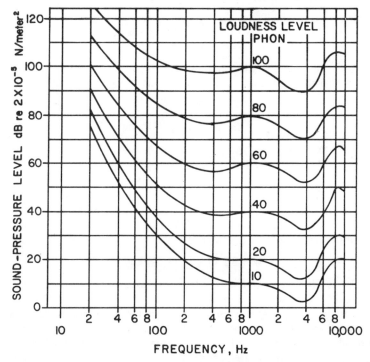

Fig. 4-5. Contours of equal loudness for pure tone sounds. Each contour is identified as the sound-pressure level at 1000 H_z and is given the designation of "phon." [16].

---

[7] There are a number of such sets of curves which have been published. The differences in values that are indicated in the various curves are due to slightly different test conditions. Reference to the original publication is suggested to determine exactly what these differences are.

The loudness level, P, in phons (dB) is related to the loudness, S, in sones, by the relation:

$$S = 2^{(P-40)/10} \text{ sones} \tag{4-4}$$

By modifying the curves of Fig. 4-5 to account for the fact that most noise of interest is not made up of only pure tones, and by adjusting the constant loudness contours to take into account the fact that for a given center frequency and band sound-pressure level a loudness index may be computed—in accordance with a procedure developed by S. S. Stevens [18]—the contours of equal loudness-index are plotted in Fig. 4-6. This plot is taken from page 9 of Reference [17]. In accordance with Reference [17] the calculated loudness-level, P, in phons, is found as follows:

1. Enter the geometric mean frequency, f_i, of each band in the abscissa of Fig. 4-6. Then from the band pressure level (ordinate in Fig. 4-5) determine the loudness index I_i for each band.
2. Find the total loudness S_t by means of the formula:

$$S_t = I_m + F\left(\sum_{i=1}^{i=n} I_i - I_m\right) \text{ sones} \tag{4-5}$$

where I_m is the largest of the loudness indexes; ΣI_i is the sum of the loudness indexes for all the bands; and F is a factor with a value dependent on the bandwidth used in the analysis of the noise, as follows:

Bandwidth	F Value
Third-Octave	0.15
Half-Octave	0.2
Octave	0.3

3. Determine the total loudness level, in phons, by reading adjacent to S_t sones, in the nomograph on the right-hand side of Fig. 4-5, the loudness level in phons, or by computation from the equation:

$$P = 40 + 33.3 \log_{10}(S_t/S_{40}), \text{ phons dB} \tag{4-6}$$

where S_{40} is the total loudness of a loudness level of 40 phons and is equal to one sone.

Loudness levels are generally used to compare the acceptability of one product with that of another, with respect to annoyance caused by noise. While it does not provide an absolute scale it can be estimated that the loudness of one noise is the same as, or twice as great, as another. That is, by a rough approximation, a noise having a total loudness of two sones is twice as loud as a noise having a total loudness of one sone. As a typical example the acceptability of the noise in the passenger compartment of automobiles is sometimes expressed (on a comparative basis) in terms of loudness levels. In Equation (4-4), for example, it may be seen that doubling the loudness in sones increases the loudness level in phons by 10 dB. Thus, a noise having a loudness-level of 40 phons would have a

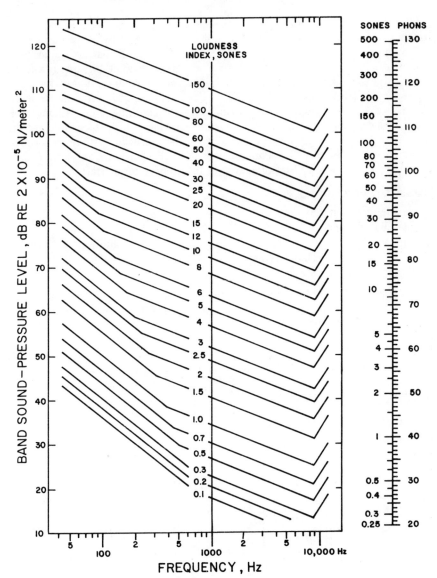

Fig. 4-6. Contours of equal loudness index to compare noise sources. [17].

loudness of 1 sone, whereas a loudness level of 50 phons would represent a loudness of 2 sones.

Perceived-Noise Level

In an attempt to improve the criterion of loudness, with respect to the acceptability of noises, K. D. Kryter [19] and his co-workers modified the procedure for collecting data by asking the subjects to compare noises on the basis of their unpleasantness or their acceptability. The results, while very similar to the subjective responses to the comparison of noises on the basis of "loudness," were sufficiently unalike to give a somewhat different rating for various sounds. This led to the development of a technique for the computation of the "perceived noise level" designated as PNL in dB or PNdB [20]. This criterion has been used by the Federal Aviation Authority for checking aircraft flying overhead. Since the use of this criterion is somewhat limited, the technique and the necessary tables will not be reproduced here; the reader may study the original Reference [20] or the 1969 Public Law 90-411, FAA regulations, amended by Part 36, to set standards for noise levels of future subsonic aircraft.

Noise Criteria Curves

As a basis for indoor noise criteria, one of the most widely used sets of values is the 1957 set of Noise Criterion Curves (NC) [21]. These were adopted by the American Society for Heating, Refrigerating and Air-conditioning Engineers (ASHRAE) [22]. The octave-band sound-pressure levels given in Table 4-3

Table 4-3. Octave-Band Sound-Pressure Levels Defining the 1957 Noise Criterion (NC) Curves as Given in Fig. 4-7

Noise Criterion	Octave-Band Center Frequency, Hz							
	63	125	250	500	1,000	2,000	4,000	8,000
	Sound-Pressure Levels, dB							
NC-15	47	36	29	22	17	14	12	11
NC-20	51	40	33	26	22	19	17	16
NC-25	54	44	37	31	27	24	22	21
NC-30	57	48	41	35	31	29	28	27
NC-35	60	52	45	40	36	34	33	32
NC-40	64	56	50	45	41	39	38	37
NC-45	67	60	54	49	46	44	43	42
NC-50	71	64	58	54	51	49	48	47
NC-55	74	67	62	58	56	54	53	52
NC-60	77	71	67	63	61	59	58	57
NC-65	80	75	71	68	66	64	63	62

represent the maximum octave-band level for a specified NC value. These values are plotted in Fig. 4-7 and are referred to as *NC curves*. The requirement is that all of the sound-pressure levels in the octave-band spectrum of a noise environment must fall below the values shown in the specified NC curve if it is to meet the NC rating specified. Usable NC rating criteria have been developed from experience and are used by ASHRAE and other groups in the acoustic design of rooms and other spaces to provide acceptable acoustic environments for the occupants. A list of criteria for a wide variety of enclosed spaces is given in Table 4-4. It should be noted that some building codes specify some of these

Ch. 4 CRITERIA FOR MACHINERY NOISE 131

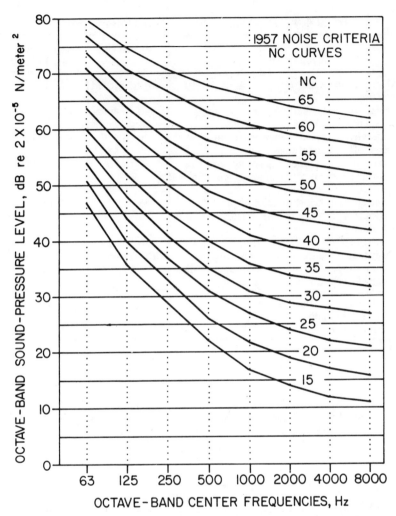

Fig. 4-7. Noise criteria curves for determination of permissible broad-band background sound levels for indoor spaces. See Table 4-4 for recommended levels for space usage.

criteria. It may be noted that the recommended NC ratings vary from a minimum of NC 20 for suburban private homes and sound reproduction studios, to a maximum of NC 60 for offices in which tabulation and computation equipment is located. There have been some significant objections to the quality of the sound associated with noise spectra that follow exactly the NC curves of 1957. The usual complaints are that the sound is both "hissy" and "rumbly." To overcome these objections, the 1957 NC curves have been modified in a paper presented by L. L. Beranek and his associates at the 1971 meeting of the Acoustical

CRITERIA FOR MACHINERY NOISE Ch. 4

Table 4-4. Recommended Indoor Noise Criteria

Type of Space	NC or PNC Curve	Approximate dBA Level
Meeting rooms, conference rooms, board rooms	Not to exceed 35	42
Private offices, small meeting rooms, executive offices	30 to 40	38 to 47
Supervisors' offices, reception rooms	30 to 45	38 to 52
Large offices, cafeterias	35 to 45	42 to 52
Laboratories, engineering rooms, drafting rooms, general office areas, open-plan office areas	40 to 50	47 to 56
Maintenance shops, office equipment rooms, computer rooms, washrooms and toilets	45 to 55	52 to 61
Control rooms, electrical equipment rooms	50 to 60	56 to 66
Manufacturing areas—foreman's office, tool crib, areas where speech communication is required	Not to exceed 60	Not to exceed 66
Manufacturing areas—assembly lines, light or heavy machinery areas, packing and shipping areas	Use OSHA* hearing-damage criteria	

*Occupational Safety and Health Act of 1971 (Reference [6]).

Table 4-5. Octave-Band Sound-Pressure Levels Defining the 1971 Preferred Noise Criterion (PNC) Curves as Given in Fig. 4-8*

Preferred Noise Criterion	Octave-band Center Frequency, Hz								
	31.5	63	125	250	500	1,000	2,000	4,000	8,000
	Sound-Pressure Levels, dB								
PNC-15	58	43	35	28	21	15	10	8	8
PNC-20	59	46	39	32	26	20	15	13	13
PNC-25	60	49	43	37	31	25	20	18	18
PNC-30	61	52	46	41	35	30	25	23	23
PNC-35	62	55	50	45	40	35	30	28	28
PNC-40	64	59	54	50	45	40	36	33	33
PNC-45	67	63	58	54	50	45	41	38	38
PNC-50	70	66	62	58	54	50	46	43	43
PNC-55	73	70	66	62	59	55	51	48	48
PNC-60	76	73	69	66	63	59	56	53	53
PNC-65	79	76	73	70	67	64	61	58	58

*These values have not yet been adopted by standardization groups.

Ch. 4 CRITERIA FOR MACHINERY NOISE 133

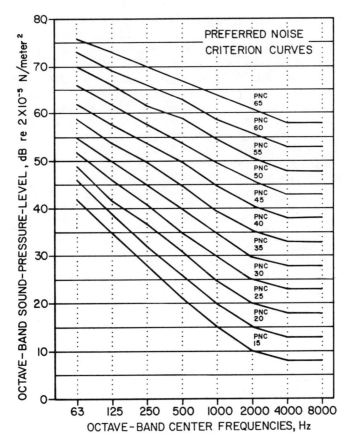

Fig. 4-8. Preferred noise criteria curves for determination of permissible broad-band background sound levels for indoor spaces. See Table 4-4 for recommended levels for space usage.

Society of America; these curves are published in Reference [23]. The resultant criteria known as the "1971 Preferred Noise Criteria Curves" are given in Table 4-5 and are plotted in Fig. 4-8. These have not yet been adopted by standardization groups; however they are at present the recommended criteria.

Criteria for Community Noise

The bases for setting realistic outdoor noise criteria for various types of communities have not been standardized to the extent that there is any wide acceptance of the various methods suggested. At the present time governmental organizations are studying the problem by making noise studies which attempt to take into account such factors as the time variations of noise in a given area, the differences in human responses to noise, the statistical distribution of noise levels, frequency, and duration, the time of day, time of year, and many other variables.

In spite of the fact that there is no widely accepted method for setting community noise criteria, the fact must be recognized that many legal criteria do exist at the federal, state, and local levels of government. In dealing with particular situations the reader should, therefore, be sure to investigate the existence of such criteria and, to the best of his ability, do what is necessary to meet them. At the present time there are criteria in several communities that are applied to external industrial noise, external transportation vehicle noise, external construction machinery and powered equipment and hand tools, powered watercraft and recreational or off-highway vehicles, and aircraft community noise. For some typical examples of such criteria and the wide range of factors involved, it is recommended that the material in pages 249 through 294 of Reference [23] be examined. Other sections of this same reference are extremely pertinent to the problems of the selection and application of noise criteria.

EXAMPLES OF APPLICATIONS OF NOISE CRITERIA

There are many factors which must be observed in selecting and applying noise criteria for a particular situation. In general, there are several reasons for applying noise criteria. The most important of these is to predict whether or not the acoustics of the system being designed will meet the criterion which is applicable, to determine by measurement whether the operating system does, in fact, meet the selected criterion and to specify changes which will produce conformity if the criterion is not met.

Example No. 1. Effect of a New Noise Source on the Hearing-Damage Risk

A new machine, which has an octave-band sound-power level spectrum as shown in Column 2 of Table 4-6, is to be installed in a shop area. The operator of this new machine is to be located a distance of two meters from the primary noise source of the new machine. The existing sound pressure levels at this location, without the new machine, are those shown in Column 4 of the table, where the "a" subscript denotes the ambient sound-pressure level. It is desired now, to determine how the new sound-pressure level at the operator's position will compare with the OSHA criterion.

Solution of Example 1

Step 1. Compute the sound-pressure levels L_{pi} which will be generated by the new machine at a 2-meter radius, assuming that the basic equation for this solution for each octave band is Equation (4-1), which may be restated as:

$$L_{pi} = L_{Wi} + C_i \qquad (4\text{-}1)$$

from Equation (4-1), where

$$C_i = \log_{10}\left[\frac{Q_{\theta i}}{4\pi r^2} + \frac{4}{R_i}\right] = 10 \log_{10} N_i$$

and from Equation (4-1a), where $R_i = S_j(\alpha_{\text{sab}})_{i,j} + 4m_i V$.

The values of the absorption coefficient, m_i, are given in Tables 4-7 and 4-8.

CRITERIA FOR MACHINERY NOISE

Table 4-6. Computation of Sound-Pressure Levels for Example 1

Col. 1	Col. 2	Col. 3	Col. 4	Col. 5	Col. 6	Col. 7	Col. 8	Col. 9	Col. 10
Octave-band center frequencies, f_i, Hz	Sound-power levels of new machine, L_{Wi} dB re 10^{-12} watt	Sound-pressure levels of new machine, L_{pi} dB re 2×10^{-5} N/meter^2	Sound-pressure levels of ambient, $L_{pa,i}$ dB re 2×10^{-5} N/meter^2	Squares of sound pressure ratios $(p_i/p_0)^2$	Squares of sound pressure ratios $(p_{a,i}/p_0)^2$	Squares of sound pressure ratios $(p_t/p_0)^2$	Total sound-pressure levels, L_{pt} dB re 2×10^{-5} N/meter^2	A-weighted sound-pressure levels, $L_{pi}(A)$, dB re 2×10^{-5} N/meter^2	A-weighted sound pressure ratios, $(p_{A,i}/p_0)^2$
31.5	103	89	97	0.794×10^9	5.011×10^9	5.805×10^9	97.6	58.2	0.007×10^8
63	101	87	90	0.501×10^9	1.000×10^9	1.501×10^9	91.8	65.6	0.036×10^8
125	100	86	86	0.398×10^9	0.398×10^9	0.796×10^9	89.0	72.9	0.195×10^8
250	99	85	84	0.316×10^9	0.251×10^9	0.567×10^9	87.5	78.9	0.776×10^8
500	100	86	84	0.398×10^9	0.251×10^9	0.649×10^9	88.1	84.9	3.091×10^8
1000	101	87	83	0.501×10^9	0.200×10^9	0.701×10^9	88.5	88.5	7.080×10^8
2000	101	87	83	0.501×10^9	0.200×10^9	0.701×10^9	88.5	89.7	9.333×10^8
4000	100	86	82	0.398×10^9	0.159×10^9	0.557×10^9	87.5	88.5	7.080×10^8
8000	101	87	80	0.501×10^9	0.100×10^9	0.601×10^9	86.1	85.0	3.162×10^8

Add values in Column 10 as shown: $\sum_{i=1}^{i=n} \left(\frac{p_{A,i}}{p_0}\right)^2 = 30.730 \times 10^8$

TABLE 4-7. Acoustic Absorption Constant, m, for Air, in feet^{-1} at 25°C (77°F)

Relative Humidity, percent	Octave-Band Center Frequency, in Hz								
	31.5	63	125	250	500	1000	2000	4000	8000	16,000
	Acoustic Absorption Constant, m									
10	...	...	...	...	0.0010	0.0020	0.0040	0.0070	0.0110	0.0140
20	...	...	...	...	0.0005	0.0010	0.0020	0.0060	0.0140	0.0190
30	...	...	...	...	...	0.0005	0.0014	0.0035	0.0090	0.0120
40	...	...	...	...	...	0.0005	0.0013	0.0030	0.0075	0.0100
50	...	...	...	...	...	0.0005	0.0012	0.0024	0.0070	0.0090
60	...	...	...	...	...	0.0005	0.0011	0.0022	0.0060	0.0080
70	...	...	...	...	...	0.0005	0.0010	0.0020	0.0060	0.0075
80	...	...	...	...	...	0.0005	0.0010	0.0020	0.0060	0.0070
90	...	...	...	...	...	0.0005	0.0010	0.0020	0.0060	0.0065

Assume, for this example, that the room dimensions and acoustic properties are as follows: length = 80 meters, width = 50 meters, and height = 10 meters; the value of $(\alpha_{sab})_{i,j}$ is 0.1 for all values of i and j; the value of $Q_{\theta i} = 2$ for all i; and the values of $4m_i$ are given in Table 4-9.

Using the values calculated below, the values of C_i in Table 4-10 may be calculated.

The value of $Q_{\theta i}/4\pi r^2 = 2/4\pi \cdot 2^2 = 0.0398$ and the value of $Sj(\alpha_{sab})_{i,j}$ = 2(80 × 10 + 50 × 10 + 80 × 50) × 0.1 = 1060 square meters for all i and j. The value of $V = 80 \times 50 \times 10 = 4 \times 10^4$ cubic meters. For all practical purposes, we may use the value $C_i = -14$ for all values of f_i. The reason that the value of C_i is nearly constant for all values of f_i in this example, is because the absorption coefficient $(\alpha_{sab})_{i,j}$, was assumed to be constant. In many instances this will not be the case and C_i will have different values for the center fre-

Table 4-8. Acoustic Absorption Constant, m, for Air, in meters^{-1} at 25°C (77°F)

Relative Humidity, percent	Octave-Band Center Frequency, in Hz									
	31.5	63	125	250	500	1000	2000	4000	8000	16,000
	Acoustic Absorption Constant, m									
10	...	...	...	...	0.0030	0.0060	0.0080	0.0200	0.0450	0.0625
20	...	...	...	...	0.0010	0.0020	0.0040	0.0150	0.0460	0.0830
30	...	...	...	...	...	0.0015	0.0028	0.0078	0.0171	0.0675
40	...	...	...	...	...	0.0010	0.0025	0.0064	0.0130	0.0500
50	...	...	...	...	...	0.0010	0.0024	0.0059	0.0111	0.0410
60	...	...	...	...	...	0.0010	0.0022	0.0055	0.0102	0.0340
70	...	...	...	...	...	0.0010	0.0021	0.0052	0.0097	0.0300
80	...	...	...	...	...	0.0010	0.0020	0.0050	0.0093	0.0260
90	...	...	...	...	...	0.0010	0.0020	0.0050	0.0093	0.0258

Table 4-9. Air Absorption Coefficient Values for Example No. 1

Octave-Band Center Frequency, f_i, in Hz	Absorption Coefficient Values, $4m_i$ (meter^{-1})*
Less than 1000	...
1000	0.0040
2000	0.0096
4000	0.0246
8000	0.0444

*These values of m_i, used above, are taken from Table 4-8. A relative humidity of 50 percent and a temperature of 25°C (77°F) was assumed.

quencies f_i. Thus the sound-pressure levels generated by the new machine may be computed by Equation (4-1) as follows:

$$L_{pi} = L_{Wi} - 14 \text{ dB}$$

The values of L_{pi} are tabulated in Column 3 of Table 4-6.

Step 2. Compute the values of the squared ratios of the root-mean-square sound pressures, p_i, produced by the new machine, to the reference pressure $p_0 = (2 \times 10^{-5} \text{ N/meter}^2)$. [Use Equation (1-30) with $p_0 = p_{\text{ref}}$.] By definition:

$$L_{pi} = 10 \log_{10} \left(\frac{p_i}{p_0}\right)^2$$

or

$$\left(\frac{p_i}{p_0}\right)^2 = \text{antilog} \left(\frac{L_{pi}}{10}\right)$$

The values of $(p_i/p_0)^2$ are tabulated in Column 5.

Table 4-10. Computation of C_i for Example No. 1.

	Octave-Band Frequency, f_i, in Hz				
Computation Elements	Less than 1000	1000	2000	4000	8000
$4m_i V$ (meter^2)	...	160	384	976	3440
R_i (meter^2)	1060	1220	1444	2036	4500
$4/R_i$ (meter^{-2})	0.378 × 10^{-2}	0.328 × 10^{-2}	0.277 × 10^{-2}	0.197 × 10^{-2}	0.089 × 10^{-2}
$N_i = \left[\frac{Q_{\theta i}}{4\pi r} + \frac{4}{R_i}\right]$ (meter^{-2})	4.378 × 10^{-2}	4.328 × 10^{-2}	4.277 × 10^{-2}	4.192 × 10^{-2}	4.089 × 10^{-2}
$C_i = \log_{10}^{-1}(N_i)$	−13.6	−13.6	−13.7	−13.8	−13.9

Step 3. Compute the values of the squared ratio of the root-mean-square sound pressures, $p_{a,i}$ produced by the ambient alone, to the reference pressure $p_0 = (2 \times 10^{-5} \text{N/meter}^2)$. As in Step 2, the squared pressure ratios are:

$$(p_{a,i}/p_0)^2 = \text{antilog} (L_{p_{a,i}}/10)$$

where the values of $L_{p_{a,i}}$ are found in Column 4 and the values of $(p_{a,i}/p_0)^2$ are tabulated in Column 6.

Step 4. Compute the squared ratio of the total of the root-mean-square pressures, p_i, produced by the ambient and the new machine, to the reference pressure $p_0 = (2 \times 10^{-5} \text{ N/meter}^2)$. Thus,

$$(p_t/p_0)^2 = (p_i/p_0)^2 + (p_{a,i}/p_0)^2$$

where p_t denotes the total sound pressure as a result of the ambient and the machine sound-pressure. These values are tabulated in Column 7.

Step 5. Compute the total sound-pressure levels L_{pi}, from the expression:

$$L_{pi} = 10 \log_{10} (p_i/p_0)^2$$

These values are tabulated in Column 8.

Step 6. Compute the A-weighted sound-pressure levels $L_{pi}(A)$ by the expression

$$L_{pi}(A) = L_{pi} = L_{pi} + C_{a,i}$$

where the $C_{a,i}$'s are the A-weighting relative response values for the octave-band center frequencies as given in Table 2-3 of Chapter 2. The octave-band values of $C_{a,i}$ are reproduced in Table 4-11, for convenience. The values of $L_{pi}(A)$ are tabulated in Column 9 of Table 4-6.

Table 4-11. A-weighted Relative Response Values for Octave-Band Sound-Pressure Levels*

Octave-Band Center Frequency, Hz	31.5	63	125	250	500	1000	2000	4000	8000
$C_{a,i}$	−39.4	−26.2	−16.1	−8.6	−3.2	0	+1.2	+1.0	−1.1

*These values are from Table 2-3 in Chapter 2.

Step 7. Determine the overall A-weighted sound level. $L_p(A)$ from the expression:[9]

$$L_p(A) = 10 \log_{10} \sum_{i=1}^{i=n} (p_{A,i}/p_0)^2$$

where the values of $(p_{A,i}/p_0)^2$, which are tabulated in Column 10, are determined from the values in Column 9 by:

$$(p_{A,i}/p_0)^2 = \text{antilog} (L_{p_{A,i}}/10)$$

---

[9] The A-weighted sound-pressure level could be obtained by combining the values in Column 9 of Table 4-6, two at a time, by using Fig. 2-3, of Chapter 2.

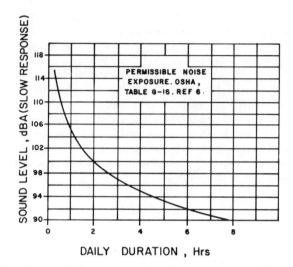

Fig. 4-9. OSHA daily permissible exposure times for continuous A-weighted sound-pressure levels.

In this example, the value of $\sum_{i=1}^{i=n} (p_{A,i}/p_0)^2$ is found by the summation of the values in Column 10, which is 30.730×10^8; therefore, the value of $L_p(A)$ is found to be:

$$L_p(A) = 10 \log_{10} (p_{A,i}/p_0)^2$$
$$= 10 \log_{10} 30.730 \times 10^8$$
$$= 94.9 \text{ dB(A)}.$$

If this value is entered in the plot of the permissible exposure times of the OSHA regulation (Fig. 4-9) it is found that a daily exposure time of 4 hours at 94.9 dBA must not be exceeded unless engineering controls are available to reduce the value of $L_p(A)$ below 90 dBA.

It should be noted that for this example, after Step 6, the values of L_{pi} of Column 8 could have been plotted in the OSHA Graph indicated in Fig. 4-3. The point of highest penetration would have been seen to be the L_{pi} value of 89.7 at 2000 Hz. This would indicate an equivalent A-weighted sound level of 95 dB, which is very close to the calculated value of 94.9.

Example No. 2. Effect of a New Noise Source on the Speech-Interference Level (PSIL)

Assume that the existing ambient octave-band sound-pressure levels $L_{pa,i}$ for the frequencies of interest ($f_5 = 500$, $f_6 = 1000$, $f_7 = 2000$) at the point of interest in an office were measured and found to be as shown in Column 2 of Table 4-14. It is apparent that the existing speech interference level is, by Equation (4-2):

140 CRITERIA FOR MACHINERY NOISE Ch. 4

PSIL = $(\frac{1}{3})[59 + 50 + 50]$ = 53 dB (PSIL) which, according to Fig. 4-1, would permit normal-voice speech communication at distances up to 6.2 feet (1.9 m) from the point of interest. It is desired to install a new machine (noise source) which has octave-band sound-power levels L_{Wi} as shown in Column 3 of Table 4-14 at a position, $r = 1$ meter (3.28 ft), from the point of interest; it is also desired to determine what effect this new noise source will have on the speech-interference level (PSIL) at the point of interest. Let it be assumed that the room has the geometrical and acoustical absorption properties shown in Table 4-12, Columns 1, 2, and 3. These values must be obtained from absorption data for the actual room surfaces in any given situation.

The equations for computing the octave-band sound-pressure levels from the octave-band sound-power levels will be:

$$L_{pi} = L_{Wi} + C_i \quad (4\text{-}1)$$

where:

$$C_i = 10 \log_{10} \left[\frac{Q_{\theta i}}{4\pi r^2} + \frac{4}{R_i} \right]$$

and:

$$R_i = \sum_j S_j (\alpha_{sab})_{i,j} + 4m_i V \quad (4\text{-}1a)$$

Table 4-12. Surface Areas, Absorption Coefficients, and Room Constants for Example No. 2

		Col. 1	Col. Group 2			Col. Group 3		
			Octave-Band Center Frequency, Hz					
Row No.	Description, Surface*	Area S_j meter^2	500	1000	2000	500	1000	2000
			$(\alpha_{sab})_{i,j}$			$S_j (\alpha_{sab})_{i,j}$		
1	Ceiling (Sound absorbing)	100	0.55	0.80	0.75	55	80	75
2	Floor (Asphalt Tile)	100	0.1	0.1	0.1	10	10	10
3	Wall Rough Plaster	30	0.04	0.06	0.05	1.2	1.8	1.5
4	Wall Rough Plaster	30	0.04	0.06	0.05	1.2	1.8	1.5
5	Wall Rough Plaster	30	0.04	0.06	0.05	1.2	1.8	1.5
6	Wall Rough Plaster	30	0.04	0.06	0.05	1.2	1.8	1.5

*Sound absorption coefficients for other materials are given in Chapter 5, "Acoustic Absorption and Transmission-Loss Materials."

Ch. 4 CRITERIA FOR MACHINERY NOISE 141

In some references the equations (above) are shown in the form of graphs from which the data are obtained. They are given in the mathematical form here because a better understanding of their application will result. Moreover, it may, in some instances, be desirable to solve those equations by computer.

Step 1. Compute the values of C_i, shown tabularized in Table 4-13. The resulting value of C_i is approximately -7 for all three octave-bands; therefore, this value (-7) can be used. Thus, Equation (4-1) becomes:

$$L_{pi} = L_{Wi} - 7 \text{ dB} \qquad (4\text{-}1c)$$

Step 2. Compute the values of the octave-band sound-pressure levels, L_{pi}, that would be produced by the new machine at a distance of $r = 1$ meter (3.28 ft), using the absorption properties given in Table 4-12. The values of L_{Wi} (Table 4-14, Column 3) are the measured values of the new noise source, or they are furnished by the supplier. In Column 4, Table 4-14, the computed values of L_{pi} are given.

Step 3. Compute the values of the squared ratios of the root-mean-square sound pressures, $p_{a,i}$, of the ambient, to the reference pressure ($p_0 = 2 \times 10^{-5}$ N/meter^2). As in Example 1,

$$(p_{a,i}/p_0)^2 = \text{antilog } (L_{p_{a,i}}/10)$$

These values are tabulated in Column 5 of Table 4-14.

Step 4. Compute the values of the squared ratios of the root-mean-square sound pressure, p_i, produced by the new machine, to the reference $p_0 = 2 \times 10^{-5}$ N/meter^2.

Table 4-13. Computation of Values of C_i for Example No. 2.

Computed Quantities	Frequency, f_i, Hz		
	500	1000	2000
$4m_i V$*	0	0	2.9
$S_j(\alpha_{sab})_{i,j}$ $j = 1$	55	80	75
(From Table 4-12) $j = 2$	10	10	10
$j = 3$	1.2	1.8	1.5
$j = 4$	1.2	1.8	1.5
$j = 5$	1.2	1.8	1.5
$j = 6$	1.2	1.8	1.5
$R_i = \sum S_j(\alpha_{sab})_{i,j} + 4m_i V$	69.8	97.2	93.9
$Q_{\theta,i}/4\pi r^2$ **	0.159	0.159	0.159
$4/R_i$	0.057	0.041	0.043
$Q_{\theta i}/4\pi r^2 + 4/R_i$	0.216	0.200	0.202
C_i	-6.7	-7.0	-7.0

*The value of m_i is obtained from Table 4-8 for 50% R.H. and 25°C (77°F) temperature. The value of V is obtained from the room dimensions which are 10m X 10m X 3m = 300m^3.

**The value of r is given as 1 meter, and a value of $Q_{\theta i} = 2$, for all frequencies was assumed on the basis that the machine is on the floor and sufficiently far enough away from the walls so that they have a small effect on the directivity factor. This value of $Q_{\theta i}$ is from Fig. 4-1.

142 CRITERIA FOR MACHINERY NOISE Ch. 4

Table 4-14. Computation of Sound-Pressure Levels for Example No. 2

Col. 1	Col. 2	Col. 3	Col. 4	Col. 5	Col. 6	Col. 7	Col. 8
Octave-Band Center Frequencies, f_i, Hz	Sound-pressure levels of ambient, $L_{pa,i}$ dB re 2×10^{-5} N/meter^2 1 meter from proposed location of new source, $L_{pa,i}$	Sound-power levels of new source, $L_{W1,i}$, dB re 10^{-12} watt. (Measured)	Sound-pressure levels of new source, $L_{p1,i}$ at 1 meter from center of noise source, dB re 2×10^{-5} N/meter^2	Squares of sound-pressure ratios of ambient pressure to the reference pressure $(p_{a,i}/p_0)^2$	Squares of sound-pressure ratios of new source at 1 meter (3.28 ft) from center of noise source, $(p_i/p_0)^2$	Total of squares of sound-pressure ratios, $(p_t/p_0)^2$ (sum of values from Columns 5 and 6)	Total sound-pressure levels, L_{pt}, dB, re 2×10^{-5} N/meter^2
500	59	75	68	0.79 × 10^6	6.3 × 10^6	7.100 × 10^6	68.5
1000	50	67	60	0.100 × 10^6	1.000 × 10^6	1.100 × 10^6	60.4
2000	50	67	60	0.100 × 10^6	1.000 × 10^6	1.100 × 10^6	60.4

As in the case of Step 3:

$$(p_{a,i}/p_0)^2 = \text{antilog}\ (L_{pi}/10)$$

These values of $(p_{1,i}/p_0)^2$ are tabulated in Column 6 of Table 4-12.

Step 5. Compute the squared ratio of the total root-mean-squared sound-pressure p_t, produced by the ambient and the new machine, to the reference pressure, $(p_0 = 2 \times 10^{-5}$ N/meter$^2)$. Thus:

$$(p_t/p_0)^2 = (p_{a,i}/p_0)^2 + (p_i/p_0)^2$$

These values, tabulated in Column 7 of Table 4-14, are obtained by direct addition of the values in Columns 5 and 6, for each octave-band center frequency.

Step 6. Compute the total sound-pressure levels, L_{pt}, from the expression

$$L_{pt} = 10 \log_{10}\ (p_t/p_0)^2$$

These values are tabulated in Column 8 of Table 4-14.

Step 7. Compute the speech-interference level from Equation (4-2). Thus:

$$\text{PSIL} = (\tfrac{1}{3})[68.5 + 60.4 + 60.4]\ \text{dB}$$

$$\text{PSIL} = 63.1\ \text{dB (PSIL)}$$

Step 8. Determine the voice level required for speech communication at a distance of one meter (3.36 ft) from the center of the new sound source.

Enter the graph of Fig. 4-4 with a value of 63.1 dB (PSIL) on the abscissa and note that normal voice levels would be satisfactory for a 2-foot (0.7 meter) separation between two people. A somewhat raised voice level would be required if the two people were 3.36 feet (approx. 1 meter) apart.

It should be noted that the computations were made for only the octave-band center frequencies at 500, 1000, and 2000 Hz. The other octave-band center frequencies were not used since only the sound-pressure levels at the three frequencies mentioned enter into computations for speech-interference levels.

Example No. 3. Computations for Specifying Sound-Power Levels of a Machine to be Purchased

It is desired to specify the maximum octave-band sound-power levels which must not be exceeded by an office machine to be purchased if the total sound-pressure levels at a point 2 meters from the machine must not exceed the values shown for the NC50 curve criterion, given in Fig. 4-7 and Table 4-3. These values are repeated in Column 2 of Table 4-15. In the symbol $L_{pd,L}$, the subscript, d, denotes that sound-pressure level is a design criterion.

The following assumptions are to be made:

1. Acoustic and geometrical conditions of the room are such that Equation (4-1) may be written:

$$L_{Wi} = L_{pi} + 13 \text{ dB} \tag{4-1d}$$

Table 4-15. Computations of Sound-Power Levels for Example No. 3

Col. 1	Col. 2	Col. 3	Col. 4	Col. 5	Col. 6	Col. 7	Col. 8
Octave-Band Center Frequencies, f_i, Hz	Sound-pressure levels of NC 50 design curve $L_{pd,i}$, dB re 2×10^{-5} N/meter²	Initial existing sound-pressure levels at point of interest. $L_{pa,i}$, dB re 2×10^{-5} N/meter²	Squared sound-pressure ratios of NC 50 design curve $(p_{a,i}/p_0)^2$	Squared ratios of initial sound pressures $(p_{a,i}/p_0)^2$	Squared ratios of sound pressure attributed to new machine $(p_i/p_0)^2$	Sound-pressure levels attributed to new machine. L_{pi}, dB re 2×10^{-5} N/meter²	Maximum sound-power levels for new machine L_{Wi}, dB re 2×10^{-12} watt
63	71	43	126.0 × 10⁵	0.200 × 10⁵	125.8 × 10⁵	71	84
125	64	46	25.15 × 10⁵	0.398 × 10⁵	24.752 × 10⁵	64	77
250	58	47	6.31 × 10⁵	0.501 × 10⁵	5.809 × 10⁵	57.6	71
500	54	43	2.515 × 10⁵	0.200 × 10⁵	2.315 × 10⁵	53.6	66
1000	51	39	1.26 × 10⁵	0.079 × 10⁵	1.181 × 10⁵	50.7	64
2000	49	34	0.794 × 10⁵	0.025 × 10⁵	0.769 × 10⁵	48.8	62
4000	48	31	0.613 × 10⁵	0.013 × 10⁵	0.600 × 10⁵	47.8	61
8000	47	29	0.501 × 10⁵	0.008 × 10⁵	0.493 × 10⁵	46.9	60

for computing sound-power levels at the point of interest, where C_i = 13 dB.
2. The sound-pressure levels existing at the point of interest, before installation of the new machine, are those indicated in Column 3 of Table 4-15. The subscript, a, in $L_{pa,i}$ denotes the ambient sound-pressure level.

Step 1. Determine values of sound-pressure levels $L_{pa,i}$, for the NC50 curve from Fig. 4-7. These values are entered in Column 2, of Table 4-15.

Step 2. Compute the values of the squared ratios of the sound pressures, $p_{d,i}$ of the NC 50 criterion to the reference pressure, p_0, by:

$$(p_{d,i}/p_0)^2 = \text{antilog}\,(L_{pd,i}/10)$$

These values are tabulated in Column 4 of Table 4-15.

Step 4. Compute the values of the squared ratios of the sound pressure p_i, that the machine can be permitted to generate at the point of interest, to the reference pressure ($p_0 = 2 \times 10^{-5}$ N/meter^2) by the expression

$$(p_i/p_0)^2 = (p_{d,i}/p_0)^2 - (p_{a,i}/p_0)^2$$

These values are tabulated in Column 6, of Table 4-15.

Step 5. Compute the values of the octave-band sound-pressure levels L_{pi} which the new machine must not exceed if the NC 50 criterion is to be met at the point of interest. This computation can be made by using the expression:

$$L_{pi} = 10 \log_{10}\,(p_i/p_0)^2$$

These values are tabulated in Column 7, of Table 4-15.

The octave-band sound-power levels to be specified for the new machine are those given in Column 8, of Table 4-15. This will provide the manufacturer with a criterion he will have to meet in designing for the acoustic performance of the machine.

Manufacturers often will avoid providing sound-power levels for machinery or simply prefer not to design to octave-band sound-power levels. There are various reasons for this, among them being the facilities and instrumentation required to obtain sound-power values. In the event that octave-band sound-power levels cannot be obtained, an approximate method for specifying the design criterion for the manufacturer is to provide sound-pressure levels at some distance from the mechanical equipment. The procedure should be to compute the octave-band sound-power levels as in the examples above, and then convert these values to octave-band sound-pressure levels. This can be done by assuming a reflection-free acoustic space and a uniformly radiating source above a reflecting plane such as a floor. Industry practice has been to specify the sound-pressure level in octave bands three feet from the equipment and five feet above the floor. For this condition the octave-band sound-pressure levels are computed from:

$$L_p = L_W - 7\text{ dB}$$

where:

L_p = Octave-band sound-pressure levels dB re 2×10^{-5} N/meter^2
L_W = Octave-band sound-power levels dB re 10^{-12} watt

Table 4-16. Sample of Equipment Noise Data Sheet for Noise Specification Sent to Suppliers (Values given are from Example No. 3.)

Equipment Description: _____ Type: _____ Item No.: _____

Octave-Band Center Frequency, Hz	Col. 1 Specified Sound-Power Level, L_{W_i} (dB re 10^{-12} watt)	Col. 2 Approximate Sound-Pressure Level, L_p. $L_p = L_W - 7$dB (dB re 2×10^{-5} N/meter^2)	SUPPLIER TO COMPLETE*		
			Col. (A) Actual (L_p or L_W, identify)	Col. (B) Special Design (L_p or L_W, identify)	Col. (C) Acoustic Treatment (L_p or L_W, identify)
63	84	77			
125	77	70			
250	71	64			
500	66	59			
1000	64	57			
2000	62	55			
4000	61	54			
8000	60	53			

*Instructions to supplier:
1. Octave-band sound-power levels are preferred, as in Column 1.
2. If sound-pressure levels are used to approximate sound-power levels, the measurements shall not be greater than three feet (0.9 meter) from the equipment and five feet (1.5 meters) above grade or floor. The measurement method must be described if Column 2 is utilized and the circumferential position for *maximum* levels is used.
3. Complete Column A for actual levels of equipment; identify L_p or L_W.
4. Complete Column B for special design for low noise; identify L_p or L_W.
5. Complete Column C for acoustic treatment such as enclosure; identify L_p or L_W.
6. Equipment will meet specified noise levels without modification: _____.
7. Additional costs required:
       For Column B: _____.
       For Column C: _____.

It is recommended that an equipment noise-data sheet be utilized to inform suppliers of the noise requirements for which bids are requested. A sample noise-data sheet is given in Table 4-16, for which the data in Example 3 are used. Some care is required in specifying proper and acceptable noise levels. Arbitrarily specifying lower noise-level requirements than are actually required may place an undue burden on the supplier and may significantly add to the cost of purchased equipment.

5

Acoustic Absorption and Transmission-Loss Materials

**WILLIAM SIEKMAN
and
JOSEPH W. SULLIVAN**

The engineer wishing to apply the information provided in this chapter must realize that there is no such thing as a class of "acoustical materials." Essentially any material, even the air carrying the human voice, can become an acoustical material when it is utilized in such a manner that the movement of sound is altered or its intensity changed.

It is the designer and applicator who, by combining their knowledge and skills, select and position appropriate materials so that a desired acoustical function is performed. The familiarity with certain physical properties of materials is a mandatory part of the designer's knowledge before any tables of values can be appreciated and a modest discussion of the characteristics of environments is a prerequisite to the establishment of practical skills. It then should become apparent that the responsibility for success or failure of any acoustical material will usually lie with the designer or the applicator; not the manufacturer. The reader should completely familiarize himself with Chapter 2, "Measurement of Sound," in order to prepare himself for the ensuing discussions, particularly those relating to the determination of acoustical properties.

I. MECHANISMS OF SOUND ABSORPTION

The process of removing sound power from a particular space—in lowering the sound pressure level within a particular volume—is known as *Sound Absorption*. This process is concerned with a lowering of the energy content of the pulsating air as it encounters a porous or flexible medium; it should be recognized as being completely different from most other types of "absorption." A blotter, for ex-

Ch. 5 ACOUSTIC MATERIALS

ample, can absorb liquids when only a corner is immersed into the fluid, but sound-absorbing materials cannot pull sound out of the air. Sound is a form of energy which can only be reduced when it is actively directed into a dissipating medium. The designer must remember that his sound-absorbing materials will work only if the sound gets to them.

The Sabin

The *sabin* is a very important sound absorption unit (not a coefficient) that expresses the total amount of sound absorption that can be obtained from a given area of a material, from a component such as a door or window, from a room divider, or from a room, etc. Mathematically, the sabin, in square feet, can be expressed for a panel made from a single material as follows:

$$A = \alpha S \quad \text{sabins}$$

where: A is the absorption of the panel in sabins, (ft^2), α is the absorption coefficient, and S is the surface area of the panel in (ft^2). The confusion concerning the sabin is that it has dimensions of square feet but it is not related to the amount of absorption of a one-square-foot area. The amount of sound absorption, in sabins, for a material is defined by a mathematical expression and obtained from a specified measurement technique, a point of great importance when utilizing acoustic materials, as shall be seen.

The absorption of a system composed of a number of materials with different sound absorption coefficients, α, can be determined from the relation:

$$A = \alpha_1 S_1 + \alpha_2 S_2 + \alpha_n S_n = \sum_{i=1}^{i=n} \alpha_i S_i$$

where:

A = total absorption in sabins
α_i = absorption coefficient of various materials
S_i = surface area of various materials

For example, the total absorption of an empty room would be computed from:

$$A = \alpha_{\text{floor}} S_{\text{floor}} + \alpha_{\text{walls}} S_{\text{walls}} + \alpha_{\text{windows}} S_{\text{windows}}$$
$$+ \alpha_{\text{ceiling}} S_{\text{ceiling}} + \alpha_{\text{door}} S_{\text{door}}$$

If there are other materials in the room, the absorption must be included. Individual objects such as chairs, room dividers, or even people, are routinely reported as so many sabins. Because α, the absorption coefficient, varies with frequency, the total absorption will vary with frequency. The usual computational methods utilize octave-bands or one-third octave bands; therefore, the total absorption, A, would normally be computed for the frequency bands of interest.

It is important to realize that the absorption coefficient, α, cannot be calculated; it must be determined experimentally. The methods for determining the absorption coefficient are treated in Section III, "Measuring Sound Absorption," in this chapter. Values of this coefficient can be found in Table 5-1 and in other

ACOUSTIC MATERIALS

Table 5-1. Typical Values and Ranges for Absorption Coefficients (α)

Materials	Frequency, Hz					
	125	250	500	1000	2000	4000
	Absorption Coefficients					
Cellulose Fiber Ceiling Tiles, #1 Mounting	.05 to .10	.20 to .25	.50 to .80	.50 to .99	.45 to .90	.30 to .70
Same, #7 Mounting	.30 to .40	.30 to .45	.35 to .65	.45 to .99	.50 to .90	.50 to .75
Mineral Fiber Ceiling Tiles, #1 Mounting	.02 to .10	.15 to .25	.55 to .75	.60 to .99	.60 to .99	.50 to .90
Same, #7 Mounting	.40 to .70	.40 to .95	.45 to .95	.55 to .99	.65 to .99	.50 to .95
Heavy Carpeting	.02 to .06	.05 to .25	.15 to .55	.35 to .70	.50 to .70	.60 to .75
Light Carpeting	.01	.05	.10	.20	.45	.65
Draperies	.03 to .15	.04 to .35	.10 to .55	.15 to .75	.20 to .70	.35 to .65
Unpainted Concrete Block	.05 to .35	.05 to .45	.05 to .40	.05 to .40	.05 to .35	.05 to .25
Painted Block, Brick, Concrete, Plaster	.01 to .02	.01 to .02	.01 to .03	.02 to .04	.02 to .05	.02 to .07
Spray-on Insulations	.03 to .08	.10 to .30	.20 to .75	.25 to .99	.25 to .95	.20 to .75
Glass, Plywood, and Other Panels	.05 to .35	.05 to .25	.04 to .18	.03 to .12	.02 to .10	.01 to .10

NOTE: The data in this table represent the unadjusted data observed on many 8- by 9-foot laboratory specimens. They must be reduced when much larger and extended areas of the same materials must be considered. A suggested method for doing this is to multiply the laboratory data by the factors given below. Note that the factors are different for the two common sample sizes, and the designer must therefore learn what size is being reported by the laboratory or manufacturer.

Common Sample Sizes, ft	Frequency, Hz					
	125	250	500	1000	2000	4000
	Absorption Coefficient Multipliers					
8 by 9	0.49	0.66	0.79	0.88	0.94	0.97
6 by 8	0.43	0.60	0.75	0.86	0.92	0.96

In very large rooms and at high frequencies, the sound absorption of the air itself can be of significance. This absorption varies with relative humidity and, in general, will be greater than the following values, which are given in sabins per 1000 cubic feet at 50 percent relative humidity:

Frequency, hertz	125	250	500	1000	2000	4000
Absorption, sabins	0.0	0.1	0.3	0.9	2.3	7.2

sources such as suppliers' catalogs. A very comprehensive source of these data is a bulletin entitled: *Performance Data-Architectural Acoustical Materials*, published by the Acoustical Materials Association.[1]

Sound Absorption Methods

Sound can be absorbed in two ways: 1. by distributing the energy over a larger volume; or 2. by converting it into another, nonaudible form of energy. An example of the first method is the effect obtained when an operating machine is moved from a reverberant room to the out-of-doors. The machine is emitting the same amount of sound energy is either environment, but the loudness within the smaller, confined environment is substantially greater than that in the nearly infinite volume of the out-of-doors. It is pertinent to point out that the outdoor or *free field* condition represents the maximum possible amount of absorption, and that therefore no sound-absorbing material can reduce the loudness of any noise source below that which would be obtained if that source were out-of-doors and away from reflecting surfaces.

There are three commonly used methods by which sound can be converted into another form of energy. First, it can be directed into and through a medium in which it experiences frictional resistance, in which its kinetic energy is converted to heat. There is an optimum *air-flow resistance* for the absorption of sound at any particular frequency which, unfortunately, is not subject to simple computation and hence, beyond the scope of this handbook. A very porous material with a very low air-flow resistance (intentionally so for its use) is the common household furnace filter. Sound will pass through such a filter with negligible reduction, as can readily be demonstrated by listening to any source (such as a radio speaker) before and after interposing such a filter. At the other extreme, if a material has a very high flow resistance; for example, a piece of sheet metal, the sound cannot readily penetrate to be absorbed and must of necessity reflect back and add energy to the space from which it came, as can be demonstrated by oscillating such a specimen near the ear and listening to the increase in intensity as the sound is reflected toward the ear. (We are not concerned at this moment with the sound which was not transmitted by the sheet; that will be considered when discussing transmission loss.) The air-flow resistance of acoustical materials is precisely measured by the standard method of ASTM: C 522-69. (See Ref. [1].) In addition to their importance to theoreticians, the results are also of importance to manufacturers, for they provide a sensitive check in quality control. For example, a doubling of the fiber diameter in a glass wool can result in a sixteen-fold change in the air-flow resistance. As mentioned, the computations which correlate air-flow resistance with sound absorption are quite involved but fortunately are not essential for the ordinary design of acoustical materials. It will be adequate for our purposes to remember this mechanism and avoid designing any interference with it.

A second method of converting sound energy is by the flexure of a thin panel or diaphragm. Although the variations in air pressure associated with sounds are normally microscopic in magnitude, they can produce a minuscule movement of

---

[1] Acoustical Materials Association, 205 West Touhy Avenue, Park Ridge, Illinois 60068.

any lightweight surface. If the materials below this surface are less than a perfectly elastic medium, some of this kinetic energy will be internally *damped* and converted into heat; some may also be conducted away as a mechanical motion or *structure-borne* sound and transmitted into another medium. Because most materials are relatively elastic at audible frequencies, this method is not ordinarily as efficient at higher frequencies as the method utilizing air-flow resistance, but it has very important applications at lower frequencies where the thickness of the acoustical material may be limited.

The third method of converting sound energy is specifically appropriate for use with coherent sound; that is, sound of a continuous frequency, like the hum of a power transformer or the whine of a fluorescent lamp ballast. It consists of a highly resonant mechanism which collects energy during a portion of each cycle and releases it after an appropriate interval so that the following incident energy is partially cancelled. The process is analogous to the non-reflective coatings applied to optical surfaces, and is technically called "destructive interference" rather than "absorption," but common usage and the fact that many such mechanisms are filled with sound-absorbing materials has established reference to them as "absorbers."

Each of these methods is illustrated in Section V, dealing with the use of acoustical materials. Each method has in common the ability to reduce the amount of sound which reaches a surface and, consequently, is not reflected back into the space from which it came.

II. MECHANISMS OF SOUND TRANSMISSION LOSS

In contrast to sound absorption (principally concerned with the reduction of sound within a space by dissipating it at the room boundaries), *sound transmission loss* is concerned with the prevention of transmission through these boundaries. The two processes are mutually incompatible: an open window is a nearly perfect sound absorber but no barrier at all; a thick piece of steel is a worthless sound absorber but a relatively good barrier; most materials lie between these extreme examples. The designer will frequently require both absorption and transmission loss in a partition, which obviously will require composite structures.

As illustrated in Fig. 5-1 the transmission of sound through a partition occurs because the pulsating air on the *source side* transfers some of its kinetic energy

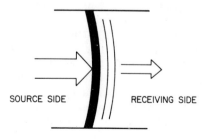

Fig. 5-1. Sound transmission by a barrier which is impervious to air.

Ch. 5 ACOUSTIC MATERIALS 151

to the partition material, causing it to ripple and twist throughout its extent. The portion of this motion which is conducted to the opposite face will generate sound in the air on the *receiving side*. The ratio of the incident energy to the radiated energy, when measured under specified conditions and expressed in decibels, is called the *transmission loss*. It is a value incapable of direct measurement with the current state of the art, but is computed from the *noise reduction*, or difference in sound levels between the spaces divided by the partition, provided that the spaces have certain defined characteristics; these will be discussed later.

There are only three common methods of increasing the transmission loss through a partition. First, the weight per exposed-square-foot can be increased by raising the density of the material or by increasing its thickness; this amounts to a change in the mass of the partition. Such improvements become limited by the cost of additional or different material, by load-bearing capabilities of the substructure, or by available space. The second method is to isolate the opposite faces of the partition so that a minimum of bridges through the partition exist to conduct the sound energy. The ultimate isolation exists when all such bridges are eliminated, as in a double-wall system, but some transmission will still take place—through the intervening air space—in a manner to be discussed below. The incorporation of a vacuum between panels has been attempted, for this would provide infinite isolation, but practical problems have prevented use of this theoretically ideal method. These concepts are illustrated in Fig. 5-2. The third and final method for increasing the sound transmission-loss is to apply a damping material to the panel surfaces such that energy is dissipated as the panel vibrates. Damping treatments are considered in Chapter 6, "Damping Materials for Vibration and Sound Control."

In general, the changes in transmission loss due to variations in density or thickness are well understood and predictable. The changes due to isolation can be predicted for most simple materials but not for complex systems. An in-

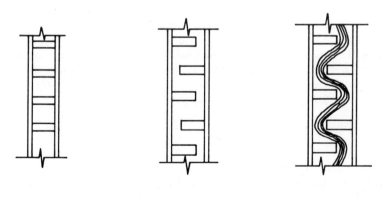

(a) STANDARD DESIGN   (b) BETTER DESIGN   (c) BEST DESIGN

Fig. 5-2. Design of double panel sound-barrier systems. a. Straight-through studs; b. Staggered studs; c. Staggered studs with sound-absorbing blanket.

crease in sound transmission resulting from the application of damping materials can be predicted from experimentally determined properties of the damping material.

III. MEASURING SOUND ABSORTION

1. Impedance Tube Method of Sound Absorption Measurement

Three standard methods exist for the measurement of sound absorption. The first, known generally as the "impedance tube method," restricts itself to sound directed perpendicularly upon a specimen. The standard method of testing is ASTM C384-58 [2]. This method has serious limitations, since there is at present no way of accurately calculating from impedance tube data the absorption of a specimen subject to randomly incident sound, which is the common situation. Nevertheless, it has the merit that relatively small samples are required, and the method permits determination of the *normal specific acoustic impedance*. This normal specific acoustic impedance is the complex ratio of sound pressure to the component of particle velocity normal to the surface of the specimen, and in equation form:

$$\text{normal specific acoustic impedance} = \frac{\text{complex sound pressure}}{\text{complex particle velocity}}$$

or

$$z = r + ix \tag{5-1}$$

where: z = the normal specific acoustic impedance, r = the acoustic resistance, x = the acoustic reactance, and i = the square root of negative one.

The fact that the normal specific acoustic impedance z, can have both a real and an imaginary part indicates that the acoustic pressure and the acoustic particle velocity are not exactly in phase at the surface of the specimen.

Since materials are usually used in various sizes and areas over extended surfaces, it is more convenient to refer to their absorption per unit area than to their sabins per specimen. This *normal sound absorption coefficient* or "α_n," is thus the fraction of the incident sound energy absorbed by a material of any size. We must note at once that α_n may vary with size; this will be discussed below. The normal absorption coefficient is related to the impedance according to the formula given below and in Reference [3]:

$$\alpha_n = 1 - \left| \frac{z - \rho c}{z + \rho c} \right|^2 \tag{5-2}$$

where: ρc is the characteristic acoustic impedance of free air = 41.5 cgs units, ρ being the density of air and c the velocity of sound; the vertical bars indicate the absolute value. It is customary to express z, r, and x in terms of their ratio to ρc.

The absorption coefficient can be determined directly from the maximum and minimum pressure amplitudes in the standing wave-pattern set up in the tube (see Fig. 5-3) using the equation given below and in Reference [2]:

$$\alpha_n = 1 - \left[\frac{\log_{10}^{-1}\left(\frac{L}{20}\right) - 1}{\log_{10}^{-1}\left(\frac{L}{20}\right) + 1} \right]^2 \quad (5\text{-}3)$$

where: L is the level difference.

The tube method permits measurements over a frequency range of about 10 to 1 with any given diameter of tube, and over an absorption coefficient range of about 1.00 to 0.05 when care is utilized. The wall of the tube must be massive and rigid, and the area must be uniform from end to end within 0.2 percent. The product of the tube length, in feet, and the *lowest* frequency, in hertz, at which measurements are to be made should be about 1000. The product of the tube inside diameter (in inches) and the *highest* frequency (in hertz) to be used, should not be greater than 8000. The sample must be inserted into the end of the tube so that the surface is as perpendicular as possible to the central axis. The sample edges must be very snug or sealed with a material such as grease or modeling clay. The measurement of samples with rough or textured faces is practical only at frequencies near the lower limit of the tube length; "cross modes" in the tube with such samples exhibits itself as a cyclical variation in the successive minimum levels as the tube length is probed. The test signal fed to the loudspeaker must be essentially free of all harmonics, and enormous precautions must be taken to prevent the probe tube from contacting the speaker or the surrounding tube. The workability of a tube is usually tested by substituting a heavy steel plate for the specimen and measuring the maximum possible stand-

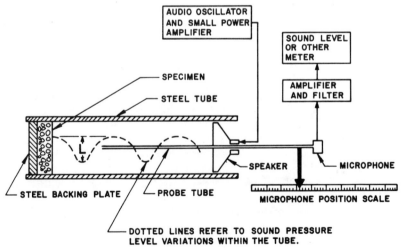

Fig. 5-3. Impedance tube arrangement for measuring normal acoustic impedance of materials.

ing wave ratio; in general, a 45 dB ratio is the minimum acceptable, while 50 dB or better, is a reasonable limit.

2. Reverberation Room Method of Sound Absorption Measurement

The second method for measuring sound absorption is known as the "reverberation room method" and is detailed in ASTM C423-72 [4]. This method is concerned with the performance of a specimen exposed to randomly incident sound, which technically occurs only when the specimen is in a *diffuse sound field*. In ASTM Standard C 384 the absorption coefficient obtained by the reverberation room method is given the symbol α_{rev} to distinguish it from the normal incidence sound absorption coefficient α_n described above. This notation will be utilized here. The creation of a diffuse sound field is so difficult that it is essentially restricted to special laboratory conditions. One might question the usefulness of data which can only be obtained in such special facilities, if the sample is to be used in a different kind of environment. A discussion of this point follows.

In a diffuse sound field, there is equal probability that the same amount of energy will pass through a given point in any direction, so the net energy flow is the same in all directions. (The intensity vector is zero.) If, by some process, energy can be equally distributed throughout a reverberant room and an absorbing material placed somewhere in that room, there will be a net energy flow in the direction of that material, thus violating the definition of a diffuse sound field. Therefore, in order to make any kind of measurement the entire surface of the room must have the same absorption as the sample, or the sample size must be so small that the net energy flow toward it is undetectable. There is no known technique for creating a diffuse sound field for the first condition, so the second is used, with the understanding that the finite sample size introduces a slight error. Note that a reverberant room is called a *reverberation room* only if it contains a diffuse sound field. The only proven technique for creating this kind of field at low, as well as other, frequencies is to make the room as reverberant as possible and simultaneously incorporate very large rotating reflectors or "vanes." These vanes must be as large as space permits, inclined to the wall surfaces, and able to move at speeds which are neither too slow for averaging processes nor so fast as to introduce noticeable doppler shift of the testing frequencies. Typically, this would mean that the vanes would have areas of about 100 square feet, or larger, and rotate somewhere between 2 and 10 rpm. Oscillating vanes have been used with success to increase the free space within the room. Moving microphones have proven themselves worthless except to average the variations within a room; however, moving *both* the microphone and the sample can theoretically obtain the same result as is obtained with a diffuse field.

In a reverberation room, the absorption is *defined* by the equation

$$A = 0.9210 \, Vd/c \qquad (5\text{-}4)$$

where: A is the absorption (in sabins), V is the volume of the room (in cubic feet), d is the rate at which sound decays in the room when measured in decibels per second, and c is the velocity of sound in the gasseous media, in feet per sec-

Ch. 5 ACOUSTIC MATERIALS 155

ond. Since the *sabin* is defined by this equation, there is no need to justify the use of this equation as long as the measurements are conducted in a reverberation room. In other environments, especially those which are heavily absorbent, the decay process changes and other equations are sometimes recommended.

The measurement is made by introducing a source of sound into the room (usually a narrow [$1/3$ octave] band of random noise) and sustaining this source until the sound field has reached a uniform level throughout; a matter of about one second. The source is then quickly extinguished and the rate at which the sound-pressure level decreases or "decays" in the room, is measured. This can be done by reading the slope of a curve obtained on a high-speed graphic level recorder, or by timing with clocks the interval between two voltage levels. If an absorbent material is introduced into the reverberation room, the decay rate will be faster than it was for the empty room. The difference between the two absorption values obtained in these measurements can normally be assumed to be that of the sample material, as the absorption of the area covered by the sample can usually be considered negligible. In equation form, the absorption coefficient is:

$$\alpha_{rev} = \frac{0.9210\, V(d_2 - d_1)}{Sc} \tag{5-5}$$

where:

α_{rev} = Acoustic absorption coefficient
V = Room volume, ft^3
c = Speed of sound in air, ft per sec
d_2 = Decay rate for the room with the sample, dB per sec
d_1 = Decay rate of empty room, dB per sec
S = Area of sample material tested, ft^2.

It has become a custom to make reference to the decay over a 60-decibel range as *the decay time;* we frequently encounter the equation

$$A = \frac{0.049\, V}{T_{60}} \tag{5-6}$$

where: A and V have values as above, and T_{60} is the time for a 60-decibel decay.

Absorption can be measured to within a few tenths of a sabin by this method provided that decays over a range of at least 40 decibels are possible, and that temperature and relative humidities can be held within a few percent throughout the course of measurement in both the empty and sample rooms. Sample sizes are specified at 8 by 9 feet, or larger, except for very unusual circumstances. The volume of the reverberation room limits the frequencies at which measurements can be made. The lowest frequency is determined by the equation:

$$V = 4\lambda^3 \tag{5-7}$$

when V is in cubic feet and λ is the wavelength, in feet. Larger sizes are recommended. However, at high frequencies the absorption of sound by the air within the room becomes appreciable and the establishment of a diffuse sound field becomes impossible.

At present there is no specification for a maximum-size room for a given frequency, but a practical limit for measurement at 5000 hertz appears to be 10,000 cubic feet. A volume of 7063 cubic feet (200 cubic meters) has been recommended for all new reverberation rooms in order to establish some uniformity.

Relation Between α_n and α_{rev}

There is no direct relation between the normal sound absorption coefficient, α_n, and the sound absorption coefficient determined by the reverberation method, α_{rev}, and it is always best to measure directly the parameter desired. Direct measurements of α_{rev} are expensive, however, and require an appropriate reverberation facility. On the other hand, the measurement of α_n is relatively easy and does not require an expensive facility. Methods to approximate α_{rev} from α_n have been determined and compared to measured values. Table 5-2 is useful in estimating the relation between α_{rev} and α_n. The value of α_n is obtained by the proper combination of the values in Row 1 and Column 1 of the table. The value of α_{rev} is obtained as the intersection of these two values in the table. For example, if α_n is 45, the intersection of 40 and 5 gives a value of $\alpha_{rev} = 71$ (from Table 5-2).

Table 5-2. Approximate Conversion of Normal Incidence Absorption Coefficients to Random Incidence

Normal Incid., α_n →	0	1	2	3	4	5	6	7	8	9
↓	Absorption Coefficients, Random Incidence, α_{rev}									
0	0	2	4	6	8	10	12	14	16	18
10	20	22	24	26	27	29	31	33	34	36
20	38	39	41	42	44	45	47	48	50	51
30	52	54	55	56	58	59	60	61	63	64
40	65	66	67	68	70	71	72	73	74	75
50	76	77	78	78	79	80	81	82	83	84
60	84	85	86	87	88	88	89	90	90	91
70	92	92	93	94	94	95	95	96	97	97
80	98	98	99	99	100	100	100	100	100	100
90	100	100	100	100	100	100	100	100	100	100

3. Steady State Method of Sound Absorption Measurement

The third method for measuring sound absorption is known as the "steady state" method. It consists of a precision measurement of the sound pressure-level within a room when a steady, fixed sound-power source is introduced into the room. The method is fundamental to the definition of sound absorption, but may be imprecise with present instrumentation and has consequently usually been reserved for situations where the other two methods will not work; in particular, those rooms which contain such a high degree of absorption that even a slight resemblance to a diffuse sound field is impossible. The technique is *not* standardized, but it is described in ASTM E336-71 (see Ref. [5]). A particularly useful steady-state sound source is commercially available from the ILG Electric Ventilating Co., 2850 N. Pulaski Rd., Chicago, Ill. The model number is ILG DSN 10910, commonly known as the "Ilg generator" or "Ilg reference sound source."

To use this method, one places the sound source in the room, preferably not near or directly over an absorbing surface, and explores the sound pressure not too near or too far from the source. The sound level meter must be equipped with one or one-third octave bandwidth filters, and the exploration is made at each frequency of interest. The sound source should be moved to other positions or other orientations and the measurements repeated. The total absorption for the room is then computed from the equation:

$$A = \log_{10}^{-1}\left[\frac{\overline{L}_W - \overline{L}_p + 16.5}{10}\right] \qquad (5\text{-}8)$$

where: A is in sabins, $\overline{L}_W$ is the average sound power output of the source in decibels re 10^{-12} watts (as provided by the manufacturer, or otherwise determined), and $\overline{L}_p$ is the average sound-pressure level measured in the area of interest. In practice, it is necessary to know $\overline{L}_p$ within small fractions of a decibel in order to measure coefficients with acceptable accuracy; special, long-time averaging methods are sometimes employed. The method is especially useful when making transmission loss measurements in field conditions, as will be discussed below.

The absorption of any material usually varies markedly with frequency, and it is customary to measure a minimum of at least six standard frequencies: 125, 250, 500, 1000, 2000, and 4000 hertz. The arithmetic average of the four frequencies, 250 through 2000, is called the *noise reduction coefficient* or NRC. Its usage is widespread; however, it is only an approximate method of summarizing the properties of a material and is not recommended except as a rough comparison of different materials.

As mentioned above, the absorption of a specimen varies with its area. To a first approximation, the specimen behaves as if it were a half-wavelength wider and longer than its measured dimensions. (The exact behavior is still in the process of being analyzed.) Since lower frequencies have longer wavelengths, this *diffraction effect* will be most noticeable there. On a typical laboratory specimen (8 by 9 feet in size) the effect at 4000 hertz is negligible; adds about 10 percent to the values at 1000 hertz; and nearly doubles the values at 125 hertz. Test laboratories are urged by the test standards *not* to make any corrections for this effect in their reports, so the designer must be alerted to this increase and make necessary allowances in his designs.

Typical values for absorption coefficients in octave bands are given in Table 5-1. Also given are correction coefficients to convert laboratory absorption coefficients obtained from an 8- by 9-foot sample to absorption coefficients for large surface areas which would exist when treating a room.

4. Summary

In summary, coefficients obtained by all three methods must be used with caution, for rarely are the conditions of measurement comparable to the conditions of use, and field installations all differ markedly. Rank ordering of different specimens, however, is very accurate, and designers have very little trouble in treating architectural spaces with use of the data obtained by these methods.

IV. MEASURING SOUND TRANSMISSION-LOSS

The sound insulating property of a material is customarily expressed in terms of its *transmission loss*, or TL, which is defined as the ratio of incident sound power to the transmitted and radiated sound power. Since the decibel is used to express the ratio of two powers, TL is appropriately expressed in decibels. In addition, by definition the incident sound power is due to the presence of a diffuse sound field, which implies correctly that reverberation rooms must be used for correct measurements as was the case for absorption measurements.

The Standard Method

The standard test method is ASTM E90-72 [6], and a sketch of a typical set-up is shown in Fig. 5-4. The specimen, usually mounted in the test opening of the source room, must be representative in size and, if constructed with mechanically movable components such as hinges, fully operable in every sense. The reverberation rooms are normally constructed with thick, solid masonry walls and rest on carefully isolated footings to eliminate *flanking paths* which

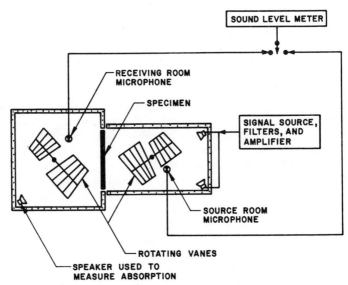

Fig. 5-4. Transmission suite arrangement for measuring the acoustic transmission-loss of materials.

could carry acoustical energy from room to room by other paths than through the specimen. The two rooms are usually designed so that no two dimensions in either room are the same; this serves to eliminate the possibility of resonances developing between the rooms. As is true for the measurement of sound absorption the use of rotating vanes is usually mandatory. Room volumes must also be selected to fit the desired range of test frequencies, which is also from 125 to 4000 hertz (at least), but measurements must be made at each one-third octave interval over this range, using one-third octave bandwidths of random noise. Some of the more sophisticated laboratories use a continuously variable technique, permitting them to make a sweep which passes through every frequency between the limits.

The difference in sound-pressure level between the rooms is called the *noise reduction*, or NR (not to be confused with the *noise reduction coefficient*, or NRC, which applies to absorption measurements). This noise reduction is dependent not only upon the insulating properties of the specimen but also upon its area (for more sound will pass through a larger area) and upon the rate at which energy leaves the receiving room (a highly absorbent room will have a lower sound-pressure level). We can eliminate these two variables by multiplying the energy ratio of the NR by the area of the specimen and dividing by the (equivalent) area of absorption. Since we are already expressing the power ratio in logarithmic form when we measure in decibels, this process is a simple arithmetic one:

$$\text{Transmission Loss, TL} = 10 \log_{10} \left(\frac{\text{incident sound power}}{\text{transmitted sound power}} \right)$$

The above definition must be used along with the acoustic properties of the receiving room to obtain:

$$TL = NR + 10 \log_{10} (S) - 10 \log_{10} (A) = L_{p1} - L_{p2} + 10 \log_{10} \left(\frac{S}{A}\right) \text{ dB} \quad (5\text{-}9)$$

where: S is the specimen area in square feet, and A is the absorption in the receiving room, in sabins. The absorption must be determined by the reverberation method, using the same microphone position used in the noise reduction measurement, and a speaker, which must be left in the receiving room during the noise reduction measurement.

The STC Rating System

The result of the transmission-loss measurements is a set of at least 16 values of transmission losses, in decibels, one for each of the prescribed one-third octave-band test frequencies. These values are useful under certain conditions for making noise reduction calculations; however, there is an obvious limitation to numerous values whenever it is necessary to make a comparison of the transmission losses of two or more materials or components.

It is easier to use single-number ratings, and for more than thirty years the arithmetic average of nine test frequencies was used to label specimens. This system has the inherent weakness of all simple averaging systems, for two quite different specimens can have identical averages; or worse, a specimen which is perfectly adequate at appropriate frequencies may be rejected by a specification based entirely upon averages because of deficiencies at an insignificant portion of its frequency range.

A classification system has been invented to overcome these difficulties and has been in use for about a decade. This is the *sound transmission class* or STC rating system given in Tables 5-3A and 5-3B. It is defined and explained in ASTM E413-70T [7]. Basically, it is a system by means of which the curve of transmission loss (TL) versus frequency for a specimen is compared against a standard contour. This contour resembles the frequency response of the human ear, and it has been adjusted to correlate with the sensitivity to common household noises.

The STC contour may be shown as a curve of the $1/3$-octave-band sound-transmission losses or it may be given in tabular form, as in Table 5-3A. In this table, the horizontal rows provide numerical values, in decibels, of the STC contours at the center frequencies given at the heading of the vertical columns. The actual STC rating is the transmission-loss value given in the 500 hertz column of Table 5-3A. The contour described by each row is identified by this single 500-hertz value.

To determine the STC value of a material, the transmission losses for the $1/3$-octave bands listed at the head of each column in Table 5-3A must be measured. These values are shown in Table 5-3B as the "Specimen TL" horizontal rows. A trial-and-error procedure is then used to determine STC value of the sample material, by comparing its transmission-loss values to those given in Table 5-3A. Select the STC contour in Table 5-3A that has the same or nearest transmission loss at the 500-hertz center frequency. Then compare the transmission-loss contour of the specimen with the selected contour from Table 5-3A by subtracting the specimen transmission-losses from the corresponding STC values for each center frequency. This procedure is shown in Table 5-3B.

In the evaluation of the comparison, the following criteria must be observed:

1. The maximum STC rating that meets the following conditions is selected to characterize the specimen.
2. In the subtraction, negative values are recorded as zero.
3. The maximum difference in any $1/3$-octave band must not exceed 8 dB.
4. The sum of the differences at all of the center frequencies must not exceed 32 dB.
5. The STC rating is the numerical value of 500 hertz center frequency of the contour from Table 5-3A that meets the above criteria.

The 500 hertz value is selected because an analysis has shown that the STC number thus obtained would roughly agree with the majority of "averages" obtained by the old single-number rating system, thus smoothing the transition to complete use of the STC. It is now the *only* single-number rating permitted; averages are specifically forbidden. Unlike the averages, the STC rating is a pure number: it is incorrect to say a specimen has, for example, an STC of 40 dB.

Using appropriate precautions, the ASTM E90 method is capable of determining and reproducing transmission-loss values within ±1 decibel, except at the lowest frequencies where the large statistical variations in random noise signals

ACOUSTIC MATERIALS

Table 5-3A. Sound Transmission Class (STC) Ratings*

\(^1/_3\)-Octave-Band Center Frequencies, hertz														
125	160	200	250	315	400	500	630	800	1000	1250	1600	2000	2500	3150	4000

Transmission Losses for Each Contour, dB

125	160	200	250	315	400	500	630	800	1000	1250	1600	2000	2500	3150	4000
39	42	45	48	51	54	55	56	57	58	59	59	59	59	59	59
38	41	44	47	50	53	54	55	56	57	58	58	58	58	58	58
37	40	43	46	49	52	53	54	55	56	57	57	57	57	57	57
36	39	42	45	48	51	52	53	54	55	56	56	56	56	56	56
35	38	41	44	47	50	51	52	53	54	55	55	55	55	55	55
34	37	40	43	46	49	50	51	52	53	54	54	54	54	54	54
33	36	39	42	45	48	49	50	51	52	53	53	53	53	53	53
32	35	38	41	44	47	48	49	50	51	52	52	52	52	52	52
31	34	37	40	43	46	47	48	49	50	51	51	51	51	51	51
30	33	36	39	42	45	46	47	48	49	50	50	50	50	50	50
29	32	35	38	41	44	45	46	47	48	49	49	49	49	49	49
28	31	34	37	40	43	44	45	46	47	48	48	48	48	48	48
27	30	33	36	39	42	43	44	45	46	47	47	47	47	47	47
26	29	32	35	38	41	42	43	44	45	46	46	46	46	46	46
25	28	31	34	37	40	41	42	43	44	45	45	45	45	45	45
24	27	30	33	36	39	40	41	42	43	44	44	44	44	44	44
23	26	29	32	35	38	39	40	41	42	43	43	43	43	43	43
22	25	28	31	34	37	38	39	40	41	42	42	42	42	42	42
21	24	27	30	33	36	37	38	39	40	41	41	41	41	41	41
20	23	26	29	32	35	36	37	38	39	40	40	40	40	40	40

*A portion of the table of transmission-loss values for corresponding STC contours is given above; it can be extended lower or higher as required. Immediately below, various contours are compared with the measured transmission loss of the resilient gypsum-board wall shown in Fig. 5-4. The STC of the wall is determined by comparing its transmission loss with progressively higher STC's, keeping track of the deficiencies as one progresses. At such time as the *sum of all deficiencies* just exceeds 32, or at any *one frequency* just exceeds 8, the next lower STC number is applied to the specimen. *The STC number for any STC contour is its transmission loss at 500 hertz.*

Table 5-3B. Example of Determining the Sound Transmission Class (STC) Rating of a Specimen

Specimen	\(^1/_3\)-Octave-Band Center Frequencies, hertz															
	125	160	200	250	315	400	500	630	800	1000	1250	1600	2000	2500	3150	4000
STC 41	25	28	31	34	37	40	41	42	43	44	45	45	45	45	45	45
Specimen TL	19	24	28	32	35	38	41	44	47	50	52	54	52	46	41	47
Deficiencies..	6	4	3	2	2	2	0	0	0	0	0	0	0	0	4	0
Sum of the deficiencies at STC 41 = 23.																
STC 42*	26	29	32	35	38	41	42	43	44	45	46	46	46	46	46	46
Specimen TL	19	24	28	32	35	38	41	44	47	50	52	54	52	46	41	47
Deficiencies..	7	5	4	3	3	3	1	0	0	0	0	0	0	0	5	0
Sum of the deficiencies at STC 42 = 31.																
STC 43	27	30	33	36	39	42	43	44	45	46	47	47	47	47	47	47
Specimen TL	19	24	28	32	35	38	41	44	47	50	52	54	52	46	41	47
Deficiencies..	8	6	5	4	4	4	2	0	0	0	0	0	0	1	6	0
Sum of the deficiencies at STC 43 = 40.																

*The specimen therefore has an STC of 42 since it meets the requirement that the deficiency at any one frequency band does not exceed 8 dB and the total deficiencies do not exceed 32 dB.

may double this error. It is a relatively simple matter to measure specimens with STC values up to about 50; at least one laboratory can make measurements on a 14- by 9-foot specimen up to about STC 80, but it should be noted to their credit that the instrumentation problems when testing such a dynamic range are formidable.

The transmission loss of a specimen can sometimes be affected by temperature and relative humidity; by the edge-mounting conditions; and by particular ratios of dimensions. The measured performance will be meaningful, but these variables must be fully reported if the results are to have meaning in other applications. As with absorption specimens, in general, a larger specimen will show poorer performance than a smaller one.

USING ACOUSTICAL MATERIALS

It is important to remember the basic difference in the use of absorbent and transmission-loss materials. Absorbing materials are used when it is desired to reduce the sound pressure levels *within* a particular environment. Transmission-loss materials are used when it is desired to reduce the sound transmission *between* two environments, i.e., to reduce the level in a different space than that where the noise is generated. One can place glass fiber on a ceiling to reduce the sound level *within* a room, but it will have only a trivial effect in reducing the sound passing through the ceiling to the room above. One can insulate a control room by installing a glass window *between* it and a studio, but this would not lower the sound level in the studio. The optimum situation will usually be obtained by utilizing both types of acoustical materials, rather than either one alone.

An absorbent material may be used to provide a sound transmission loss between two spaces in some circumstances if it is applied in a certain manner. For example, Fig. 5-5 illustrates an absorbent material applied inside a short duct to absorb the sound that is transmitted through the duct, thereby preventing the transmission of the sound between the spaces through which the duct passes. The objective of the short, sound-absorbing duct is to provide an opening between the two spaces while at the same time transmitting a minimum amount of sound. For this type of application, the absorbing material must be along the path of the transmitted sound inside the duct. The design calculation for such ducts is given in Equation (5-23).

This treatment can also be used for longer ducts in closed systems. In general, sound that is transmitted along the fluid path inside a duct can only be absorbed by an absorbent material applied inside of a duct. Sound that is being transmitted through the duct wall can be absorbed or attenuated by wrapping an absorbent material with a septum overlayer around the outside surface, as shown later in Fig. 5-18; in some cases sound attenuation is also accomplished by placing an absorbent material inside the pipe. Further details of sound propagation inducts are given in Chapter 9, "Fan and Flow System Noise."

Treating Rooms With Absorbent Materials

The manner and extent to which a room is treated with absorbent material depends principally on the source-to-receiver distance and the geometry of the

Ch. 5 ACOUSTIC MATERIALS

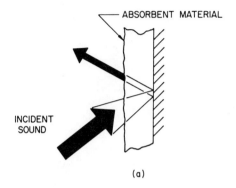

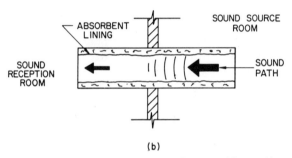

Fig. 5-5. Industrial applications for acoustic absorption materials. a. Absorption material applied to room walls, interior surfaces, or source side of barriers; b. Absorption material applied to ducts.

room. Sound reaching the ear always consists of two components; one transmitted directly from the source and one made up of all reflections from the room's surfaces. Both components are governed by the sound power-output of the source. The level of the direct sound depends only on the distance from the source and is reduced 6 dB for each doubling of distance (Inverse Square Law). The level of the reflected sound depends on room size, shape, and the absorption of the surfaces; and it may or may not depend on source distance, depending on room proportions, *regular* or *extended*.

Case 1.—Regularly shaped room, longest floor dimension is less than 5 times the ceiling height, as shown in Fig. 5-6. The sound pressure-level of the reflected sound depends only on the room absorption, A, and can be calculated, approximately, from the following equation:

$$L_p = L_W - 10 \log_{10}(A) + 16.5 \text{ dB} \tag{5-10}$$

where L_p is in decibels re 2×10^{-5} N/meter^2; L_W is in decibels re 10^{-12} watts; and A is in sabins.

As mentioned above, the total sound-pressure level is the sum of sound intensities (logarithmic sum of the decibel levels) of the direct and reflected components. The total sound-pressure level is shown graphically in Fig. 5-7. From

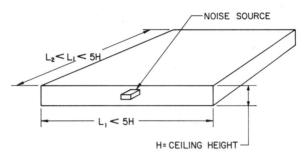

Fig. 5-6. Regularly shaped room with longest floor dimension less than five times the ceiling height.

this graph we can see that close to the source the sound-pressure level is governed primarily by distance; whereas, far from the source, the sound-pressure level is governed by the room acoustic absorption only. Thus, increasing the room acoustic absorption is not effective in reducing the sound-pressure level close to the source, but *is* effective at distances greater than about $0.5\sqrt{A}$ feet[1] from the source. The limit of reduction possible by absorptive treatment is established by the curve of direct sound level versus distance, as in Fig. 5-7. Far from the source, where reflected sound dominates, reduction in level from L_{p1} to L_{p2} by increasing room absorption from A_1 to A_2 sabins is obtained directly from Equation (5-10) and is:

$$\text{Noise Reduction} = L_{p1} - L_{p2} = 10 \log_{10} (A_2/A_1) \text{ dB} \qquad (5\text{-}11)$$

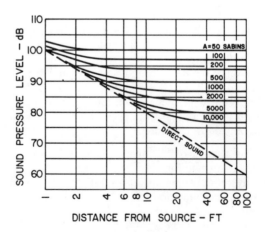

Fig. 5-7. Relation of sound level to source distance and room absorption in regularly shaped rooms. Levels cannot be reduced below the dashed line (direct sound) by absorption alone.

---

[1] The unit for the sabin (A) is square feet (ft^2); thus, the unit for the square root of the sabin must be feet.

Ch. 5 ACOUSTIC MATERIALS

We see, therefore, that a tenfold increase in absorption can reduce the level by 10 dB at a distance of $0.5 \sqrt{A}$ feet from the source. The maximum absorption ratio A_2/A_1 possible, is governed by the "hardness" of the room before treatment and the available areas for absorptive treatment. Generally, a ratio of about 10 is the practical limit.

Case 2.—Extended room, one or more floor dimensions greater than 5 times the ceiling height, as illustrated in Fig. 5-8. In this case wall reflections have a

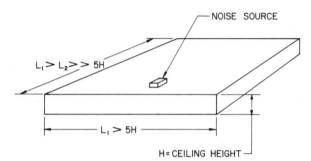

Fig. 5-8. Extended room with one or more floor dimensions greater than five times the ceiling height.

negligible effect except very near the walls and the level of the reflected sound is controlled largely by the reflections between floor and ceiling. Thus the practical limit of noise reduction by acoustic absorption is to remove all ceiling reflection and this can be approached by a ceiling treatment of $\alpha = .90$ acoustic absorption coefficient, or higher. The effectiveness of the ceiling treatment depends not only on the distance between source and receiver, as in Case 1, but also on the ceiling height. We thus define the ratio of source distance to ceiling height, D/H, and note in Fig. 5-9 how the sound is affected with changes in this ratio. Close to the source, hard ceiling reflection contributes little to the direct sound; therefore, adding absorption to the ceiling will have very little effect in reducing the sound. But far from the source, the hard ceiling reflection is about equal to the direct sound and ceiling absorption will therefore be more effective. We note that the effect of ceiling absorption increases with distance and that roughly a 10 dB reduction at 10 times ceiling height can be achieved by increasing the absorption coefficient from 0.02 to 1.00.

Figure 5-9 also suggests that for a given source distance a low ceiling is more effective than a high ceiling. In practice this can be achieved by suspended acoustical ceilings in new construction, or in existing buildings by space absorbers, if the overhead area is cluttered. For example, 2 X 4 foot glass fiber boards hung vertically on wires are very effective. A coverage of approximately 100 boards per 1000 sq ft of projected ceiling area is equivalent to a ceiling absorption coefficient of 1.00 [8]. The exact pattern is not important, however, and their major disadvantage is a potential interference with moving objects and shading of light sources.

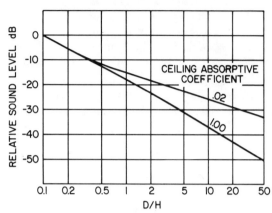

Fig. 5-9. Level of combined direct and reflected sound in an extended room (floor and ceiling reflection only) with a hard ceiling and with a completely absorptive ceiling. D/H = ratio of source distance to ceiling height.

Multiple Sources of Sound

The foregoing discussion, for simplicity, concentrated on sound from a single source yet the results can be applied effectively to more complicated situations. For example, one frequently finds extended shop areas where machining, fabrication, assembly, and packaging are carried out in one large room. If the inherently noisier machining and fabrication processes are grouped together at one end of the room, the sound level at the other end—where assembly and packaging are done—can be significantly reduced by an appropriate ceiling treatment as outlined above.

Where there is a concentration of machinery, the sound level within the area will tend to be spacially uniform particularly for repetitive processes of like machinery such as screw machines, cold-heading machines, and the like. In such cases an effective use of acoustically absorbent material would be to place the material near the machinery, for the nearer the material is to the noise source, the greater will be the angle of space covered (as seen from the source) and this will reduce the total amount of material required. In this sense it is common practice to construct free-standing barriers or partial enclosures around noisy machines, lined on the source side with absorbent materials. This is the best treatment from the standpoint of the machine operator.

Treatment of the entire room under such conditions should not be considered superfluous, however. Reduction in the overall reverberation time increases comfort and intelligibility of communication, and greatly increases the ability of a human being to locate a source of noise; a factor in safety.

Overall surface absorptive treatment is straightforward to apply in new plant construction by using absorptive roof decks and by lining structural walls at strategic locations with fibrous material protected by perforated metal facings. Figure 5-10 shows such construction using slotted structural blocks (acting as Helmholtz resonators) to enhance low-frequency absorption. For existing industrial rooms, wall and ceiling surfaces can be treated with absorbers con-

Ch. 5 ACOUSTIC MATERIALS 167

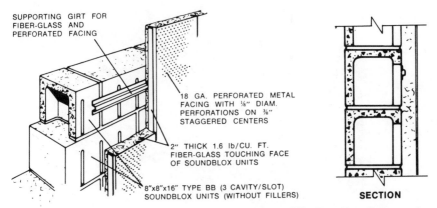

Courtesy of The Proudfoot Company, Inc.

Fig. 5-10. Acoustically absorbent wall treatment for new construction.

sisting of glass fiber batts contained in perforated metal pans. A typical installation is illustrated in Fig. 5-11.

Protection of Absorbent Material

Since the majority of absorbent materials are fragile and can shred or dust, it is often necessary to cover them in some manner. A perforated covering will have little effect on the absorption efficiency of the materials provided that it has only trivial air-flow resistance. Its open area should not be less than 10 percent of the total area, and the distance between openings must not be more than one-quarter wavelength at the highest frequency of interest (about one-half inch, for example, at 5000 hertz). Impermeable coverings, which may be required to prevent absorption of oil vapors or other dirt, or for sanitary purposes, must have a very low weight per square foot. As a rule, the weight should be less than 0.02 lb/ft^2 for good high-frequency absorption. Plastic films, 2 mils thick, or less, are commonly used, sometimes completely enclosing the absorbent material like a pillowcase, as shown in Fig. 5-12. Although this may slightly reduce high-frequency absorption it will improve low-frequency absorption. The variety of possible coverings is limited only by the imagination of the designer, ranging from perforated masonry, to wire meshes, to exotic fabrics. For industrial applications, perforated metal or hardboard are most practical. If an impermeable plastic film is used together with a perforated covering, great care must be taken so that the film is not compressed against the perforations, for this would create little drumheads, essentially destroying the transmission of sound into the absorbing material.

Effect of Air Spacing

Whenever possible, there is a definite advantage in mounting the absorbent material with an air space between it and any impervious backing. The ability of a material to absorb low frequencies is usually limited in ratio to its thickness. Some of the benefits of thickness without actually using too much material can

Courtesy of Barry Controls Div. of Barry Wright Corp.

Fig. 5-11. Installation of acoustically absorbent panels on walls and ceiling of an existing room to reduce reflected sound.

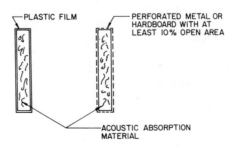

Fig. 5-12. Acoustically absorbent material covered with protective film or a perforated retainer.

be achieved by the incorporation of such an air space. The most obvious opportunity for such air spaces is found in ceilings, where it is desirable to leave a plenum space for access to utility devices, but even the small air space provided by mounting over furring may be advantageous. Table 5-4 indicates the variation in absorption coefficients for one-inch-thick glass fiber on three types of mountings:

Table 5-4. Absorption Coefficients for 1-inch Glass Fiber

Type of Mounting	Frequency, Hz						
	125	250	500	1000	2000	4000	NRC
	Absorption Coefficients						
No air space:	.08	.24	.68	.91	.96	.94	.70
3/4-inch air space:	.13	.37	.85	.99	.99	.97	.80
15-inch air space:	.69	.94	.75	.98	.99	.99	.90

The anomalous reduction at 500 hertz as the air space is increased from 3/4 inch is a resonance phenomenon. There is an optimum air space for any particular noise frequency, but seldom is the noise source a single frequency. The designer usually tries to get a maximum amount of absorption at all frequencies, except for very special environments such as concert halls. It is pertinent to recognize that a 10 percent change in the absorption coeffients may not be detectable by ear, but a 50 percent change will be substantial.

TRANSMISSION-LOSS MATERIALS

Noise reduction by sound absorption alone will seldom prove completely satisfactory. If the source is very loud, it will be necessary to isolate it with a total enclosure or an insulating barrier. Whereas reduction of 10 decibels represents about an optimum amount for absorptive treatments, 10 to 60 decibels or more of noise reduction may be feasible for barriers ranging from simple screens to full enclosures. The cost of such barriers is greater, proportional to the attenuation, thus the designer must carefully evaluate his needs as well as his products. For industrial applications, it is usually sufficient to lower the unwanted noise down to the background level; assuming, of course, that the background level is at least below legislative requirements. Ideally, it should be about 10 decibels more quiet than the background as measured with the noise source shut off. Legislation and economics will often set the limits rather than good design practice, unfortunately.

Technical Considerations

a. Limp Mass Law—The major single contributor to sound insulation is mass; specifically, the weight per square foot is the most important variable available to the designer. Because denser materials tend to transmit less sound than lighter materials, higher frequencies are more difficult to transmit than lower frequencies. This relationship follows from the laws of motion and is known as the *Limp Mass Law*, implying that the predictions will only be accurate for limp materials. The law can be written in equation form for any angle of incident sound. For use in this handbook, and in most architectural situations, only the so-called *Field Incidence Limp Mass Law*, which restricts itself to the angles likely to be encountered under typical conditions, is needed. In equation form, this relationship is:

$$TL = 20 \log_{10} W + 20 \log_{10} f - 33 \text{ dB} \qquad (5\text{-}12)$$

where: TL is the transmission loss, in decibels; W = the weight in pounds per square foot of the specimen; and f = the frequency, in hertz. Since no material is perfectly limp, this equation is used principally as the first order approximation of transmission loss. It is a handy reference as an "ideal goal" when examining the characteristics of new systems.

b. Sound Transmission Class—The sound transmission (STC) rating provides a practical means of specifying materials and designs for walls, ceilings, floors, windows, and other features used in the construction of a partition for sound control or an enclosure within a building. Some typical STC ratings are given in Table 5-5.

c. Insulation at Low Frequencies—The laws of physics preclude the design of simple walls with large transmission losses at low frequencies. In addition, these frequencies are conducted efficiently through side walls and floors, "flanking" the direct path. When confronted with rumbly noise sources, it may be neces-

Table 5-5. Representative Sound Transmission Classes of Common Materials or Systems*

Common Material or System	Sound Transmission Class (STC)
1¾-inch-thick, hollow-core wood door	19
¼-inch-thick plate glass, single pane	26
½-inch gypsum board, each side of 2- by 4-inch wood studs	32–38
1¾-inch-thick, solid-core wood door, well gasketed	35
Wood floor, 2- by 8-inch joists, gypsum-board ceiling	40
Plastered walls with resilient suspension	40–45
4-inch, hollow cement-block wall, 45 lb per cu ft, painted	40
Simple, dual-glazed windows, ¼-inch panes	38–42
8-inch, hollow cement-block wall, 45 lb per cu ft, painted	45
Staggered-stud drywall partition	45
3-inch-thick poured concrete wall, 148 lb per cu ft	46
4-inch-thick poured concrete wall, 144 lb per cu ft	50
8-inch, dense cement-block wall, 60 lb per cu ft	50
6-inch-thick, poured-concrete wall, 152 lb per cu ft	54
8-inch, solid cement-block wall, 142 lb per cu ft	56
12-inch-thick (triple thick) brick wall	59
16-inch-thick, solid cement-block wall, 142 lb per cu ft	62
Double masonry walls, glass fiber filled	73–74
Limit of Riverbank, 14- by 9-foot test facility	84

*For STC ratings of additional systems see Table 5-13.

Ch. 5 ACOUSTIC MATERIALS 171

sary to provide complete isolation of adjoining rooms; this may require installation of separate footings in severe cases. Right-angle turns in walls provide an impedance mismatch and substantial reduction can sometimes be obtained by changing wall-thicknesses at intersections.

When the noise source contains *only* low-frequency sounds, the use of very stiff walls will be beneficial. An example of such a wall would be paper or metal honeycomb sandwiched between gypsum-board or sheet-metal faces. Later, we will see this principle applied to a machine enclosure. The property of stiffness as we ordinarily think of it is limited to static or very low frequencies. At middle frequencies, the stiff core becomes a bridge across the partition, transmitting sound with efficiencies far below those predicted by the limp mass law. At still higher frequencies, the efficiency is restored, as can be seen in Fig. 5-13, which is the transmission loss of an idealized sandwich of gypsum-board faces and expanded-foam core. At present, most such walls are designed by trial and error processes, and the designer should refer to sources of data published by manufacturers of such sandwiches unless he has opportunity for testing in commercial, or other, laboratories.

d. Multiple Panel Construction—If, for impact resistance or some other reason, such stiff constructions are desired, then the acoustical performance may be improved by adding an additional partition, usually with an air space between the two. The use of two walls in close proximity creates additional problems. At low frequencies, the air between such "double walls" (acting as a spring) couples them so that they behave like a single wall. At high frequencies, this coupling changes and the transmission loss may reach large values, usually displaying a

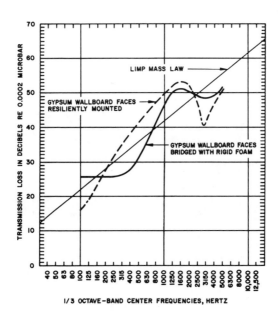

Fig. 5-13. Sound transmission loss as a function of frequency for gypsum wallboard.

slope increasing at about 12 decibels per octave. At the junction of these two regions, however, the reaction between the faces and the air spring becomes a resonant condition, and the two walls combined will transmit more sound than either one alone. This resonant frequency can be calculated from the equation

$$f_0 = 170 \sqrt{\frac{W_1 + W_2}{dW_1 W_2}} \text{ hertz} \tag{5-13}$$

where: W_1 and W_2 are the surface weights, in pounds per square feet, for each of the walls and d is the air space, in inches. When the wall is symmetric (each face identical) this reduces to:

$$f_0 = \frac{240}{\sqrt{dW}} \text{ hertz} \tag{5-14}$$

where: W is the weight per square foot of one of the walls. An idealized curve of the performance of a single and double wall of identical *total* weights per square is shown in Fig. 5-14.

Symmetric constructions are to be used with caution, however, since they encourage the generation of resonances. As an example, two identical sheets of plywood paneling, resiliently attached to each side of a concrete-block wall will lower its sound transmission class (STC) several numbers; in this case the lightweight panel easily accepts energy from the incident sound, translates it into a mechanical vibration on the elastic concrete which is readily conducted to the opposite side, where the lightweight face regenerates the sound. The addition of intermediate layers into air spaces may also reduce performance by increasing

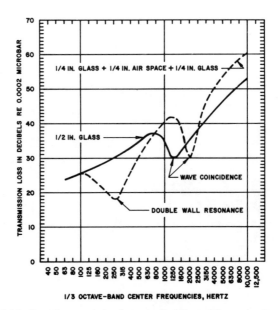

Fig. 5-14. Sound transmission loss as a function of frequency for glass.

ACOUSTIC MATERIALS

the frequency of the double-wall resonances and the range over which the double wall behaves as a single wall. This latter case is apt to arise when the available thickness of a partition is limited and the designer must add materials to existing constructions. In any case, resonances can be reduced by adding absorbent material in the air space to increase air damping, as was shown in Fig. 5-2.

e. Coincidence Frequency—Note in Fig. 5-14 the large dip in transmission loss at 1250 hertz for the half-inch glass. This dip is characteristic of all materials; its *critical frequency* is only dependent upon surface density and bending stiffness, not overall dimensions. More commonly known as the "coincidence frequency," this anomaly has a fascinating explanation. In air, the velocity of sound is the same at all frequencies. In solids, however, the bending or flexural wave has a velocity proportional to the square root of the frequency. Thus, below a certain frequency, bending waves will travel more slowly in the solid material than in the adjacent air; above this frequency, they will travel faster. At the critical frequency where the velocity is the same in air and the material, energy is efficiently conducted into, through, and out of the solid. A perfectly elastic wall of infinite area would theoretically have a transmission loss near zero at this point, becoming acoustically transparent.

The coincidence frequency can be computed from the equation:

$$f_c = \frac{c^2}{2\pi} \sqrt{\frac{M}{B}} \text{ Hz} \qquad (5\text{-}15)$$

where:

f_c = Coincidence frequency, Hz
M = Surface density of panel, slug/in.^{2}
$B = \dfrac{Et^2}{12}$ = Flexural rigidity per unit width, lb/in.
E = Young's Modulus, lb/in.^{2}
t = Panel thickness, in.
c = Sound wave speed in air, in./sec

Table 5-6 shows the density, the product of the surface density, and the critical frequency for a number of common building materials. For steel sheet the coincidence frequency, f_c, can be estimated from:

$$f_c = \frac{500}{\text{Thickness in inches}}, \text{ Hz} \qquad (5\text{-}15a)$$

The designer has several possible ways of avoiding or minimizing the coincidence dip: 1. The thickness of the materials can be so chosen that the dip will occur at an inconsequential frequency; 2. A material can be coated with substances having damping characteristics; 3. Two materials with different coincidence frequencies can be combined so that the dips will cancel each other. An example of Method No. 1 is the use of thin sheets of metal or thick concrete walls; 16-gage steel has its dip at 8000 hertz, and an 8-inch concrete wall at 115 hertz, which may both lie outside of the frequencies of interest. An example of

Table 5-6. Density, and Products of Surface Density and Critical Frequency for Common Building Materials

Material	ρ_m, lb/ft³	ρ_m, kg/m³	$w f_c$, Hz-lb/ft³	$M_s f_c$, Hz-slug/ft²	$M_s f_c$, Hz-kg/m²	c, m/s	η, at 1000 Hz
Aluminum	170	2,700	7,000	217	34,700	5,100	10^{-4}–10^{-2}
Brick	120–140	1,900–2,300	7,000–12,000	217–373	34,700–58,600	...	0.01
Concrete dense poured	150	2,300	9,000	279	43,000	3,400	.005–02
Concrete (clinker) slab plastered on both sides, 2 in. thick	100	1,500	10,000	310	48,800	...	.005–02
Masonry block:							
Hollow cinder (nominal 6-in.-thick)	50	750	4,750	148	23,200	...	.005–02
Hollow cinder, ⁵/₈ in. sand plaster each side (nominal 6-in.-thick)	60	900	5,220	162	25,500	...	.005–02
Hollow dense concrete (nominal 6-in.-thick)	70	1,100	4,720	147	23,000	...	0.007–0.02
Hollow dense concrete, sand filled voids (nominal 6-in.-thick)	108	1,700	8,650	269	42,200	...	variable
Solid dense concrete (nominal 4-in.-thick)	110	1,700	11,100	345	54,100	...	0.012
Fir timber	40	550	1,000	31	4,880	3,800	0.04
Glass	156	2,500	7,800	242	38,000	5,200	0.001–0.01
Lead: Chemical or tellurium	700	11,000	124,000	3,850	605,000	1,200	0.15
Antimonial (hard)	700	11,000	104,000	3,240	508,000	1,200	0.002
Plaster, solid, on metal or gypsum lath	108	1,700	5,000	165	24,500	...	0.005–0.01
Plexiglas or Lucite	70	1,150	7,250	225	35,400	1,800	0.002
Steel	480	7,700	20,000	621	97,500	5,050	10^{-4}–10^{-2}
Gypsum board (¹/₂ to 2 in.)	43	650	4,500	140	20,000	6,800	0.01–0.03
Plywood (¹/₄ to 1¹/₄ in.)	40	600	2,600	81	12,700	...	0.01–0.04
Wood waste materials bonded with plastic, 5 lb/ft²	48	750	15,000	466	73,200	...	0.005–0.01

method No. 2 is the lamination of two pieces of glass to a layer of viscoelastic plastic. Method No. 3 is frequently used in dual-pane window designs, where one pane is of a different thickness than the other; in using this method, a ratio of 2 to 1 is recommended for thicknesses.

Above the coincidence frequency, the transmission loss is controlled by the internal elasticity and damping of the material. A limp material such as a sheet of leaded vinyl will exhibit essentially the characteristics expected from the limp mass law, but an elastic material such as steel or concrete may perform 6 or more decibels poorer than predicted from the mass law alone. Calculations to estimate the transmission loss of a panel above coincidence can be quite involved and will not be dealt with here. The interested reader may study Reference [10] for a more detailed account.

f. Insertion Loss—One method of comparing systems simply without actually determining the acoustic transmission loss is to utilize a parameter called the acoustic *"Insertion loss,"* IL." The insertion loss is the difference in sound-pressure level with an added noise control system compared to the sound-pressure level without the added system. The insertion loss is a method for determining the change in acoustic performance without the need of finding the actual acoustic transmission loss of the total system which may be difficult and time-consuming in many cases.

g. Drumhead Resonances and the Kettledrum Effect—Every panel has natural or "drum-head" resonances, determined by its dimensions, at which the transmission loss will be reduced. The lowest frequency at which such resonances can exist is calculated from the equations:

$$f_{11} = \frac{\pi}{2}\left(\frac{1}{a^2} + \frac{1}{b^2}\right)\sqrt{\frac{B}{M}}; \quad \text{or } f_{11} = \frac{144\, c^2}{f_c}\left(\frac{1}{a^2} + \frac{1}{b^2}\right) \quad (5\text{-}16)$$

where:

f_{11} = Natural frequency of panel, or drumhead resonance, Hz
M = Surface density of panel, slug per in.^{2}
B = Flexural rigidity per unit width, $\frac{Et^3}{12}$, lb/in.
a = Panel width, in.
b = Panel length, in.
c = Speed of sound in the material, ft/sec
f_c = Coincidence frequency, Hz [Equation (5-15) or (5-15a)]
E = Young's modulus for material, lb/in.^{2}
t = Panel thickness, in.

Many multiples (harmonics) of these lowest frequencies will exist and are responsible for most of the wiggles in the transmission-loss curves. This frequency is of special interest, however, for below it the transmission loss is determined by the stiffness of the panel, usually increasing with *decreasing* frequency at about 6 decibels per octave, a sort of "reverse limp-mass law."

Equation (5-16) assumes that the panel is not in close proximity to a rigid surface or enclosed on one side by a cavity. If a cavity is present it represents an air spring, or an additional stiffness to the panel, causing the lowest resonance frequency to increase according to Equation (5-13).

$$f_0^2 = f_{11}^2 + \frac{\rho c^2}{4\pi^2 M d} \qquad (5\text{-}17)$$

where ρ is the density of air in the cavity and d is the cavity depth. This "kettledrum effect" can greatly influence the lowest resonant frequency when the cavity is small. For example, Fig. 5-15 shows the theoretical insertion loss of a single panel of surface weight 3 lb/ft^2 of two different materials (lead .052-in.-thick, plywood 1-in.-thick) spaced 3 inches from a rigid, vibrating surface (see Ref. [11]). We observe that even though these two materials represent

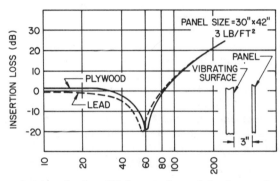

Fig. 5-15. Theoretical insertion loss (IL) for a single panel closely spaced to a rigid vibrating surface.

extremes in stiffness-to-weight ratio for commonly available materials, the lowest resonance frequency for both panels is about the same. In this case, the cavity stiffness predominates and is about 5½ times the stiffness of the plywood panel, and many times greater than that for the lead panel.

The theoretical insertion loss at or below the lowest resonance frequency is given by Reference [12] and Equation (5-18):

$$\text{IL} = 10 \log_{10} \left[\left(1 + \frac{R}{\rho c}\right)^2 \sin^2 kl + \left(\cos kl - \frac{x}{\rho c} \sin kl\right)^2 \right] \text{dB} \qquad (5\text{-}18)$$

where:

    IL = Insertion loss, dB

    k = Wave number, $\dfrac{2\pi f}{c}$, in.$^{-1}$

    c = Speed of sound in air, in. per sec

    l = Cavity depth, in.

    $x = 2\pi f M - \dfrac{S}{2\pi f}$

    S = Panel bending stiffness, slug per in.^{2} per sec^2

M = Surface density, slug per in.^{2}
R = Panel resistance, $f_0 M\eta$, slug per in.^{2} per sec
ρ = Density of air in cavity, slug per in.^{3}
η = Damping factor
f_0 = Resonant frequency, Equation (5-17), Hz

PRACTICAL CONSIDERATIONS

The good designer will always check the coincidence frequency and the panel resonant frequencies then design the system so that these frequencies are not critical to the acoustical performance of the unit. In descending order of importance, a good design will incorporate the maximum practical amount of mass, isolation between opposite sides, and internal damping. Almost every design, unfortunately, will be a first order approximation and will require verification (or justification) by actual test in a laboratory or on the sample situation. A conservative approach is almost certain to insure satisfactory performance even in this type of situation.

The design, the construction, and the installation of insulating partitions can each contribute to inadequate performance, and it is most desirable that the persons responsible for each of these stages be aware of the others' efforts, to some extent. For example, a good design consisting of glass fiber sandwiched between metal skins can be damaged in assembly by overcompression of the blanket, especially near the framework; the addition of electrical, telephone, or plumbing connections during installation can introduce bridging of the faces and thus destroy the desirable isolation. All partitions which are expected to have STC's in excess of 40 must be completely sealed at all joints (peripheral and other) with a nonhardening mastic or dense foam rubber. A swinging or rolling door not equipped with airtight gaskets and automatic threshold-sealing devices will usually be limited at STC 35, or lower, regardless of its original design potential.

When situations arise which require partitions in excess of STC 60, it may be desirable to examine alternative treatments, such as relocation of the noise source or of the quiet room, unless cost and space are of no concern. It should be noted that even a 16-inch-thick, solid concrete wall has an STC of only 62.

NOISE CONTROL BY BARRIERS

If the sound source is out-of-doors, or in a room with a large amount of absorption, blocking the direct (visual) path with a partial barrier may be adequate for noise control. Masonry walls are frequently erected for the same purpose around outdoor machinery and power transformers, and berms are sometimes constructed along highways passing through residential neighborhoods. One may expect an insertion loss of 10 to 20 dB for these cases. Indoors, however, even a modest amount of reverberation will destroy the effectiveness of such a shield, for reflected paths may contribute as much or more sound energy as was stopped by the shield. A barrier to be *fully* effective should be impervious and have a surface weight of not less than 4 lb/ft^2; for multipanel construction the weight can be correspondingly less. Based on

178　　ACOUSTIC MATERIALS　　Ch. 5

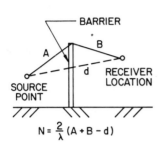
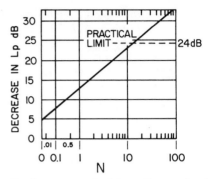

Fig. 5-16. Sound attenuation by barriers. a. Barrier geometry; b. Attenuation as a function of the parameter, N.

theoretical considerations, the decrease in sound-pressure level in the shadow zone of the barrier is given in Reference [9] and is related to Equation (5-19), see Fig. 5-16, and Equation (5-20), which are given below:

$$N = \frac{2}{\lambda}(A + B - d) \qquad (5\text{-}19)$$

$$\text{Decrease in } L_p = 20 \log_{10} \left[\frac{\sqrt{2\pi N}}{\tanh \sqrt{2\pi N}} \right] + 5 \text{ dB} \qquad (5\text{-}20)$$

where:

N = Geometric factor
$A + B$ = Shortest path length of wave travel over the barrier between the source and the receiver, ft or meters (See Fig. 5-16.)
d = Straight line distance between source and receiver, ft or meters (See Fig. 5-16)
λ = Wavelength of sound = $\dfrac{1128}{\text{frequency (Hz)}}$ ft = $\dfrac{344}{\text{frequency (Hz)}}$ meters
L_p = Sound-pressure level, dB

The equations above hold for ideal conditions where there are no reflections from nearby surfaces and the length of the barrier is sufficiently greater than the height so that end effects are negligible. Even under these ideal conditions the maximum reduction attainable is about 24 dB. One would expect, therefore, that for realistic conditions, the actual reduction in sound level would fall under the curve of Equation (5-19) shown in Fig. 5-16.

Example No. 1

It is desired to construct a free-standing barrier to be placed in front of a machine to reduce its noise level, as illustrated in Fig. 5-17. The machine is located near a wall and the wall and ceiling have been heavily treated with absorbent material. Under these conditions let us assume the noise-level contribu-

Ch. 5 ACOUSTIC MATERIALS 179

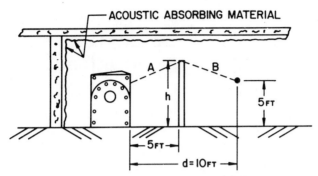

Fig. 5-17. Barrier geometry for Example No. 1.

tion from the machine is 92 dBA at a point 10 feet in front of the machine and 5 feet above the floor, and it is desired to reduce this to 85 dBA. It has been determined that the source of noise in the machine is concentrated at a point 5 feet above the floor and that the barrier cannot be closer than 5 feet from this point. The problem is: What should be the minimum barrier height?

The procedure for calculating the barrier height is given below in a step-by-step manner. Because separate calculations must be made for each octave-band center frequency, it is usually convenient to tabularize the calculations, as done in Table 5-7.

Step 1: Measure the sound-pressure level in octave bands at the operator position. These values are shown in Column 2 of Table 5-7.

Step 2: Add the A-weighting corrections reproduced in Column 3, to the values in Column 2, and record this summation for each octave band in Column 4. The A-weighting corrections are found in Table 2-3 of Chapter 2. They convert the octave-band data to the A-scale criteria prescribed by OSHA.

Step 3: Select an octave band for which the noise-level reduction is to be made in the first estimate. The design of a suitable barrier requires a trial-and-error solution because the noise reduction in each octave band resulting from the installation of a barrier will be different. By inspection of Column 4, it can be seen that the greatest octave-band sound-pressure levels occur in the 500 and 1000 Hz octave bands. For this reason, select a required noise reduction of 10 dB at 1000 Hz. This will be the first estimate for the barrier design for which the new sound-pressure levels will be computed.

Step 4: Calculate the sum of the sound-pressure levels in Column 4; however, these values must be added logarithmically. Figure 2-3, in Chapter 2, can be utilized to perform this addition. When this method is used, the addition must be done by adding two values, then adding their sum to the next value, etc., until the entire sum is obtained. This method will be illustrated in Example No. 2. A more precise method

Table 5-7. Problem on Barrier Design, Example No. 1

Col. 1	Col. 2	Col. 3	Col. 4	Col. 5	Col. 6	Col. 7
Octave-Band Center Frequency, Hz	Existing Level, dB (Measured)	A-weighting Factors, dB	Corrected Level, dB, L_{pi}	$\dfrac{L_{pi}}{10}$	$\left(\dfrac{p}{p_{\text{ref}}}\right)_i^2 = \log_{10}^{-1}\left(\dfrac{L_{pi}}{10}\right)$	Calculation of dBA
63	80	−25	55	5.5	.32 × 10^6	
125	76	−16	60	6.0	1.00 × 10^6	
250	78	−9	69	6.9	7.94 × 10^6	Equivalent A-weighted sound-pressure level: 10 log$_{10}$ (1460.99 × 10^6) = 91.6 dBA; round off to 92 dBA
500	89	−3	86	8.6	39.81 × 10^6	
1000	91	0	91	9.1	1258.92 × 10^6	
2000	80	+1	81	8.1	125.89 × 10^6	
4000	73	+1	74	7.4	25.12 × 10^6	
8000	64	−1	63	6.3	1.99 × 10^6	
			$\sum \log^{-1} \dfrac{L_p}{10} = \sum_{i=1}^{i=n}\left(\dfrac{p}{p_{\text{ref}}}\right)_i^2 = 1460.99 \times 10^6$			

which is readily acceptable to solution with an electronic calculator is to apply the following Equation (1-30) from Chapter 1. The results are shown in Columns 5, 6, and 7, in Table 5-7, and the mathematical equations from which these results are obtained are repeated below:

$$L_{pi} = 10 \log_{10} \left(\frac{p}{p_{\text{ref}}}\right)_i^2 \qquad (1\text{-}30)$$

$$\frac{L_{pi}}{10} = \log_{10} \left(\frac{p}{p_{\text{ref}}}\right)_i^2 \qquad (1\text{-}30\text{A})$$

$$\log^{-1} \frac{L_{pi}}{10} = \left(\frac{p}{p_{\text{ref}}}\right)_i \qquad (1\text{-}30\text{B})$$

$$\sum \log^{-1} \frac{L_{pi}}{10} = \sum_{i=1}^{i=n} \left(\frac{p}{p_{\text{ref}}}\right)_i^2 \qquad (1\text{-}30\text{C})$$

Using Equation (1-30A), the values of L_{pi} for each octave band in Column 4 are divided by 10 to find $L_{pi}/10$ and these values are recorded in Column 5. Equation (1-30B) indicates that the logarithm for each value of $L_{pi}/10$ given in Column 5 must be determined; the results are shown in Column 6 for each octave band. Equation (1-30C) merely directs that a summation of the values in Column 6 be obtained, which in Table 5-7, is shown to be 1460.99×10^6. The A-weighted sound-pressure level is recalculated by using Equation (1-30) again; however, in this case the *sum of the squares* of the pressure is used as

ACOUSTIC MATERIALS

Table 5-7 (con't). Problem on Barrier Design, Example No. 1

Col. 1	Col. 8	Col. 9	Col. 10	Col. 11	Col. 12	Col. 13	Col. 14
Octave-Band Center Frequency, Hz	Wavelength $\lambda = \dfrac{1128}{f(Hz)}$, ft	$N = \dfrac{\lambda}{2}[A+B-d]$	Noise Reduction, dB	New Level, dB	$\dfrac{dB}{10}$	$\left(\dfrac{p}{p_{ref}}\right)^2 = \log_{10}^{-1}\left(\dfrac{dB}{10}\right)$	Calculate dBA with barrier
63	17.9	.026	5.5	49.5	4.95	.89 × 10⁵	
125	9.00	.052	5.9	54.1	5.41	2.57 × 10⁵	Equivalent A-scale sound-pressure level 10 log₁₀ (2019.8 × 10⁵) = 83.05 dBA, round off to 83 dBA
250	4.51	.105	6.7	62.3	6.23	16.98 × 10⁵	
500	2.26	.211	8.0	78.0	7.80	630.96 × 10⁵	
1000	1.13	.423	9.9	81.1	8.11	1288.25 × 10⁵	
2000	.56	.839	12.4	68.6	6.86	72.44 × 10⁵	
4000	.28	1.708	15.3	58.7	5.87	7.41 × 10⁵	
8000	.14	3.416	18.3	44.7	4.47	.30 × 10⁵	
						sum = 2019.80 × 10⁵	

shown below:

$$L_p = 10 \log_{10}\left[\sum_{i=1}^{i=n}\left(\frac{p}{p_{ref}}\right)_i^2\right] \text{ dBA}$$

$$= 10 \log_{10}(1460.99 \times 10^6)$$

$$= 91.6 \text{ dBA} \qquad (5\text{-}21)$$

This is usually rounded to the nearest whole decibel, which in this case is 92 dBA, as given in Column 7 of Table 5-7.

Step 5: Calculate the required barrier height, h, that should result in a 10 dB noise reduction for the 1000 Hz octave band. From the geometry of Fig. 5-17:

$$A^2 = B^2 = \left(\frac{d}{2}\right)^2 + (h-5)^2 \qquad (5\text{-}22)$$

Thus:

$$A + B = 2\left[\left(\frac{d}{2}\right)^2 + (h-5)^2\right]^{1/2}$$

From Fig. 5-16:

$$A + B = \frac{N\lambda}{2} + d$$

Combining:

$$2\left[\left(\frac{d}{2}\right)^2 + (h-5)^2\right]^{1/2} = \frac{N\lambda}{2} + d$$

From which:

$$h = 5 + \frac{1}{2}\left[\left(\frac{N\lambda}{2} + d\right)^2 - d^2\right]^{1/2}$$

From the graph in Fig. 5-16, N is approximately 0.35 and λ is $\frac{1128}{1000}$ = 1.13 ft. Thus,

$$h = 5 + \frac{1}{2}\left\{\left[\frac{(0.35)(1.13)}{2} + 10\right]^2 - 10^2\right\}^{1/2}$$

$$= 6.1 \text{ ft.}$$

Step 6: Determine the acceptability of the calculated barrier height (6.1 ft) by calculating a new sound-pressure level at each octave-band level.

    a. Calculate the wavelength, λ:

$$\lambda = \frac{1128}{f} \text{ (Hz) ft} \qquad (1\text{-}2)$$

The resulting values are given in Column 8, Table 5-7.

    b. Calculate N (see Fig. 5-16) for each octave band from Equation (5-19):

$$N = \frac{2}{\lambda}(A + B - d) \qquad (5\text{-}19)$$

The resulting values are shown in Column 9.

    c. Calculate the noise reduction caused by the barrier, using Equation (5-20). These values are recorded in Column 10.

    d. Determine the A-weighted octave-band sound-pressure levels with the barrier in place by subtracting the values in Column 10 from those in Column 4, and recording these results in Column 11.

Step 7: Calculate the A-scale sound-pressure level with the barrier in place. This procedure is identical to the procedure in Step 4. The results are given in Columns 12, 13, and 14. The sound-pressure level with the barrier will be approximately 83 dBA.

Step 8: Steps 3 through 7 are repeated, if necessary, until the required A-weighted sound-pressure level is attained. Succeeding calculations are made by changing the selected noise reduction in Step 3 to a higher value and repeating the calculations. In this example the required sound level was obtained by the first trial and no further calculations are required.

The result of these calculations yields a sound-pressure level of 83 dBA which is close enough to the design goal of 85 dBA to show that a barrier height of 6 ft will be adequate.

Ch. 5 ACOUSTIC MATERIALS 183

Example No. 2

A large air supply fan in one corner of a shop area is generating an A-scale sound pressure level of 94 dBA in the work area some distance away. This is above the OSHA limit of 90 dBA and is preventing a full 8 hours per shift use of the space. In addition, it is desired to create an employees' relaxation room adjoining this shop area, which will possibly mean an improved common wall. The wall construction of this room will be painted block. It is necessary to both reduce the sound level to meet the OSHA limit in the shop area and to calculate the requirements for a common wall which can provide an NC-40 background level within the adjacent relaxation room to be constructed.

The shop area is 48 feet wide by 90 feet long by 40 feet high, thus having an area of 18,700 square feet and a volume of 172,800 cubic feet. Because of existing access doors and windows, only the ceiling and one large wall are capable of being treated. The common wall with the relaxation room will be 10 feet high and 20 feet wide; the relaxation room will be 30 feet long and will contain hard furniture only; the ceiling alone can receive acoustical treatment.

The first task is to obtain the octave-band sound-pressure level due to the fan in the shop area. Using a sound-level meter with an octave-band filter set, the sound-pressure-level spectrum was measured; the results of these measurements are presented in Table 5-8.

Table 5-8. Measured Sound-Pressure Level in Untreated Shop Area Due to an Air Supply Fan (for Example No. 2).

Octave-Band Center Frequency, Hz							
63	125	250	500	1000	2000	4000	8000
Sound-Pressure Level, L_p, dB re 2 X 10^{-5} N/meter^2							
79	99	95	91	88	82	74	52

For this sample problem the sound-pressure level in the 63 Hz and 8000 Hz octave bands was observed to be considerably lower than the other octave bands. A comparison with the A-scale weighing factors and the NC-40 criteria will confirm that these two octave bands do not significantly contribute to the analysis. Therefore, the following analysis will proceed, utilizing only the 125, 250, 500, 1000, 2000, and 4000 Hz octave bands. In general, this may not be adequate for a solution and additional octave bands may be required in the analysis; however, the steps in the solution would proceed in exactly the same manner if additional octave bands are included.

The acoustic absorption within the shop area must be determined by calculation or by measurement techniques. For a calculation procedure, the absorption coefficients for the various room surfaces could be estimated from the values in Table 5-1. The computational method will be the least accurate because the actual room-surface absorption coefficients will not be known with confidence.

ACOUSTIC MATERIALS Ch. 5

A measurement method for an existing space is always preferable and results in a more precise evaluation. Two experimental methods can be utilized for the measurement data from which the room absorption can be determined. The decay-rate method was described earlier in this chapter, in Section III, "Measuring Sound Absorption." This method requires a strip chart recorder, a broadband noise source, an octave-band filter set, a sound-level meter which can provide an electrical voltage proportional to decibels, and a means of calibrating the entire system. Many industrial plants will not have such a collection of noise analysis equipment available; therefore, an alternative method will be described.

The absorption can be obtained with a portable sound-level meter and a noise source for which the octave-band sound-power-levels, L_W, are known. One commercially available noise source is the ILG reference sound source described in Chapter 2. It is necessary to make sound measurements with the ILG source operating in the room and all other noise sources off. The absorption can then be calculated from the formula:

$$A = \log_{10}^{-1} \left[\frac{L_W - L_p + 16.5}{10} \right] \text{ sabins} \qquad (5\text{-}8)$$

where: L_W is the known sound power, in decibels, of the source in the desired frequency band and L_p is the measured sound-pressure-level, in decibels, for the corresponding frequency band. The use of Equation (5-8), above, assumes the measured sound-pressure level, L_p, is in the reverberant acoustic field of the room. To satisfy this condition the measurements must be taken at a distance greater than $0.5 \sqrt{A}$ feet from the sound source as indicated earlier in this chapter by Fig. 5-7, where: A is the absorption, in sabins. Since A is unknown, it is not immediately evident how far from the sound source the measurements should be taken. In general, measurements should be taken as far as possible from the source but not close to walls or other reflective surfaces. For most industrial spaces a distance of 50 ft from the source to the measurement location is sufficient. After calculating the room absorption, A, it can be determined by the use of $0.5 \sqrt{A}$ if the measurement location was in the reverberant field. If it is impossible to separate the source and the measurement location sufficiently, another method to determine the room absorption should be utilized, such as the decay-rate method.

The results for the absorption, in sabins, computed from Equation (5-8) are given in Table 5-9. The measurement location was 50 ft from the ILG reference sound source. A check of the final absorption values for the room gives $0.5 \sqrt{A} = 0.5 \sqrt{2239} = 24$ ft, therefore, the measurements were all in the reverberant sound field.

The sound power of the calibrated ILG source is given in row 1, and the measured octave-band sound-pressure levels with the ILG source operating, is given in row 2 of Table 5-9.

The quantity $L_W - L_p + 16.5$ is computed from rows 3 and 4 and is given in row 5. The value for the total absorption, A, in the room is obtained by taking $\log_{10}^{-1}$ which follows from the solution for A from Equation (5-8). From the resulting absorption values in row 6, the absorption of the air in the room must be subtracted to obtain the absorption of the surfaces only. The air absorption

ACOUSTIC MATERIALS

Table 5-9. Determination of Room Absorption for Example No. 2

Row No.	Procedure	Octave-Band Center Frequency, Hz					
		125	250	500	1000	2000	4000
1	Sound-Power, L_W, of ILG Source (See Chapt. 2, Table 2-4)	81	81	81	81	81	79
2	Measured Sound-Pressure Level, L_p, in Room, Caused by ILG Source, dB	66	68	67	65	65	62
3	$L_W - L_p$ (Row 1, minus Row 2), dB	15	13	14	16	16	17
4	Add 16.5 dB [As in Equation (5-8)]	+16.5	+16.5	+16.5	+16.5	+16.5	+16.5
5	$10 \log_{10} A = L_W - L_p + 16.5$	31.5	29.5	30.5	32.5	32.5	33.5
6	$A = \log_{10}^{-1}\left(\dfrac{L_W - L_p + 16.5}{10}\right)$ sabins	1413	891	1122	1778	1778	2239
7	A due to Air in Room, sabins (See Table 5-11)	0	17	52	155	397	1240
8	Absorption, A_0, due to Room Surfaces, sabins	1413	874	1070	1623	1381	999

is obtained from the room volume, 172,800 cu ft, and the air absorption per 1000 cu ft, as given in Table 5-1. The resulting air absorption is given in row 7. Subtracting row 7 from row 6 yields the total absorption of the room surfaces, in sabins, as given in row 8 of Table 5-9.

From the results of the room absorption calculations in Table 5-9 the effect of wall treatment can be determined in the shop area where the fan is located.

We can assume that 100 percent of the ceiling can be covered with a suspended ceiling system (usually called a #7 mounting) and 80 percent of the one 40- by 48-ft wall covered with the same material mounted against ⅛-inch-thick spacers (usually called a #1 mounting). A manufacturer of mineral-fiber ceiling board has advertised a material, compatible in other respects, which has the laboratory absorption coefficients with an 8- by 9-foot specimen shown in row 1 and row 3 of Table 5-10.

The values in rows 1 and 3 of Table 5-10 must be corrected for diffraction effects to account for the variation in acoustic performance between an 8- by 9-foot laboratory sample and the large area to be treated in the shop area. The corrections are made by using the factors given in the footnote to Table 5-1 and shown in rows 2 and 4 in Table 5-10. The corrected absorption values are shown in rows 5 and 6 of Table 5-10.

To obtain the absorption, in sabins, of the treated surfaces, multiply the ceiling absorption coefficient values of row 5 by the area of the ceiling, 4320 sq ft, and multiply the wall-treatment absorption values of row 6 by 80 percent of the wall area which is 1536 sq ft. The resulting absorption values are given in rows 8 and 9, respectively.

The absorption of the untreated surfaces must be determined. If the assumption is made that the original absorption was equally distributed, 12844/18700 sabins of the total area remains untreated. Taking this ratio of the values in row 8 of Table 5-9 results in the absorption of the untreated surface given in row 10 of Table 5-10.

Table 5-10. Absorption Calculations and the Resulting Sound-Pressure Level for Example No. 2

Row No.	Procedure	Octave-Band Center Frequency, Hz					
		125	250	500	1000	2000	4000
1	#7 Mounting Absorption Coefficient	.80	.75	.78	.93	.86	.64
2	#7 Mounting Correction Factor (From Table 5-11, 8 X 9-Foot Sample)	.49	.66	.79	.88	.94	.97
3	#1 Mounting Absorption Coefficient	.12	.34	.83	.99	.79	.55
4	#1 Mounting Correction Factor (From Table 5-11, 8 X 9-Foot Mounting)	.49	.66	.79	.88	.94	.97
5	#7 Mounting Corrected Absorption Coefficient (Row 1 X Row 2)	.39	.50	.62	.82	.81	.62
6	#1 Mounting Corrected Absorption Coefficient (Row 3 X Row 4)	.06	.22	.66	.87	.74	.53
7	Air Absorption per 1,000 cu ft (From Table 5-11)	.0	.1	.3	.9	2.3	7.2
8	Treated Ceiling Absorption, sabins (Row 5 X Ceiling Area)	1685	2160	2678	3542	3499	2678
9	Treated-Wall Absorption, sabins (Row 6 X 8 X Wall Area)	92	338	1014	1336	1137	814
10	Absorption of Untreated Surface, sabins (Row 8, Table 5-9 X Ratio Treated to Untreated Surfaces)	971	600	735	1115	949	686
11	Air Absorption, sabins $\left(\dfrac{\text{Room Volume}}{1000} \times \text{Row 7}\right)$	0	17	52	156	397	1240
12	New Total Absorption, A_1 (Sum of Rows 8, 9, 10, and 11)	2748	3115	4479	6149	5982	5418
13	A_0 (From Table 5-9)	1413	874	1070	1623	1381	999
14	$\dfrac{A_1}{A_0}$	1.94	3.56	4.19	3.79	4.54	5.42
15	Fan L_p, dB (From Table 5-8)	99.0	95.0	91.0	88.0	82.0	74.0
16	$10 \log_{10} \dfrac{A_1}{A_0} = L_{p1} - L_{p2}$	2.9	5.5	6.2	5.8	6.6	7.3
17	New Fan L_p, dB (Row 15 minus Row 16)	96.1	89.5	84.8	82.2	75.4	66.7
18	A-weighting Factors (From Table 2-3, Chapter 2)	-16.1	-8.6	-3.2	0.0	+1.2	+1.0
19	A-weighted Octave-Band Sound-Pressure Level (Sum of Rows 17 and 18)	80.0	80.9	81.6	82.8	76.6	67.7
20	Summation of Values (*Note:* The addition must be performed logarithmically. It is shown here performed by adding two values at a time, using Fig. 2-3, in Chapter 2.)						

```
      83.5
          ┐
           ├── 85.7
          ┘        ┐
                    ├── 87.5
                   ┘        ┐
                             ├── 87.8
                            ┘
```

A-weighted Sound-Pressure Level = 87.8 dBA

Ch. 5 ACOUSTIC MATERIALS 187

The air absorption of the room must also be included. The volume of the room, 172,800 cu ft, is divided by 1000 and multiplied by the air absorption in sabins per 1000 cu ft as given in the footnote to Table 5-1 and row 7 in Table 5-10. The absorption value, in sabins, for air is given in row 11 of Table 5-10. Finding the sum of the values in rows 8, 9, 10, and 11 results in the total absorption, in sabins, for the treated shop area. These values are given in row 12 of Table 5-10. The reduction in sound-pressure level was given by Equation (5-11) and repeated here:

$$L_{p_1} - L_{p_2} = 10 \log_{10}\left(\frac{A_1}{A_0}\right) \qquad (5\text{-}11)$$

where: A_1 is the absorption after treatment and A_0 is the original absorption of the shop area. The values of A_0 are given in Table 5-9 row 8 and in Table 5-10, row 13. The values of A_1 are given in row 12 of Table 5-10. Taking the ratio of these values results in the ratios given in row 14 of Table 5-10. Taking $10 \log_{10}$ of the ratios given in row 14 results in the values in row 16 which are decibel sound-pressure-level reductions in the shop area. Subtracting these reductions from the corresponding values in Table 5-8, which are repeated in row 15 in Table 10, results in the sound-pressure levels in octave bands after treatment of the shop area. These resulting levels are listed in row 17 of Table 5-10. The A-weighted sound-pressure level can be obtained for the treated shop area by adding the octave-band A-weighting factors from Table 2-3, Chapter 2. These are given in row 18 Table 5-10, and the sum of values in rows 17 and 18 are given in row 19.

The A-weighted sound-pressure level is obtained by combining the values of row 19, two at a time, utilizing Fig. 2-3 of Chapter 2. The calculation is shown in row 20 to result in 87.8 dBA. This value, rounded to the nearest decibel, will yield an A-weighted sound-pressure level of 88 dBA which is within the OSHA eight-hour specification of 90 dBA. This completes the treatment of the work area.

Selection of Wall Between Shop and Relaxation Room

The noise reduction required between the work area and the relaxation room is obtained by subtracting the sound-pressure level in the relaxation area from that in the source room. The process is tabulated in rows 1, 2, and 3 of Table 5-11. The resulting values in row 3 are the required noise reduction, in octave bands, which must be achieved by a combination of wall transmission-loss and acoustic absorption in the relaxation room.

We will design for the same acoustic material and a #7 mounting on the ceiling of the relaxation room as was used in the shop area. The absorption coefficients, correction factors, and the corrected absorption coefficients are given in rows 1, 2, and 3 of Table 5-12. The ceiling area for the relaxation room is 600 sq ft; the remaining surfaces have a total area of 1100 sq ft, and will be assumed to have absorption coefficients in octave bands equivalent to painted cement block as given in Table 5-1.

Given in row 6 of Table 5-12 is the total absorption of the ceiling of the relaxation room. It was computed by multiplying the ceiling area, 600 sq ft by the ceiling absorption coefficients of row 3. The total absorption of the remain-

ACOUSTIC MATERIALS Ch. 5

Table 5-11. Required Noise Reduction, in Decibels, for Relaxation Room in Example No. 2

Row No.	Procedure	Octave-Band Center Frequency, Hz					
		125	250	500	1000	2000	4000
1	Sound-Pressure-Level, L_p, in Treated Shop Area, dB (Row 17, Table 10)	96.1	89.6	85.1	82.8	77.2	70.9
2	NC-40 Criteria, From Table 4-3, Chapter 4	56.0	50.0	49.0	41.0	39.0	38.0
3	Required Noise Reduction, dB, (Row 1, minus Row 2)	40.1	39.6	36.1	41.8	38.2	32.9

ing surfaces is obtained by multiplying the remaining surface area, 1100 sq ft, by the absorption coefficients of row 4, which are obtained from Table 5-1, for painted block. These values are given in row 7 of Table 5-12. The absorption of the air in the room is computed by multiplying the values at the bottom of

Table 5-12. Calculations of Required Transmission-Loss for Wall Between Shop Area and Relaxation Room, for Example No. 2

Row No.	Procedure	Octave-Band Center Frequency, Hz					
		125	250	500	1000	2000	4000
1	#7 Mounting Absorption Coefficient	.80	.75	.78	.93	.86	.64
2	#7 Mounting Correction Factor (See Table 5-11, 8 X 9-Foot Sample)	.49	.66	.79	.88	.94	.97
3	#7 Mounting Corrected Absorption Coefficient (Row 1 X Row 2)	.39	.50	.62	.82	.81	.62
4	Wall and Floor Absorption Coefficient (From Table 5-1, Painted Block)	.02	.02	.03	.04	.05	.07
5	Air Absorption per 1000 cu ft (From Table 5-11)	.0	.1	.3	.9	2.3	7.2
6	Ceiling Absorption, sabins (Row 3 X Ceiling Area)	234	300	372	492	486	372
7	Absorption of Remaining Surfaces, sabins (Row 4 X Area of Remaining Surfaces)	22	22	33	44	55	77
8	Air Absorption, sabins (Row 5 X $\frac{60,000}{1,000}$)	0	6	18	54	138	432
9	Total Absorption, A, sabins (Sum of Rows 6, 7, and 8)	256	328	423	590	679	881
10	$10 \log_{10} S$ (S = Common Wall Area = 200 sq ft)	23.0	23.0	23.0	23.0	23.0	23.0
11	$10 \log_{10} A$ (A is given in Row 9)	24.1	25.2	26.3	27.7	28.3	29.4
12	$10 \log_{10} S - 10 \log_{10} A$ (Row 11 minus Row 10)	-1.1	-2.2	-3.3	-4.7	-5.3	-6.4
13	Required Noise Reduction, dB (From Table 5-11)	40.1	39.6	36.1	41.8	38.2	32.9
14	Required Transmission-Loss, dB-Rounded Off. (Sum of Rows 12 and 13)	39	37	33	37	33	26

Table 5-1 (repeated in row 5) by 60, since the room volume is 60,000 cu ft. These results are tabulated in row 8.

The total absorption, in sabins, for the relaxation room is given in row 9, which is the sum of the values in rows 6, 7, and 8. Equation (5-9), repeated below,

$$TL = NR + 10 \log_{10} S - 10 \log_{10} A \qquad (5\text{-}9)$$

is used to compute the required transmission-loss of the common wall between the shop area and the relaxation room. S is the area of the common wall, 200 sq ft, and A is the total absorption in the relaxation area, given by row 9 Table 5-12. The calculation for the required transmission-loss is done in rows 10, 11, 12, 13, and 14, of Table 5-12. The values in row 14 are the required transmission-loss for the common wall. The STC contour for the wall construction must be equal to, or larger than, the required transmission-loss. This is determined by selecting a contour from Table 5-3A for which the transmission-loss for each center frequency is equal to, or greater than, the corresponding value of the required transmission-loss as given in Table 5-12, row 14. Examining the tabular values of various STC contours from Table 5-3A we note that the STC-55 contour has the following values:

Frequency, Hz	125	250	500	1000	2000	4000
STC-55 Contour	39	48	55	58	59	59

This means we should look for walls having STC ratings of 55, or greater, for the common wall between the shop and the relaxation room. From Table 5-5, two possible wall constructions are a 6-inch poured-concrete wall with STC-54 which will meet the design requirements within 1 dB, or an 8-inch, solid cement-block wall with STC-56. Since this is a modification to an existing building, the wall easiest to construct would be the 8-inch, solid cement-block, therefore this construction is selected for the common wall between the shop area and the relaxation room. To achieve the desired sound transmission in the installed condition, doors in the common wall must be acoustical and have the same STC rating as the wall. No straight-through openings such as electrical outlets can exist in the common wall. If air-conditioning ducting penetrates the common wall, precautions must be taken to prevent sound transmission through the duct; sound traps and silencers may also be required. If a suspended ceiling exists in the relaxation room, care must be taken to prevent sound from entering from the shop area via the ceiling path, thus circumventing the acoustical wall. In general, for a successful system, all sound paths must be treated to obtain the desired transmission-loss from the shop to the relaxation room.

NOISE CONTROL BY ENCLOSURES

When a reduction in noise of 20 dB, or more, is required, it is generally necessary to use a full enclosure. Sometimes a partial enclosure of two or more sides, with or without top may provide the needed attenuation without heavy absorption treatment in the room as demanded by barriers. Partial enclosures have obvious advantages over full enclosures in terms of cost, accessibility, and ventilation, but design and construction should be done carefully and there should be the option of closing them completely in marginal cases. Partial enclosures will be treated in a later section of this chapter.

For purposes of discussion enclosures can roughly be divided into two types: first, those which are "loose fitting," that is, where there is sufficient space within the enclosure so that negligible coupling occurs between machine and enclosure panels. Such enclosures obey the ordinary rules of room acoustics as discussed previously. The second type may be called "close fitting," and here coupling does occur and must be attended to in order to achieve satisfactory performance.

Wrappings

Enclosures of the close-fitting type often take the form of wrappings where the barrier material is literally wrapped around the noise source. Figure 5-18 shows such treatment for a noisy pipe [see Ref. 13]. The leaded vinyl provides the barrier; and it is supported by and decoupled from the pipe by a layer of glass fiber. Note the use of tape to hold the assembly together and seal the joints. In general, one can expect a reduction in noise of 20 to 30 dB above 500 Hz for a single wrapping of 1 lb/ft^2 lead or lead-vinyl, or the equivalent; for additional treatment a second blanket/septum layer can be added. Any material which is limp, massive, and impervious will make a good septum, for example, linoleum, asphalt paper, and thin sheet-steel. The blanket should be firm enough to support the septum, but not too firm as to prevent its serving as a vibration bridge. Besides glass fiber, open-cell foams are good for this purpose. Properly done, septum treatments of this type are relatively easy to apply and are very effective. Figure 5-19 shows another application where the cylinder of an engine in a recreational vehicle was wrapped to determine whether the noise radiating from the cooling fins were significant. In this case the material was a commercially available composite consisting of lead sheet 1 lb/ft^2 sandwiched between layers of $1/4$-inch open-cell polyurethane foam. Such material is very easily formed as shown in the illustration, where the "snout," projecting downward, was formed

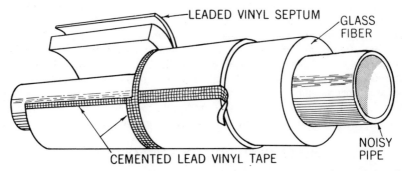

Courtesy of Technical Services, Lead Industries Association, Inc.

Fig. 5-18. Exploded view showing the construction of a glass fiber with a leaded vinyl septum wrapping for noise reduction of piping systems.

ACOUSTIC MATERIALS

Fig. 5-19. Application of a lead-foam wrapping system for noise control.

from the same material to serve as a lined duct or exit silencer for the forced air coolant. Although this was an experimental application, such treatments will often lead directly to a practical, production fix of a noise problem.

Free-Standing Enclosures

More conventional, free-standing enclosures can be constructed from a variety of commonly available materials, for example, as shown in Table 5-13. All stud walls given in the table could use the staggered-stud construction technique to boost the STC in excess of 50, if required. Often, it is advantageous to use commercially available enclosures based on module construction, such as that shown in Fig. 5-20. Such structures take advantage of standard-size barrier panels which are locked together with a variety of sealed joints to form a structure rigid enough to be transported, should the need arise. Partial enclosures, production-equipment enclosures, engine test cells, production test rooms, and in-plant supervisory offices, etc., are all possible, using modular construction.

Lined Ducts

Total enclosures often require openings for ventilation ducts, conveyors, etc., which if left untreated would destroy the benefits of the enclosure. Treatment generally consists of using some type of lined duct or silencer to "trap" the noise yet still permit an opening, as seen previously in Fig. 5-5. For ventilation ducts, there exists a large variety of commercially available silencers that are engineered

Table 5-13. Transmission-Loss of Sound Barrier Walls*

DESCRIPTION OF CONSTRUCTION			Nom Wt. (psf)	TRANSMISSION-LOSS (TL) AT STATED FREQUENCY (dB)								STC†
One Face	Frame or Core	Other Face		125	170	250	350	500	700	1000	2000	4000	
WOOD STUD WALLS													
½" Gypsum Board	2 x 4 Wood Studs	½" Gypsum Boards	5½	17	22	27	32	35	37	41	43	39	40
½" Gypsum Board plus 3 lb Sheet Lead	2 x 4 Wood Studs	½" Gypsum Board plus 3 lb Sheet Lead	11½	28	35	36	45	43	47	51	56	61	47
STEEL STUD WALLS													
⅜" Gypsum Lath plus ½" Plaster	1⅝" Steel Studs	⅜" Gypsum Lath plus ½" Plaster	13¼	27	32	37	41	43	44	46	39	47	41
⅜" Gypsum Lath plus ½" Plaster plus 3 lb Sheet Lead	1⅝" Steel Studs	⅜" Gypsum Lath plus ½" Plaster	16¼	35	38	43	46	45	47	47	49	58	43
⅜" Gypsum Lath plus ½" Plaster plus 3 lb Sheet Lead	1⅝" Steel Studs	⅜" Gypsum Lath plus ½" Plaster plus 3 lb Sheet Lead	19¼	36	40	45	47	47	49	53	61	47	
⅜" Gypsum Lath plus ½" Plaster plus 4 lb Sheet Lead	1⅝" Steel Studs	⅜" Gypsum Lath plus ½" Plaster plus 4 lb Sheet Lead	21¼	39	40	44	48	48	48	50	55	60	50
CINDERBLOCK													
Paint	4" Lightweight Cement Block	Paint	–	40	38	40	38	40	45	48	55	56	44
None	4" Lightweight Cement Block	4 lb Sheet Lead on ¼" Plywood on 1 x 2 Furring	–	41	43	46	44	47	54	56	63	67	49
PERIMETER FRAMED PARTITIONS AND MACHINERY ENCLOSURES													
⅜" Plywood	Wood Perimeter Frame	None	1¼	18	18	16	21	21	24	25	27	22	24
⅜" Plywood plus 3 lb Sheet Lead	Wood Perimeter Frame	None	4¼	33	33	34	35	37	38	41	44	45	40
18 Gage Steel	Steel Angle and Channel Frame	None	2	26	25	27	28	30	32	35	38	41	33
18 Gage Steel plus 3 lb Sheet Lead	Steel Angle and Channel Frame	None	5	33	32	34	35	38	39	41	44	46	40
DOORS													
Existing	Hollow Core Wood (With Gasketing)	Existing	2	14	12	12	15	12	19	24	27	28	16
2 lb Sheet Lead plus ⅛" Plywood	Hollow Core Wood (With Gasketing)	Existing	4.3	23	20	21	27	23	28	29	31	32	26
2 lb Sheet Lead plus ⅛" Plywood	Hollow Core Wood (With Gasketing)	2 lb Sheet Lead plus ⅛" Plywood	6.6	21	23	26	30	28	29	32	33	35	31
Existing	Solid Wood Core Door	Existing	–	23	31	23	26	28	21	29	27	35	23
LEADED VINYL SHEET													
Free-Hanging Leaded Vinyl Sheet			3.0	26	26	28	28	30	32	35	41	45	34
Free-Hanging Leaded Vinyl Sheet			1.5	22	22	23	23	25	29	30	35	42	28
Free-Hanging Leaded Vinyl Sheet			.75	16	16	17	16	18	21	24	30	35	22
Free-Hanging Leaded Vinyl Sheet			.50	13	13	14	13	15	18	21	26	32	18

*Excerpted from "Practical Application of Sheet Lead Barriers," courtesy of the Lead Industries Association.
†For STC rating of additional systems see Table 5-5.

to adapt to almost every conceivable situation. Sometimes these stock silencers can be applied to conveyor lines or other situations where a straight-duct section is needed. Often this is not the case, however, and one must resort to designing and fabricating one's own. Figure 5-21 shows a lined duct used in a test cell

Ch. 5　　ACOUSTIC MATERIALS　　193

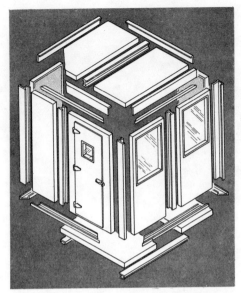

Courtesy of Industrial Acoustics Company, Inc.

Fig. 5-20. The use of module panels in construction of total enclosures.

where a dynamometer powered a disk brake setup via a gear-belt drive. The noise from the dynamometer interfered with the brake tests and consequently it was placed behind a barrier and the drive belt and all necessary control cables were routed through the duct. The duct, about 4 feet long, was made from $3/4$ inch plywood lined on two sides with 2-inch, open-cell foam.

Fig. 5-21. Lined duct silencer for noise reduction in a test cell.

Fig. 5-22. Lined duct silencer used to quiet a noisy blower in a slide projector.

Figure 5-22 shows another lined-duct type silencer fitted to a slide projector to reduce the blower noise. The duct was attached to the rear of the projector over the air outlet opening and was designed to permit its easy removal for storage of the projector. Aside from the aesthetics, the duct was curved to take advantage of the additional attenuation offered by a bend in a lined duct. In general, in attacking noise problems of this sort, care must be taken not to unduly restrict the air flow. In this case no problems were encountered, and the reduction in noise, diffuse field, was 5 dBA.

The attenuation in dB/ft for a rectangular duct is given by reference [14]

$$\text{Duct Attenuation} = 12.6 \frac{P}{S} \alpha^{1.4} \text{ dB/ft} \quad (5\text{-}23)$$

where:

P = Acoustically lined perimeter of duct, in.
S = Cross-sectional open area of duct, in.^{2}
α = Reverberant room determined absorption coefficient for the duct liner material.

The above equation is applicable within 10 percent accuracy for ducts having cross-sectional dimensions in the ratio 1:1 to 2:1, absorption coefficient between .2 and .4, and frequencies between 250 and 2000 Hz [15]. For a more detailed account of silencer design and selection see References [15]-[18].

Sample Problem No. 3

There are certain types of equipment which, because of the nature of the noise they radiate, require special enclosures. Examples of such equipment are rotating electrical machinery, transformers, and vibrating mixers. Each of these radiates predominantly pure tones at very low frequencies. Figure 5-23 shows a vibratory mixer which oscillates at 60 Hz, at amplitudes as high as $1/4$ inch [19]. The

Ch. 5   ACOUSTIC MATERIALS   195

Fig. 5-23. Vibratory mixer in an enclosure with front and top panels removed.

hopper of the mixer is, in effect, a crude, low-frequency loudspeaker. At these low frequencies, acoustic modes in a large room can be easily excited, causing amplification of the sound waves resulting in intolerable noise levels, and even rattling of windows and light fixtures.

A decision was made to enclose such a machine even though requirements were quite stringent: the enclosure had to be close-fitting and light enough for two men to lift, and had to have access panels on top and front for loading and unloading the hopper. It was estimated that a 10-dB reduction in noise at 60 Hz would suffice. This would require a 10-dB insertion loss (IL) from the enclosure. A consideration of the available space made it necessary to limit the average spacing between the machine and enclosure panels to 3 inches; thus the enclosure would measure approximately 24 × 40 × 32 inches.

Step 1: From previous discussion (see Fig. 5-12) it is known that the enclosure panels will have to be very stiff; that is, the lowest resonance (kettledrum) frequency of each panel will have to be much greater than 60 Hz. Calculate first, the required flexural rigidity of the largest panel. In the stiffness region (below resonance), Equation (5-18) can be simplified to:

$$\text{IL} = 10 \log_{10}\left(1 + \frac{2ls}{\rho c^2}\right) = 10 \log_{10}\left(1 + \frac{8\pi^2 Ml}{\rho c^2} f_{11}^2\right) \quad (5\text{-}24)$$

Combine this with Equation (5-16) to get the flexural rigidity of the panel per unit width:

$$B = \frac{Et^3}{12} = \frac{\rho c^2}{2\pi^4 l}\left(\frac{a^2 b^2}{a^2 + b^2}\right)^2 \left(\log_{10}^{-1}\left[\frac{IL}{10}\right] - 1\right) \qquad (5\text{-}25)$$

$$B = \frac{(.0024)(1130)^2}{(2)(144)(3)\pi^4}\left(\frac{40^2 \cdot 32^2}{40^2 + 32^2}\right)^2 \left(\log_{10}^{-1}\left[\frac{10}{10}\right] - 1\right) = 1.28 \times 10^5 \text{ lb-in.}$$

Thus, the required flexural rigidity of the largest panel is 1.28×10^5 lb-in.

Step 2: Calculate the required thickness for a homogeneous panel. For a 32 × 40 inch panel the required thickness and weight of a single homogeneous panel is calculated using Equation (5-25), $B = Et^3/12$, for some common materials. These are listed in Table 5-14. Except for plywood, it can be seen from Table 5-14 that because of their excessive weight and cost, homogeneous panels are not practical for this application. For an insertion loss in excess of 10 dB, a plywood panel would also be unacceptable because it would be too thick.

Table 5-14. Thickness and Weight Data for Homogeneous Panels

Material	E, lb/in.^{2}	Panel thickness, in.	Panel weight, lb/ft^2	Enclosure total weight, lb
Steel	30 × 10^6	0.37	14.9	521
Aluminum	10 × 10^6	0.54	7.6	264
Plywood	1 × 10^6	1.15	3.5	122

Step 3: Thus, since a homogeneous panel is not acceptable, a composite panel must be used. Therefore, we must determine the required thickness for a composite panel. Assume the composite panel consists of stressed face sheets stabilized by a lightweight core such as honeycomb or foam. The selected panel weight is 1.5 lb/ft^2 for a total enclosure weight of 52.5 lb. The flexural rigidity for a composite is given approximately by:

$$B_{\text{composite}} = \left(\frac{t^3}{6} + \frac{t(D-t)^2}{2}\right)E_{\text{face}} = 1.28 \times 10^5 \text{ lb-in.} \qquad (5\text{-}26)$$

The required face thickness (t) is calculated using surface density of .75 lb/ft^2 per sheet and the panel thickness (D) is then determined from Equation (5-24). The results are tabulated in Table 5-15.

Any one of these panels would be satisfactory. We are now in a position to trade off between sheet thickness, panel thickness, material, and cost as governed by available commercial sizes. Based on these considerations, a honey-

Ch. 5 ACOUSTIC MATERIALS

Table 5-15. Thickness and Weight Data for Composite Panels

Face Material	Face thickness, in.	Panel thickness, in.	Panel weight,* lb/ft^2	Enclosure weight,* lb
Steel	.018	.693	1.5	52.5
Aluminum	.053	.728	1.5	52.5
Plywood	.250	1.250	1.5	52.5

*Core weight neglected.

comb plywood panel having a thickness of 1.250 in. and faced with .250 in. plywood was selected, as shown in Fig. 5-23.

ESTIMATION OF NOISE REDUCTION WITH PARTIAL ENCLOSURES

In many instances, it is not possible to completely enclose a machine or system because of accessibility requirements and the need to transfer materials in and out of the machine. Noise reduction can be obtained with partial enclosures, although the reduction in sound level will not be as great as with a complete airtight enclosure. Equations are available by which an estimate of the noise reduction from a partial enclosure can be obtained. In those instances where noise reductions not in excess of 20 decibels are required, a partial enclosure may be acceptable.

Sound Level Inside an Enclosure

The sound level inside an enclosure can be estimated in the same manner as in estimating the sound level in a room. Calculations can be made using the method described in Chapter 4, "Criteria for Machinery Noise," and illustrated on page 134. One must realize that if a noise source is placed within an enclosure with little or no sound-absorbing material on the inside, the sound level will be greater within the space defined by the enclosure than the sound level in the same space without the enclosure. The increase in sound level inside the space defined by the enclosure is the result of reflected sound waves from surfaces inside the enclosure. If it is assumed that a homogeneous sound field exists inside a complete enclosure without openings, the sound pressure inside the enclosure can be computed from Equation (5-27), [10]:

$$p_j^2 = \frac{4\rho c W_j}{\sum_i \alpha_{i,j} S_i} \qquad (5\text{-}27)$$

where:

p_j^2 = Average squared sound-pressure inside the enclosure in some frequency band, j.
W_j = Sound power, in watts, for the noise source in frequency band, j.
ρ = Density of air

c = Speed of sound in air
$\alpha_{i,j}$ = Acoustic absorption coefficient for i^{th} surface in j^{th} frequency band
S_i = Surface area of the ith surface inside the enclosure to which material with absorption coefficient is applied.

Often, the inner surfaces of enclosures are treated with the same material and $\alpha_{i,j}$ becomes the acoustic absorption of this material for the jth frequency band.

For an enclosure with openings, an approximation to the equivalent absorption of the enclosure surfaces can be made by assuming that the sound absorption coefficient of the opening is 1.0. The total absorption of an enclosure with openings would be computed for one frequency band from:

$$\sum_i \alpha_i S = \alpha_1 S_1 + \alpha_2 S_2 + \cdots \cdot 1.0 S_o \qquad (5\text{-}28)$$

where: $\alpha_1 S_1$ is the absorption, in sabins, for surface 1, $\alpha_2 S_2$ is the absorption for surface 2, etc., and S_o is the total open area of the enclosure.

The amount of acoustic power escaping the enclosure through an opening is the *acoustic power incident* on the open area. Since the sound field was assumed to be homogeneous inside the enclosure, the ratio of acoustic power incident on the opening to the sound power generated by the enclosed noise source becomes equal to the absorption ratio which reduces to the relation in Equation (5-29):

$$\frac{W_{\text{incident}}}{W_{\text{source}}} = \frac{S_{\text{open}}}{S_{\text{open}} + \sum_i \alpha_i S_i} \qquad (5\text{-}29)$$

or

$$\frac{W_I}{W_S} = \frac{S_o}{S_o + \sum_i \alpha_i S_i} \qquad (5\text{-}30)$$

where:

S_o = The open area of the enclosure
$\sum_i \alpha_i S_i$ = The absorption, in sabins, for the treated inside surface areas
W_I = Sound power incident on openings
W_S = Sound power of the source.

The sound power incident on the open areas must be corrected by means of a radiation factor to account for the directivity and diffraction effects of the opening location compared to the field point of interest. Each opening has a radiation coefficient, η, depending on the location of the opening as illustrated in Fig. 5-24, and given in Table 5-16. With the observer in front of the partial enclosure, the location of the opening will influence the effective sound power radiated to the observation location such that the power radiated is equal to the power incident on the opening, times the radiation coefficient:

$$W_R = \eta W_I \qquad (5\text{-}31)$$

Ch. 5 ACOUSTIC MATERIALS 199

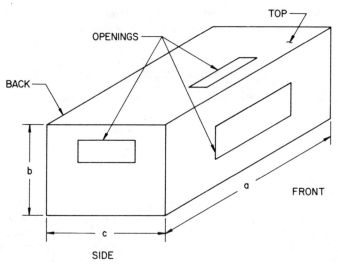

Fig. 5-24. Geometry of a partial enclosure having openings for access. Operator position is assumed to be in front of the enclosure.

where: W_R is the effective radiated sound power reaching an observer in front of the enclosure, W_I is the sound power incident on the inside of an opening in the partial enclosure, and η is the radiation coefficient found in Table 5-16. Using the values of η, the radiated sound power from an opening can be related to the sound power of the enclosed source by:

$$\frac{W_R}{W_S} = \frac{\eta W_I}{W_S} = \frac{\eta S_o}{S_o + \sum_{i=1}^{n} \alpha_i S_i} \quad (5\text{-}32)$$

Table 5-16. Acoustic Radiation Coefficient (η) as a Function of Opening Location. (Assumes observation point in front of enclosure)

Location of Open Area	Acoustic Radiation Coefficient, η
Front	1.00
Side*	0.30
Top*	0.30
Back*	0.15

*If the side, top, or back of the enclosure is near a reflective surface, the corresponding value of η in the table should be doubled.

If more than one opening exists for the partial enclosure, the resulting effective sound power at the front of the enclosure is obtained by summation:

$$\frac{\text{Total effective power}}{\text{Total source power}} = \frac{\sum\limits_{\text{openings}}^{m} W_R}{W_S} = \frac{\sum\limits_{k=1}^{m} \eta_k W_I}{W_S}$$

$$= \frac{\sum\limits_{k=1}^{m} \eta_k S_{ok}}{S_{oT} + \sum\limits_{i=1}^{n} \alpha_i S_i} \qquad (5\text{-}33)$$

In the final form, η_k accounts for the radiation coefficient for the various openings in the enclosure, S_{ok} is the area of the various openings, S_{oT} is the total open area of the enclosures, α_i is the acoustic absorption coefficient for the ith material inside the enclosure, and S_i is the surface area of the ith material inside the enclosure.

In most machinery applications, one material will be utilized on the inside of the enclosure and $\Sigma_i \alpha_i S_i$ will reduce to αS.

The noise reduction, NR, for a source enclosed with a partial enclosure can be estimated from:

$$\text{NR} = -10 \log_{10}\left(\frac{W_R}{W_S}\right) \text{ decibels} \qquad (5\text{-}34)$$

The above form assumes the noise source originally was located in a free space not close to reflecting walls. If the source initially is located close to a wall or large reflecting surface, the noise reduction is computed from the difference between the initial condition and the treated condition.

$$\text{NR} = 10 \log_{10}\left(\frac{W_R}{W_S}\right)_{\text{initial}} - 10 \log_{10}\left(\frac{W_R}{W_S}\right)_{\text{treated}} \qquad (5\text{-}35)$$

Example No. 4—Partial Enclosure

Consider an enclosure for a production machine which must have openings on each side for material flow. The machine is located in a factory, far from walls or other reflecting surfaces (except the floor). The dimensions of the proposed partial enclosure are $a = 10$ feet, $b = 5$ feet, $c = 6$ feet (see Fig. 5-24). An opening 2 feet by 1 foot is required on each side of the enclosure. No openings are required on the front, top, or back surfaces. The operator is located in front of the machine. It is desired to determine the noise reduction, in octave bands, which can be achieved with a partial enclosure.

Step 1: Determine the absorption coefficients for the proposed enclosure. Assume the machine will be enclosed with reinforced sheet steel such that the acoustic absorption coefficients are given by row 2 of Table 5-17. These values are estimates taken as the minimum values from Table 5-1.

Ch. 5   ACOUSTIC MATERIALS   201

Table 5-17. Calculations For Partial Enclosure, From Example No. 4

Row No.	Procedure	125	250	500	1000	2000	4000	8000
					Octave-Band Center Frequencies, Hz			
1	Absorption Coefficients, Top; Front; Back; and Sides	.05	.04	.03	.02	.02	.01	.01
2	Absorption Coefficients, Floor	.01	.01	.01	.01	.02	.02	.02
3	Absorption, sabins—Front (50 ft^2) αS	2.5	2.0	1.5	1.0	1.0	.5	.5
4	Absorption, sabins—Side (28 ft^2) αS	1.4	1.1	.8	.6	.6	.3	.3
5	Absorption, sabins—Side (28 ft^2) αS	1.4	1.1	.8	.6	.6	.3	.3
6	Absorption, sabins—Top (60 ft^2) αS	3.0	2.4	1.8	1.2	1.2	.6	.6
7	Absorption, sabins—Back (50 ft^2) αS	2.5	2.0	1.5	1.0	1.0	.5	.5
8	$\sum \alpha_i S_i$	10.8	8.6	6.4	4.4	4.4	2.2	2.2
9	Total Absorption, sabins = 4 ft^2; S_o + $\sum \alpha_i S_i$	14.8	12.6	10.4	8.4	8.4	6.2	6.2
10	η, For Side Opening = 0.3 $\sum \eta S = 0.3(2) + 0.3(2) = 1.2$	1.20	1.20	1.20	1.20	1.20	1.20	1.20
11	$\dfrac{W_R}{W_S} = \dfrac{\eta S_o}{S_o + \sum \alpha_i S_i}$ (Row 10, divided by Row 9)	.081	.095	.115	.143	.143	.192	.192
12	Noise Reduction, dB = $-10 \log_{10} \left(\dfrac{W_R}{W_S}\right)$	10.9	10.2	9.4	8.5	8.5	7.1	7.1
13	Absorption Coefficients for Acoustic Foam Treatment on Inside of Top and Front Panels	.23	.54	.60	.98	.93	.94	.96
14	Absorption, sabins—Treated Top (60 ft^2) αS	13.8	32.4	36.0	58.8	55.8	59.4	57.6
15	Absorption, sabins—Treated Front (50 ft^2) αS	11.5	27.0	30.0	49.0	46.5	49.5	48.0
16	Total Absorption, sabins S_o + $\sum \alpha_i S_i$ Sum of Rows 4, 5, 7, 14, and 15, plus S_o, where: S_o = 4 ft^2	34.6	67.6	73.1	114.0	108.5	114.6	111.3
17	$\dfrac{W_R}{W_S} = \dfrac{\eta S_o}{S_o + \sum \alpha_i S_o}$ (Row 10, divided by Row 16) For Treated Enclosure	.0347	.0178	.0164	.0105	.0111	.0105	.0108
18	Noise Reduction For Treated Enclosure, dB $-10 \log_{10} \left(\dfrac{W_R}{W_S}\right)$, Equation (5-34)	14.6	17.5	17.8	19.8	19.6	19.8	19.7

Step 2: Determine the absorption, in sabins, for each surface. This is obtained by multiplying the surface area of the respective surface, times the absorption coefficient of that surface. The resulting values are given in rows 3, 4, 5, 6, and 7 of Table 5-17. (The area and absorption of the openings is deleted from these calculations.) The absorption of the floor is not included in this case since it can be assumed that the machine will cover the largest portion of the floor area inside the enclosure. As can be seen from row 2, Table 5-17, the absorption coefficients for the floor are small; therefore, the total floor absorption is negligible. For a larger enclosure, however, the floor absorption must be included in the calculation. In this case, the calculated noise reduction for the enclosure will be conservative because the floor absorption is omitted.

Step 3: Sum up the absorption in each octave band. The sum of the values in rows 3, 4, 5, 6, and 7 are given in row 8.

Step 4: Determine the absorption of the total system, including the open areas. Since the total open area is 4 sq ft, and the absorption coefficient of the openings is assumed to be 1.0, 4 ft^2 is added to the values of row 8, as indicated in row 9 of Table 5-17.

Step 5: Find the product of radiation coefficient and open area. For this example, both openings are on the side of the proposed enclosure. From Table 5-16, the value of η is 0.3. Each open area is multiplied by 0.3 and the sum for both areas is obtained as given by row 10. The value is the same for all octave bands.

Step 6: Find the ratio of sound power radiated by the openings to the sound power of the source, W_R/W_S. This ratio is given in row 11, as the ratio of the values in row 10 to the values in row 9, of Table 5-17.

Step 7: Calculate the noise reduction, in decibels, for the operator located in front of the machine. This is done by using Equation (5-35), above. In this case the machine was initially located at a relatively great distance from walls or other reflecting surfaces. For this reason the first term of Equation (5-35) is equal to zero and the equation is reduced to the following form:

$$NR = -10 \log_{10} \left(\frac{W_R}{W_S}\right)_{treated}$$

Thus, the $\log_{10}$ of the values in row 11 must be multiplied by minus ten (-10), resulting in positive numbers for the values noise reduction given in row 12.

It is anticipated that the addition of acoustic absorption material inside the partial enclosure will further reduce the sound level in front of the enclosure, since the use of sheet steel has low absorption coefficients as presented in the above calculations. Assume that to the above enclosure a one-inch treatment of acoustical foam will be added to the inside top and inside front panels. Estimate

Ch. 5 ACOUSTIC MATERIALS 203

the noise reduction of this new configuration compared to no enclosure. The acoustic absorption coefficients for one-inch foam are given in row 13, of Table 5-17. These values must be obtained from the supplier of the acoustical material.

Step 8 : Calculate the absorption, in sabins, for the treated inside top and front panels. This is obtained by multiplying the absorption coefficients for acoustic foam by the surface areas of the top and front panels, respectively. The results are given in rows 14 and 15 of Table 5-17.

Step 9 : Determine the total absorption for the enclosure by adding the values in rows 4, 5, 7, 14, and 15 where $S_o = 4$ ft^2. The sums are given in row 16.

Step 10: Compute the ratio W_R/W_S from the values in rows 10 and 16. These ratios are given in row 17.

Step 11: Calculate the noise reduction for the enclosure with acoustical material relative to no enclosure. Again, the first term in Equation (5-35) is equal to zero, reducing this equation to the form given in Step 7. The $\log_{10}$ of the values in row 17 are multiplied by minus ten (-10) to obtain the noise reduction in row 18.

The values given in row 18 of Table 5-17 are estimates of the noise reduction, in octave bands, for the partial enclosure with absorption material on the inside top and front panels compared to no enclosure. In designing partial enclosures, the transmission loss of the enclosure material must be selected to be greater than the computed noise reduction by at least ten decibels in every octave band in order to achieve the desired result. Transmission of sound through the walls of the partial enclosure will reduce the effectiveness of the noise-control system.

To achieve greater noise reduction with a partial enclosure, the openings can be fitted with acoustically lined ducts, as described previously in this chapter in the section entitled "Noise Control by Enclosures—Lined Ducts." For an enclosure treated with a lined duct, the duct attenuation is obtained from Equation (5-23), duct attenuation = $12.6 \frac{P}{S} \alpha^{1.4}$ dB/ft, times the length of the lined duct which will give the attenuation in decibels.

For this case a duct transmission coefficient is found from:

Duct Transmission Coefficient

$$\beta = \log_{10}^{-1} \left[\frac{12.6 \frac{P}{S} \alpha^{1.4} L}{10} \right] \qquad (5\text{-}36)$$

where:

$12.6 \frac{P}{S} \alpha^{1.4}$ = the duct attenuation, dB/ft

L = length of the lined duct.

From this duct transmission coefficient the effective power radiated from the openings, as given by Equation (5-33), becomes:

$$\frac{\sum\limits_{openings} W_R}{W_S} = \frac{\sum\limits_{k=1}^{m} \beta_k \eta_k W_I}{W_S} = \frac{\sum\limits_{k=1}^{m} \beta_k \eta_k S_{ok}}{S_{oT} + \alpha_i S_i} \qquad (5\text{-}37)$$

and the noise reduction is the same form as given by Equation (5-34). The effect of the lined duct is to reduce the sound power incident on the opening by a factor, β, before it reaches the radiating end of the duct.

Example No. 5—Partial Enclosure with Lined Ducts

Assume that the attenuation for the partial enclosure of Example No. 4, given by row 18, Table 5-17, is not sufficient to meet the sound-pressure-level requirements at the worker position. Additional sound-absorption material inside the enclosure will reduce the sound level at the worker position and thus provide a greater noise reduction than given by row 18, of Table 5-17. Since there is a limit to the amount of sound-absorbing material which can be applied to the interior of the enclosure and since the openings are the source of worker-position sound-pressure level, the addition of short lined ducts at each opening will be considered. This procedure has the advantage that the sound-absorbing material is applied at the most effective location and thus a greater noise reduction is anticipated with the equivalent amount of material.

Determine the noise reduction of the partial enclosure of Example No. 4, if a lined duct two-feet-long is added to both 1 ft × 2 ft openings, one at each end of the enclosure. The duct is to be lined with one-inch-thick acoustic foam identical to the treatment used on the inside of the enclosure of Example No. 4. This solution will utilize a number of the calculations from Example No. 4.

Step 1: Determine the attenuation of the lined duct. This calculation is illustrated in rows 1, 2, and 3 of Table 5-18. In row 1 the acoustic absorption coefficients for the one-inch-foam are repeated from row 13, Table 5-17. In row 2 the absorption coefficients are raised to the 1.4 power and in row 3 the attenuation is computed for the lined duct. The perimeter, P, of the 1 ft × 2 ft duct is 72 inches. The cross-sectional area, S, of the duct is 288 sq in., and the length, L, is 2 feet. The final values for the attenuation, in decibels, of the duct is given in row 3.

Step 2: Calculate the duct transmission coefficient from Equation (5-36). The calculations are shown in rows 4 and 5.

Step 3: Determine the sound power radiated from the open duct to the front of the enclosure. The quantity $\beta \eta S$ is computed in row 7, where β is the duct transmission coefficient; η the radiation coefficient from Table 5-16, which is equal to .30 for both openings of this example; and S is the total open area for the two side openings.

Step 4: Compute the ratio of sound power radiated to the front of the enclosure, to the sound power of the unenclosed source, W_R/W_S, in row 9.

Table 5-18. Calculations for a Partial Enclosure with Lined Duct Openings. From Example No. 5

Row No.	Procedure	Octave Band Center Frequencies, Hz							
		125	250	500	1000	2000	4000	8000
1	Absorption Coefficients, α, for Acoustic Foam Treatment for Duct Liner $\alpha^{1.4}$	.23 .13	.54 .42	.60 .49	.98 .97	.93 .90	.94 .92	.96 .94
2	Duct Attenuation, dB $12.6\dfrac{P}{S}\alpha^{1.4}L$ $P = 72$ in., $S = 288$ in.^{2}, $L = 2$ ft [Equation (5-23)]	.41	1.32	1.54	3.06	2.84	2.90	2.96
3	Duct Attenuation/10 $\dfrac{12.6\dfrac{P}{S}\alpha^{1.4}L}{10}$	.041	.132	.154	.306	.284	.290	.296
4	Duct Transmission Coefficient, β $$\beta = \dfrac{1}{\log_{10}^{-1}\left[\dfrac{12.6\dfrac{P}{S}\alpha^{1.4}L}{10}\right]}$$ [Equation (5-36)]	.91	.74	.70	.49	.52	.51	.51
5	$\sum \eta S = 0.3(2.0) + 0.3(2.0) = 1.20$ From row 10, Table 5-17 $\sum \beta \eta S = \beta \eta S_1 + \beta \eta S_2 = \beta \eta(4.0)$	1.20 1.092	1.20 .888	1.20 .840	1.20 .588	1.20 .624	1.20 .612	1.20 .612
6 7 8	Total Absorption of Enclosure, sabins $S_O + \sum \alpha_i S_i$ From row 16, Table 5-17	34.6	67.6	73.1	114.0	108.5	114.6	111.3
9	$\dfrac{W_R}{W_S} = \dfrac{\beta \eta S}{S_O + \sum \alpha_i S_i}$ row 7, divided by row 8	.0316	.0131	.0115	.0052	.0058	.0053	.0055
10	Noise Reduction of Enclosure with Lined Duct $NR = -10\log_{10}\left(\dfrac{W_R}{W_S}\right)$ dB [Equation (5-34)]	15.0	18.8	19.4	22.9	22.4	22.7	22.6

Step 5: Calculate the noise reduction of the partial enclosure with lined ducts using Equation (5-35). Since in this case the machine was initially located far from walls and other reflecting surfaces, the first term is equal to zero and the equation is modified as given below and explained in Step 7, Example No. 4.

$$NR = -10 \log_{10} \frac{W_R}{W_{S \text{ treated}}}$$

To find the values in row 10, take $\log_{10}$ of the values in row 9 and multiply the answer by minus ten (-10).

A comparison of row 18, Table 5-17, and row 10, Table 5-18, indicates that the addition of the 2-foot-long lined duct to each side opening resulted in approximately 2 decibels increase in the noise reduction in front of the enclosure except at the 125 Hz octave band which had an increase of less than 1 decibel. Increasing the length of the lined ducts will increase the noise reduction at the front of the enclosure. For this example, the increase in noise reduction is equal to the attenuation of the lined ducts. For an enclosure where the openings are on several sides, the resulting increase in noise reduction will be some combination of each duct attenuation factor and each radiation factor.

6

Damping Materials for Vibration and Sound Control

DAVID I. G. JONES

Introduction

Structures and machines are normally designed to meet innumerable criteria which are defined by the use for which the structure or machine is intended, and which include: (a) geometrical and tolerance factors, (b) external static loading and strength considerations, (c) external dynamic loads and loads resulting from movement within and/or of the structure or machine, and (d) dynamic response characteristics. The first two of these factors (a) have little to do, directly, with noise and vibration phenomena although their influence on the second pair of factors (b) can be very marked. The third set of factors (c), can produce serious noise and vibration problems, as, for example, the noise emanating from meshing gear teeth in an engine transmission. However, it is very often the dynamic response characteristics (d), of the machine or structure which amplify and aggravate these sources of noise and/or vibration and thereby cause even more serious problems.

Care must be taken in the design to avoid phenomena which include:

(a) Sonic fatigue, where the resonant response characteristics of aircraft skin structure coupled with intense noise from jet engines leads to high stresses and early failure

(b) Resonances of jet engine fan or turbine blades which can couple with multiples of the shaft rotation speed to generate undesirable vibration induced stresses, unless these resonances are carefully avoided

(c) Rotating shafts, where critical speeds can occur at or near design rotation speeds and lead to shaft whirling problems

(d) Gear systems, which tend to resonate at certain speeds unless careful geometrical design or the use of more appropriate gear materials correct these conditions

(e) Thin panels, plates, and shells in machines and structures which often vibrate excessively and act as serious secondary sources of noise

(f) Pipes and tubes which can vibrate excessively in response to excitation to cause sound or premature failure

(g) Modern steel-frame buildings which can vibrate excessively under wind and earthquake loading.

It must be clearly understood that noise control problems can be exceedingly difficult to resolve because, although resonance amplified vibration may be the root cause, possibly in an inaccessible location, many other parts of the structure will serve to amplify, transmit and/or re-radiate the output of the original source. Vibration control problems are usually somewhat simpler to resolve because the vibration problem is usually identifiable as directly associated with some abnormal dynamic behavior of the part affected, and not with some far distant point which may be the source in the case of some other noise problems. It follows, therefore, that if a noise problem is directly and intimately linked with local abnormal dynamic behavior of a part of a structure, some noise reduction may be achieved by controlling the vibration; if this is not the case then a solution will be far more difficult to achieve. The available technology for control of vibration problems in structures and machines will now be discussed in greater depth, bearing in mind that this technology must be utilized with knowledge and care, and recognizing that it also can be used to solve many, but certainly not all problems of noise control. The one phenomenon which dominates the dynamic behavior of structures and machines is resonance. To understand what this means the simple mass-spring-damper system, illustrated in Fig. 6-1, should be considered.

Single Degree of Freedom System

This system consists of a mass M connected through a spring of modulus K to a rigid surface. A force is applied to the mass at a finite frequency of oscillation.

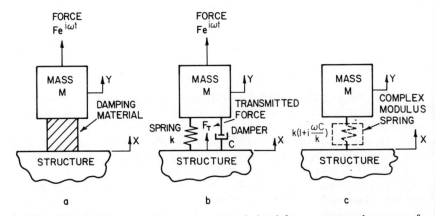

Fig. 6-1. Mass-spring-damper system. a. A mass isolated from a structure by means of a damping material which has spring-like properties; b. An equivalent spring-mass-damper system; c. An equivalent system where the spring and damper have been replaced by an equivalent spring with a complex modulus.

At very low frequencies, the mass simply moves steadily as the force varies with time. As the frequency increases, the inertia force $-M\ddot{X}$ gradually increases until it is eventually exactly equal and opposite to the force KX of the spring, and the amplitude of the vibration $|X|$ can then build up indefinitely to an infinitely large value, at least in theory. The particular frequency at which this occurs is known as the *resonance frequency*. At higher frequencies, the inertia force $-M\ddot{X}$ becomes extremely large and the mass M tends to "stand-still." So much for a very simple structure. For a more complex structure or machine, similar phenomena occur, but these will have more than one resonance.

Complex Structures

Each resonance frequency of the system is associated with a very high amplitude (infinite in theory but not in practice) of vibration at all points of the structure; the pattern of vibration amplitudes over the surface, nondimensionalized as necessary, is known as the corresponding *normal mode of vibration*. Vibration problems occur when one, or more, of these resonant frequencies occur at frequencies where significant excitation exists, so that high amplitudes arise and cause failure or malfunction.

Vibration Control Technology

Vibration control is the technology available for:

(a) Moving the resonance peaks to less sensitive values
(b) Reducing the magnitude of the exciting forces at the source
(c) Changing the mode shapes in a favorable manner, and/or
(d) Reducing the height of the resonance peaks.

Technique (a) is widely used to design shafts, turbines, and other rotating machines so that resonant peaks do not fall directly over frequencies of significant excitation. The procedure is not always perfect because it is difficult to make a large number of resonant frequencies consistently avoid a large number of excitation frequencies under all speed conditions; moreover, this is not possible at all if the excitation is of significant magnitude over a broad frequency range and not merely at a number of discrete frequencies. Technique (b) is clearly desirable at all times, but definite limits exist as to the amount of reduction in the excitation that can be achieved. Technique (c) is difficult to apply since it demands a remarkably precise knowledge of the dynamic response of the structure, but it is certainly a possible approach in some instances, usually in conjunction with technique (a). Technique (d) will consistently reduce the severity of a vibration problem, but demands a totally different technology from the first three techniques, which essentially involve either reducing the magnitude of the excitation or making geometrical changes in the structure. In order to reduce the amplitude at each resonance to a low value, relative to that customarily measured, energy must be dissipated in the structure before it has the opportunity of generating high resonant amplitudes. This is what is meant by "damping." Useful damping, however, can be introduced into a structure only by using materials or techniques which dissipate much larger amounts of energy than resonating structures customarily do.

When a specific vibration problem is encountered by an engineer, therefore, his first task will be to identify what part of his structure or system is vibrating, and whether or not a simple design modification might not resolve the problem. If he decides that damping is, in fact, worth investigating further he must next determine the temperature of the vibrating component because any damping materials he chooses to use will be temperature sensitive; employing an inappropriate damping material will prove useless. With the temperature known, he must next examine the vibrating component in more detail to decide whether or not its mode of vibration is essentially "plate like" or "beam like." If it is plate like, he might consider a free layer treatment consisting of a homogeneous layer of an appropriate elastomeric material bonded to the surface by means of a suitable stiff adhesive. If the damping material is chosen correctly, it will have its peak damping capability near the operating temperature and will be applied initially in a thickness about equal to that of the plate. If the damping so introduced is significant but not sufficient, he can add more thickness, or he can try a more complicated treatment such as applying alternate layers of a viscoelastic adhesive and metal—constraining or stiffening layers—otherwise known as "damping tapes." The advantage aimed for by this change might be a reduced weight or net thickness. If, however, the original free layer treatment application is not successful, the engineer must ask himself whether this is because the plate motion is not really plate like or whether the vibrations are true resonances and not mere forced motions of the structure. If true resonances are involved, he might, at this point, recognize that the strain distributions on the surface are such that the layered treatments are not really effective, and instead, might then consider using a tuned device of some sort. One example, in its most ideal form, might be a shallow cylindrical specimen of a suitable damping material attached at one flat surface to an appropriate point on the structure, and at the other, to a mass. The link and mass would be so chosen that the resonant frequency of the damper is equal to that of the structure in the most important mode to be damped. This *ad-hoc* approach can be very effective in solving fairly simple problems, but it should be borne in mind that there still must be some rational basis for selecting the appropriate materials and designing the treatments—which is the subject of this chapter.

Some of the fundamental aspects of damping from the point of view of materials characterization, response of damped structures, and applications will be reviewed. Design-oriented data are discussed where available, including the damping behavior of several materials. It should be clearly recognized that the data on materials, while extremely useful for materials selection and dynamic behavior prediction, are only valid for the specific materials tested and changes in processing and composition could change the properties. Measurement of the damping properties of the specific material, or materials, to be used in each specific application, is recommended.

SYMBOLS

The following symbols are used in the equations in this chapter:

A resonance amplification factor or dimensionless constant as in text

Ch. 6 DAMPING MATERIALS 211

	a	acceleration
A', B', C'		constants
	B	dimensionless constant
	C	damping coefficient
	D	flexural rigidity or subscript denoting damping materials according to test
	E	Young's modulus of elastic material
	$\overline{E}$	complex Young's modulus of elasticity
	E_D	real part of Young's modulus
	E_c	constraining layer modulus
	E_e	effective Young's modulus
	exp	exponential function
	e	E_D/E – modulus ratio
	f	frequency (Hz)
	f_n	nth natural frequency
	f_{nm}	nmth natural frequency
	f_r	resonant frequency of damped system
	F	force
	G	shear modulus
	G_D	real part of shear modulus
	h	thickness of plate or beam
	h_D	thickness of damping material
	h_c	thickness of constraining layer
	i	$\sqrt{-1}$ (imaginary number operator)
	K	spring stiffness or radius of gyration
	$\overline{K}$	complex spring stiffness
K_S, K_T		shape factors
L, l		lengths
	M	mass
	M_s	mass of structure
	n	h_D/h – thickness ratio
	N	number of layers
	R	radius
	r	ratio of λ_n to half-wavelength of plate in normal direction
S, S'		areas
	t	time
T, T_o		absolute temperatures
W, W_o		displacements
	$\dot{X}$	velocity
	x	coordinate along beam
X, Y		displacement amplitude
$\ddot{X}, \ddot{Y}$		accelerations
	y	instantaneous displacement
	α_n	dimensionless constant
	α_T	temperature shift factor
β, β_{nm}		dimensionless constant
	α	non-dimensional parameter
	δ	y/l

Δ x/L
η_D loss factor in tension-compression
η_D' shear loss factor
η_e effective loss factor
η, η_s, η_s' loss factors—see text
$\lambda, \lambda_n, \lambda_{nm}$ modal half-wavelengths
ν, ν_D Poisson's ratio
ξ_{nm} eigenvalue
ρ, ρ_o, ρ_D densities
σ stress
ϕ_{nm} normal mode. Also shear parameter
ψ_e effective mass ratio
ψ_{nm} inverse shear parameter
ω frequency (rad/sec)
ω_D damped natural frequency
$\hat{\epsilon}$ maximum strain
ϵ, ϵ_o strains

DAMPING BEHAVIOR OF MATERIALS

Sources of Damping

Most materials used for structural purposes are not noteworthy for the amount of energy which they are capable of dissipating under cyclic strain and, indeed, such behavior would be regarded as unacceptable for most design purposes unless dynamic response criteria were of primary importance. However, it is true that all materials, when strained cyclically, do dissipate some energy no matter how little it may be and therefore all materials normally possess damping in some degree. Introducing the concept of damping in a useful way is in terms of the various types of stress-strain curves which can be observed in simple specimens of various types of material that are uniaxially strained, cyclically, to various amplitude levels. Once a steady state has been reached, after several cycles, the stress-strain curve of a material will not be a straight line, as for a perfectly elastic material, but will be instead, a closed loop known as a *hysteresis loop*. The shape of this loop will depend on the material itself, environmental factors such as temperature; frequency; and the strain amplitude. No matter what the shape may be, the area enclosed by the stress-strain loop will be a direct measure of the energy dissipated per unit volume per cycle, and hence, a measure of the damping. This damping behavior arises from a wide variety of internal sources including grain boundary effects, point defect relaxations, thermoelastic effects, eddy currents in metals and alloys, and interactions between chains of large molecules in elastomeric materials. Further details may be followed up in the literature, particularly in the work of Lazan [1].

Damping of Elastomers

As a broad rule, the damping behavior of most elastomeric, or rubberlike, materials is the most readily amenable to characterization and analysis as compared with, say, the metallic class of materials. In the case of most elastomeric

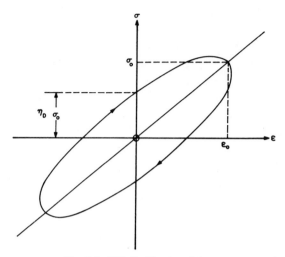

Fig. 6-2. Elliptical hysteresis loop.

material specimens, subjected to uniaxial cyclic strain, the steady state stress-strain curve is essentially an ellipse, as illustrated in Fig. 6-2.

Complex Modulus Representation

The degree of simplicity which evolves from this particular fact is best realized by writing down the equation of the ellipse, then inserting a sinusoidally varying strain $\epsilon = \epsilon_o \sin \omega t$ and calculating, from the equation of the ellipse, the stress necessary to produce the motion indicated thereby. The result [2, 3] is an exceedingly simple representation of the material behavior in which the real moduli E or G of a perfectly elastic material are replaced by complex moduli $E_D(1 + i\eta_D)$ or $G_D(1 + i\eta'_D)$ which are, of course, meaningful in a direct sense only if the strain and stress vary as $\exp(i\omega t)$ in the time domain. This complex arithmetic approach then allows one to directly apply any solutions which are obtainable for the perfectly elastic case by merely changing over to complex arithmetic. Those who are less familiar with complex numbers can work equally well with a real number representation. These two representations may be written as:

$$\sigma = E_D(1 + i\eta_D)\epsilon \tag{6-1}$$

or

$$\sigma = E_D\epsilon + \frac{E_D\eta_D}{\omega}\frac{d\epsilon}{dt} \tag{6-2}$$

This form of the stress-strain relation applies for a dynamic situation where the modulus of elasticity includes both the "springiness" and the energy dissipation of the material, as illustrated in Fig. 6-1. For a time-dependent loading, there will be a phase difference between the stress and the strain of the form,

$$\sigma = Ee^{i\phi}\epsilon \tag{6-3}$$

In complex number notation

$$e^{i\phi} = \cos\phi + i\sin\phi = \cos\phi \left(1 + i\frac{\sin\phi}{\cos\phi}\right)$$

$$= \cos\phi\,(1 + i\tan\phi) \tag{6-4}$$

therefore

$$\sigma = E\cos\phi\,(1 + i\tan\phi)\,\epsilon \tag{6-5}$$

by defining:

$$E\cos\phi = E_D$$

$$\tan\phi = \eta_D$$

therefore

$$\sigma = E_D(1 + i\eta_D) \tag{6-6}$$

If one makes the analogy of stress σ as force per unit area, and strain ϵ as displacement per unit length, the model for damping material becomes similar to a single degree of freedom vibration system as shown in Fig. 6-1b. For a time dependent applied force, $F\exp(i\omega t)$, the resulting displacement is not in phase with the applied force because the damper force is related to velocity, and spring force is related to displacement. The exponential notation for a sinusoidal force and a sinusoidal displacement are:

$$\text{Applied Force} = F\exp(i\omega t) \tag{6-7}$$

$$\text{Resulting Displacement} = Y\exp(i\omega t) \tag{6-8}$$

Using these equations, the differential equation for the system in Fig. 6-1b can be written as:

$$M\frac{d^2y}{dt^2} + C\frac{dy}{dt} + K = F\exp(i\omega t) \tag{6-9}$$

if the structure is assumed rigid.

Using the exponential form for $y = Y\exp(i\omega t)$ the above equation becomes:

$$(-\omega^2 M + i\omega C + K)\,Y = F \tag{6-10}$$

This result can be written in terms of a complex modulus spring if one defines

$$\overline{K} = K + iK_i = K\left(1 + i\frac{K_i}{K}\right) \tag{6-11}$$

where:

$$K_i = \omega C$$

then Equation (6-10) becomes:

$$(-\omega^2 M + \overline{K})\,Y = F \tag{6-12}$$

Ch. 6 DAMPING MATERIALS

If the mass is eliminated from the system and the force is applied directly to the isolation material, the system equation is

$$F = \overline{K}Y \qquad (6\text{-}13)$$

If this is written in terms of force per unit area = σ, and displacement per unit length, ϵ, the equation becomes:

$$\sigma = \overline{E}\epsilon \qquad (6\text{-}14)$$

where $\overline{E}$ represents the complex modulus of the material

$$\overline{E} = \frac{\overline{K}}{\text{unit area}} = \frac{K\,[1 + i(K_i/K)]}{\text{unit area}} = E_D \left(1 + i\,\frac{\omega C}{K}\right) \qquad (6\text{-}15)$$

and we define $\eta_D = \omega C/K$.

The complex modulus accounts for energy storage as strain energy and energy loss due to the loss factor η_D. The real part of the complex moduli are often called the *storage* moduli and the imaginary part, the *loss* moduli.

It must be realized that E_D and η_D are, in general, functions of frequency, temperature, strain amplitude, and pre-strain for any given material. Another commonly used representation, namely

$$\sigma + A'\,\frac{d\sigma}{dt} = B'\epsilon + C'\,\frac{d\epsilon}{dt} \qquad (6\text{-}16)$$

is equivalent to (6-1), $\epsilon = \epsilon_o \exp(i\omega t)$ if A', B', and C' are chosen appropriately to simulate as closely as possible the actual variation of E_D and η_D with frequency. By means of these simple stress-strain relationships, one can obtain some idea of the meaning of E_D and η'_D as applied to the vibrations of a simple system.

One Degree of Freedom Viscoelastic System

For a simple one degree of freedom system, as in Fig. 6-1b, with the structure motion $x = X \exp(i\omega t)$, the motion of the mass can be obtained from Newton's second law. In this instance the forcing force, $F \exp(i\omega t)$, will be assumed to be zero, resulting in a base exitation problem where the input motion is the motion of the structure (x).

$$\sum F = Ma \qquad (6\text{-}17)$$

Spring Force + Damper Force = Inertia Force

$$K(x - y) + C\left(\frac{dx}{dt} - \frac{dy}{dt}\right) = M\,\frac{d^2 y}{dt^2} \qquad (6\text{-}18)$$

Using the concept of the complex spring modulus, as provided in Equation (6-11), the equation of motion (6-18) may be written in the following manner:

$$K\left(1 + i\,\frac{\omega C}{K}\right)(x - y) = M\,\frac{d^2 y}{dt^2} \qquad (6\text{-}19)$$

Rearranging and using relations from Equation (6-15) enables us to transform Equation (6-19) as follows:

$$M \frac{d^2y}{dt^2} + \frac{S}{h} E_D(1 + i\eta_D) [y - X \exp(i\omega t)] = 0 \qquad (6\text{-}20)$$

If the solution $y(t)$ is expressed as $Y \exp(i\omega t)$, the ratio of the motion of the mass (Y) to the motion of the structure (X) becomes:

$$Y/X = \frac{(1 + i\eta_D)(E_D S/h)}{-M\omega^2 + (E_D S/h)(1 + i\eta_D)} \qquad (6\text{-}21)$$

The magnitude of the complex quantity Y/X gives the equation of the response curve of $|Y/X|$ versus ω or, $M\omega^2 h/E_D S$. At the peak amplitude point $M\omega^2 h/E_D S = 1$ and:

$$|Y/X|_{max} = \frac{\sqrt{1 + \eta_D^2}}{\eta_D} = A \qquad (6\text{-}22)$$

where A is the resonance amplification factor.

From Fig. 6-3 the response curve for $|Y/X|$ is shown as a function of frequency, ω. The "peak" or maximum response occurs at the damped natural frequency of the system denoted by ω_D. The maximum response, A, is defined as the resonance amplification factor. From Equation (6-22):

$$\eta_D = \frac{1}{\sqrt{A^2 - 1}} \qquad (6\text{-}23)$$

so that η_D is a direct measure of the resonant amplification transmitted through a simple specimen at resonance and is, hence, a direct measure of the damping in the material.

The resonant amplification factor A is seen to increase when the amplitude of the mass (Y) is large for a given amplitude of the structure (X). On the other

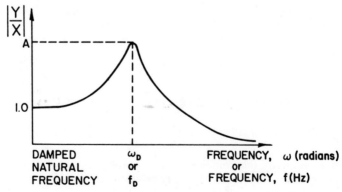

Fig. 6-3. Response curve of the displacement amplitude ratio, $|Y/X|$, as a function of frequency. X is the motion of the structure and Y is the motion of the mass to be isolated, as shown in Fig. 6-1.

hand, the loss factor, η_D, will be smaller when Y is large for a given value of X. Thus, a large value of η_D is indicative of desirable vibration dampening material or a good vibration dampening system.

Effects of Temperature, Frequency, and Strain

So much for characterization. The way in which the large number of available elastomeric materials differ from each other is in the variation of E_D and η_D with frequency, temperature, strain amplitude, and prestress; temperature being by far the most important factor, since E_D can vary by as much as four orders of magnitude over a narrow temperature range in some cases. The variation of Young's Modulus E_D and loss factor η_D with temperature for an elastomer, at fixed frequency and cyclic strain amplitude, are typically of the form shown in Fig. 6-4. Three distinct temperature regions are observed, namely the glassy region, the transition region, and the rubbery region. In the glassy region, E_D is high and η_D is low; in the transition region, E_D varies rapidly with change in temperature and η_D is high; in the rubbery region, E_D varies more slowly with change in temperature and η_D is lower than in the transition region, although not as low as in the glassy region. At the very highest temperatures, irreversible thermal decomposition of the material occurs.

If the effects of frequency are also to be taken into account, one of the most useful techniques for plotting the data is by means of the temperature-frequency equivalence concept [4]. In this approach $(T_o \rho_o / T\rho) E_D$ and η_D are plotted against the so-called reduced frequency parameter $f\alpha_T$, where α_T is a function of the absolute temperature T, and T_o is a reference temperature, again on the absolute scale. Usually ρ_o/ρ is regarded as 1.0 over a wide temperature range and is ignored. The preparation of "master curves" such as these is extremely useful

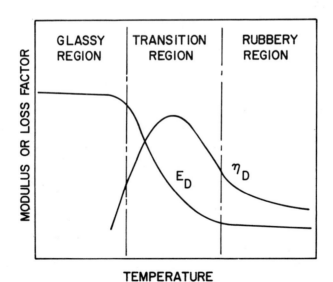

Fig. 6-4. Variation of E_D and η_D with temperature.

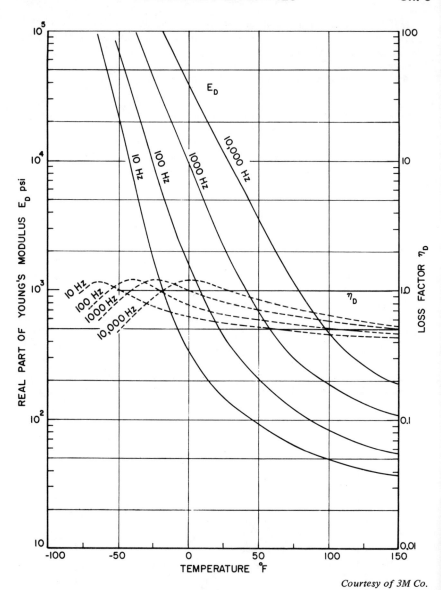

Courtesy of 3M Co.

Fig. 6-5. Complex modulus data for 3M-428 (AFML data).

for extrapolating test results obtained under a wide, but still limited, set of conditions. For example, in a test series one may have data over the frequency range 100 to 1000 Hz and a temperature range from 0 to 200 F and wish to estimate the properties at 100 F and 2 Hz. In order to achieve this, the available data to produce a best fitting set of master curves are first used and only then is

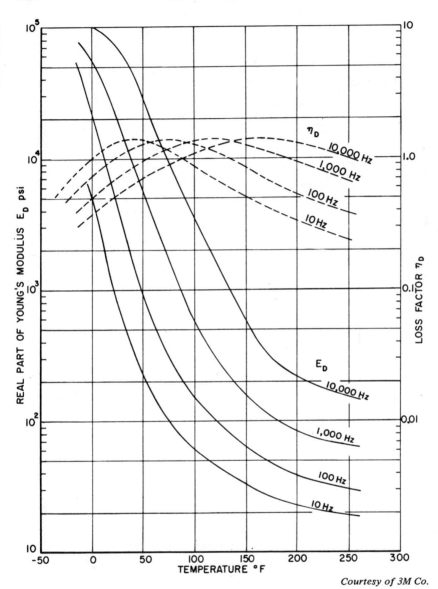

Fig. 6-6. Complex modulus data for 3M-467 (AFML data).

the estimate made. The process is most satisfactorarily accomplished empirically by judging the factor α_T on the basis of the shift needed to make the curve of log $(T_o E_D/T)$ versus log (frequency) at temperature T, match as closely as possible the curve of log (E_D) versus log (frequency) at temperature T_o, while at the same time matching the curves of η_D versus log (frequency) at temperatures

220 DAMPING MATERIALS Ch. 6

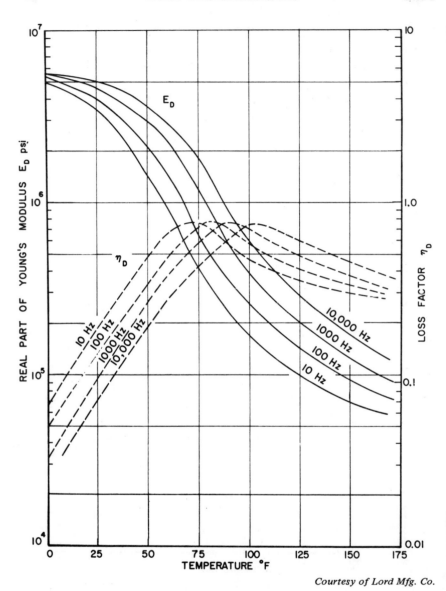

Courtesy of Lord Mfg. Co.

Fig. 6-7. Complex modulus data for LD-400 (AFML data).

T and T_o. The data given in Figs. 6-5 to 6-19 were obtained in this manner. In this way, the limitations of the measuring technique or techniques can be partly compensated for, at least. Measurement techniques will not be discussed here as the subject is well referenced in the literature [5, 11].

Ch. 6   DAMPING MATERIALS   221

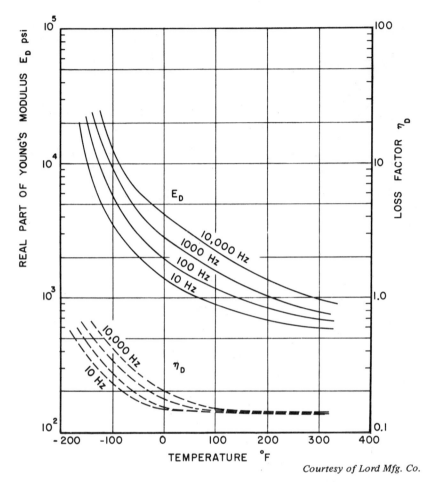

Courtesy of Lord Mfg. Co.

Fig. 6-8. Complex modulus data for BTR (AFML data).

    The real part of Young's modulus, or the Complex modulus, and the loss factor for many commonly used damping materials are given as a function of temperature in Figs. 6-5 to 6-19. It is important to realize that both E_D and η_D are temperature dependent; thus, the damping material must be selected on the basis of the temperature at which it is to be used. This is a most important consideration because a damping material that is effective at one temperature may be totally ineffective when used at another temperature. Often the damping material properties are given only for normal room temperatures by the manufacturer. When these materials are to be used at other temperatures, their damping behavior will vary, and it is well to consult the manufacturer or Figs. 6-5 to 6-19, beforehand.

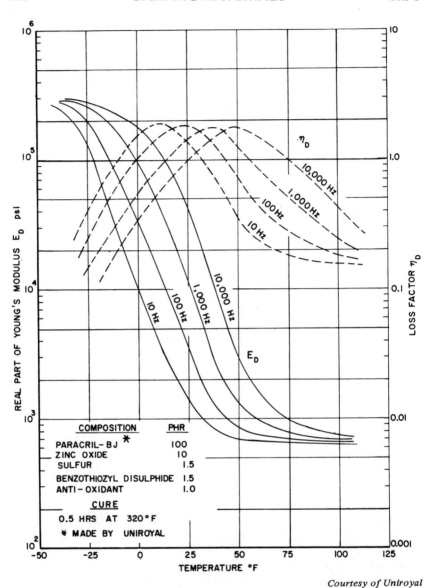

Courtesy of Uniroyal

Fig. 6-9. Complex modulus data for Paracril-BJ (AFML data).

Damping of Metals

For a typical structural alloy, the hysteresis loop is usually thin and cusped, as sketched in Fig. 6-20, so that the damping is very low. As the material is strained into the plastic range, the loop will thicken and hence the damping will

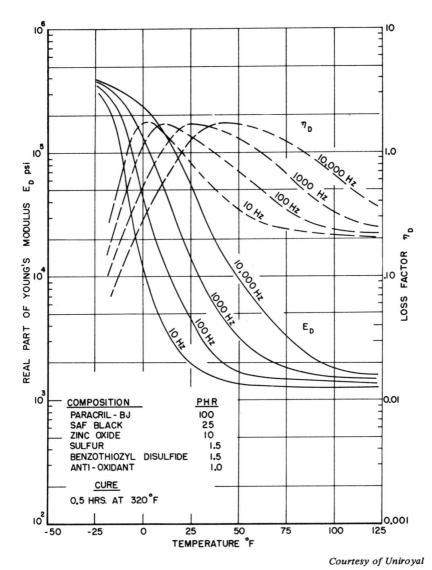

Fig. 6-10. Complex modulus data for Paracril-BJ with 25 PHR carbon (AFML data).

increase, but only at the expense of operating the material in a region where it may fail in a relatively small number of cycles.

This behavior is characteristic of most commonly used structural alloys and one has to examine somewhat less common alloy systems before one may encounter higher levels of damping at low strain levels, relative to the point at which plasticity effects become significant. Alloy systems which exhibit high

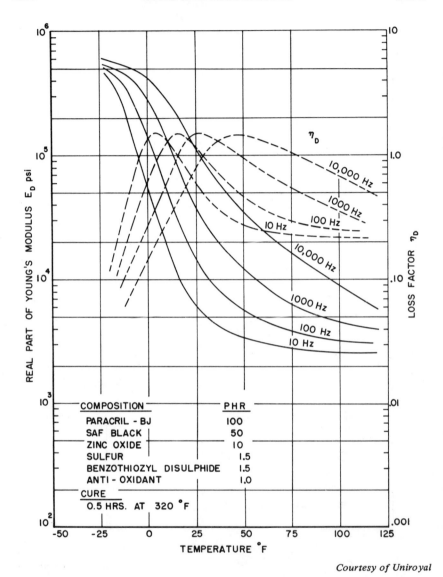

Courtesy of Uniroyal

Fig. 6-11. Complex modulus data for Paracril-BJ with 50 PHR carbon (AFML data).

damping behavior include some of the manganese-copper alloys, iron-aluminum alloys, and magnesium alloys [1, 12, 14].

Typical Damping Behavior

Resonant vibration tests on cantilever beams are often used to obtain a measure of the damping properties of the beam material. The results of a vibration

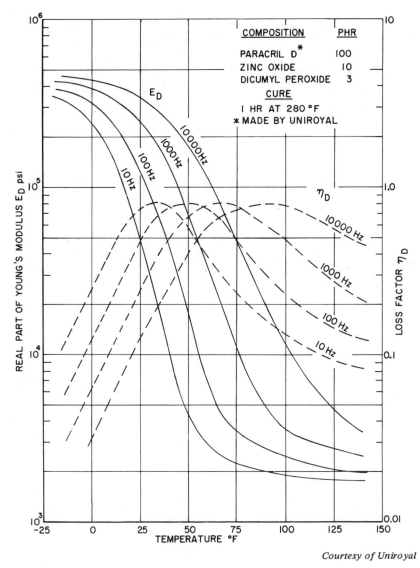

Courtesy of Uniroyal

Fig. 6-12. Complex modulus data for Paracril-D (AFML data).

test are illustrated in Fig. 6-21, where f_n is the resonant frequency of the nth mode and Δf_n is the "half-power" bandwidth. The "half-power" bandwidth, Δf_n, is determined by the points that are at an amplitude of 0.707 times the maximum amplitude on either side of the response peaks (see Fig. 6-21). If the peak acceleration at the tip of the beam is $\ddot{Y}$ and the input acceleration occurring at the root of the beam is $\ddot{X}$, data such as these will provide a measure of the

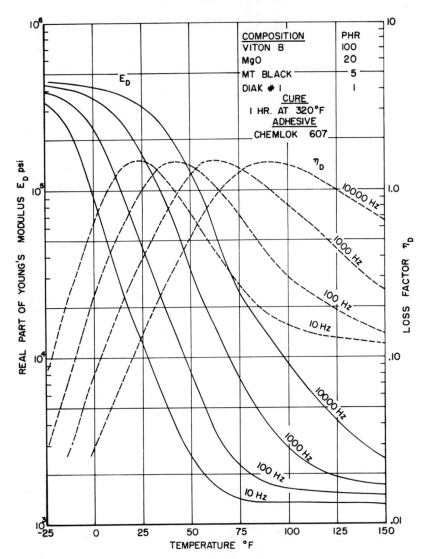

Fig. 6-13. Complex modulus data for Viton-B (AFML data).

damping of the nth mode, which is expressed by two numbers, or loss factors, η_s and $\eta_{s'}$.

$$\eta_s = \frac{\Delta f_n}{f_n} \tag{6-24}$$

$$\eta_{s'} = \frac{\alpha_n}{\sqrt{\left|\frac{\ddot{Y}}{\ddot{X}}\right|^2 - 1}} \tag{6-25}$$

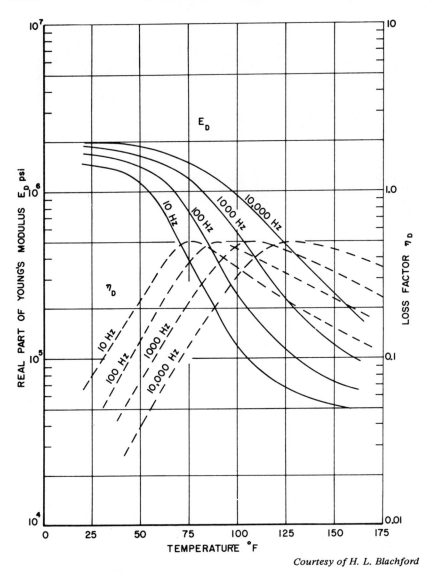

Courtesy of H. L. Blachford

Fig. 6-14. Complex modulus data for Aquaplas F-70 (AFML data).

and

$$\alpha_n = \frac{\int_o^L \phi_n(x)\, dx}{\int_o^L \phi_n^2(x)\, dx}$$

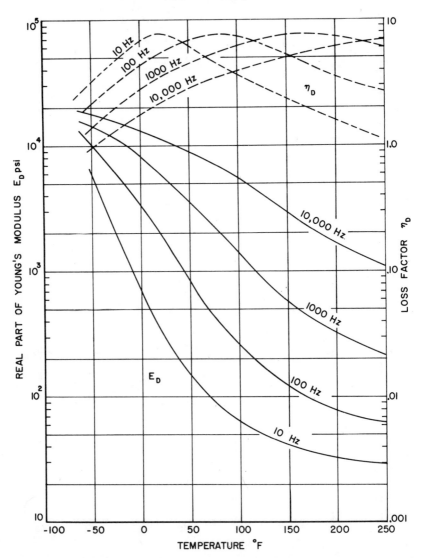

Courtesy of Borden Chemical Co.

Fig. 6-15. Complex modulus data for Mystik 7402 (AFML data).

where $\phi_n(x)$ is the nth normal mode of the cantilever beam and L is the length of the beam.

These quantities can be used to obtain a simple measure of the variation of the beam damping in each mode as a function of peak strain amplitude at the root of the beam, on the surface, and/or frequency, and temperature. The quantities η_s and $\eta_{s'}$ are not a direct measure of the damping behavior of the material itself

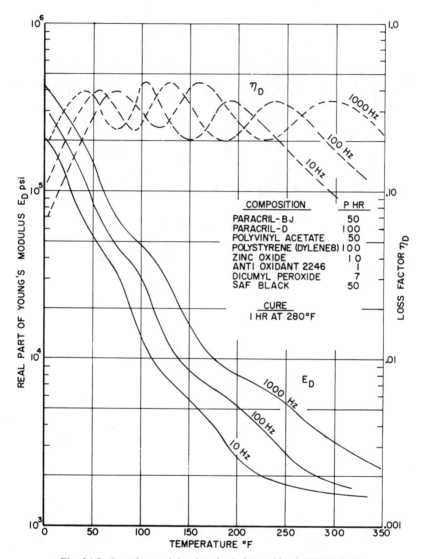

Fig. 6-16. Complex modulus data for Polymer Blend (AFML data).

unless the material is linear, i.e., η_s and η_s' are independent of strain amplitude, but do give a qualitative insight into the suitability of a given alloy for damping purposes.

To illustrate the damping behavior of a typical structural material, Fig. 6-22 shows the variation of η_s with frequency for a beam machined carefully from a 7075-T6 aluminum block. The beam was 7 inches long, 0.45 inch wide, and 0.05 inch thick. Results are given for the beam in air at atmospheric pressure,

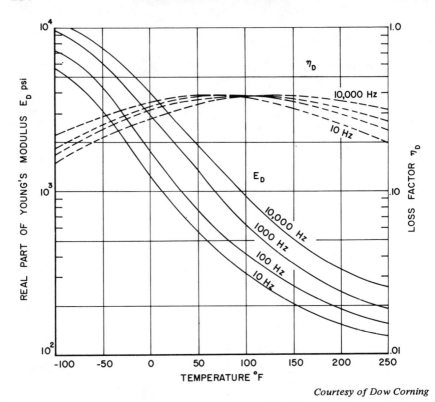

Fig. 6-17. Complex modulus data for Sylgard 188 (AFML data).

and at a pressure equivalent to an altitude of 125,000 feet. They show that, for this particular geometry and frequency range, the contribution to the damping from atmospheric sources, such as acoustic radiation, is negligible. It was also found that the damping was independent of strain, i.e., it was linear, and hence a direct measure of the damping of the material as well as of the specimen. However, in view of the low damping levels, these results are of little value in estimating the damping of a complex structure.

In contrast to this, the behavior of a highly damped manganese-copper alloy, known as Sonoston [12], is highly dependent on strain amplitude. Results of tests on cantilever beams of width, 0.50 inch; thickness, 0.04 inch; and of various lengths are shown in Fig. 6-23. In this case, the graphs of η_s versus peak strain are representative of the specimens only, and not of the material *per se*, except in a qualitative manner.

Damping of Enamels

Vitreous enamels have been used for decorative and protective purposes for thousands of years but it is only in the very recent past that attention has been paid to the damping behavior of enamels. While the mechanism whereby energy

Ch. 6 DAMPING MATERIALS 231

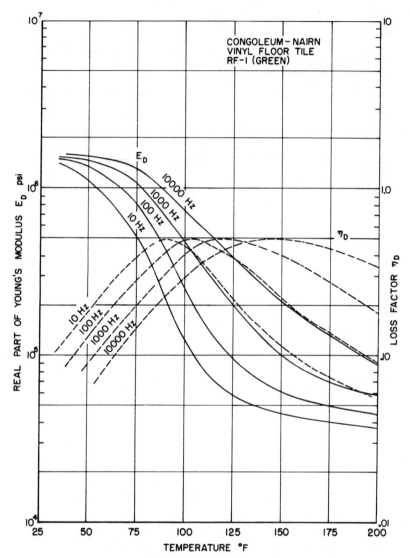

Fig. 6-18. Complex modulus data for Vinyl floor tile (AFML data).

is dissipated is somewhat different from those operating in metals and elastomers, the damping behavior of enamels is phenomenologically very similar to that of the elastomers.

Effects of Temperature and Frequency

At constant frequency, the complex modulus representation of enamel behavior as a function of temperature is typically as illustrated in Fig. 6-24. It is

232   DAMPING MATERIALS   Ch. 6

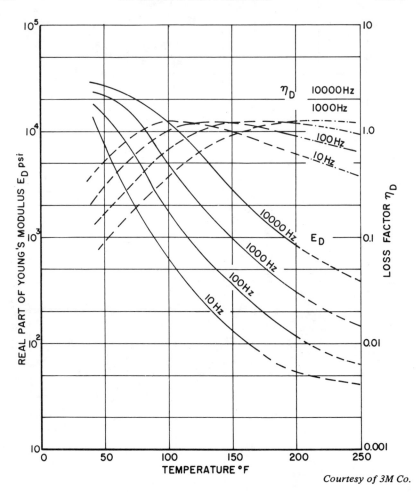

Courtesy of 3M Co.

Fig. 6-19. Complex modulus data for ISPD-110 (AFML data).

obvious that the only real behavioral difference between an enamel and an elastomer is that at the high end of the "rubbery region" the enamel will melt while the elastomer will break down.

With regard to variation with frequency, the temperature-frequency equivalence principle is again applicable to the behavior of at least some enamels, so that graphs of E_D and η_D versus $f\alpha_T$ can be obtained from experimental data. Note that T_o/T is here assumed equal to 1.0, since the transition temperature is usually very high on an absolute scale and the temperature band width of useful damping is at most 200°F wide.

Figure 6-25 shows the variation of E_D and η_D with changes in temperature for a typical enamel [15].

Ch. 6 DAMPING MATERIALS 233

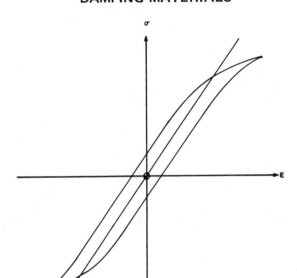

Fig. 6-20. Cusped hysteresis loop.

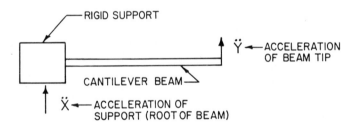

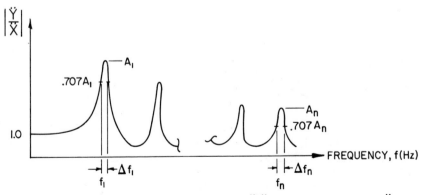

Fig. 6-21. Response curve of the acceleration ratio, $|\ddot{Y}/\ddot{X}|$, for a cantilever beam. $\ddot{Y}$ is the acceleration of the tip of the beam and $\ddot{X}$ is the applied acceleration at the root of the beam.

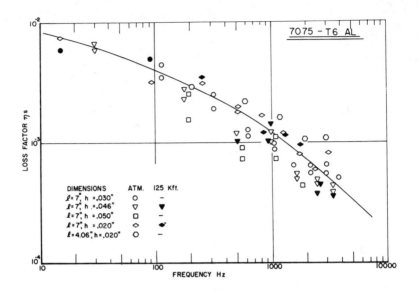

Fig. 6-22. Variation of η_S with frequency for aluminum beam.

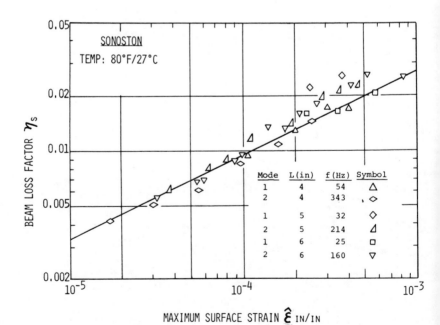

Fig. 6-23. Variation of η_S with strain for Sonoston.

DAMPING MATERIALS

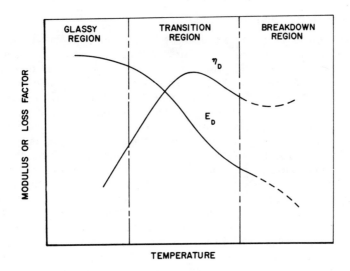

Fig. 6-24. Typical damping behavior of enamel as function of temperature.

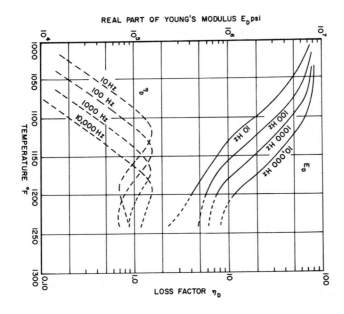

Fig. 6-25. Complex modulus data for Chicago Vitreous 325 Enamel (AFML Data).

RESPONSE OF DAMPED STRUCTURES

Nominally Undamped Structures

Most built-up structures possess damping in each mode of vibration far in excess of the intrinsic damping of the structural material itself. This is because many sources of energy dissipation exist in a built-up structure that are not found in a simple specimen. These additional damping sources can include *acoustic damping* arising through sound radiation from the vibrating surfaces of the structure, *joint damping* due both to slipping between interfaces and to "air-pumping" caused by transverse relative motion of overlapping metal sheets [16], and to air pumping through holes in bounding surfaces enclosing finite volumes of air, as structural surfaces surrounding the volume, vibrate. One aspect of the art of vibration control must be to enhance, as far as possible this natural damping within a structure without in any way reducing its strength, stiffness, or fatigue life characteristics. This, however, is a difficult art since the phenomena involved are extremely sensitive to details of structural geometry and construction and are by no means well-understood, quantitatively. For the present, therefore, we are obliged to take the natural damping in each mode of a structure very much as it comes. Even more than this, the variability of the natural damping in any given structure geometry is immense. If, for example, one attempts to build several nominally identical structures using normal fabrication techniques, the measured values of the modal loss factor, Equation (6-8), in any given mode may vary by a factor of as much as ten, typically varying from $\eta_s = 0.002$ to $\eta_s = 0.02$. This unpredictability, as much as anything else, makes reliance on natural damping a rather uncertain affair, because the displacements and stresses at resonance under sinusoidal excitation are inversely proportional to η_s, and the stresses and displacements under random excitation are inversely proportional to $\sqrt{\eta_s}$. See Ref. [17].

Unconstrained Layer Treatments

The free or unconstrained layer treatment is by far the simplest way of introducing damping into a sheet-metal type of structure. The treatment consists of a layer of an appropriate elastometric material bonded, by a suitable adhesive, to

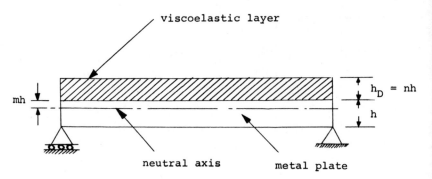

Fig. 6-26. Unconstrained layer treatment.

those vibrating surfaces of the structure which are vibrating primarily in a bending type of mode as shown in Fig. 6-26. As these surfaces bend, the treatments on the surfaces are deformed cyclically in tension-compression and so dissipate energy.

Unconstrained Damping of Unstiffened Beams and Plates

The damping loss factor, η_s, for a given mode of vibration in an unstiffened beam covered by a layer of viscoelastic material can be expressed by the following equations:

$$\eta_s = \frac{\eta_D en}{1+en}\left[\frac{3+6n+4n^2+2en^3+e^2n^4}{1+2en(2+3n+2n^2)+e^2n^4}\right] \quad (6\text{-}26)$$

$$\left[1+\frac{n\rho_D}{\rho}\right]\left[\frac{f_r}{f_n}\right]^2 = \frac{1+2en(2+3n+2n^2)+e^2n^4}{1+en} \quad (6\text{-}27)$$

where:

$\eta_s = \dfrac{\Delta f_n}{f_n}$ [see Equation (24)]

$n = \dfrac{h_D}{h}$

h = thickness of the beam
h_D = thickness of the viscoelastic layer

$e = \dfrac{E_D}{E}$

E_D = real part of Young's modulus
ρ = density of the beam material
ρ_D = density of viscoelastic material
f_r = resonant frequency of the nth mode of the damped beam
f_n = nth natural frequency of the undamped beam.

These now classic equations were originally derived by Dr. H. Oberst and Dr. K. Frankenfeld, in 1952 [18].

If, instead of an unstiffened beam, the treatment is applied to an unstiffened plate, these equations still apply, provided that the Poisson's ratio ν_D of the viscoelastic material is approximately equal to that of the beam or plate material, ν.

The author has found an alternative form [19] of the above equations to be of considerable value, particularly in view of the possibility of extending them to apply to stiffened plates. These equations are:

$$\eta_s = \frac{\eta_D}{1+\dfrac{A}{Be}} \quad (6\text{-}28)$$

$$\left[1+\frac{ne_D}{\rho}\right]\left[\frac{f_r}{f_n}\right]^2 = \frac{A+Be}{2} \quad (6\text{-}29)$$

where:

$$A = \frac{(1 - n^2 e)^3 + (1 + [2n + n^2]\, e)^3}{(1 + ne)^3} \tag{6-30}$$

and

$$B = \frac{(2n + 1 + n^2 e)^3 - (1 - n^2 e)^3}{(1 + ne)^3} \tag{6-31}$$

Needless to say, the two sets of equations are identical, as may readily be verified.

Design Procedure

The plate geometry is given, the undamped resonant frequencies f_n are known, and E_D and η_D are assumed known for the damping material as functions of temperature and frequency. For given thickness ratio n, $\left(\text{Fig. 6-26}, n = \dfrac{h_D}{h}\right)$, the mode to be considered is then chosen and a first guess is made at the frequency f_r of that mode for the damped system. Using Equation (6-29), a new value of f_r is calculated using values of E_D and η_D corresponding to the initial value of f_r. The process is repeated until the two values of f_r converge, then, from Equation (6-28), the value of η_s is calculated. The process is repeated over the desired temperature range. If the damping behavior so achieved is not suitable, either n may be increased or another material with a more suitable peak-damping temperature is chosen. The most important criterion, as Equation (6-26) clearly shows, is that the product $E_D \eta_D$ is as large as possible at the temperature where maximum damping is required.

Sample Calculation No. 1

Find the loss factor η_s and the resonant frequency in the fundamental mode of a rectangular steel plate, clamped at all four edges, of thickness ¼ inch, if the plate is damped by means of a layer of viscoelastic material, ¼-inch thick. The temperature is 75°F, $E_D = 10^6$ lb/in.² and $\eta_D = 0.4$. E for steel is 3×10^7 lb/in.² and the fundamental frequency for the undamped plate is 200 Hz. The density of steel is 0.3 lb/in.³ and is 0.06 lb/in.³ for the damping material. For the given data:

$$n = h_D/h = 1$$
$$e = 10^6/3 \times 10^7 = 1/30$$
$$\eta_D = 0.4$$
$$f_{11} = 200 \text{ Hz}$$
$$\frac{n\rho_D}{\rho} = 0.06 \times 1/0.3 = 0.20$$

∴ From Equation (6-26):

$$\eta_s = 0.115$$

and from Equation (6-27):

$$(1 + 0.2)(f_r/200)^2 = 1.422$$

$$\therefore f_r = 218 \text{ Hz}$$

Alternately, using Equations (6-28) and (6-29):

$$A = 2.03$$

$$B = 24.6$$

$$\therefore \eta_s = \frac{0.4}{1 + \frac{2.03 \times 30}{24.6}} = 0.115$$

$$(1 + 0.2)(f_r/200)^2 = (2.03 + 24.6/30)/2 = 1.422$$

$$f_r = 218 \text{ Hz, as before.}$$

Damping of Stiffened Plates

The presence of stiffeners in a sheet-metal type of structure will serve to reduce the effectiveness of a free layer treatment. The prediction of the effect of such treatment on modal damping is, in general, very difficult and beset with many assumptions. If, as is often the case for structures having thin, stiffened, skin type construction, the treatment is applied over all the surfaces which deform primarily in bending, the damping in each mode can be estimated from the Equation (6-19):

$$\eta_s \approx \eta_D/[1 + (A - 2 + \beta_{nm})/Be] \tag{6-32}$$

$$(1 + \rho_D n/\rho)(f_r/f_{nm})^2 \approx 1 + (A - 2 + Be)/\beta_{nm} \tag{6-33}$$

where A and B are defined as before and:

$$\beta_{nm} = \frac{2\xi_{nm}^4 \iint_S \phi_{nm}^2 \, dxdy}{\iint_D \left[\left(\frac{\partial^2 \phi_{nm}}{\partial \Delta^2}\right)^2 + \left(\frac{L}{l}\right)^4 \left(\frac{\partial^2 \phi_{nm}}{\partial \delta^2}\right)^2 + 2\nu \left(\frac{L}{l}\right)^2 \left(\frac{\partial^2 \phi_{nm}}{\partial \Delta^2}\right)\left(\frac{\partial^2 \phi_{nm}}{\partial \delta^2}\right)\right] dxdy}$$

$$\tag{6-34}$$

Estimation of β_{nm}

In general it is exceedingly difficult to calculate β_{nm} because the detailed information on mode shapes is usually not available. One can estimate β_{nm} in some idealized cases, but in the event of encountering a practical problem, perhaps the best procedure is to apply a treatment consisting of a known, suitable, viscoelastic material in known thickness and over the entire surface involved in the vibration modes of interest. Measured response records can then be used to estimate η_s and Equations (6-32) and (6-33) would then be used in reverse to estimate β_{nm}. From this point onward, the equations could then be used directly, to estimate the effects of other treatment thicknesses and/or other treatment materials.

Sample Calculation No. 2

The plate of sample calculation No. 1 is stiffened while the same damping treatment is retained over the whole surface. The loss factor is reduced to 0.04. Estimate β_{nm} for the fundamental mode, and the added thickness of material needed to return η_s to its unstiffened plate value.

From Equation (6-32), using the same values of n and e:

$$0.04 = \frac{0.4}{1 + (2.03 - 2 + \beta_{nm})/(24.6/30)}$$

$$\therefore \frac{30}{24.6}(0.03 + \beta_{nm}) = 9.0$$

$$\beta_{nm} = \frac{9.0 \times 24.6}{30} - 0.03 = 7.37$$

We must now guess at a new value of n needed to raise η_s to a value of 0.115. By trying several values, it is eventually found that the value of n must be equal to 1.84 to raise the loss factor to the desired level.

Multiple Constrained Layer Treatments

Figure 6-27 illustrates the various geometries of multiple layer damping treatments which have been examined analytically and experimentally. The very simplest constrained layer treatment consists only of a thin layer of viscoelastic

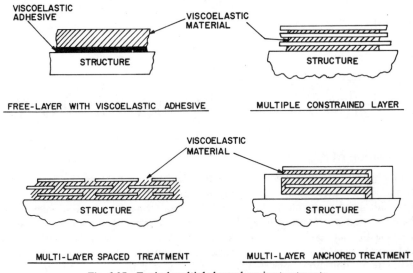

Fig. 6-27. Typical multiple layer damping treatments.

material attached to the surface of the structure and constrained by a stiff layer, usually of metal, on the outer surface. Multiple layers are easily produced by repeating the pattern. The advantages of this type of treatment are manifold. For example, since the mechanism of energy dissipation is primarily through shear deformation of the viscoelastic material as the surface deforms in bending, the viscoelastic material need not have an exceptionally high shear modulus to be effective, as is the case for a free layer treatment [20]. Furthermore, by choosing very stiff constraining layers one can further enhance the damping capability and, finally, by choosing appropriate different materials for each viscoelastic layer, one can optimize the damping to some degree to be effective over a wider temperature or frequency range than one usually achieves with a single viscoelastic material between each constraining layer [21, 22].

Damping of Unstiffened Beams and Plates

Unlike the free layer treatment, the analysis of constrained layer treatments is generally difficult and analysis and experimental data are not always in good agreement, although the qualitative behavior of these systems seems to be well understood [23, 25]. For a designer's purposes semi-empirical data, based on experiments, are often preferable although the price paid for this approach is the limited number of parameters which can be varied. A comprehensive program to evaluate multiple constrained layers on simple beams, for example, would have to consider wide variations in the number of layers N, the complex shear modulus $G_D(1 + i\eta'_D)$ of the viscoelastic layers [or $E_D(1 + i\eta_D)$], the modulus E_c of the constraining layer, the viscoelastic material thickness h_D, the constraining layer thickness h_C, the modal half-wavelength λ_n in the nth mode, the length L and the thickness h of the beam, the modulus E of the beam, and the boundary conditions. This total of at least ten parameters can easily lead to the necessity for conducting an enormous number of response tests and would become highly impractical if a large number of values of each parameter had to be examined. Fortunately, however, analysis can help greatly in showing which nondimensional groups of parameters are the most promising to look into. In fact, elementary theory seems to show that many of the parameters can be combined into a single "shear parameter" defined by $E_D \lambda_n^2 / E_C h_C h_D$ which, when combined with the geometrical ratios h_D/h_C and h_D/h, the modulus ratio E_C/E, and N, reduces the total number of independent parameters from ten to five. Even further reductions can be achieved for treatments comprising five or more pairs of layers. In this case, one can approximate the behavior of the multiple layer treatment quite adequately in terms of an equivalent free layer treatment having an equivalent, complex Young's modulus $E_e(1 + i\eta_e)$, and for which E_e/E_c and η_e/η_D depend strongly only on $E_D \lambda_n^2 / E_C h_C h_D$, and h_D/h_C, thereby eliminating the ratios h_D/h, E_C/E, and N, at least to a degree. The number of significant parameters is thereby reduced to two and experimental investigations are greatly simplified.

Equivalent Complex Moduli

Extensive tests [26] have been carried out on a clamped-clamped aluminum beam, seven inches long, one inch wide, and 0.050 inch thick for a number of different multiple constrained layer treatments. Typical measured graphs of η_s and $(1 + \rho_D n/\rho)(f_r/f_n)^2$ versus $E_D \lambda_n^2$ are plotted in Fig. 6-28. From these graphs, using equations (6-28) and (6-29) in reverse, the values of E_e and η_e can be deduced for each selected value of $E_D \lambda_n^2$. E_e and η_e represent the equivalent Young's modulus and loss factor, respectively, of a single free layer of damping material which would provide the same results as the multiple layer damping treatment. Typical graphs of E_e/E_C and η_e/η_D versus $E_D \lambda_n^2/E_C h_D h_C$ are shown in Fig. 6-29. To extend the data to unstiffened plates, it is necessary only to replace the beam shear parameter $\phi_n = E_D \lambda_n^2 / E_c h_c h_D$ by a plate shear parameter $\phi_{nm} = E_D \lambda_{nm}^2 /(r^2 + 1) E_c h_c h_D$ where λ_{nm} is the half-wavelength of the nmth mode and r is the ratio of λ_{nm} to the distance between the other two node lines bounding a segment of a mode. For example, for a 10 by 8 in. plate pinned on all four edges, λ_{11} for the fundamental mode is 8 in. and $r = 8/10$.

Design Procedure

For a plate of given dimensions and classical boundary conditions, i.e., clamped, pinned, or free, first λ_{nm} and r are estimated for the mode of interest, on the basis of undamped modal patterns, assumed to be known. At each temperature, the resonant frequency f_r of the nmth mode of the damped plate is guessed and E_D and η_D are read off from an appropriate data sheet for the viscoelastic material selected. For given constraining layer modulus E_C, adhesive thickness h_D and constraining layer thickness h_c, one can then calculate ϕ_{nm}

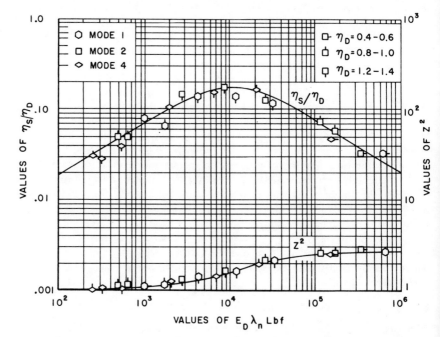

Fig. 6-28. Graphs of η_s and Z^2 versus $E_D \lambda_n^2$ for $N = 4$, $h_c = 0.002''$, $h_D = 0.002''$.

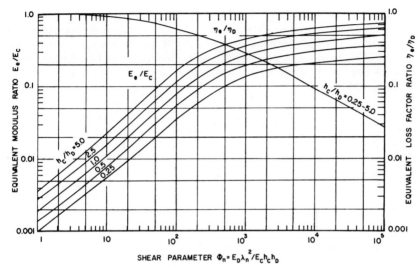

Fig. 6-29. Graphs of E_e/E_c and η_e/η_D versus $\phi_n = E_D \lambda_n^2/E_c h_c h_D$.

and read E_e/E_c and η_e/η_D from Fig. 6-29. Then, as E_c and η_D are known, E_e and η_e are calculated. For a given number of layer pairs N, in the treatment, the ratio n of treatment thickness to plate thickness can be estimated. Now calculate η_s and $(1 + \rho_D n/\rho)(f_r/f_{nm})^2$, and hence, η_s and f_r, from Equations (6-28) and (6-29), using the determined value of E_e and η_e. If f_r, as calculated, differs significantly from the initial value, this should be repeated.

Sample Calculation No. 3

A clamped-clamped steel beam 2 ft in length and 0.25-in. thick is to be damped by a multiple layer constrained damping treatment over the whole surface. If the adhesive properties are $h_D = 0.004$ in., $E_D = 360$ lb/in.^{2}, $\eta_D = 1.4$, and the constraining layer properties are $E_c = 10^7$ lb/in.^{2} and $h_c = 0.020$ in.; (a) find the loss factor of the damped beam in the fundamental mode when five (5) layer pairs are used; (b) estimate the effect of increasing E_c to 3×10^7 lb/in.^{2}; (c) how many layers for $E_c = 10^7$ lb/in.^{2} would be needed to achieve an equal increase in damping?

(a) For the fundamental mode of a clamped-clamped beam of length L:

$$\lambda_1 = \frac{\pi L}{\xi_1} = \frac{\pi \times 24}{4.73} = 15.95 \text{ in.}$$

where ξ_1 is the first eigenvalue of a clamped-clamped beam [27].

The eigenvalue is related to the natural or resonant frequency of the beam by: $\xi_1 = \left(\frac{\omega_1^2 A \rho}{EI}\right)^{1/4}$, where ω_1 is the fundamental mode natural frequency in radians per second, A is the cross-sectional area of the beam, ρ is the density of the beam material, E is the modulus of elasticity of the beam material, and I is the area moment of inertia for the beam cross-section. The eigenvalue is determined by the differential equation and the corresponding boundary conditions for beam vibration. Only certain parameters will satisfy both the differential equation and the boundary conditions, simultaneously. Beam vibration solutions are beyond the level of this material. For a discussion of this topic see Reference [2].

$$\therefore \phi_1 = \frac{\lambda_1^2 E_D}{E_c h_D h_c} = \frac{(15.95)^2 \times 360}{10^7 \times 0.004 \times 0.020} = 114$$

∴ From Fig. 6-29 for $h_c/h_D = 5.0$:

$$E_e/E_c = 0.18$$
$$\therefore E_e/E = 0.18$$

and

$$\eta_e/\eta_D = 0.57$$
$$\therefore \eta_e = 0.57 \times 1.4 = 0.796$$

Therefore, from Equations (6-26) or (6-28), with $n = 5(.004 + .020)/0.25 = 0.48$:

$$\eta_s = 0.259$$

(b) If we now increase E_c to 3×10^7, then ϕ_1 becomes:

$$\phi_1 = 114/3 = 38$$
$$\therefore E_e/E_c = 0.078$$
$$\therefore E_e/E = 0.078 \times 3 = 0.234$$

and

$$\eta_e/\eta_D = 0.72$$
$$\therefore \eta_e = 0.72 \times 1.4 = 1.01$$

Therefore, by Equations (6-28) or (6-29):

$$\eta_s = 0.372$$

(c) In order to estimate the number of layers with $E_c = 10^7$ lb/in.² needed to achieve the above loss factor we must simply repeat the calculation procedure (a) with $N = 6, 7, 8$—until this value of η_s is equalled or exceeded. For $N = 8$ we have:

$$\eta_s = 0.388$$

Sandwich Panels

Sandwich panel construction is widely used with honeycomb cores wherever high stiffness is required. Cores having viscoelastic properties are also attractive, although boundary attachment problems are not readily overcome. Some typical sandwich constructions are illustrated in Fig. 6-30.

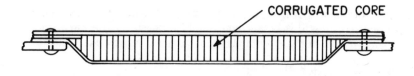

CORRUGATED CORE

HONEYCOMB SANDWICH

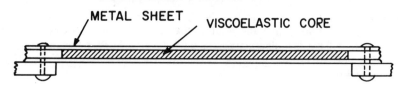

METAL SHEET   VISCOELASTIC CORE

THREE LAYER SANDWICH

METAL SHEET   VISCOELASTIC CORE

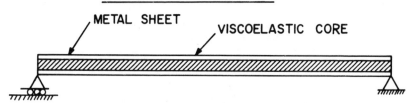

THREE LAYER PINNED SANDWICH

Fig. 6-30. Typical sandwich constructions.

Damping in Symmetric Viscoelastic Sandwich Panels

If panel boundaries are assumed to be pinned or simply supported, their damping and stiffness characteristics [28, 29] can be estimated from the Equations:

$$\eta_s = \frac{3\eta_D \psi_{nm}(1+n)^2}{(1+\psi_{nm})^2 + \eta_D^2 + 3(1+n)^2 (1+\psi_{nm} + \eta_D^2)} \tag{6-35}$$

$$\left[1 + \frac{\rho_D h_D}{2\rho h}\right]\left(\frac{f_r}{f_o}\right)^2 = 1 + \frac{3(1+n)^2 (1+\psi_{nm}+\eta_D^2)}{(1+\psi_{nm})^2 + \eta_D^2} \tag{6-36}$$

$$\psi_{nm} = \frac{(r^2+1)}{(1-\nu^2)} \frac{\pi^2}{2} \frac{E}{G_D} \left(\frac{h}{\lambda}\right)^2 n \tag{6-37}$$

where $n = h_D/h$, λ is the half-wavelength in one direction, r is the ratio of λ to the half-wavelength in the other direction, and ν is Poisson's ratio for the outer sheets.

If the boundaries are clamped—which would be a closer approximation to what might be used in a practical application—the size of the equivalent pinned plate of the same aspect ratio can be estimated by simply varying the pinned plate dimensions until the observed natural frequencies are the same as those for the clamped plate in the appropriate mode, in lieu of a more exact analysis [25].

Sample Calculation No. 4

A steel sandwich plate is 12 inches long by 20 inches wide by $3/16$ inch net thickness. The other sheets are of $1/16$ inch thickness each, of Young's modulus $E = 3 \times 10^7$ lb/in.^{2}, and the core layer is $1/16$-inch thick with Young's modulus of 360 lb/in.^{2} and a loss factor of 1.4. The plate is pinned along all four sides. Calculate the damping and the resonant frequency for the fundamental mode. The density of the damping material is 0.04 lb/in.^{3}. For the undamped plates, each $1/16$ inch thick, the equation of motion for the vibration of a plate has been found to be:

$$\frac{Eh^3}{12(1-\nu^2)} \nabla^4 W - \rho h \frac{\delta^2 W}{\delta t^2} = 0 \tag{6-38}$$

where:

$$\nabla^4 W = \frac{\delta^4 W}{\delta x^4} + 2 \frac{\delta^4 W}{\delta x^2 \delta y^2} + \frac{\delta^4 W}{\delta y^4}$$

For the fundamental mode $W = \sin(\pi x/12) \sin(\pi y/20)$ if the x and y axes are defined as being along the 12-inch and 20-inch sides, respectively, with the origin at a corner. Substituting into the Equation of motion given, for $\nu = 1/3$: Poisson r

$$\frac{Eh^3}{12(1-\nu^2)} \left[\frac{1}{12^2} + \frac{1}{20^2} \right]^2 - \rho h \omega_{11}^2 = 0 \tag{6-39}$$

whence

$$\omega_{11}^2 = \frac{3 \times 10^7 \times (1/16)^2 \times 386(400 + 144)}{0.3 \times 12(8/9)(400 \times 144)}$$

$$\therefore f_{11} = 184 \text{ Hz}$$

For the damped plate, we note that $n = 1$ and $\eta_D = 1.4$. Therefore, by Equation (6-37):

$$\therefore \psi_{11} = \frac{[(12/20)^2 + 1] \pi^2 \times 3 \times 10^7 \times (1/16)^2 \times 1}{(1 - 1/9) \, 2 \times (360/3) \times 12^2}$$

$$= 50.9$$

From Equation (6-35):

$$\eta_s = \frac{3 \times 1.4 \times 50.9(2)^2}{(51.9)^2 + (1.4)^2 + 3 \times 4(51.9 + 1.96)}$$

$$= 0.252$$

and, from Equation (6-37):

$$\left[1 + \frac{0.04 \times 1}{2 \times 0.3}\right] \left(\frac{f_r}{184}\right)^2 = 1 + \frac{3(2)^2 (1 + 50.9 + 1.96)}{(51.9)^2 + 1.96}$$

$$\therefore 1.067(f_r/184)^2 = 1.12$$

$$\therefore f_r = 184 \times 1.025 = 189 \text{ Hz}$$

Tuned Damping Devices

The various layered damping treatments operate effectively only in structures and modes of vibration for which high surface strains occur. For structures which have low surface strains involving nonplate-like behavior, as in the case of very highly curved elements, the tuned damper concept may be of greater utility. For a tuned damper to be of value the essential prerequisites are that the damper be located at a point of high amplitude response, such as an antinode of a mode, and that the frequency spectrum of the response have a single resonance, a number of widely separated resonances, or a number of widely separated groups of modes.

Let us first consider the damper itself, in isolation, in order to understand the principle of action.

Damper Behavior

Figure 6-1 shows an idealized sketch of a tuned damper, consisting of a mass attached to a spring exhibiting viscoelastic damping behavior. This simple tuned damper is essentially a simple, single-degree-of-freedom system. From Equation (21), the force exerted on the structure by the damper, resulting from an excitation $X \exp(i\omega t)$ at the point of attachment, is:

$$F = -K(Y - X)(1 + i\eta) \exp(i\omega t) \qquad (6\text{-}40)$$

$$= \frac{-M\omega^2 X}{1 - \frac{M\omega^2}{K}(1 + i\eta)} \qquad (6\text{-}41)$$

At the frequency for which $M\omega^2/K = 1$, we see that:

$$F = \frac{-ixK}{\eta} = \frac{\dot{x}K}{\eta\omega} \qquad (6\text{-}42)$$

indicating that the force F is opposing the velocity $\dot{x}$, i.e., it is a damping force. By proper choice of M, K, and η, in principle a damper can be designed to introduce a high degree of damping into a number of structures [30, 31].

Tuned Dampers in Simple Structures

If the structure is relatively simple, to the extent that it either has only a single resonance or a number of very widely separated resonances, a tuned damper can be designed very readily simply by ensuring that the damper frequency $\omega_D = \sqrt{K/M}$ is close to the frequency of the mode to be damped.

The effect of the damper on the response of the structure is to split the original single mode into two [32]. The lower frequency branch corresponds to the damper mass M and the structure surface moving essentially in-phase while the higher frequency peak corresponds to the mass and surface moving essentially out-of-phase. The effect of moving the damper frequency ω_D relative to the frequency ω_n of the mode in question, to emphasize one of the peaks at the expense of the other, with an optimum damping case occurring when the two response peaks are of equal amplitude.

For simple structures meeting the requirements of a single resonance peak, or widely separated resonances, the response behavior of the damped system near the lowest frequency can be written in the form:

$$\frac{W}{W_o} = \frac{1}{1 - \left(\frac{\omega}{\omega_1}\right)^2 - \frac{\psi_e(\omega/\omega_1)^2}{1 - (\omega/\omega_D)^2/(1 + i\eta)}} \tag{6-43}$$

where W_o is the static displacement under the applied forces at zero frequency and ψ_e is an effective mass given by:

$$\psi_e = \left(\frac{M}{M_s}\right)\gamma_1$$

where M_s is the mass of the structure and γ_1 is a nondimensional parameter depending on the mode shape and the damper location. Calculation of γ_1 will be difficult, in general, although it is easily worked out for simple cases.

Equation (6-43) shows that the response W/W_o at any frequency ω and any point in the structure, depends only on the effective mass ψ_e, the frequency

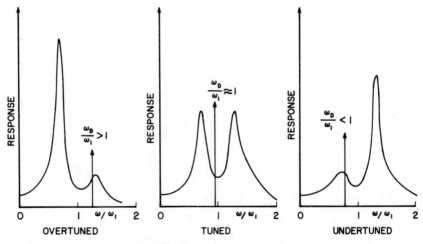

Fig. 6-31. Typical response spectra.

Ch. 6   DAMPING MATERIALS

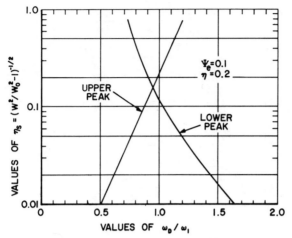

Fig. 6-32. Variation of $\eta_s = (|W_0^2/W^2| - 1)^{-1/2}$ with ω_D/ω_1.

ratio ω_D/ω_1, and the damper loss factor η. For given η and ψ_e, therefore, varying ω_D/ω_1 will alter the tuning of the damper and vary the relative amplitude of the two response peaks generated by the damper, as in Fig. 6-31. Figure 6-32 shows the variation of the value of $\eta_s = (|W_0^2/W^2| - 1)^{-1/2}$ with ω_D/ω_1 for typical values of η and ψ_e.

Design Procedure

If η and ψ_e are not small, the estimated value of ω_D/ω_1 for optimum damping can be predicted from the formula:

$$\left(\frac{\omega_D}{\omega_1}\right)_{opt} = \frac{1}{(1 + \psi_e)^{1/2}(1 + \eta^2)^{1/4}} \qquad (6\text{-}44)$$

For a structure having essentially plate-like geometry, γ_1 can be estimated from:

$$\gamma_1 = \frac{\sum_{j=1}^{J} \phi_1^2 (\Delta_j, \delta_j)}{\int_S \phi_1^2(\Delta, \delta) d\Delta \, d\delta} \qquad (6\text{-}45)$$

where ϕ_1 is the mode shape (assumed known), S is the surface area of the plate-like structure, and Δ_j, δ_j are the coordinates of the jth damper, if the number of separate identical dampers is J.

With γ_1 and, hence, ψ_e known, an appropriate damper material for which η is known, would be selected. For this value of ψ_e and the calculated value of ω_D/ω_1 for optimized damping, the value of $|W_o/W|$ can then simply be read off the curves in Fig. 6-33. If this level of damping is adequate, the calcula-

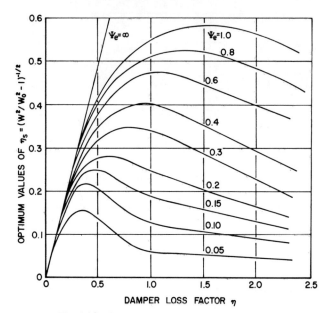

Fig. 6-33. Optimum value of η_s versus η and ψ_e.

tion need not be repeated. If temperature variability is important, the calculation should be repeated at several different temperatures.

Sample Calculation No. 5

A structure has a single resonance peak at 400 Hz. The mass of the structure is 10 lb and the tuned damper is to be placed at a location for which γ_1 is estimated to be 3.0. Using a viscoelastic material having a Young's modulus of 1000 lb/in.^{2} and a loss factor of 0.2, design a damper that will reduce the peak response to less than seven times the static displacement under any system of forces.

From Fig. 6-33, ψ_e must be at least equal to 0.1 for the value of $|W_o/W|$ to be greater than 0.143.

From Equation (6-44), for $\psi_e = 0.1$, and $\eta = 0.2$:

$$(\omega_D/\omega_1)_{opt} = 0.94$$

$$\therefore (\omega_D)_{opt} = 400 \times 0.94 = 376 \text{ Hz}$$

Now

$$\psi_e = \frac{M}{M_s}\gamma_1$$

$$\therefore 0.1 = \frac{M}{10} \times 3$$

$$\therefore M = 0.33 \text{ lb}$$

But

$$(\omega_D)^2_{opt} = \frac{K}{M} = \frac{E_D S K_T}{hM}$$

where $E_D = 1000$ lb/in.^{2}, S is the cross-sectional area of the damper (assumed of cylindrical shape as in Fig. 6.1), h is the length, and K_T is a shape factor [2] given by:

$$K_T = 1 + \beta(S/S')^2$$

where S' is the nonload-carrying area of the specimen and β is a nondimensional quantity equal to about 2.0 for an unfilled elastomer and 1.5 for a filled elastomer. For a cylindrical specimen of radius R, therefore:

$$(376 \times 2\pi)^2 = \frac{1000 \pi R^2 (1 + R^2/4h^2)}{h (0.33/386)}$$

$$\therefore \frac{R^2}{h^2} \left\{ 1 + \frac{R^2}{4h^2} \right\} = \frac{1.52}{h}$$

Any values of h and R which satisfy this equation will therefore be acceptable dampers. For $R/h = 1$, for example:

$$\frac{1.52}{h} = 1 \left(1 + \frac{1}{4}\right) = \frac{5}{4}$$

and

$$\therefore h = 1.21 \text{ in.}$$

$$R = 1.21 \text{ in.}$$

If, on the other hand, a shear type damper of the type illustrated in Fig. 6-34 were to be designed, the procedure would be as before:

$$(\omega_D)^2_{opt} = \frac{K}{M} = \frac{G_D S K_S}{hM}$$

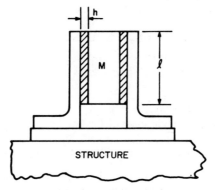

Fig. 6-34. Tuned damper in shear.

where G_D is the shear modulus (usually about one third ($\frac{1}{3}$) the Young's modulus E_D), S is the load-carrying area, h is the thickness of the shear layers [33], and K_S is now given by:

$$K_S = \frac{1}{1 + h^2/3\,6\,K^2}$$

and K is the radius of gyration of the shear layer cross section about the neutral axis of bending. For a rectangular shear layer, $K = l/\sqrt{12}$.

$$\therefore (376 \times 2\pi)^2 = \frac{333\,(2\,\text{lb})}{h(0.33/386)\,(1 + h^2/3l^2)}$$

$$\therefore \frac{h}{l}\left(1 + \frac{h^2}{3l^2}\right) = 0.1406$$

For $h/l = 0.1$,

$$b = 0.1\,(1 + 0.01/3)/0.140$$
$$= 0.717\,\text{in}.$$

if we now let $l = 1$ in., $h = 0.1$ in.; and let $l = 2$ in., $h = 0.2$ in., and so on.

Tuned Dampers in Complex Structures

When tuned dampers are applied to complex structures possessing many closely spaced resonances, the simplicity of the foregoing is lost and the effect of the dampers on the structural response is dependent on the exact nature of the structural geometry, so that no relatively general design concepts can be formulated.

It can be said, however, that if the modes of vibration of interest extend over more than an octave band of frequency, have widely dissimilar energies associated with them, and/or have many node lines, then tuned dampers are not likely to be of much value in reducing structural response. If, however, the modes of vibration of interest do fall within an octave band, have similar energies, and few node lines, then it may be said that tuned dampers will be very effective indeed. This type of response behavior is exactly that which is seen in the case of the skin-stringer class of structure.

Tuned Dampers in Skin-Stringer Structures

A typical skin-stringer structure is illustrated in Fig. 6-35. Here the modes of vibration are typically divided into groups or bands of modes corresponding to:

(a) no nodes within the bays;
(b) one, two, or more node lines within the bays parallel to the stiff frames; and
(c) one, two, or more node lines within each bay parallel to the stringers.

The frequency limits of each band depend critically on the aspect ratio of each bay; namely, the ratio of frame separation to stringer separation. If the

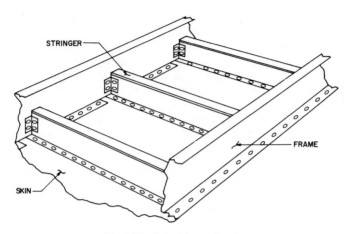

Fig. 6-35. Skin-stringer structure.

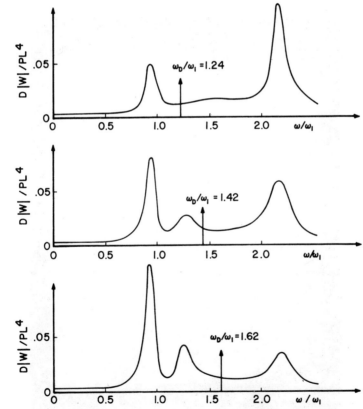

Fig. 6-36. Damped response spectra for skin-stringer structure (center bay, $\eta = 0.5$, $\psi_e = 0.1$).

aspect ratio is small, the first band having no mode lines within each bay will be separated from the others, whereas for high-aspect ratios the bands will all tend to overlap. In practice this means that tuned dampers are not apt to be of much utility for aspect ratios greater than about two. For aspect ratios less than two, however, several investigations [34, 35] have shown that tuned dampers can be very effective. Figure 6-36 shows the effect in a typical case [34]. No general design rules are available even in this case, however, beyond the fact that the damper should be tuned so that its resonant frequency lies near the center of the band and the damper loss factor η should be as high as possible.

APPLICATIONS

The foregoing information can be used to assist a designer in selecting the correct damping techniques and materials to solve a vibration problem once he has determined that increasing the damping is the appropriate approach. However, it cannot be said too often that a combined analytical-experimental approach is really necessary because of the great complexity of real structures

Fig. 6-37. Dispenser center web.

Ch. 6 DAMPING MATERIALS 255

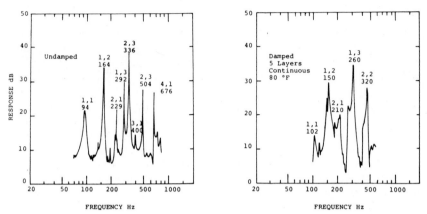

Fig. 6-38. Typical response spectra.

and the resulting need to verify that the analysis really does represent the dynamic behavior of the system. In order to illustrate this approach, some examples of practical problems arising in Air Force systems will be discussed presently. Problems encountered by various other investigators and organizations are described in various publications, and merit further reading [36, 45].

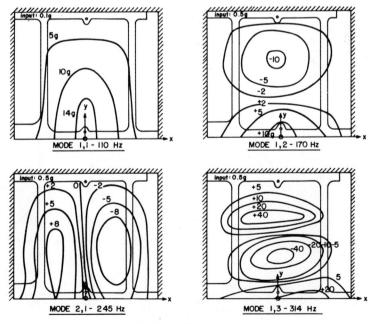

Fig. 6-39. Mode shapes.

Weapons Dispenser Problem

A plate-type vibration problem [3, 17] encountered in a weapons dispenser is of interest because it illustrates the way one designs layered damping treatments. The geometry of the center webs of the dispenser is illustrated in Fig. 6-37, and it is seen to consist essentially of a lightly stiffened, nearly square plate, clamped on three edges and free on the other. A typical measured response spectrum is illustrated in Fig. 6-38. Many resonances can be seen to have occurred over the entire frequency range examined, and the problem was that one or more of these modes was excited by cavity resonances in the dispenser, causing very rapid failure. Tuned damping devices were not appropriate, both because of the large number of well-separated resonances and the very tight clearances involved. As the mode shapes in Fig. 6-39 show, however, strains were high on the surface which vibrated essentially in plate-like modes for which layered damping treatments are appropriate, and of these, free or multiple-constrained layer treatments were found to be the most appropriate in order to avoid the major redesign and refabrication of the very large number of structures involved.

Free Layer Treatment

The problem involved is clearly that of a stiffened plate covered over most of the surface by a free layer treatment. In order to estimate β_{nm}, therefore,

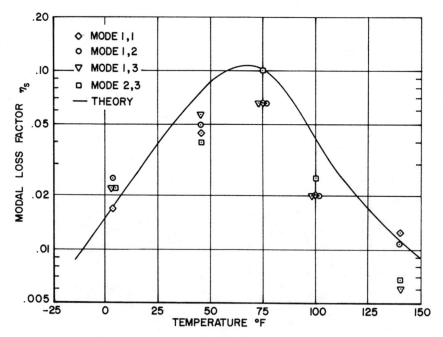

Fig. 6-40. Variation of η_s with temperature for LD-400.

the procedure of first applying a known treatment and measuring η_s for each mode must be followed. Figure 6-38 illustrates a typical damped response spectrum. Figure 6-40 shows the variation of η_s with temperature for several modes, for a layer of LD-400, 0.055-inch thick, bonded to the center area, and bounded by the integral stiffeners by means of 3M-410 adhesive tape. (This was for convenience in removing the treatment after testing.) Using the properties of this treatment, determined from cantilever beam tests, and Equations (6-32) and (6-33), the value of β_{nm} was estimated to be about 8.0. Since the temperature of peak damping was required to be near 0°F, this particular treatment was not used to solve this problem. Its use, however, was vital both to demonstrate that a layered treatment was an effective way of introducing significant amounts of damping into each and every mode at minimal cost and complexity, and to determine β_{nm}.

Multiple Constrained Layer Treatment

Since a lower temperature of peak damping was required, a material having suitable properties at 0°F was sought. The material selected was 3M-428A damping tape, consisting of a 0.002-inch adhesive layer on a 0.005-inch soft-aluminum sheet. The two-layer sheets were then built up, one upon the other,

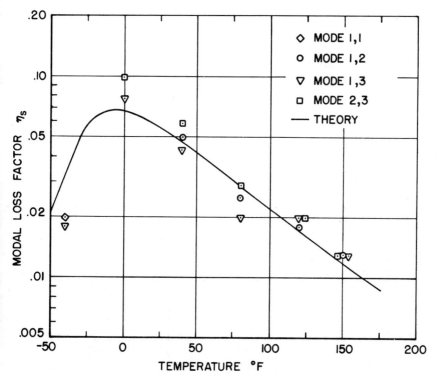

Fig. 6-41. Variation of η_s with temperature for 3M-428A.

to make a multiple-layer constrained damping treatment in the center area of the panel.

The design procedure which can be adopted here is, for each mode, to assume an initial guess as to the frequency. Then E_D and η_D are read off the complex modulus data for 3M-428. Using an estimate of λ_n for the mode, which of course involves some uncertainty (the effects can be assessed by trying several estimates for λ_n), and knowing E_C, h_C, and h_D. The values of E_e and η_e can be for the equivalent free layer, from Fig. 6-29. These values can then be used in conjunction with Equations (6-32) and (6-33) to estimate f_r and η_s. If the calculated value of f_r is markedly different from the initial estimate, the procedure should be repeated.

Calculated graphs of η_s versus temperature for five-layer pairs of 3M-428A are shown in Fig. 6-41, for an assumed value $\beta_{nm} = 8.0$, λ_n assumed to be about 8.0 inches for all the modes examined and $r \simeq 2/3$. It is seen that the design procedure outlined is in reasonable agreement with the experimental data. It is clear, therefore, that this treatment was acceptable for introducing significant amounts of damping from $-30°F$ to $+50°F$, or possibly, $+75°F$. At higher temperatures, the damping would probably prove to be inadequate. Subsequently it developed that this higher temperature damping was not needed to solve the service problem, but tests on multiple-layer treatments were car-

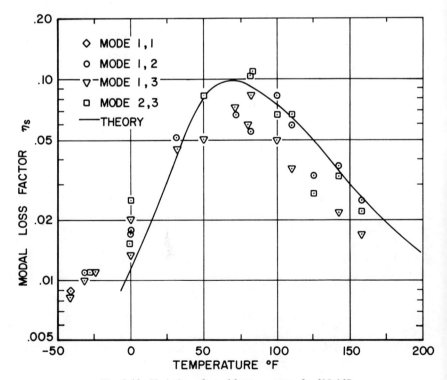

Fig. 6-42. Variation of η_s with temperature for 3M-467.

Fig. 6-43. IFF antenna.

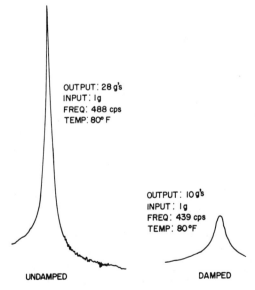

OUTPUT: 28 g's
INPUT: 1g
FREQ: 488 cps
TEMP: 80° F

OUTPUT: 10 g's
INPUT: 1g
FREQ: 439 cps
TEMP: 80°F

UNDAMPED   DAMPED

Fig. 6-44. Response spectra under shaker excitation.

ried out using 3M-467 adhesive instead of 3M-428. The results of analysis and experiment are shown in Fig. 6-42.

IFF Antenna Problem

This exercise [3, 17, 46] illustrates the procedure for solving a problem for which a layered damping treatment is totally unsuitable. The antenna geometry is illustrated in Fig. 6-43. The surface was highly curved with very small plate-like areas. Investigation of the response spectra near the center, illustrated in Fig. 6-44, showed a single resonance peak near 485 Hz, at which frequency the antenna vibrated like a diaphragm. No other resonance peaks were observed up to 2000 Hz. This problem was associated with high response amplitudes and low surface strains except near the rim, where early failure occurred under gunfire-induced excitation over a broad frequency range. Since a tuned damper is one of the few devices in which damping depends on displacement, or more precisely, acceleration, rather than surface strain, it was concluded that such a device must be designed to solve this particular problem.

The damper geometry was designed mainly on the basis of the need to attach the damper to the electrical connector. The geometry finally selected is shown in Fig. 6-45. The final geometry had a number of machined slots designed to

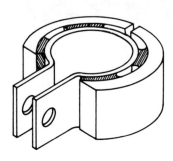
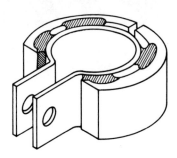

UNSLOTTED UNITS SLOTTED UNITS

Fig. 6-45. Damper geometry.

optimize bond strength and stiffness of the damper. The material stiffness was also varied through control of compounding and composition parameters until acceptable overall damping and strength levels were achieved over a broad temperature range. Figure 6-46 shows some measured values of the amplification factor, that is, the ratio of output response near the damper location to the input acceleration at resonance, as a function of temperature at various input levels. The material used was Paracril-BJ with 25 parts per hundred Carbon Black. Comparison with the low strain level damping behavior of the material, as in Fig. 6-10, shows a qualitative dependence of the damping on the closeness of the temperature to the transition temperature. This clearly shows

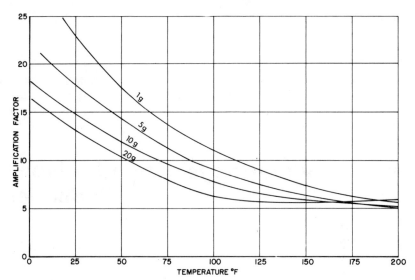

Fig. 6-46. Variation of amplification factor with temperature.

that the transition temperature should be as low as possible for good damper performance over a broad temperature range. The Carbon Black filter was used to increase the strength of the elastomer, but the stiffness of the elastomer also changed, as Figs. 6-9 and 6-10 show. The slots were machined in order to compensate somewhat for this increase in stiffness.

7

Vibration Isolation for Noise Reduction

JAMES F. HAMILTON
and
MALCOLM J. CROCKER

Introduction

Noise reduction by vibration isolation is concerned with reducing the forces and motions which result in vibration and noise radiation. When a vibrating machine is rigidly attached to its base the force or motion is directly transmitted to the base and can result in undesirable noise or vibration. A vibration isolator is placed between the source of the vibratory effect and the base, to produce a reduction in the transmitted force or motion.

The general concept of steady-state, harmonic vibration[1] isolation can be illustrated by a single-degree-of-freedom system[2] as shown in Fig. 7-1. The system consists of a rigid body representing the isolated piece of equipment which is connected to the base structure through the isolator.

There are two general types of isolation problems that the noise reduction engineer is faced with:

1. The source of the noise is known and is feeding into the base structure where it can produce undesirable radiated noise or motion. Most effectively the isolator is placed immediately between the source and the base to reduce the transmitted force into the base. Fig. 7-1A.
2. A portion of a structure (i.e., some panel or floor) is radiating undesired noise and the source of the noise is from this vibrating component; see

---

[1] Steady-State harmonic vibration assumes that the force, F, has a constant amplitude and is at discrete frequency (i.e., sinusoidal in time) and that the corresponding motion of the mass, X, also has a constant amplitude at the same discrete frequency.

[2] A single-degree-of-freedom system is one which is assumed to be free to move in one direction only (e.g., the X direction).

Ch. 7 VIBRATION ISOLATION FOR NOISE REDUCTION 263

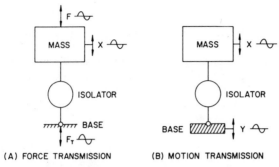

Fig. 7-1. Single-degree-of-freedom vibration isolation systems.

Fig. 7-1B. Isolation of the noise-radiating portion of the structure is necessary to reduce the motion being transmitted to it from the base. As was shown in Chapter 6, damping treatments on the radiating portions of the structure can also be utilized and are most effective when the panel or structure is in a resonant vibration condition; i.e., when vibrating at a natural frequency.

The effectiveness of the isolator is measured by the reduction of the transmitted force or motion from that of the input force or motion. One of the most used measures of the effectiveness of an isolator is the transmissibility. The *transmissibility* is defined as the ratio of the magnitude of the transmitted quantity to the magnitude of the excitation quantity. (Ref. 1)[3]

$$\text{Force Transmissibility} = \frac{\text{Force Transferred}}{\text{Force Input}},$$

$$T_F = \frac{F_T}{F} \tag{7-1}$$

$$\text{Motion Transmissibility} = \frac{\text{Motion Transferred}}{\text{Motion Input}},$$

$$T_M = \frac{X}{Y} = \frac{\dot{X}}{\dot{Y}} = \frac{\ddot{X}}{\ddot{Y}} \tag{7-2}$$

where motion can refer to displacement, velocity, or acceleration.

While motion transmissibility is defined for displacement, velocity, and acceleration, the velocity transmissibility is generally used because of the sound radiation dependency on the surface velocity of the radiating panel.

The *transmissibility* is the proportion of the vibratory effect transferred through the isolator. The *efficiency* of the isolator is the proportion of the vibratory effect that is *not* transmitted as determined by:

$$\eta_T = (1 - T) \cdot 100\% \tag{7-3}$$

---

[3] Reference number in parentheses refers to the bibliography at the end of the book.

264 VIBRATION ISOLATION FOR NOISE REDUCTION   Ch. 7

where:

T is T_M, or T_F, as applicable for a given situation.

Another useful measure of isolator effectiveness is the ratio of the magnitude of the dynamic displacement of the isolated machine to the magnitude of the displacement which would be caused by a static force of the same magnitude of the dynamic force. This ratio is called the *Dynamic Magnification Factor* and a small value is necessary to ensure a small motion of the machine.

$$DFM = \text{Dynamic Magnification Factor} = \frac{\text{Displacement of Isolated Mass}}{\text{Displacement if Excitation Force is Applied Statically}},$$

$$DMF = \frac{X}{X_0} = \frac{X}{F/K} \tag{7-4}$$

where:

$F/K = X_0$, the static deflection of the mass (magnitude of applied force divided by the spring rate).

SIMPLE VIBRATION–ISOLATION THEORY
Description of Theoretical Model

For the case of a machine resting on its suspension system on a floor, a simple single-degree-of-freedom system is often a good theoretical model as shown in Fig. 7-2. The isolators in the suspension system will possess both stiffness, K, and damping, R, and for simplicity, the actual isolators have been replaced with an equivalent spring and an equivalent damper. The base is assumed to be perfectly rigid compared to the stiffness of the isolator spring, K.

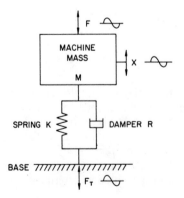

Fig. 7-2. Simple single-degree-of-freedom vibration isolation system with isolator having damping properties.

Ch. 7 VIBRATION ISOLATION FOR NOISE REDUCTION 265

For vertical displacement, the undamped natural or *resonance frequency*, f_n, of the machine resting on its suspension is given by:

$$f_n = \frac{1}{2\pi}\sqrt{K/M} = \frac{1}{2\pi}\sqrt{\frac{Kg}{W}} \qquad (7\text{-}5)$$

or

$$f_n = \frac{1}{2\pi}\sqrt{\frac{g}{\Delta_{static}}}$$

where:

f_n = machine suspension resonance frequency, (Hz),
K = equivalent spring stiffness, $\frac{lb}{in.}$, or $\frac{N}{m}$
M = machine mass, $\frac{lb}{in./sec^2}$, or Kg
W = weight; lb (no metric equivalent: in SI given as Mg)
g = gravitational acceleration; $\frac{ft}{sec^2}$, or $\frac{m}{s^2}$
Δ_{static} = static deflection of the mass on the spring.

Even if there is no exciting force present, the machine mass, M, will vibrate vertically at this frequency if displaced vertically and released.

With damping included in the system the natural frequency becomes:

$$f_d = f_n\sqrt{1 - (R/R_c)^2} \text{ Hz} \qquad (7\text{-}6)$$

This frequency, f_d, is called the damped resonant frequency; it is the free vibration frequency of oscillation when damping exists in the system, as shown in Fig. 7-2.

The damping coefficient, R, is the constant of proportionality equating the damper force and damper velocity, as given below:

$$R = \frac{F}{V}; \frac{lb}{in./sec} \quad \text{or} \quad \frac{N}{m/s} \qquad (7\text{-}7)$$

Critical damping is the minimum amount of damping to prevent any oscillation from occurring when a spring and damper-connected mass, as in Fig. 7-2, is displaced from its equilibrium position and released. This will occur when the damping is equal to:

$$2\sqrt{MK}$$

For each mass and spring rate this term is a constant which is defined as the critical damping coefficient, R_c; thus:

$$R_c = 2\sqrt{MK} \qquad (7\text{-}8)$$

266 VIBRATION ISOLATION FOR NOISE REDUCTION Ch. 7

To obtain a dimensionless number involving damping that will apply to all systems regardless of their dimensions, the damping coefficient is divided by the critical damping coefficient and this ratio is called the damping ratio, δ; thus:

$$\delta = \frac{R}{R_c} \qquad (7\text{-}9)$$

This ratio conveniently expresses the amount of damping in a system as compared to the critical damping.

Simple Harmonic Force or Motion Input

If the exciting force $F(t)$ is caused by out-of-balance, magnetic forces or similar excitations it is likely to be simple harmonic (sinusoidal in time). The vibration amplitude of the mass, X, can be found from:

$$DMF = \frac{X}{F/K} = \sqrt{\frac{1}{[1 - (f/f_n)^2]^2 + 4(f/f_n)^2 (R/R_c)^2}} \qquad (7\text{-}10)$$

where:

X = machine displacement, (in.), or meters
F = exciting force, lb, or N
f = frequency of excitation force (often the running speed of the machine) Hz
R = damping coefficient, $\frac{\text{lb}}{\text{in./sec}}$, or $\frac{N}{m/s}$
R_c = critical damping coefficient; $\frac{\text{lb}}{\text{in./sec}}$, or $\frac{N}{m/s}$
K = spring rate of the isolator; lb/in.
f_n = natural frequency as defined by Equation (7-5), Hz

Equation (7-10) is plotted in Fig. 7-3.

If the input time-varying force, $F(t)$, occurs at the resonance frequency f_n, then very large amplitudes of the mass motion occur. In fact, for zero damping, $R = 0$, an infinite amplitude is predicted. In practice, there is always some damping present and the amplitude of the machine displacement is always finite, even though it may become quite large at resonance.

Equation (7-10) is given in a non-dimensional form, by dividing the vibration amplitude X by F/K. The value, F/K, is the *static* displacement which would occur if the machine were subjected to a *static* force, F. This ratio $X/(F/K)$ is called the *dynamic magnification factor, DMF*.

Force Transmissibility

If the base is assumed to be perfectly rigid then, of course, it will not move. However there will still be a time varying force transmitted to the base $F_T(t)$ equal to the summation of the spring force $Kx(t)$ and the damper force $R\dot{x}(t)$.

Ch. 7 VIBRATION ISOLATION FOR NOISE REDUCTION 267

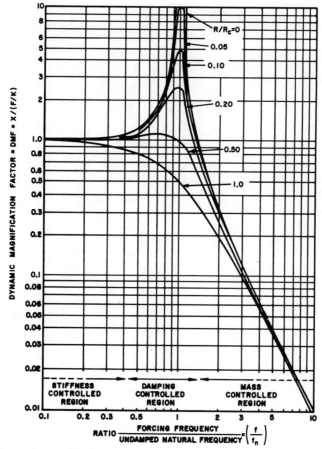

Fig. 7-3. Dynamic magnification factor with respect to the ratio of the forcing frequency and the natural frequency of a simple single-degree-of-freedom system.

The ratio of the amplitude of the force transmitted to the base, F_T, to the amplitude of the exciting force, F, is given by (Ref. 3):

$$T_F = \frac{F_T}{F} = \sqrt{\frac{1 + 4(f/f_n)^2(R/R_c)^2}{[1 - (f/f_n)^2]^2 + 4(f/f_n)^2(R/R_c)^2}}. \quad (7\text{-}11)$$

This equation is plotted in Fig. 7-4.

Stiffness, Damping, and Mass-Controlled Regions

If the exciting frequency, f, is close to the resonance frequency, f_n, then the frequency ratio $f/f_n \approx 1$ and provided the damping ratio $R/R_c = \delta$ is small,

268 VIBRATION ISOLATION FOR NOISE REDUCTION Ch. 7

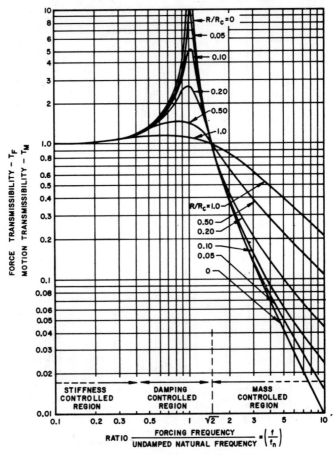

Fig. 7-4. Force transmissibility and motion transmissibility with respect to the ratio of the forcing frequency to the natural frequency of a simple single-degree-of-freedom system.

Equations (7-10) and (7-11) give:

$$DMF \approx \frac{1}{2\delta} > 1,$$

$$T_F \approx 1 + \frac{1}{2\delta} > 1. \tag{7-12}$$

In the region where the forcing frequency is nearly equal to the resonant frequency, or $f/f_n \approx 1$, both the force transmitted and the motion of the machine mass, M, are large and inversely proportional to the damping ratio, δ. In this region increasing the damping decreases both the force transmitted and the motion of the mass, while, inversely, decreasing the damping increases them. For this reason the region where $f \approx f_n$ is known as the *damping controlled region* (see Figs. 7-3 and 7-4).

Ch. 7 VIBRATION ISOLATION FOR NOISE REDUCTION 269

In the region where the forcing frequency, f, is much less than the resonance frequency f_n, f/f_n approaches zero and Equations (7-10) and (7-11) give:

$$DMF = \frac{X}{F/K} \approx 1$$

$$T_F = \frac{F_T}{F} \approx 1 \qquad (7\text{-}13)$$

In this case it can be seen that $F_T \approx F$; i.e., the transmitted force is approximately equal to the exciting force. Also, $X \approx F/K$, which shows that the motion of the mass is inversely proportional to the isolator stiffness, K. For this reason, the frequency region where $f \ll f_n$, is called the *stiffness controlled region*. If the exciting frequency, f, is large compared to the natural frequency, f_n, ($f \gg f_n$) then the motion can be shown to be inertia or mass controlled. All three regions are marked on Figs. 7-3 and 7-4.

In Fig. 7-4 it can be seen that the force transmitted to the base is always equal to or greater than the applied force in the region where f/f_n is less than $\sqrt{2}$. Considering the machine mass and the isolator as a system (see Fig. 7-2), this system should be so designed that it will not operate in the region where f/f_n is less than $\sqrt{2}$. While damping will reduce the transmitted force in this region, it will always equal or exceed the applied force to the mass. If a steady-state operation of the machine in this region cannot be avoided, it would be best not to use an isolator and to bolt the machine firmly to the floor, provided that the floor is rigid. In this event the force transmissibility, T_F, is equal to one, which is less than it would be if damping were present. This practice is common for certain types of machinery; e.g., machine tools. However, it should be emphasized again that whenever possible the system should not be designed to operate in the region where $f/f_n < \sqrt{2}$.

As shown in Fig. 7-4, when the exciting frequency is increased so that $f/f_n > \sqrt{2}$ the mass-controlled region is entered where $T_F < 1$; in fact, if $f/f_n \gg 1$, then $T_F \ll 1$, which is highly desirable from the force isolation point of view. Thus, this region is also called the vibration isolation region. By noting the effect of the damping ratio, R/R_c, in the vibration isolation region (or mass-controlled region) in Fig. 7-4, it is evident that the force transmissibility, T_F, is adversely affected by damping. Considering the system in Fig. 7-2 in which the isolator has damping, for use on a system operating in the vibration isolation region, the isolator selected should have as little damping as possible to obtain the maximum transmissibility. For operation in this region it is desirable to design the isolation system so that $f/f_n \gg 1$ and R/R_c is at a minimum. All isolators do have some damping, as shown in Table 7-1, and the curve in Fig. 7-4 for which $R/R_c = 0$, is theoretical only.

If a machine or engine is operating in the vibration isolation region a dangerous situation could arise when passing through the damping controlled region on coming up to or down from the operating speed. If the damping control region is passed rapidly it is unlikely that a high amplitude or transmissibility will be reached; however, if this region is passed through more slowly the amplitude and transmissibility can be very large. When this is likely to occur the isolator must

Table 7-1. Typical Damping Ratios

Material	Damping Ratio $\delta = R/R_c$	T_F (max.)
Spring Steel	0.005	100.0

Elastomers
  Natural Rubber
  Neoprene
  Silicone Base
  Low-Temperature
	
	0.05
	0.05
	0.15
	0.12
		
		10.0
		10.0
		3.5
		4.5

Friction-Damped Springs Metal Mesh Air Damping Felt and Cork	0.33 0.12 0.17 0.06	1.5 4.0 3.0 8.0
Air Mounts	Variable by means of flow orifices and surge tanks.	100.0

be designed to have a higher damping ratio at some sacrifice to its performance at the steady-state operating speed.

Displacement, Velocity, and Acceleration Transmission

As discussed at the beginning of the chapter, it is sometimes desired to isolate a portion of a structure against undesired motion. Again, a simple single-degree-of-freedom system with harmonic input is a good theoretical model as shown in Fig. 7-5. The ratio of the displacement amplitude transmitted to the mass, X, to the input displacement amplitude of the base, Y, is given by (Ref. 3):

$$T_M = \frac{X}{Y} = \sqrt{\frac{1 + 4(f/f_n)^2 (R/R_c)^2}{[1 - (f/f_n)^2]^2 + 4(f/f_n)^2 (R/R_c)^2}} \qquad (7\text{-}14)$$

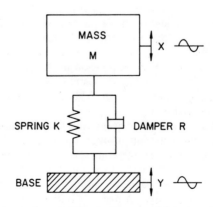

Fig. 7-5. Simple one-degree-of-freedom system with isolator having damping.

Ch. 7 VIBRATION ISOLATION FOR NOISE REDUCTION 271

where:

T_M = Motion transmissibility
X = Amplitude of the mass (or machine) motion; inches, or mm.
Y = Amplitude of the base motion; inches, or mm.
(Other symbols defined on page 266.)

This result is the same as that for force transmissibility as given by Equation (7-11). The ratios of mass velocity to base velocity ($\dot{X}/\dot{Y}$) and acceleration amplitudes ($\ddot{X}/\ddot{Y}$) are also given by Equation (7-14); i.e., in solving for these ratios, the identical equation results. Hence, Fig. 7-4 for force transmissibility may be used interchangeably if the vertical axis is read as any of the motion transmissibilities T_M. See Equations (7-2) and (7-14). Additionally, all the factors which apply in the reduction of force transmissibility, T_F, apply equally to the reduction of motion transmissibility, T_M.

All three of the performance indices, transmissibility, isolation efficiency, and dynamic magnification factor, are dimensionless to allow general application to different size systems.

Most noise problems involving machinery have a variety of mechanisms that can produce the exciting forces, i.e., inertia forces, magnetic forces, unbalance, misalignment, periodic impacts, gear drives, etc. The vibration isolator designer usually cannot settle for reduction at a single frequency but must consider a broad range of isolation frequencies determined by the range of frequencies of the input force or motion, the range of important resonance frequencies of the structure, and the audible frequency range. Some machines or structures have exciting inputs which are primarily shock inputs causing additional complication in the isolation treatment, as will be discussed later in this chapter.

Practical Difficulties

It is highly desirable for low force or motion transmissibility to ensure that $f/f_n \gg 1$. If f/f_n is to be made large, relative to $\sqrt{2}$, this usually means that f_n must be made small, since the frequency of the exciting force, f, is normally fixed by the operational requirements of the machine. Equation (7-5) suggests that either the machine mass, M, must be made large or the spring stiffness, K, must be made small. In most practical cases it is easier to make K small. However, a practical difficulty can arise with the stability of the suspension system in the lateral direction. If the vertical deflection is large, the machine may be unstable in a horizontal plane and "tip" from side to side while operating. This condition is of most concern for a lightweight device such as a fan, blower, pump, compressor, etc. The usual practice in designing an isolation system for such applications is to provide a concrete base for the machine which is at least ten times heavier than the machine and then to select suitable springs to suspend the machine plus the isolation base. This approach provides acceptable isolation with smaller, lateral deflection of the springs.

The static deflection, X_s, of the mass, M (i.e., the amount the spring deflects when the mass is placed on it) is:

$$X_s = \frac{Mg}{K}, \text{ or } \frac{W}{K} \qquad (7\text{-}15)$$

where: g = gravitational constant, 386, $\left(\dfrac{\text{in.}}{\text{sec}^2}\right)$ substituting for K/M from Equation (7-4) gives:

$$X_s = \dfrac{g}{4\pi^2 f_n^2} \qquad (7\text{-}16)$$

The static deflection X_s plotted against f_n from Equation (7-16) is shown in Fig. 7-6. The static deflection can be one way, in principle, of experimentally determining the resonance frequency f_n of the system from the relation:

$$f_n = \dfrac{1}{2\pi} \sqrt{\dfrac{386}{\Delta_{\text{static}}}} \text{ Hz} \qquad (7\text{-}5)$$

where Δ_{static} is the static deflection of the system on the isolator, in inches. Values for f_n are in Table 7-4, page 301, for various ranges of static deflection.

In practice it is common to specify an isolation system by the required static deflection of the machine on the isolator. This is a concept easily understood; to give the weight of the machine and the desired static deflection is the usual American practice.

In order to achieve appreciable vibration-isolation when the exciting frequency, f, is small, f_n must be very small so that $f/f_n \gg 1$ is maintained. If

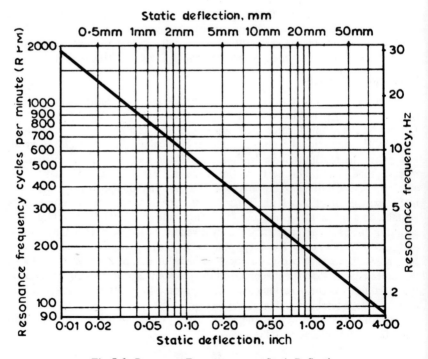

Fig. 7-6. Resonance Frequency versus Static Deflection.

Ch. 7 VIBRATION ISOLATION FOR NOISE REDUCTION 273

f_n is very small, as discussed above, K must be very small, or springs having a very low spring constant must be used. However, as Fig. 7-6 shows, this can result in an extremely large static deflection which it may not be possible to tolerate in practice. Springs having a very low spring constant may result in an unstable machine which could be tipped over. When this condition exists, the usual practice is to mount the machine on a heavy base, usually concrete, and then to select isolators for the total weight of the machine and base. If the base has a relatively large surface area and a low center of gravity, the "tipping" problem can be minimized.

Example

Consider a machine where the exciting frequency, f, is equal to 20 Hz. If the damping ratio, $R/R_o = \delta$, is equal to 0.2 and the desired transmissibility, $T_F \approx 0.05$, determine the static deflection of the isolators. From Fig. 7-4 the frequency ratio, f/f_n, is found to be approximately 9.

$$\frac{f}{f_n} = 9$$

and

$$f_n = \frac{f}{9} = \frac{20}{9} = 2.2 \text{ Hz}$$

From Fig. 7-6 with f_n = 2.2 Hz, the static deflection is found to be approximately 2 inches. This much static deflection may be too large to be tolerated for stability in the horizontal directions. To minimize horizontal instability a heavy base could be used which will lower the center of gravity of the system. Another remedy is to provide horizontal bumper stops to limit the horizontal motion of the system. A third possible solution is to use air mounts. These are very costly compared to steel-spring type isolators. Although more costly, in some instances air mounts are the better solution.

Typical Isolators

Isolators may be divided into the following types:
Springs.—Metal springs are the most common form of isolator in use. They have several advantages including the wide range of stiffnesses available and their freedom from creep. However, they have the disadvantage of low inherent damping. Damping is normally added by using friction dampers or other devices such as air chambers with orifices or wire-mesh inserts. Most spring isolators are normally used in compression and have linear deflection rates unless several springs are combined in one design. A typical coil-spring isolator and load/deflection curve is shown in Fig. 7-7.

Elastomers.—Elastomers such as natural rubber, neoprene, and butyl are often used in the construction of isolators. They can be used either in compression or shear. They have an advantage in that they normally possess sufficient internal damping that permits them to be operated at the machine resonance frequency for short periods of time. Special, higher cost elastomers can be produced to operate at temperatures up to 400°F. Elastomers have a disadvantage in that

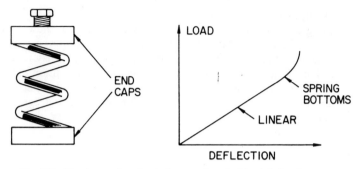

Fig. 7-7. Simple metal spring isolator and typical load-deflection curve.

they may creep excessively over a period of time. For that reason they should not be subjected to continuous strains exceeding 10 or 15 percent in compression or 25 to 50 percent in shear. Typical elastomeric isolators used in compression and shear are shown in Fig. 7-8. Further information may be found in (Refs. 5 and 6).

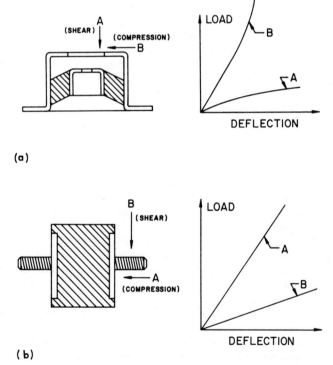

Fig. 7-8. Simple elastomeric isolators: a. Strip isolator (extending normal to plane of paper); b. Solid elastomeric isolator.

Ch. 7 VIBRATION ISOLATION FOR NOISE REDUCTION

Pneumatic Isolators

Pneumatic isolators, air springs, or air mounts are particularly useful when low frequency forces (e.g., 5 or 10 Hz) are present. Noise at such a low frequency is not normally a problem from the hearing point of view although it can result in unpleasant physical sensations, such as nausea. More often it is important to protect delicate instrument packages from such low-frequency excitations. In this case, very low frequency isolators are needed with resonance frequencies in the range, $0.5 < f_n < 3.0$ Hz. A spring isolator would have a static deflection of 1.1 inch for a 3.0 Hz resonance frequency and 3.3 feet for a 0.5 Hz resonance frequency! However, when the machine rests on an air spring, zero static deflection can be achieved. This is partly obtained by further inflation of the isolator when under the machine weight. Air springs also have the advantage that the damping is large at resonance but small at the high frequencies where it offers very good isolation. Figure 7-9 illustrates a simple, sealed pneumatic isolator.

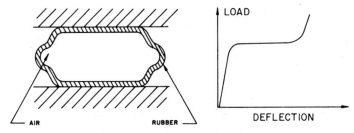

Fig. 7-9. Simple pneumatic isolator.

The stiffness of the air cavity of the pneumatic isolator is given by (Ref. 6):

$$K = \frac{\gamma M g A}{V_o}, \quad (7\text{-}17)$$

and the resonance frequency is

$$f_n = \frac{1}{2\pi}\sqrt{\frac{K}{M}} = \frac{1}{2\pi}\sqrt{\frac{\gamma g A}{V_o}}, \quad (7\text{-}18)$$

where:

γ = ratio of specific heats = 1.4 for air,
A = load supporting area of air spring, in.^{2},
V_o = air cavity volume, in.^{3}
g = gravitational constant, 386 in./sec.^{2}

Other Materials

Several other materials are used as vibration isolators, such as wool felt, cork, glass fibers, foam, wire mesh, die-cut elastomer parts and others. These are often used in pad or blanket form. These materials have their uses in vibration-isolation; however, their properties, including durability, are not as well documented as the spring, elastomeric, and pneumatic isolators discussed above.

Manufacturers' specifications should be obtained for selecting any type of isolator system. Up-to-date information is readily available for accurate analysis. Table 7-1 gives typical values of damping and maximum values of force transmissibility allowed for different materials.

High-frequency Effects

Although the simple single-degree-of-freedom theory described earlier under "Simple Vibration," is adequate for the design of vibration isolators in most cases, in some cases it produces less isolation than expected. This problem is due to a variety of causes and primarily results in more transmissibility at high frequencies than the simple theory predicts. Since the ear is more sensitive to high-frequency noise (greater than about 1000 Hz) this is sometimes important.

The main reasons for the failure of the simple theory at high frequencies are:

1. The foundation cannot always be assumed to be perfectly rigid. Sometimes it is not much stiffer than the isolators used and causes additional higher frequency resonances to occur with an increase in transmissibility at these frequencies (see Fig. 7-10).
2. The machine, itself, cannot always be assumed to behave as a concentrated, rigid mass. Its mass is actually spatially distributed with inherent internal flexibility. This effect causes additional resonances to occur at the higher frequencies with associated increased transmissibility at these frequencies (see Fig. 7-11).
3. The vibration isolators do not have zero mass as assumed previously. The isolator mass is distributed through the isolator and at high frequencies standing waves will occur, again causing increased transmissibility.

Foundation Flexibility

The effect of flexibility of the foundation on the isolator transmissibility must be considered in the selection of practical vibration isolation mountings. In the simplified earlier analysis, the foundation was assumed to be completely rigid. It was also assumed that the function of the isolator system was to allow the mass of the machine to move in such a manner that most of the excitation force was balanced by the inertia force of the machine mass. Generally, the relatively low stiffness of the isolator system permits this effect while allowing the foundation to remain substantially motionless. However if the stiffness of the isolator is allowed to become comparable to the foundation stiffness (or greater), the deflection of the isolator will become smaller and the foundation will also deflect with increased transmissibility and decreased isolator efficiency.

In a dynamic sense, supporting foundations or floors should have natural frequencies as high and be as stiff as possible compared to the system being isolated. If the natural frequencies of the machine-isolator system and the foundation are far apart, ideally in the order of at least 10 to 1, the two natural frequencies of the combination will be essentially the same as the two separate natural frequencies and the transmissibility curve will be similar to Fig. 7-10 as

Ch. 7 VIBRATION ISOLATION FOR NOISE REDUCTION 277

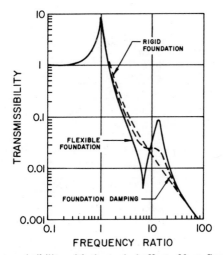

Fig. 7-10. Transmissibility with the typical effect of base flexibility added.

taken from (Ref. 1). Note that this is the same as Fig. 7-4 with the effect of foundation flexibility added.

Good design practice calls for designing the isolators assuming a rigid foundation with the stipulation that the selected machine-isolator system frequency should be well below the foundation frequency. The first resonance for wooden floors is generally in the 20 to 30-Hz region while for concrete floors it is in the 30 to 100-Hz region depending on the base under the concrete floor. Additionally, to reduce base resonance effects, the machine should not have any operating frequencies within 0.8 to 1.3 times the natural frequencies of the base. As shown in Fig. 7-10, damping in the foundation is very helpful.

Devices Attached to Lightweight Panels

In the previous section, vibration isolation was considered for the case of a machine supported by an isolation system on a foundation. This assumption is reasonable in many cases where the foundation is massive compared to the machine to be isolated or if the foundation is supported by the ground. In situations such as a motor, fan, linkage, flow valve, etc., attached to a lightweight panel, the panel cannot be assumed to be rigid. This case is of great concern in noise control since the panel may have an imposed motion due to the transmitted force and result in radiation of sound.

When a device is attached to a nonrigid panel the motion of the panel is the concern, since it is the element which will radiate sound in most cases rather than the motion of the device attached to the panel. If the damping is neglected ($R = 0$), the motion of the panel at the connection point can be approximated as

$$X_{panel} = \frac{F(t)}{(M_1 + M_2)(2\pi f)^2 \sqrt{1 - \left(\frac{f}{f_r}\right)^2}} \qquad (7\text{-}19)$$

278 VIBRATION ISOLATION FOR NOISE REDUCTION Ch. 7

where:

$$f_r = \frac{1}{2\pi} \sqrt{\frac{K}{M_1 M_2/(M_1 + M_2)}} \text{ Hz} \qquad (7\text{-}20)$$

$F(t)$ = forcing function, lb or N
f_r = resonant frequency of the system, Hz
M_1 = mass of device to be isolated, lb/in./sec^2, or Kg
M_2 = mass of the panel, lb/in./sec^2, or Kg
K = spring rate of isolator, lb/in., or Kg/m

For this system the ratio of transmitted force to applied force is given by

$$\frac{F_T}{F} = \frac{M_2}{M_1 + M_2} \sqrt{\frac{1}{[1 - (f/f_r)^2]^2}} \qquad (7\text{-}21)$$

Note that except for the constant $M_2/(M_1 + M_2)$ the magnitude of the force ratio has the same form as for the case of a rigid foundation with damping, R, equal to zero as given by Equation (7-11). This means that Figs. 7-3 and 7-4 are valid for the case of a nonrigid panel if the magnitude is multiplied by $M_2/(M_1 + M_2)$ and f_r is used instead of the natural frequency for a rigid panel, f_n. For a rigid or nonrigid panel the ratio of forcing frequency to natural frequency should be large compared to 1, for example:

$$\frac{f}{f_n} \gg 1, \quad \text{or} \quad \frac{f}{f_r} \gg 1$$

If the panel mass is much greater than the mass to be isolated, the approximation that the foundation is rigid is quite good; for example, if $M_2 = 10 M_1$

$$\frac{M_2}{M_1 + M_2} = \frac{10 M_1}{M_1 + 10 M_1} = \frac{10}{11} = .91 \quad \text{(See Equation 7-21.)}$$

and

$$\frac{M_1 M_2}{M_1 + M_2} = \frac{10 M_1 M_1}{M_1 + 10 M_1} = \frac{10}{11} M_1 = .91 M_1 \quad \text{(See Equation 7-20.)}$$

Thus, both the natural frequency and the ratio of transmitted force to applied force will be approximately equal to the corresponding values for the rigid panel assumption. If, on the other hand, the panel is light such that $M_1 = M_2$ then,

$$\frac{M_2}{M_1 + M_2} = \frac{M_1}{M_1 + M_1} = \frac{1}{2} \quad \text{(See Equation 7-21.)}$$

and

$$\frac{M_1 M_2}{M_1 + M_2} = \frac{M_1 M_1}{M_1 + M_1} = \frac{1}{2} M_1 \quad \text{(See Equation 7-20.)}$$

therefore

$$f_r = \frac{1}{2\pi} \sqrt{\frac{K}{\frac{1}{2} M_1}} = \frac{\sqrt{2}}{2\pi} \sqrt{\frac{K}{M_1}} = \sqrt{2} f_n \qquad (7\text{-}22)$$

Ch. 7 VIBRATION ISOLATION FOR NOISE REDUCTION 279

The resulting natural frequency for the nonrigid panel, f_r, is greater than the natural frequency for the rigid panel, f_n, and hence, an isolator with a larger static deflection is required to obtain the same ratio of forcing frequency to natural frequency, $\frac{f}{f_n}$ or $\frac{f}{f_r}$, and to obtain the same magnitude of transmitted force.

Internal Resonance Effects

As with the effect of flexibility in the foundation, internal flexibility and resonances in the machine can change the transmissibility curve. Small magnitude excitations can be amplified by internal resonances causing increases in the transmissibility from that of the rigid machine curve as shown in Fig. 7-11.

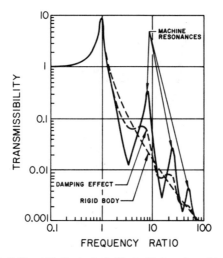

Fig. 7-11. Transmissibility with the typical effect of internal machine resonances added.

Care should be taken with systems having many flexible members and broad-band frequency excitation. Stiffening the machine to raise the frequency of the internal resonances and/or damping material applied to the machine members will substantially reduce the effect of the resonances. Generally, little is done concerning these resonances unless they are too close to the isolator natural frequency or can be shown to be a major contributor to the noise or vibration problem.

Wave Effects in Isolators

Wave effects in isolators can significantly alter the transmissibility at high frequencies, particularly when the mass of the isolators is significant compared to the machine mass. The transmissibility at these higher frequencies is also dependent on the damping present in the isolators. Wave effects in isolators have been studied by several authors (Refs. 8 and 9). Theoretical results from (Ref. 9) are presented in Fig. 7-12 for three different values of the ratio μ of machine mass M to isolator mass m. Also shown is a curve resulting from a measured response for $\mu = 70$. The results are presented for two values of damping ratio ($\delta = 0$ and $\delta = 0.03$), and for three different values of the ratio $\mu = M/m$. Experimental results essentially confirm the theoretical findings (Ref. 10).

Figure 7-12 shows that standing waves in the isolators can reduce their effectiveness at frequencies starting about thirty times greater than the resonance frequency f_n of a machine-isolator system. The first standing wave resonance occurs when there is a half-wavelength of displacement in the isolator. The second harmonic occurs at twice this frequency when there is a full wavelength, the third harmonic at three times the frequency when there is a wavelength and a half, and so on.

In order to reduce the wave effects in an isolator so that high-frequency vibra-

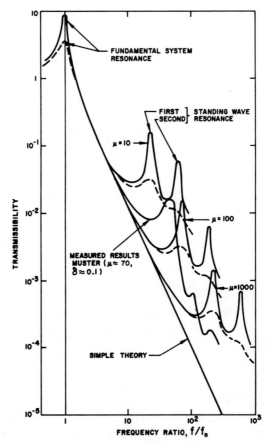

Fig. 7-12. Transmissibility with the typical wave effect of the isolators added: $\delta = 0$; $\delta = .03$ (measured results from Muster $\mu \approx 70$, $\delta \approx 0.1$ [Ref. 10]). $\delta = R/R_c$.

Ch. 7 VIBRATION ISOLATION FOR NOISE REDUCTION

Table 7-2. Characteristic Impedances of Some Common Materials (Bar Values Given Where Possible)

Material	Mass Density, ρ $\left(\dfrac{\text{lb-sec}^2}{\text{ft}^4}\right)$	Wave Speed, c $\left(\dfrac{\text{ft}}{\text{sec}}\right)$	Characteristic Impedance, ρc $\left(\dfrac{\text{lb-sec}}{\text{ft}^3}\right)$
Aluminum	5.24	16,900	8.85×10^4
Cast Iron	14.90	12,130	18.10×10^4
Lead	21.90	3,940	8.64×10^4
Steel	14.90	16,550	24.65×10^4
Glass	4.45	17,050	7.60×10^4
Concrete	5.04	10,200	5.14×10^4
Cork	0.465	1,640	0.0762×10^4
Pine	0.871	11,500	1.00×10^4
Rubber (Hard)	2.13	4,760	1.01×10^4
Rubber (Soft)	1.84	230	0.0424×10^4
Water (Fresh)	1.935	4,860	0.940×10^4
Air	0.00234	1,125	0.000263×10^4

tion and subsequent noise is more efficiently isolated, low-mass isolators should be used together with appreciable damping. As shown in Table 7-2, steel springs normally have lower damping and higher mass than elastomeric isolators. Hence, elastomeric isolators are to be preferred where high-frequency isolation is an important problem.

If spring isolators are in use with a machine, the additional use of rubber, felt, or some other soft material in series with the spring will normally reduce or eliminate the high-frequency problem due to an impedance mismatch between the isolators.

Impedance Mismatch

It is a well-known fact in electricity, acoustics, and the theory of vibration that if impedances are matched there is a maximum rate of energy transfer from one system or medium to another. If there is an impedance mismatch then there is a low rate of energy transfer. In vibration isolation, of course, one is interested in a low transfer of energy.

For simple harmonic stress waves (in solids) or pressure waves (in fluids), it may be shown that the fraction of incident energy transmitted from medium (1) to medium (2) (if the media are infinite in extent and have the same surface area) is:

$$\alpha_t = \frac{4\dfrac{\rho_1 c_1}{\rho_2 c_2}}{\left(1 + \dfrac{\rho_1 c_1}{\rho_2 c_2}\right)^2}, \qquad (7\text{-}23)$$

and the fraction of energy reflected is:

$$\alpha_r = \frac{\left(1 - \dfrac{\rho_1 c_1}{\rho_2 c_2}\right)^2}{\left(1 + \dfrac{\rho_1 c_1}{\rho_2 c_2}\right)^2} \qquad (7\text{-}24)$$

In these equations α_t is the *Transmission Coefficient* and α_r is the *Reflection Coefficient*.

The product ρc of the density ρ, and speed of sound, c, is known as the *characteristic impedance* of the medium. It is immediately obvious from equations (7-23) and (7-24) that if $\rho_1 c_1 = \rho_2 c_2$ that $\alpha_t = 1$ and $\alpha_r = 0$; all the energy is transmitted and none is reflected. If $\rho_1 c_1 \gg \rho_2 c_2$ or $\rho_2 c_2 \gg \rho_1 c_1$ then α_r approaches 1, all the energy is reflected; none is transmitted.

The second case, when $\rho_2 c_2 \gg \rho_1 c_1$ or $\rho_1 c_1 \gg \rho_2 c_2$, is the case of impedance mismatch; the case desired in vibration isolation. This simple theory is often used when machines made from dense materials are placed on felt, rubber, fiberglass, or other low-density materials to achieve vibration isolation, and is a useful concept. Unfortunately, the theory is too simple because the media under question, the metal of the machine and the felt or rubber pad are not infinite in extent. This causes severe complications in the theory because of the standing waves which will exist in the media. However, this simple theory can at least be used as a guide. Characteristic impedances of some common materials are given in Table 7-2.

Example of a Rubber Pad Under a Steel Spring

Suppose it is desired to place some rubber under the spring isolators of a machine to reduce high-frequency transmission through the steel springs to the foundation. As discussed above, this would be expected to be effective if the spring impedance and rubber impedance are sufficiently mismatched.

For this case Equation (7-23) can be used if we let $\rho_1 c_1$ equal the impedance of the spring and $\rho_2 c_2$ equal the impedance of the rubber, i.e.:

$$\rho_1 c_1 = Z_{\text{spring}} \qquad \rho_2 c_2 = Z_{\text{rubber}}$$

then:

$$\alpha_t = \frac{4 \dfrac{Z_{\text{spring}}}{Z_{\text{rubber}}}}{\left(1 + \dfrac{Z_{\text{spring}}}{Z_{\text{rubber}}}\right)^2} \qquad (7\text{-}25)$$

The impedance, Z_{spring} of a steel coil spring is given by the following relation (Ref. 8):

$$Z_{\text{spring}} = \left(\frac{1}{2\sqrt{1+v}}\right)\left(\frac{d}{D}\right)(\rho c) \qquad (7\text{-}26)$$

Ch. 7 VIBRATION ISOLATION FOR NOISE REDUCTION 283

where:

ν = Poisson's ratio
d = Wire diameter (in.)
D = Coil diameter (in.)
ρ = Steel density = 0.000725 lb/in.^{4}/sec^2
c = Speed of sound in steel = 16,550 ft/sec.

Hence, for the steel spring:

$$Z_{spring} = \left(\frac{1}{2\sqrt{1+0.3}}\right)\left(\frac{1}{5}\right)[0.000725(16,550)12]$$

$$= (0.439)(0.2)[144]$$

$$= 12.6 \frac{lb}{in.^3/sec}$$

Choosing a soft rubber for the pad, the characteristic impedance is obtained from Table 7-2 as:

$$\rho c_{rubber} = 0.0424 \times 10^4 \frac{lb\text{-}sec}{ft^3} = .25 \frac{lb\text{-}sec}{in.^3}$$

therefore

$$\frac{Z_{spring}}{\rho c_{rubber}} = \frac{12.6}{.25} = 50 \text{ (rounded)}$$

A good impedance mismatch has apparently been achieved. If the simplifying assumption that the spring and rubber are infinite in extent is made, Equation (7-23) gives

$$\alpha_t = \frac{4(50)}{(1+50)^2} \approx .08$$

This simple theory suggests that only 8 percent of the energy transmitted down the spring would be transmitted into the rubber.

Additional Effects

Although most isolation systems are designed using the simple vibration-isolation-system theory, there are additional factors that should be considered which can yield practical improvements or prevent difficulties in a given isolation system. The following discussion of certain of these factors is designed to give the reader a qualitative view for use in isolator design.

Inertia Blocks.—In some cases the use of an isolator with sufficiently low stiffness to isolate the vibration may result in an isolated machine which wobbles so badly that it may not operate properly. It is frequently desirable to mount the machine on a more massive inertia block which is, in turn, supported by isolators as shown in Fig. 7-13. There are several advantages to this arrangement. One highly desirable feature is that the stiffness of the isolator system can be

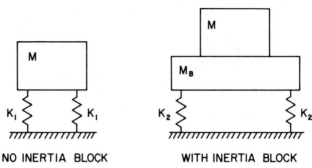

Fig. 7-13. Isolated machine with and without inertia block.

greatly increased by the addition of the inertia block. To maintain the same natural frequency and, therefore, the same transmissibility, the stiffness of the isolators must be increased in proportion to the increased mass.

1. Increasing the isolator stiffness decreases the vibration amplitude by decreasing F/K, the displacement for the force applied statically [see Equation (7-10)].
2. If the original stiffness was so low as to be physically difficult to attain, the increased stiffness due to the increased mass will be easier to attain.

The addition of an inertia block can be especially effective in the case of machines which have massive unbalanced moving members since the movement tends to be proportional to the ratio of the unbalanced moving members to the fixed masses of the machines. Typical of installations of this type are drop hammer machines. Reciprocating engines or compressors also represent a special problem in that close to one-third of the engine mass may be moving with the system fundamentally out-of-balance. Units with an odd number of cylinders will present an even more severe situation. Acceptable transmission isolation may be obtained by the use of appropriately selected isolators, but "large" amplitude motions of the isolated machine may still result from the severe unbalanced forces. To keep this motion to a level acceptable for connection to ancillary systems through flexible connections, an inertia base of two to five times the mass of the machine may be needed.

Many complete machine systems consist of a separate electric motor or engine driving a machine, for example, an electric motor and large blower connected by a semirigid drive shaft. To maintain alignment these components should never be mounted on independent isolation systems. Having the system mounted as a unit on a rigid concrete inertia block of one or more times the mass of the machine system will simplify the isolator selection and maintain small deflections.

A large concrete base will also serve as a rigid support for machines requiring a support for proper operation, such as a power shear. A base of this type will also help to isolate the machine from external equipment. For example, isolating a sensitive grinder from the disturbance of a forklift truck moving across the surrounding floor.

Ch. 7 VIBRATION ISOLATION FOR NOISE REDUCTION

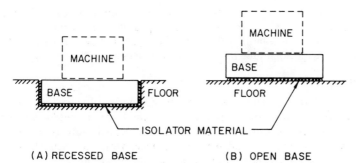

(A) RECESSED BASE   (B) OPEN BASE

Fig. 7-14. Recessed and open inertia block machine bases.

Typical concrete bases are illustrated in Fig. 7-14. Fig. 7-14A illustrates a built-in recessed base while Fig. 7-14B illustrates a nonrecessed base which can be moved with the machine. Usually, the weight of the base should be at least equal to the weight of the machine and in some cases much heavier. For lightweight machinery a base weight as much as ten times the machine weight might be used.

One of the primary advantages of an inertia block is easier control of the location of the center of gravity with respect to the machine force and the reaction forces of the isolators. This control will permit reduction of the vibration of the base to simple uncoupled motions in the coordinate directions (elemental single-coordinate modes).

Modes of Vibration

In the case of free vibrations a *mode of vibration* is a repetitive oscillation of a natural frequency following a given path. Elemental single-coordinate modes may be simple oscillatory translation or rotation along or around any of the three orthogonal axes. Complex modes consist of oscillatory motion characterized in that they may follow a path having displacement components along two or more of the orthogonal axes; this may include rotation or a combination of rotation and translation.

Machines supported by isolators have several degrees of freedom which are identifiable as: motion in the vertical direction, motion in two horizontal directions, and rotations about three axes. In general, each rigid machine on an isolation system will have six degrees of freedom; three translational and three rotational. The resulting vibrational motion of the machine will be a combination of these six coordinates which correspond to the six degrees of freedom. For vibration isolation we are generally interested in the vertical motion of the machine, however, the actual motion will be a superposition of three translational motions and three rotational motions. The placement of the isolators in relation to the center of gravity of the mass determines the resulting motion associated with the vibrational response. By various combinations of isolator attachment locations, various motions or "modes" of vibration will result. Some isolator attachment locations result in translation only or rotation only; other attachment locations result in combined translation and rotational motion.

286 VIBRATION ISOLATION FOR NOISE REDUCTION Ch. 7

When the combined motion exists the resulting motion is termed *modal coupling*. This is of importance in vibration isolation because the system may be designed acceptably for vertical motion of the machine only to have a rotational "mode" excited with large amplitude rotations of the machine while operating.

Modal Coupling

Coupling of the elemental single-coordinate modes of vibration is a physical condition that results from poor selection of the locations and/or stiffnesses of the isolators. If the isolators are selected and located properly, the elemental modes will be uncoupled and will exist independently of each other. Figure 7-15 illustrates the technique for locating the isolators properly.

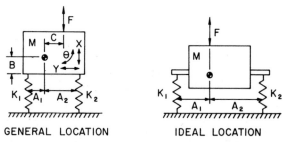

Fig. 7-15. Isolator location for modal coupling.

The points of attachment of the isolator to the machine should be in the horizontal plane of the center of gravity of the machine, as shown in Fig. 7-15B. Thus, the horizontal components of the forces in the isolator will pass through the center of gravity of the isolated mass and will not excite a vertical or rotational motion of the mass. The stiffness of the isolators is selected to be inversely proportional to the distance from the center of gravity for a rotation uncoupling effect. Thus, a vertical translation will not produce an isolator reaction moment if

$$K_1 A_1 = K_2 A_2 \qquad (7\text{-}27)$$

Obviously, equidistant isolator locations will call for equally stiff isolators. Similar approaches in the other vertical plane and the horizontal plane using the appropriate isolator stiffnesses and locations will produce complete uncoupling in the three translational and rotational elemental modes. For example, uncoupling the horizontal rotation is accomplished by making the isolator transverse stiffness inversely proportional to the distance from the center of gravity.

The elemental modes that are excited are determined by the type of excitation (force, moment) and location of the machine action. Machine excitation forces passing through the center of gravity as shown in Fig. 7-15B will excite motions only in this direction and the isolators need only be effective in this direction. However, the machine excitation forces generally consist of many types at many frequencies and the isolator designer must consider isolation of all forces for proper isolation control and noise reduction.

Ch. 7 VIBRATION ISOLATION FOR NOISE REDUCTION 287

For good vibration isolation the driving frequencies must be significantly higher than the highest modal frequency. The modal frequencies of the elemental modes may be evaluated by the following simple equations, and the simple transmissibility curve of Fig. 7-4 is used to select the isolators. For the translational modes the modal frequencies are:

$$f_i = \frac{1}{2\pi}\sqrt{\frac{K_i}{M}}, \quad i = x, y, z \qquad (7\text{-}28)$$

where:

K_i = total isolator stiffness in the direction of motion (lb/in.)
M = isolated mass (lb/in./sec^2).

For the rotational modes, the modal frequencies are:

$$f_j = \frac{1}{2\pi}\sqrt{\frac{KA^2}{J}}, \quad j = \theta, \phi, \psi \qquad (7\text{-}29)$$

where:

KA^2 = moment parameter, sum of the product of isolator stiffness and the moment arm in the direction of rotation (lb/in.)
J = rotational moment of inertia of the isolated mass (lb/in./sec^2).

Consideration should be given to the location of the center of gravity of the isolated mass to the location of the isolators. If necessary the location of the center of gravity can be changed by attaching an inertia block to the original system as shown in Fig. 7-16. For complicated systems such as an electric motor

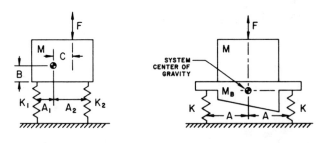

GENERAL SYSTEM   SPECIAL SYSTEM

Fig. 7-16. Inertial block design for center-of-gravity location.

and large blower on an inertia block, control of the center-of-gravity location will greatly simplify the isolator selection and location.

Dynamic Absorbers

Dynamic vibration absorbers may be of particular usefulness when a system to be isolated has a predominant excitation force at one frequency or at such a low frequency as to be difficult to control for normal isolators; for example, a plate or shell that is radiating noise strongly at a single resonance frequency. It

288 VIBRATION ISOLATION FOR NOISE REDUCTION Ch. 7

is also a useful approach to use when the system to be isolated is mounted on a flexible foundation. The use of a dynamic absorber can eliminate the motion and force transmitted to the foundation at the foundation resonance.

The basic principle is that of adding an absorbing spring-mass assembly to the original isolated system as shown in Fig. 7-17. The natural frequency, f_a, of the

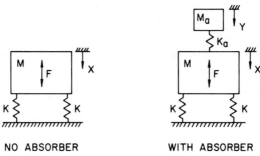

Fig. 7-17. Use of vibration absorber.

absorber assembly is designed to equal the frequency, f, of the excitation force or motion to be absorbed.

$$f = f_a = \frac{1}{2\pi}\sqrt{\frac{K_a}{M_a}} \qquad (7\text{-}30)$$

where:

f_a = natural frequency of the absorber, Hz
K_a = absorber spring rate, lb/in., N/m
M_a = mass of the absorber, lb-sec^2/in., or Kg

The absorber will develop sympathetic vibrations at the excitation frequency which will eliminate most of the original motion of the isolated system and the transmitted force at the frequency. The amplitude of the absorber mass motion will be

$$Y = \frac{F}{K_a} \qquad (7\text{-}31)$$

where F is the amplitude of the machine force.

It is important to check the amplitude of motion of the absorber mass, since at low driving frequencies the absorber spring may need to be so soft as to allow excessive motion. At this point more absorber mass will permit a stiffer absorber spring. The size of the absorber mass is normally in the order of $1/10$ to $1/4$ of the original system mass.

Isolation Bridging

The importance of avoiding unintentional bridging or "stiff" connections across the isolating system cannot be overemphasized. Vibration isolators may have

Ch. 7 VIBRATION ISOLATION FOR NOISE REDUCTION

been correctly selected and installed but inadvertent inclusion of bridging may severely reduce the effectiveness of the isolation. Even "soft" bridging may cause modal coupling which can reduce the total isolation efficiency of the system. Typical examples of isolation bridging include:

1. Use of rigid pipelines or electrical conduit instead of flexible connections; high-pressure piping may require coiling or the use of other flexible piping designs.
2. Use of rigid shafting or bracing between mechanical systems instead of flexible couplings or connections.
3. Attachment of bracing or pipe hangers to previously isolated panels or walls.

Mobility and Maintenance

Besides the obvious advantages of noise and vibration reduction there are other advantages that can be obtained in the use of vibration-isolation systems. One important advantage is the mobility of the equipment. By the use of proper vibration isolation so that the machine need not be bolted down to maintain its position, any machine may be relocated easily as production and operation scheduling are changed. Properly designed isolator mounts keep the machines from "walking" yet still allow for easy relocation. A standard forklift and machinery rollers will allow relocation of all but the very largest pieces of equipment.

Another advantage may be a general reduction of maintenance. Checking and tightening of clamps and bolts will be reduced along with wear in certain parts where deformation may be causing misalignment and stress.

SHOCK EXCITATION

Inherent in many machines are relatively large forces that are not harmonic in nature but are applied suddenly due to impacting members, engagement or braking of drive members, or other machine actions of a similar nature. The result of such actions is to:

1. Produce vibrations of the machine structure at the machine resonance frequencies and consequent radiation of sound from the machine surfaces;
2. Transmit forces to the supporting structure with subsequent problems of sound radiation elsewhere in the supporting structure.

Modern trends in the design of machinery with greater speeds and automated operation have increased the severity of these problems. Increased velocities of moving elements in high-speed machines obviously increases the available kinetic energy in the machine while automatic control mechanisms characteristically involve sudden changes with consequent impulsive motions and forces.

Reduction of impact noise generally calls for isolation of the machine from the supporting foundation or isolation of the radiating portion of the machine from the impact source within the machine. In addition, proper control of the impact source can frequently reduce the magnitude of the problem.

Shock Isolation

While the principles involved in shock isolation are very much similar to those involved in steady-state harmonic isolation, described earlier in the section on Vibration Isolation, some differences exist due to the transient nature of a shock. Reduction in the severity of a shock is obtained by the use of isolators that store the vibrational energy at the relatively high rate associated with the shock excitation and then release it at a much lower rate which is characteristic of the isolator natural frequency.

Shock isolators are usually elastic members (identical to steady-state harmonic isolators) inserted between the shock source and the system requiring protection. The elastic member (spring) stores the shock energy in deformation energy and then releases the energy causing the isolated body to vibrate at the natural frequency of the isolator system, until the energy is dissipated by the isolator internal damping.

The dynamic performance of an undamped shock-isolator system is shown in Fig. 7-18. The impact history is represented by a triangular pulse of magnitude,

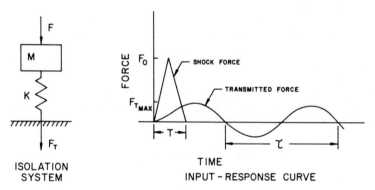

Fig. 7-18. Dynamic performance of an undamped shock isolator system.

F_o, and a time period, T. The natural period, τ, of the isolator is selected to be long compared to the shock pulse period, T, so that the shock pulse has disappeared before the isolated system has had time to respond to any great extent. The undamped natural period of the isolated system is given by

$$\tau = \frac{1}{f_n} = 2\pi \sqrt{\frac{M}{K}}. \qquad (7\text{-}32)$$

It is apparent that shock isolation needing long, natural periods is also accomplished by the use of low-frequency isolation systems.

Useful performance indices of shock isolators are generally given by the ratio of maximum response quantities to the maximum excitation quantities (Ref. 12).

$$\text{Shock Force Transmissibility} = \frac{\text{Maximum Transmitted Force}}{\text{Maximum Exciting Force}}$$

Ch. 7 VIBRATION ISOLATION FOR NOISE REDUCTION

$$T_{SF} = \frac{F_{T\,MAX}}{F_o} \qquad (7\text{-}33)$$

$$\text{Shock Velocity Transmissibility} = \frac{\text{Maximum Velocity of Machine}}{\text{Maximum Input Velocity}}$$

$$T_{SV} = \frac{\dot{X}_{MAX}}{\dot{Y}_{MAX}} \qquad (7\text{-}34)$$

$$\text{Shock Magnification Ratio} = \frac{\text{Maximum Motion of Machine}}{\text{Deflection if Max Force Applied Statically}}$$

$$SMF = \frac{X_{MAX}}{X_o} = \frac{X_{MAX}}{F_o/K} \qquad (7\text{-}35)$$

Shock Transmissibility Curve

The shock transmissibility curve is a plot of the shock transmissibility, T, as a function of the ratio of the shock period to the isolator natural period, T/τ. Figure 7-19B gives the shock transmissibility curves for three different shaped shock pulses. For this particular curve the shock transmissibility has been defined in terms of the *effective excitation input* where the effective shock transmissibility is

$$T_{SF} = \frac{F_{T\,MAX}}{F_e} \qquad (7\text{-}36)$$

and the effective input force is

$$F_e = \frac{1}{T} \int_{t=0}^{t=T} F(t)\,dt \qquad (7\text{-}37)$$

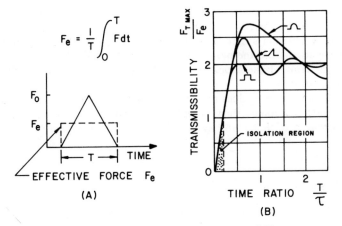

Fig. 7-19. Effective shock force transmissibility.

or the average force as shown in Fig. 7-19A. It is apparent that for shock isolation the natural period, τ, of the system must be more than $4T$ and preferably more than $10T$. Thus, for good shock isolator design

$$\tau \geq 10T \quad \text{or} \quad f_n \leq \frac{1}{10T} \tag{7-38}$$

It can be seen readily from Fig. 7-19B that when the shock pulse period is small compared to the natural period of the system, the severity of the shock is determined by the area of the shock pulse and not by its shape. The shock-transmissibility curve in the figure is given for an undamped isolation system but for damping ratios ($\delta = R/R_c$) of less than $\delta = 0.5$ the change is relatively small. Since the transmissibility will increase with damping, a low damping ratio, δ, is desirable in shock isolation problems.

Periodic Shock

Most shock pulses in machines occur with some regularity or periodicity, as in the example shown in Fig. 7-20. Generally, the period between shock pulses,

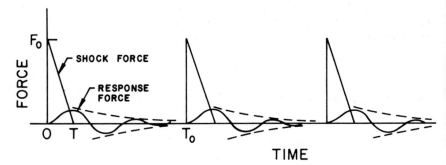

Fig. 7-20. Periodic shock pulse excitation.

T_o, is very long compared to the shock pulse period, T. If the internal damping in the shock isolation system is sufficient to "damp out" the response due to one shock pulse before the next shock pulse occurs, the isolation system will be designed in terms of the shock transmissibility curve to obtain a time period ratio $\tau/T \geq 10$.

If the excited response does not damp out before the next pulse, there will be a steady-state harmonic effect as well as the shock pulse effect. Ideally the natural period, τ, of the isolator system should be chosen so that

$$\tau \geq 10T, \quad \text{or} \quad f_n \leq \frac{1}{10T}$$

for shock isolation, and

$$f_n \ll \frac{1}{T_o}$$

for vibration isolation.

Ch. 7 VIBRATION ISOLATION FOR NOISE REDUCTION 293

The shock periodic frequency $1/T_o$ generally is very low (in the order of 0 to 2 Hz) and is difficult to isolate without large static deflections, as discussed earlier in Vibration Isolation. Since the shock periodic frequency is too low to be easily audible and the motion at this frequency is also usually small, it is normal practice to allow the isolator frequency, f_n, to lie above the shock periodic frequency, $1/T_o$, if need be. Care must be taken so that the isolator frequency, f_n, is not in the direct region of the shock periodic frequency.

Example

A machine weighing 500 lbs is to be isolated for a square pulse shock excitation of 50 lbs. Determine the static deflection of the isolator within acceptable design limits for machine stability.

Machine Weight $W = 500$ lb
Shock Input $F_o = 50$ lb (Square Pulse)
$T = 0.01$ sec.
$T_o = 0.5$ sec. (2 Hz).

Assume that $f_n = 1$ Hz, is an acceptable isolator frequency. Rearranging Equation (7-5) and solving for Δ_{static}:

$$\Delta_{static} = \frac{1}{4\pi^2} \frac{g}{f_n^2} = \frac{1}{4\pi^2} \frac{386}{1}$$

$$= 9.8 \text{ in.}$$

This static deflection is too large for steel isolators and is unacceptable for the horizontal stability of the system.

Considering the shock isolation criterion, $\tau \geq 10\,T$,
Let $\tau = 10\,T$ or .1 sec:

$$f_n = \frac{1}{\tau} = \frac{1}{.1} = 10 \text{ Hz}$$

From Equation (7-5):

$$\Delta_{static} = \frac{1}{4\pi^2} \frac{g}{f_n^2} \frac{386}{100}$$

$$\approx .1 \text{ in.}$$

This static deflection can easily be obtained with steel isolators and will provide acceptable isolation for the shock-excited machine. However, this condition is the minimum static deflection of the isolator and a larger deflection would be more desirable. The actual allowable static deflection would depend upon the stability or positioning requirements of the machine but is usually under two inches.

Calculation of Shock Force Transmissibility for .1 in static deflection.
With $T/\tau = .1$:

$$T_{SF} = \frac{F_{T\,MAX}}{F_e} \approx 0.25, \quad \text{from Fig. 7-19.}$$

Equivalent Force Input

$$F_e = \frac{1}{T} \int_0^T F_o \, dt = 50 \text{ lb}$$

Transmitted Shock Force to foundation through isolator.

$$F_{T\,MAX} = T_{SF} \cdot F_e = 0.25(50) = 12.5 \text{ lb}$$

Shock Force Reduction

Reduction of the shock excitation will inherently reduce the severity of the shock vibration problem and, as much as possible, machinery should be designed with this in mind. From the previous discussion it is evident that the following design rules should be considered:

1. The mass, M, should be as large as possible. The massiveness of the machine is an effective deterrent to shock excitation.
2. The maximum shock force, F_o, should be made as small as possible, i.e.:
   Where the force results from an impact by a mass, M_o, having an approach velocity, V_o, the force of the impact can be reduced by decreasing the magnitude of M_o or V_o, or both.
3. Cushioning or reducing the abruptness of the impact by a more flexible insert between the impacting members will reduce the excitation of the higher resonances and reduce the noise radiation. Reducing the abruptness of starting or stopping operations will have a similar effect.

Care should be taken to ensure that these changes do not increase the period, T, of the shock impulse, to the point where the transmitted force or response motion actually increases.

SOURCE REDUCTION

The internally generated forces in equipment and machines come from a number of sources such as rotating and reciprocating members, unbalance, flow of fluids, fluctuating magnetic fields, and many others. In many cases, such as stamping machines, centrifugal blowers, etc., it is inherent in the design that such excitation exists; however, the magnitude of the excitation can be significantly reduced by proper attention to design and manufacturing practice. While certain sources of excitation can be almost eliminated, most machines have so many sources of excitation, that the overall excitation level of the machine cannot be eliminated and isolation for noise reduction will still be necessary. Many times, certain noise sources are inherent in the function of the machine and cannot be reduced.

In general, the basic sources of excitation in a machine are the following: (Refs. 16 and 17).

1. Rotating and Reciprocating Unbalance
2. Friction and Rubbing
3. Impact or Cutting Forces

Ch. 7 VIBRATION ISOLATION FOR NOISE REDUCTION

4. Imbalance Between Load and Drive Forces or Moments
5. Magnetic Field Effects
6. Fluid Flow and Turbulence.

It is beyond the scope of this discussion to delineate all ways of reducing or eliminating these sources of excitation and the following discussion will serve only to indicate possible sources and source frequencies. Corrective treatments are usually implied in the knowledge of the source.

Rotary Unbalance

The excitation frequency is the speed of rotation and, usually, the higher harmonics of that speed. For an unbalanced shaft and rotor the fundamental excitation frequency is

$$f = \frac{\text{rpm}}{60}, \text{Hz} \qquad (7\text{-}39)$$

Misaligned or bent shafts may cause excitation at double this frequency and its harmonics.

Unbalance occurs when the center-of-mass of a rotor does not coincide with the axis of rotation. The residual unbalance, M_u, is related to the total rotor mass, M. The specific unbalance is equivalent to the displacement of the center of gravity by an amount equivalent to the static unbalance and is defined as:

$$e = M_u/M \qquad (7\text{-}40)$$

The *specific unbalance*, e, has been shown to vary inversely as the running speed, n, of a rotor, to achieve equivalent balance grades for machinery which leads to the following relationship:

$$en = \text{constant} \quad \text{or} \quad e\omega = \text{constant} \qquad (7\text{-}41)$$

where ω = circular frequency in rad/sec.

The unbalance of a rotor is given by the expression, $M_u e$, in ounce-inches or in gram-millimeters.

The relationship $e\omega$ = constant, gives the result that, for geometrically similar rotors rotating with equal peripheral speeds, the stresses in the rotors and in rigid bearings will be equal. Balance quality grades are based on this relationship.

Balance quality grades have been established which comprise a range of permissible residual unbalances given by certain magnitudes of the product, $e\omega$, given in millimeters per second. The balance quality grades given by the International Standard ISO 1940-1973 (E) are listed in Table 7-3 along with descriptions of rotor types. The upper limits of e plotted as a function of maximum operating speed are given by Fig. 7-21. The major balance quality grades are separated from each other by a multiple of 2.5. The intermediate lines of Fig. 7-21 are fractions of the 2.5 multiple for a finer grading, if desired. Note that the vertical scale is in lb-in./(lb of rotor weight), or in center-of-gravity displacement, in inches. Table 7-3 is a guide for the specification of balance quality for various types of rotating machinery or it may be used for selecting a special balance quality specification if unbalance forces exist in a critical machine location.

Table 7-3. Balance Quality Grades for Various Groups of Representative Rigid Rotors

Balance Quality, Grade G	$e\omega$,[1,2] mm/s	Rotor Types—General Examples
G 4 000	4 000	Crankshaft-drives[3] of rigidly mounted slow marine diesel engines with uneven number of cylinders.[4]
G 1 600	1 600	Crankshaft-drives of rigidly mounted, large, two-cycle engines.
G 630	630	Crankshaft-drives of rigidly mounted, large, four-cycle engines. Crankshaft-drives of elastically mounted marine diesel engines.
G 250	250	Crankshaft-drives of rigidly mounted, fast, four-cylinder diesel engines.[4]
G 100	100	Crankshaft-drives of fast diesel engines with six or more cylinders.[4] Complete engines (gasoline or diesel) for cars, trucks, and locomotives.[5]
G 40	40	Car wheels, wheel rims, wheel sets, drive shafts. Crankshaft-drives of elastically mounted, fast, four-cycle engines (gasoline or diesel) with six or more cylinders.[4] Crankshaft-drives for engines of cars, trucks, and locomotives.
G 16	16	Drive shafts (propeller shafts, cardan shafts) with special requirements. Parts of crushing machinery. Parts of agricultural machinery. Individual components of engines (gasoline or diesel) for cars, trucks, and locomotives. Crankshaft-drives of engines with six or more cylinders under special requirements.
G 6.3	6.3	Parts or process-plant machines. Marine main-turbine gears (merchant service). Centrifuge drums. Fans. Assembled aircraft gas-turbine rotors. Fly wheels. Pump impellers. Machine-tool and general machinery parts. Normal electrical armatures. Individual components of engines under special requirements.
G 2.5	2.5	Gas and steam turbines, including marine main turbines (merchant service). Rigid turbo-generator rotors. Rotors. Turbo-compressors. Machine-tool drives.

(See footnotes at end of Table.)

Ch. 7 VIBRATION ISOLATION FOR NOISE REDUCTION

Table 7-3 (con't). Balance Quality Grades for Various Groups of Representative Rigid Rotors

Balance Quality, Grade G	$e\omega^{1,2}$ mm/s	Rotor Types—General Examples
G 2.5 (Cont.)	2.5	Medium and large electrical armatures with special requirements. Small electrical armatures. Turbine-driven pumps.
G 1	1	Tape recorder and phonograph (gramophone) drives. Grinding-machine drives. Small electrical armatures with special requirements.
G 0.4	0.4	Spindles, disks, and armatures of precision grinders. Gyroscopes.

[1] $\omega = 2\pi n/60 \approx n/10$, if n is measured in revolutions per minute and ω in radians per second.
[2] In general, for rigid rotors with two correction planes, one-half of the recommended residual unbalance is to be taken for each plane; these values apply usually for any two arbitrarily chosen planes, but the state of unbalance may be improved upon at the bearings. For disk-shaped rotors the full recommended value holds for one plane.
[3] A crankshaft-drive is an assembly which includes the crankshaft, a flywheel, clutch, pully, vibration damper, rotating portion of connecting rod, etc.
[4] For the purposes of this International Standard, slow diesel engines are those with a piston velocity of less than 9 m/s; fast diesel engines are those with a piston velocity of greater than 9 m/s.
[5] In complete engines, the rotor mass comprises the sum of all masses belonging to the crankshaft-drive described in Note 3, above.

Example

A centrifugal water chiller has a rotor weight of 500 lbs and operates at 3600 rpm. Select a balance grade and determine the unbalance of the rotor.

Solution: From Table 7-3, this rotor type is best described by balance quality grade G 6.3. From Fig. 7-21 for G 6.3 and an operating speed of 3600 rpm, the unbalance is .0007 lb-in./lb. For a 500-lb rotor the unbalance $M_u e$ = .0007 × 500 = 0.35 lb-in., or 5.6 oz-in.

The resulting force due to the rotating unbalance is the product of the unbalance and the square of the rotational speed.

$$\text{Unbalance force} = M_u e \omega^2$$

The component of this force in the vertical direction is given by

$$F_{\text{vertical}} = M_u e \omega^2 \sin \omega t \qquad (7\text{-}42)$$

For the above example of a water chiller the vertical time dependent force due to the unbalanced rotor is given by

$$F_{\text{vertical}} = \frac{0.35 \text{ lb-in.}}{386 \text{ in./sec}^2} \left(\frac{2\pi \, 3600}{60}\right)^2 \sin\left(\frac{2\pi \, 3600}{60} t\right)$$

$$F_{\text{vertical}} = 129 \sin(120\pi t) \text{ lb}$$

298 VIBRATION ISOLATION FOR NOISE REDUCTION Ch. 7

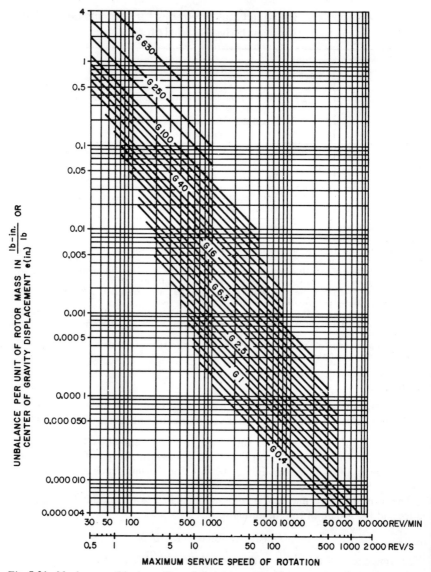

Fig. 7-21. Maximum residual specific unbalance corresponding to various balancing quality grades G.

Thus, the maximum force is 129 lbs and the exciting frequency is $\frac{120\pi}{2\pi}$ rad, or 60 Hz.

Reciprocating Unbalance

This excitation is produced by the inertia forces of reciprocating masses in the machine. Generally the reciprocating motion is not sinusoidal and will result in

Ch. 7 VIBRATION ISOLATION FOR NOISE REDUCTION 299

not only the fundamental drive frequency but relatively large amplitude higher harmonics. The magnitude of the second harmonic is usually dependent on the length of the connecting rod to the crank radius. Complete dynamic balancing is generally not possible and because of the large reciprocating masses, use of an inertia block may be necessary to reduce the system motion.

Friction Excitation

This excitation arises from a lack of, or poor, lubrication and causes a high frequency excitation due to rapid, intermittent contacting of the rubbing surfaces. Both the surface finish, and the normal loading, affect the excitation level.

Impact

Inherent in many types of equipment is the phenomenon of relatively large forces suddenly applied, or machine elements of relatively high velocity suddenly contacting other elements. Reducing the impact magnitude or "cushioning" the impact within the machine will reduce the impact effects. Additional reduction can be obtained by increasing the mass of the structure to reduce the overall acceleration of the system. The impact forces tend to excite the system at its natural frequencies which causes transmitted forces at these frequencies. In general, impact excitation effects will damp out in time, dependent on the damping in the system.

Load-Drive Imbalance

Most systems have a general imbalance between the drive force or torque, and the load force or torque at any instant of time, with only the work per cycle being constant. This difference in load-drive magnitude results in an accelerating-decelerating motion with resulting vibration. Usually, all frequencies of the fundamental and higher harmonics are generated. Reduction can generally be obtained by "flywheel" inertial effects or by larger drive systems to smooth out the difference between the load and the drive.

Magnetic Field Effects

These excitations primarily occur in electric drive motors and are principally caused by the periodic forces which exist in the stator and the rotor. The torque fluctuations about the mean drive torque occur at twice line frequency and its harmonics and are normally the main source of noise in electric motors. Although usually of a lower magnitude, the radial components produce vibrational forces which appear at harmonics of the line frequency and at frequencies related to the slip and line frequency. In two pole induction motors, these frequencies are:

$$f = n \cdot 2f_{Line} \text{ Hz} \quad (7\text{-}43)$$

where:

n = harmonic number, integer

and

$$f = N \cdot \frac{\text{rpm}}{60} \pm 2f_{Line} \text{ Hz} \quad (7\text{-}44)$$

where:

N = number of rotor slots.

Careful mounting of electric motors with soft mounts specially designed to isolate against the torque fluctuations can greatly reduce electric-motor noise.

Fluid-Flow Noise

One of the major sources of excitation from this phenomenon is called "blade noise." This excitation is the result of a fluid (frequently, air) impulse every time an impeller or fan blade passes a stationary flow interruption point (such as a post). The rate of repetition of this impulse determines the fundamental excitation frequency and its harmonics:

$$f = N \cdot \frac{\text{rpm}}{60} \cdot n \text{ Hz} \qquad (7\text{-}45)$$

where:

N = number of blades, and
n = harmonic number, integer.

This excitation occurs at any flow interruption point in the system and the total system excitation is a function of the total number of flow interruptions and their spacing during one cycle of the machine.

This same phenomenon will cause noise in electric motors where the rotor slots play the part of the blades and the air gaps in the stator play the role of flow interruption.

Bearings and Gears

Rolling bearing and gear noise are principally caused by the periodic impacts of the rollers (balls) or gear teeth. The excitation frequencies are at the rotation speeds and multiples of the rotational speeds due to the number of rollers or gear teeth. If a high spot exists on one gear, the frequency will be the product of the rotational speed and the number of teeth in the meshing gear. Broadband noise can also be produced in bearings and gears. A method for estimating the sound-pressure levels produced by gears is given in (Ref. 18).

Sleeve bearings may produce oil excitations which occur at one-half the rotational speed. For poorly lubricated bearings with large clearances, double frequency excitation can be caused by interval impact.

Transformers

The excitation effort orginates in the core due to the magnetostrictive effect and, since the effect is independent of the direction of magnetization, produces frequencies that are twice the line frequency and harmonics that are multiples of twice the line frequency:

$$f = 2 \cdot f_{\text{Line}} \cdot n \text{ Hz} \qquad (7\text{-}46)$$

where:

n = integer.

Ch. 7 VIBRATION ISOLATION FOR NOISE REDUCTION

Application of Vibration Isolation Principles

In industry, the majority of vibration-induced noise problems occur when one or more machines transmit vibrations into the floor or support structures of the building. The principles of vibration generation and transmission have been thoroughly treated in preceding sections of this chapter, and in this closing section some procedures and methods for applying and specifying vibration isolators (mounts) to industrial equipment will be shown. In almost all cases it will be found practical and economical to specify a commercial, off-the-shelf mount to satisfy isolation and other requirements. It is important, therefore, to become familiar with the various types of machinery mounts available for industrial applications (see Refs. 19 and 20). Basically, the choice of mount will depend on the static deflection of the mount under the weight of the machine. As seen from Equation (7-16) and Fig. 7-6, the static deflection of the isolator and the natural frequency of the machine mounted on the isolator are related. For a more direct comparison this relationship is given in Table 7-4 for an appropriate

Table 7-4. The Natural Frequency of an Isolator as a Function of its Static Deflection Under Load

Static Deflection	Natural Frequency		Static Deflection	Natural Frequency	
in.	Hz	cpm	in.	Hz	cpm
0.02	22.0	1320	2.7	1.9	114
0.04	15.7	940	3.0	1.8	108
0.06	12.8	770	3.4	1.7	102
0.01	10.0	600	3.8	1.6	96
0.2	7.0	420	4.4	1.5	90
0.3	6.0	360	5.0	1.4	84
0.4	5.0	300	6.0	1.3	78
0.6	4.0	240	7.0	1.2	72
0.8	3.5	210	8.0	1.1	66
1.1	3.0	180	10.0	1.0	60
1.2	2.8	168	12.0	0.9	54
1.4	2.6	156	15.0	0.8	48
1.7	2.4	144	20.0	0.7	42
2.0	2.2	132	27.0	0.6	36
2.4	2.0	120	39.0	0.5	30

range of static deflections. The recommended deflection range for the most common types of isolators is shown in Table 7-5.

Specific Mount Types

<u>Undamped Steel Springs</u>.—Designed to be used under vertical loading, the steel coil spring is generally used for natural frequencies below 6 Hz. A free-standing unit is shown in Fig. 7-22, complete with leveling bolt and noise isolation pad attached to the base.

Table 7-5. Recommended Deflection Ranges for Common Types of Industrial Isolators

Deflection Range	Isolator
0.5 inch, and greater	Steel spring or air spring
0.3 to 0.5 inch	Elastomeric mounts in tandem
0.1 to 0.25 inch	Single elastomeric mounts or 2 to 4 layers of ribbed neoprene pads.
0.1 inch, and under	Single layer of ribbed neoprene pad.

Free-standing steel springs should be used for deflections greater than 1 inch, and must have a diameter large enough to guarantee lateral stability under vertical loading. Table 7-6 gives recommended minimum spring diameters (OD) for a range of load and deflection conditions.

If space limitations necessitate, smaller diameter springs than those recommended in Table 7-6 can be used if they are used with spring housings. However, the possibility of dirt entrapment and the spring rubbing against the housing may lead to short-circuiting the spring with a loss in isolation efficiency.

If additional restriction of lateral motion of stable steel springs is required, spring mounts can be purchased with stabilizer housings as shown in Fig. 7-23.

Specially designed spring mounts are also available for overhead (hanging) suspensions as shown in Fig. 7-24.

Damped Steel Springs.—Although undamped steel springs provide optimum isolation at operating speeds above the mount natural frequency, it may be ne-

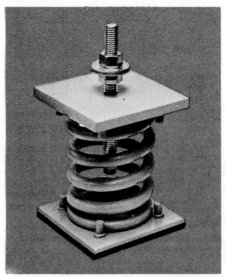

Courtesy, Consolidated Kinetics Corp.
Fig. 7-22. High-deflection spring type isolator.

Ch. 7 VIBRATION ISOLATION FOR NOISE REDUCTION

Table 7-6. Recommended Spring Diameters (inches) for Free-Standing Springs

Load Range, lb	Maximum Deflection Required, inches			
	1	1.5	2	3.5
	Recommended Spring Diameters, inches			
to 250	2.5	3.5	3.5	5.5
250 to 500	3.5	4.0	4.5	7.0
500 to 1000	4.5	4.5	5.5	7.5
1000 to 2000	5.0	5.5	7.0	8.0

cessary to compromise by including a small amount of damping in the mount to reduce the motion of the machine during startup and shutdown as the machine passes through resonance. Damped steel spring mounts are available in a range of sizes. In the type illustrated in Fig. 7-25, the damping is accomplished by adding a dry frictional surface between the top and bottom housings and adjustment of the friction force is achieved by a spring-loaded mechanism. It must be

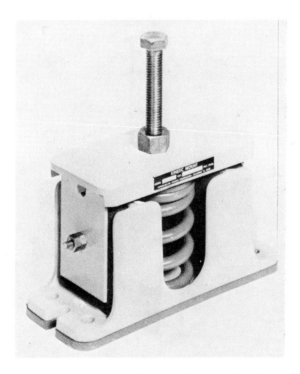

Courtesy, Consolidated Kinetics Corp.

Fig. 7-23. Spring mounts with stabilizer housings.

Courtesy, Consolidated Kinetics Corp.

Fig. 7-24. Spring mounts for overhead suspension.

Courtesy, Barry Controls

Fig. 7-25. Adjustable damped steel spring mounts.

Ch. 7 VIBRATION ISOLATION FOR NOISE REDUCTION   305

remembered that isolation is reduced at the operating speeds above resonance when damping is incorporated in the mount.

General Purpose Elastomeric Mounts.—Isolators made from elastomeric materials, such as neoprene, incorporate features which make them attractive for general purpose machinery mounts. Mounts of this type can be molded in a variety of shapes and sizes to give desired stiffness characteristics in both vertical and lateral directions. Examples of two such mounts are shown in Figs. 7-26 and 7-27. Because of its resistance to oils and washing solutions, neoprene is the

Courtesy, Consolidated Kinetics Corp.
Fig. 7-26. Neoprene unit mount.

favored material and its inherent damping is sufficient to limit transmissibility at the mount natural frequency and to attenuate high-frequency resonances in the mount.

Pad Mounts.—A wide variety of materials including rubber, felt, cork, and compressed glass fiber can be used in pad form to provide effective vibration isolation at machine mounting points. One of the most versatile is the ribbed neoprene pad illustrated in Fig. 7-28. Neoprene pads can be purchased in several thicknesses, styles, and durometer hardnesses; as well as in sheet form to allow cutting and fitting as needed.

By suitable stacking, either by nesting or insertion of steel plates between layers, a wide range of isolation requirements can be met. Simple, inexpensive, ready-made isolators are available with a single or double neoprene ribbed pad bonded to a steel plate containing a leveling screw. Typical sizes are shown in Fig. 7-29.

306 VIBRATION ISOLATION FOR NOISE REDUCTION Ch. 7

Courtesy, Barry Controls
Fig. 7-27. Low-frequency elastomeric compression mounts.

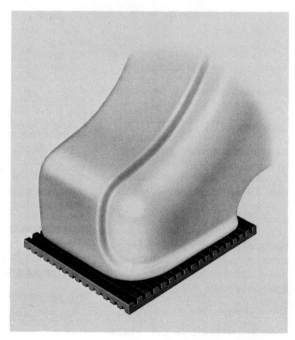

Courtesy, Consolidated Kinetics Corp.
Fig. 7-28. Ribbed neoprene mount.

Ch. 7 VIBRATION ISOLATION FOR NOISE REDUCTION 307

Courtesy, Barry Controls

Fig. 7-29. Neoprene pad mounts with leveling screws.

Pneumatic Rubber Mounts.—Air-filled mounts, typical of the one shown in Fig. 7-30, offer advantages over steel springs for special low-frequency applications. Isolation efficiencies up to 99 percent can be obtained at low natural frequencies from 0.1 to 5 Hz, without the large static deflection required of steel springs. In addition, excellent lateral stability, adequate internal damping, self-leveling by air volume adjustment, and shock protection are inherent in their design. Their disadvantages, in comparison with steel springs, are cost; limited load capacity; and the necessity of periodic inspection for proper operation.

Roller Mounts.—Horizontal reciprocating compresses, cold headers, and the like, produce very large horizontal forces at low operating speeds. Specially designed mounts are available for isolation of such forces. Figure 7-31 shows a roller mount where rollers are incorporated to provide resiliancy in the horizontal direction at a mount natural frequency as low as 1 Hz.

Mount Selection Procedures

The recommended mount types for a wide variety of machinery are presented in Table 7-7. In considering the mount system design and selection, it is necessary to obtain some or all of the following information about the equipment and its environment:

1. Location where equipment is to be mounted.
2. Equipment operating characteristics: power, rpm, strokes per minute, starting torque.

308 VIBRATION ISOLATION FOR NOISE REDUCTION   Ch. 7

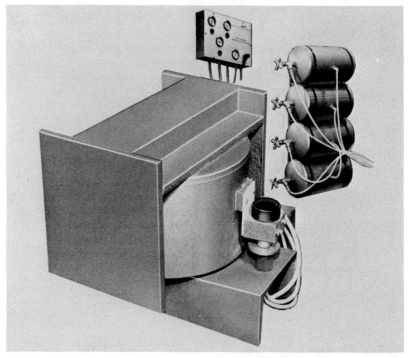

Courtesy, Consolidated Kinetics Corp.
Fig. 7-30. Pneumatic rubber mount.

3. Direction and magnitude of principal forcing frequencies.
4. Equipment weight and weight distribution.
5. Number, size, and location of mounting feet relative to the center of gravity.
6. Equipment attachments such as ducts, pipes, and electrical conduits.
7. Type of surface and structure to which equipment is to be mounted.
8. Required isolation efficiency.

The amount of information actually needed for a given situation depends on the complexity of the equipment and the seriousness of the vibration problem. The minimum information needed is the driving frequency, the weight per mount, and the desired isolation efficiency. For many machines, the major driving frequency occurs from unbalanced forces at the operating speed of the machine. The weight of the machine is either distributed approximately equally at its mounting points, or if not, one can judge the distribution of weight from the area of the mounting feet since most machinery is designed for 50 psi floor loading. When in doubt, of course, either the manufacturer should be consulted or the machine weighed, if practicable.

Ch. 7 VIBRATION ISOLATION FOR NOISE REDUCTION 309

Fig. 7-31. Roller mount.

Courtesy, Barry Controls

For machines that fit into the category above, the mount selection procedure is as follows:

1. Choose an isolation efficiency. For noncritical installations 85 percent efficiency is usually adequate. From Fig. 7-3, 85 percent isolation is equivalent to .015 transmissibility; this yields a ratio of driving frequency to natural frequency of 3 (0.05 critical damping).
2. Divide the driving frequency by the frequency ratio to obtain the natural frequency of the mount. From Table 7-4 (or Fig. 7-6) determine the static deflection of the mount.
3. If the weight of the machine is equally distributed, divide the weight by the number of mounting points to obtain the weight supported, per mount. As a guide in selecting a type of isolator use Table 7-7, then determine the size from the manufacturer's rated load versus deflection curves.
4. If the weight of the machine is unequally distributed, the mounts should be selected on the basis of equal static deflection per mount. Each mount is then selected as in Step 3. Generally, this means using different rate springs or dual (parallel) or tandem (series) spring installations to achieve equal deflection. If this is not practical use subframes or rails and/or inertia blocks to shift the mounting points on the center of gravity, or both, to achieve an even weight distribution.

Table 7-7. Recommended Mount Types for Various Applications*

	Leveling Mount	Pneumatic Rubber Mount	Pad Mount	Roller Mount	Undamped Spring Mount	Damped Spring Mount	General Purpose Elastomeric Mount	"Active" Pneumatic Isolation Mount
Air Compressors (vert. & Y type)								
Air Compressors (horizontal)								
Air Conditioners	●	●						
Balancers	●	●			●	●	●	●
Bending Machines	●					●	●	
Blowers	●	●					●	
Boring Mills	●							
Brakes	●	●					●	
Broaches								
Clickers	●			●				
Cold Headers	●	●						
Die Casting Machines						●		
Diesel Engines	●	●			●	●	●	
Drill Press	●		●					
Drop Hammers	●	●				●	●	
Drug Machines	●	●					●	
Fans	●					●	●	
Food Processing Equipment	●							
Forging Hammers	●							
Four Slides	●							
Gear Hobbers	●							
Grinders	●	●						
Injection Molding Machines	●							●
Jig Borers	●							●
Lathes	●							●
MG Sets		●			●	●	●	
Milling Machines	●							
Nail Formers	●			●				

Ch. 7 VIBRATION ISOLATION FOR NOISE REDUCTION 311

	Packaging Machinery	Paper Working Machinery	Printing Machinery	Punch Presses	Riveters	Sanders	Saws	Screw Machines	Shakers	Shapers	Shears	Shock Test Machines	Shoe Machinery	Swaging Machines	Transformers	Typewriters	TWX Machines	Vacuum Pumps	Vibrators	Vibratory Feeders	Welders	Wirewinding Machines	Woodworking Machinery
						●		●															
	●	●			●												●	●	●		●		
																	●	●		●			
		●		●	●					●	●	●			●		●		●		●		
			●				●		●	●	●				●	●			●				
	●	●	●	●	●		●	●		●	●			●	●		●			●	●		

*Courtesy of Barry Controls.

312 VIBRATION ISOLATION FOR NOISE REDUCTION Ch. 7

Table 7-8. A Guide for Selecting Vibration Isolating Systems for Some Mechanical and Electrical Equipment

Equipment	Power Range	Speed Range, rpm	AREA[6] Mount Type (See Table 7-9)			AREA[6] Weight Ratio (See Note 1, below)			AREA[6] Deflection, in. (See Note 2, below)		
			Min	Normal	Critical	Min	Normal	Critical	Min	Normal	Critical
Centrifugal and Axial Flow Fans	<3 hp	All	NN[3]	2	3	...	...	2	...	1.0	1.5
	3-25 hp	<600	2	3	3	...	2	3	1.0	1.5	2.0
		600-1200	2	2	3	...	...	2	0.5	1.0	1.5
		>1200	2	2	3	...	...	2	0.5	1.0	1.0
	>125 hp	<600	2	3	3	...	2	3	1.5	2.0	3.0
		600-1200	2	3	3	...	2	2	1.0	1.5	2.0
		>1200	2	3	3	...	...	2	0.5	1.0	1.0
Motor-Pump Assemblies	<20 hp	450-900	2, 3	3	3	1.5	2-3	3-4	1.5	1.5	3.0
		900-1800	2, 3	3	3	2.5	1.5-2.5	2-3	1.5	1.0	2.0
		>1800	2, 3	3	3	2.5	1.5-2.5	2-3	1.0	0.75	1.5
	20-100 hp	450-900	3	3	3	2-3	2-3	3-4	1.5	1.5	4.0
		900-1800	3	3	3	1.5-2.5	1.5-2.5	2-3	1.0	1.0	3.0
		>1800	3	3	3	1.5-2.5	1.5-2.5	2-3	0.75	1.0	2.0
	>100 hp	450-900	3	3	3	2-3	3-4	3-4	2.0	2.0	5.0
		900-1800	3	3	3	2-3	2-3	2-3	1.5	1.5	4.0
		>1800	3	3	3	1.5-2.5	2-3	2-3	1.0	1.0	3.0
Internal Combustion Engines and Engine-Driven Equipment	<25 hp	All	2	2	3	...	2-3	3-4	0.35	0.35	2.5
	25-100 hp	All	2	2	3	...	2-3	3-4	0.35	1.75	3.5
	>100 hp	All	2	2	3	...	2-3	3-4	0.35	2.50	5.0
Reciprocating Type Air Compressors, 1 to 2 Cylinders	<20 hp	300-600	4	NR[4]	NR	4-8	...	...	4.0	4.0	...
		600-1200	4	4	4	2-4	3-6	3-6	2.0	2.0	4.0
		>2400	4	4	4	1-2	2-3	...	1.0	...	...
	20-100 hp	300-600	4	NR	NR	6-10	...	...	5.0	...	...
		600-1200	4	NR	NR	3-6	3-6	3-6	3.0	3.0	4.0
		>1200	4	4	NR	2-3	...	...	1.5	...	...
Refrigeration Reciprocating Compressors and Package-Chiller Assemblies	10-50 Tons	600-900	2	3	3	...	2-3	3-4	2.0	2.0	3.0
		900-1200	2	3	3	...	2-3	3-4	1.5	1.5	2.0
		>1200	2	3	3	...	2-3	2-3	1.0	1.5	2.0
	>50 Tons	600-900	3	3	4	2-3	3-4	4-6	2.0	3.0	3.0
		900-1200	3	3	3	...	2-3	3-5	1.5	2.0	2.0
		>1200	2	3	3	...	2-3	3-4	1.5	2.0	2.0

Ch. 7 VIBRATION ISOLATION FOR NOISE REDUCTION

Equipment	Size	Weight Ratio[1]	Min. Area Layers	Normal Area Layers	Critical Area Layers	Weight Ratio	Min. Area Defl.	Normal Area Defl.	Critical Area Defl.[2]
Refrigeration Rotary-Compressor, and Package-Chiller Assemblies	100-500 Tons	>3000	2	2	3	2-3	0.75	1.0	1.5
	500-1000 Tons	>3000	2	2	3	3-5	1.0	1.5	1.5
Absorption Refrigeration Assemblies	All Sizes	...	1	2	2	...	0.25	0.5	1.0
Cooling Towers, Propeller Type (See Note 5, below) <25 hp	150-300	NN	2	2	2	...	...	5.0	5.0
	300-600	NN	2	2	2	...	...	3.0	3.0
	>600	NN	2	2	2	...	...	3.0	3.0
25-150 hp	150-300	NN	2	2	2	...	...	6.0	6.0
	300-600	NN	2	2	2	...	...	4.0	4.0
	>600	NN	2	2	2	...	...	3.0	3.0
>150 hp	150-300	NN	2	2	2	...	...	6.0	6.0
	300-600	NN	2	2	2	...	...	5.0	5.0
	>600	NN	2	2	2	...	...	4.0	4.0
Cooling Towers Centrifugal-Type <25 hp	450-900	NN	2	2	2	...	...	1.0	1.5
	900-1800	NN	2	2	2	...	...	0.75	1.0
	>1800	NN	2	2	2	...	...	0.75	0.75
25-150 hp	450-900	NN	2	2	2	...	...	1.5	2.0
	900-1800	NN	2	2	2	...	...	1.0	1.5
	>1800	NN	2	2	2	...	...	0.75	1.0
>150	450-900	NN	2	2	2	...	...	2.0	3.0
	900-1800	NN	2	2	2	...	...	1.5	1.5
	>1800	NN	2	2	2	...	...	1.0	1.0
Transformers	<10 kva	...	1	1	2	...	0.13	0.13	0.25
	10-100 kva	...	1	2	2	...	0.13	0.25	0.5
	>100 kva	...	1	2	2	...	0.25	0.25	0.5

[1] Weight Ratio = Weight of Equipment Mounted on Inertia Block or Base / Weight of Inertia Block or Base

[2] Deflection of the combined equipment and inertia block on the mount. For normal and critical areas these deflections may have to be increased, depending on the floor flexibility. For normal areas, the deflection should be three to six times the floor deflection and critical areas six to ten times the floor deflection (see text).

[3] NN denotes not needed; mount machine directly on the floor.

[4] NR denotes not recommended, machine cannot be suitably isolated.

[5] Minimum location is on a slab outside of building. To minimize waterfall noise from rooftop location over critical areas, place two or three layers of neoprene pad between tower and springs, or between tower and roof if the propeller and drive assembly are vibration-isolated from the tower.

[6] 1 – *Minimum Area:* Basement or on-grade slab location.

2 – *Normal Area:* Upper floor location, but not above or adjacent to a critical area, private offices, classrooms, conference rooms, operating rooms, etc.

3 – *Critical Area:* Upper floor location above or adjacent to private offices, classrooms, conference rooms, operating rooms, etc.

314 VIBRATION ISOLATION FOR NOISE REDUCTION   Ch. 7

Note: There is no standard way to present isolator performance rating. Most manufacturers will present load versus deflection, or load versus natural frequency curves, but some will only give the natural frequency at the maximum rated load. To find the natural frequency at an intermediate load, use the formula:

$$f'_n = f_n \sqrt{\frac{W}{W'}} \tag{7-47}$$

where f_n is the rated natural frequency at the rated maximum load, W, and f'_n is the desired natural frequency at the intermediate load, W'.

More complicated and involved design and selection procedures are called for 1. when the required isolation efficiency is very high (say, 98 percent, or greater); 2. when the driving frequencies are such as to cause angular as well as translational deflections of the equipment; 3. when the forcing is due to combined shock and periodic forces; and 4. when flexible floor structures are involved in the installation.

Table 7-9. Mount Types

Mount Type	Schematic Illustration	Description
1		Pad mounts of sufficient layers to obtain desired static deflection at load rates of 25–60 psi (45 durometer neoprene), or 145–150 psi (65 durometer neoprene). Floor should be level or self-leveling pads used to insure rated loading.
2		Assembly of equipment bolted to stiff steel frame to maintain alignment of components. Frame mounted on springs, general purpose elastomer mounts, or pads, depending on the amount of static deflection required.
3		Assembly of equipment bolted to inertia block which is supported on steel springs with the top of the springs mounted as high as possible on the block. Bottom of springs may rest on pedestals or block can can be recessed into the floor.
4		Same as Type 3 except that tops of springs are close to the height of the center of gravity of the combined block and equipment assembly.

Ch. 7 VIBRATION ISOLATION FOR NOISE REDUCTION

Sufficient vibration data have been generated for mechanical equipment room installations to make mount selection for this equipment rather straightforward. Recommended mountings for a variety of such equipment is given in Table 7-8. Mount types specified in this table are described in Table 7-9. For critical installations, where vibration transmission must be kept to an absolute minimum, attention to detail is extremely important. It involves consideration of where in the building the isolated equipment is to be located, and the isolation of attachments to the equipment.

Machine Location

The location of equipment in a building can influence the isolator selection in two ways: first, the proximity to critically occupied areas will determine the isolation efficiency, and second, upper floor structural flexibility may require mounts softer than that prescribed by rigid floor theory.

Classification of Areas.—As with other aspects of noise control in confined spaces, acceptable noise intrusion due to vibration transmission depends on the use of the space. For example, one would expect the required isolation efficiency to be much higher for a machine located above a conference room as compared to that needed for the same machine located above a drafting room. As a guide in isolator selection, it is convenient to identify three classifications of area: 1. *Minimum Area*—Basement or on-grade slab location; 2. *Normal Area*—Upper floor location, but not above or adjacent to a critical area; and 3. *Critical Area*—Upper floor location above or adjacent to private offices, classrooms, conference rooms, operating rooms, etc.

In factories, warehouses, and other noncritical environments, upper floor locations can be considered minimum condition areas. It must be stressed that these are only guidelines, and each situation must be carefully examined on the basis of present and anticipated future use.

Examples of the Use of Table 7-8.

A. A 10 hp centrifugal fan which operates at 1750 rpm is to be located on ground level or a concrete slab floor in a mechanical equipment room for a process air system. Select the mount type for the application.

Solution. Since this is a noncritical area, we will use the *Minimum Area* classification from Table 7-8. For a centrifugal fan from 3 to 25 hp and over 1200 rpm, in a minimum area, a type 2 mount from Table 7-9 is recommended with isolators having static deflections of 0.5 inches, as given in Table 7-8. From the number of isolators (presumably 4) and the weight of the fan, the proper isolators can be selected from a vendor.

B. A 15 hp, 750 rpm reciprocating air compressor is to be located on a second floor of a factory. If there are no critical areas near, select the isolator system.

Solution. This location is classified as a *Normal Area* and, as indicated in the fourth column of Table 7-8, the recommended mounting from Table 7-9 is mount type 4, and the recommended static isolator deflection is 4 inches.

Floor Conditions.—As stressed earlier, the transmissibility theory given by Equation (7-11) and illustrated in Fig. 7-4, holds for an infinitely rigid foundation. In practice, a ground-floor slab can be considered rigid but an elevated floor cannot. Depending on span and thickness, concrete floors have natural frequencies in the range of 30 to 100 Hz, and wooden floors from 20 to 30 Hz. To obtain adequate isolation in critical areas, the mount natural frequency should not be greater than $1/10$ to $1/6$ the driving frequency, nor should the mount static deflection be less than 6 to 10 times the floor static deflection. For less critical areas, the mount natural frequency can be raised to $1/6$ to $1/3$ the driving frequency and the static mount deflection reduced to 3 to 6 times the floor deflection.

Example

A motor-generator set with a driving frequency of 1800 cpm is to be installed in a critical area. It has been determined that the floor static deflection will be 0.2 inch at the mounting location. How should the mounts be selected?

By the first floor condition, above, the mount natural frequency should be $1/10$ to $1/6$ of 1800 cpm, or 180 to 300 cpm. From Table 7-4, the corresponding mount deflection should be 0.4 to 1.1 inches.

By the second criterion, the mount deflection should be 6 to 10 times the floor deflection, or 1.2 to 20 inches.

Therefore, the mounts should be selected based on the larger deflection requirement, 1.2 to 2.0 inches, to obtain the desired isolation.

Equipment Attachments

To maintain the desired vibration isolation, attachments to a piece of equipment, such as ducts and pipes, must be flexibly connected or mounted 1. to permit freedom of motion of the equipment, and 2. to reduce vibration transmission through the attachment mounts.

Ducts should be attached to fans via flexible rubber or canvas bellows and electrical conduits should be very flexible. In addition, for very large fans, resilient supports for the duct of $1/4$-inch deflection should be provided in the vicinity of the fan.

Pipe runs to and from mechanical equipment must receive special consideration. As a rule, the first three pipe supports from the equipment should have a static deflection of at least $1/2$ the deflection of the equipment to which it is mounted. Beyond that distance, resilient supports of up to $1/2$-inch or more may be needed for an additional distance depending on the proximity of the pipe run to critical areas. It may be necessary in some cases, such as with steam lines and cooling-tower pipes, to use resilient mounts over the entire pipe run.

Resilient pipe supports may be attached to floor or ceiling and may be either steel springs or other types, depending on the required deflection (see Table 7-5). If springs are used, neoprene pads or washers must be placed in series with the spring to prevent high-frequency transmission.

For smaller hydraulic and pneumatic pumps, motors, and compressors, it is usually satisfactory to make connections via braided neoprene hose looped in a loose "c" or "s" configuration so that machine motion is not transmitted along the axis of the hose to the connecting pipe run.

Ch. 7 VIBRATION ISOLATION FOR NOISE REDUCTION 317

Detail Considerations

1. When using steel springs always use at least one ribbed neoprene pad in series with the spring either at the top or base of the spring.
2. If hold-down bolts are necessary to prevent machine creeping make sure they do not short-circuit the mount.
3. Periodic inspections should be scheduled for isolated equipment to be sure that trash does not collect in the isolators or under the equipment to short-circuit the isolators.
4. Machines with high centers of gravity, or subject to accidental collision with aisle traffic may require lateral stabilizing devices to keep them in place. If used, such devices must not short-circuit the isolators.
5. Machines with high starting torques requiring high isolation efficiencies may need to be mounted on inertia blocks to control motion at startup.
6. Grouting, shims or leveling isolators should be used on uneven floors to assure equal loading of each isolator.
7. Neoprene and other pad-type isolators should be loaded to the manufacturer's recommendations. Underloaded pads will not give desired performance and overloaded pads will have reduced life. This last is true for all elastometric mounts.
8. For a given style and size isolator, the static deflection can be doubled or the natural frequency cut in half by stacking two isolators (series or tandem installation). To halve the static deflection, or double the natural frequency, put two isolators side-by-side (parallel or dual installation) connected by a bridge plate to the mounting hole of the machine.

Equipment Maintenance

Worn or malfunctioning equipment is frequently manifested in increased vibration levels at the equipment mounts, as well as an increase in the directly radiated noise levels from the equipment. In this respect, proper equipment maintenance is imperative. A guide to vibration identification is given in Table 7-10. In most cases, a vibration measurement, with suitable filtering of the signal, is necessary to detect causes of excessive vibration. A guide to determining how serious a vibration is, based on the amplitude of the vibration, is shown in Fig. 7-32.

Typical Vibration-Isolation Problems

Problem 1.

A hydraulic power supply consists of an 1800 rpm motor driving a hydraulic pump; both are bolted to a sheet metal pan containing an integral oil reservoir. It is desired to isolate the motor-pump unit from the pan which acts as a sounding board. The motor-pump unit weighs 250 lb and the isolation efficiency desired is 95% minimum, or a transmissibility of .05.

Assume $R/R_c = 0.05$. From Fig. 7-4, $f/f_n = 5$.

Therefore

$$f_n = \frac{f}{5} = \frac{1800}{5} = 360 \text{ cpm} = 6 \text{ Hz}$$

Table 7-10. Vibration Identification Guide*

Cause	Frequency Relative to Machine rpm	Phase-Strobe Picture	Amplitude	Comments
Unbalance	1 × rpm	Single steady reference mark	Radial—steady proportional to unbalance	Common cause of vibration.
Defective anti-friction bearing	10 to 100 × rpm	Unstable	Measure velocity 0.2 to 1.0 in./sec radial	Velocity largest at defective bearing; as failure approaches velocity signal will increase, frequency will decrease.
Sleeve bearing	1 × rpm	Single reference mark	Not large	Shaft and bearing amplitude about the same.
Misalignment of coupling or bearing	2 × rpms, sometimes 1 or 3 rpms	Usually 2 steady reference marks, sometimes 1 or 3	High axial	Axial vibration can be twice radial. Use dial indicator as check.
Bent shaft	1 or 2 × rpm	1 or 2	High axial	...
Defective gears	High rpm × gear teeth	...	Radial	Use velocity measurement.
Mechanical looseness	1 or 2 × rpm	1 or 2	Proportional to looseness	Radial vibration largest in direction of looseness.
Defective belt	Belt rpm × 1 or 2	...	Erratic	Strobe lite will freeze belt.
Electrical	Power line frequency × 1 or 2 (3600 or 7200 rpm)	1 or 2 rotating marks	Usually low	Vibration stops instantly when power is turned off.
Oil whip	Less than rpm	Unstable	Radial—unsteady	Frequency may be as low as one-half rpm.

Ch. 7 VIBRATION ISOLATION FOR NOISE REDUCTION 319

Aerodynamic	1 × rpm or number of blades on fan × rpm	...	Variable at beat rate	May cause trouble in case of resonance.
Beat frequency	1 × rpm	Rotates at beat rate		Caused by two machines running at close rpm.
Resonance	Specific criticals	Single reference mark	High	Phase will change with speed. Amplitude will decrease above and below resonant speed. Resonance can be removed from operative range by stiffening.

*Courtesy of Production Measurements Corp.

320 VIBRATION ISOLATION FOR NOISE REDUCTION Ch. 7

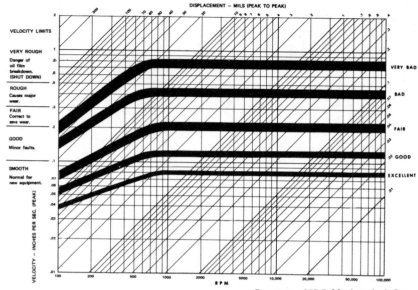

Fig. 7-32. Guide to determine how serious a vibration is, based on amplitude of vibration. *Courtesy of IRD Mechanalysis Inc.*

weight per mount = 250/4 = 62.5 lb for 4 mounts equally loaded. Choose a general-purpose elastomeric mount.

From Fig. 7-26 choose an RDA-55 type mount which has a rated deflection of 0.34 inch at 62.5 lb. Since this exceeds the minimum deflection of 0.30 inch the design is safe. The design assumes that the sheet metal pan does not deflect appreciably with respect to the static deflection of the mounts. In this case it should be less than $1/5$ the static deflection, or less than 0.070 inch. To make this design work, hydraulic and electric lines to and from the unit would have to be suitably flexible and not hinder the motion of the unit on its mounts.

Problem 2.

A 15 hp rooftop exhaust-fan unit weighing 400 lb and operating at 1200 rpm is to be located over a noncritical area. It has been determined that the roof deflection at the point of mounting would be no greater than 0.1 inch. From Table 7-8, the recommended static deflection for the mounts is 1.0 inch. From Fig. 7-6 or Table 7-4, the natural frequency will be 180 cpm or 3 Hz. The frequency ratio is

$$\frac{f}{f_n} = \frac{1200}{180} = 6.66$$

and the mount deflection to roof deflection ratio is 1.0/0.1 = 10. Since the deflection ratio is greater than the frequency ratio, the mounts should be sufficiently soft, relative to the roof, to do the job. For a four-mount support, the weight per mount is 100 lb. To guard against horizontal motion due to wind gust loading choose a housed steel-spring mount as in Fig. 7-23. A type SL-120

Ch. 7 VIBRATION ISOLATION FOR NOISE REDUCTION

A mount is rated at 120 lb for a 1.18-inch deflection. The deflection at 100 lb would be

$$\frac{100}{120} (1.18) = 0.983 \approx 1.00 \text{ inch.}$$

This spring meets the deflection requirements.

APPLICATION OF PRINCIPLES TO MACHINE-NOISE PROBLEMS

The engineering principles already described have been applied to a variety of small and large machines from typewriters to punch presses. Substantial noise reductions have been achieved in those cases where the dominant noise is produced by vibration energy flowing from a machine to a structure where it is reradiated as sound. Some actual cases of vibration isolation applied to machines for the purposes of noise reduction are given below.

Vibration Isolation of a Book-Trimming Machine

In many cases the noise in printing plants is very high. In this particular problem (Ref. 19), eight book-trimming and stacking machines were installed on the third floor of a building and the resulting noise level in a drafting room below was thought to be unacceptable. Believing that most of the noise came from forces transmitted through the direct mountings of the machines, it was decided to isolate that part of the machines containing the book-trimming mechanisms.

Each mechanism weighed 3,472 lb. The time history of machine cutting forces resembled that shown in Fig. 7-33 with an impulsive period, $T = 0.02$ sec,

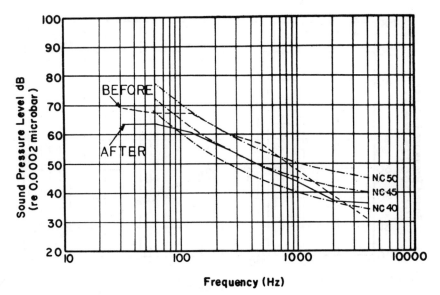

Fig. 7-33. Octave-band sound-pressure levels measured in drafting room before and after installation of isolators under book-trimming machine. [Ref. 19.]

322 VIBRATION ISOLATION FOR NOISE REDUCTION Ch. 7

and a repetition period T_o = 0.4 sec, corresponding to impulsive and periodic frequencies of 50 Hz and 2.5 Hz, respectively. Choosing the natural frequency, f_n to be 10 Hz (corresponding to a static deflection of 0.1 inch), and assuming the weight symmetrically distributed on four isolators, each isolator had to support 868 lb and exhibit a stiffness of 8,680 lb/inch. Any rigid connections between the trimming machine portion and other portions were flexibly connected to avoid "short-circuiting" the isolators.

Octave-band sound-pressure-level readings (re 0.0002 microbar) in the drafting room, before and after the isolators were installed, are shown in Fig. 7-33. With approximately 5 dB noise reduction throughout the frequency range, the final noise level in the room (NC 45) was just acceptable. A further reduction to NC 40 could easily have been achieved by adding acoustic absorption material to the room.

In another room under the machines, the noise level was reduced approximately 10 dB throughout the frequency range. Furthermore, the annoying cyclic thump every 0.4 second was replaced by a rumbling sound in both rooms below the machinery room. Some of this sound was caused by auxiliary equipment which also could have been isolated from the floor.

Vibration Isolation of Printing Presses

Vibration isolation has been used to reduce noise in printing plants (Ref. 20). The problem of isolating printing presses on the third floor of a building was examined to achieve noise reduction. The operating frequency of the presses ranged between 0.8 to 1.4 Hz and the floor was found to have a resonance fre-

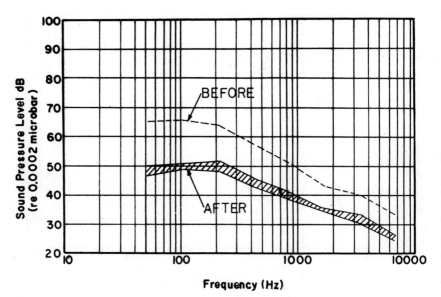

Fig. 7-34. Octave-band sound-pressure levels measured in an office before and after installation of isolators under printing presses. [Ref. 20.]

Ch. 7 VIBRATION ISOLATION FOR NOISE REDUCTION 323

quency in the region of 12 to 15 Hz. The isolators were selected to give coupled machine-floor resonance frequencies that were as low as possible but out of the operating frequency range of the presses. Fig. 7-34 shows the noise reduction (between 8 and 15 dB) that was achieved using felt and rubber isolation pads.

Vibration Isolation of an Air-Handling System

Vibration isolation was used for noise reduction of an air-handling system installed on the first floor of a two-story building (Ref. 21). Although the machine supports rested on thin neoprene pads, they were also bolted directly to a 6-inch, concrete structural slab. The fan operated at 1000 rpm, weighed approximately 2000 lb, and supplied 15,900 cfm of air to the ventilation system of the building. The spaces directly under the fan installation were used as storage rooms and no attention was given to the problem of noise in the area. However, in a conference room adjacent to the storage area, the noise levels were unacceptably high, even though this room was fitted with a heavy suspended ceiling and was at a distance of 40 feet from the fan. Furthermore, the noise in the conference room was easily demonstrated to come from the fan installation.

The fundamental resonance frequency of the supporting floor was measured by shock excitation and found to be 12 Hz. Accordingly, the fan system was mounted on six spring-type vibration isolators chosen to have a static deflection of one inch, thereby providing a system natural frequency of 3.2 Hz.

The measured sound-pressure levels in the conference room before and after the installation of vibration isolators are shown in Fig. 7-35. The decrease in noise levels above 125 Hz are significant (between 3 and 17 dB) and clearly

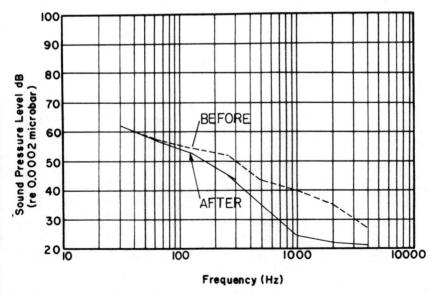

Fig. 7-35. Octave-band sound-pressure levels measured in a conference room before and after installation of isolators under air-handling system. [Ref. 21.]

324 VIBRATION ISOLATION FOR NOISE REDUCTION Ch. 7

demonstrate the importance of having adequate vibration isolation in mechanical equipment rooms to reduce the flow of vibrational energy into the building structure. Such energy can travel relatively great distances and then be reradiated as noise in some distant part of the building.

Vibration Isolation of a Nail-Making Machine

Excessive noise generated by ten nail-making machines in a machine shop was considered (Ref. 21). The nail-making machines were originally rigidly bolted to the concrete floor. Essentially, each machine operated as follows. Wire was continually fed into the machine where it was then chopped and headed to form the desired length nail. This was achieved by a guillotine cutting action within the machine which created a periodic history of shock pulses. These shock pulses were readily transmitted into the concrete floor, thereby exciting structural modes and causing noise to be radiated acoustically.

As previously discussed, the usual practice is to choose an isolator that provides a natural period much greater than the shock pulse duration but less than the period of application of the force. In this case the machine made 300 nails per minute (i.e., pulse period T_o of 0.2 sec) with a measured pulse time duration, T, of approximately 0.01 sec. Elastomeric vibration isolators were selected to provide a natural period τ of 0.1 sec (corresponding to a 10-Hz natural frequency) with a static deflection of 0.1 inch. Figure 7-36 shows the sound-pressure level measured in the machine shop before and after the installation of the isolators. The overall sound level was reduced by approximately 9 dBA, and the remedial action taken was considered successful.

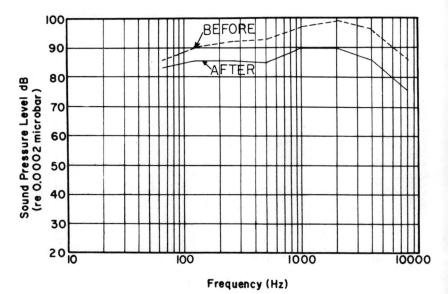

Fig. 7-36. Octave-band sound-pressure levels measured in a machine shop with ten nail-making machines before and after installation of isolators. [Ref. 21.]

Ch. 7 VIBRATION ISOLATION FOR NOISE REDUCTION 325

Vibration Isolation of an Air Conditioner

Excessive noise was generated by a 27,500 Btuh residential central air conditioner (Ref. 22), the main concern being the 120-Hz hum at twice line-frequency. This hum was caused by ¼ and ⅓ hp electric motors used to drive the two blowers. By improving the design of the motor vibration isolators (reducing their stiffness), the sound-pressure level was reduced by approximately 25 dB at 120 Hz and 12 dB at 240 Hz, as shown in Fig. 7-37. The measurements shown were made five feet from the air conditioner installed in an anechoic room. In the experimental arrangement, four insulated ducts were fitted to remove the air flow and much of the broad-band fan noise from the anechoic room. The slight decrease in the broad-band fan noise shown in Fig. 7-37 resulted from shortening the ducts midway through the program to eliminate the masking of the 120- and 240-Hz pure tones.

Vibration Isolation of a Typewriter

A common noise source in typical offices is the typewriter. One solution is to place the typewriter on rubber feet or on a piece of thick felt. This is clearly a

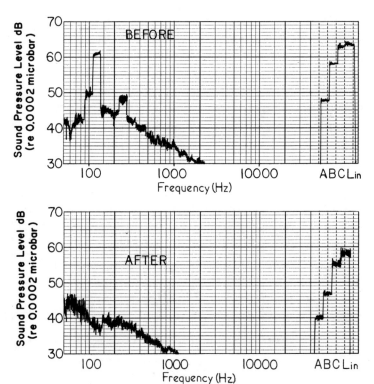

Fig. 7-37. Third-octave-band sound-pressure levels measured 5 feet from a central air-conditioning unit before and after improved isolation of the blower electric motors. [Ref. 22.]

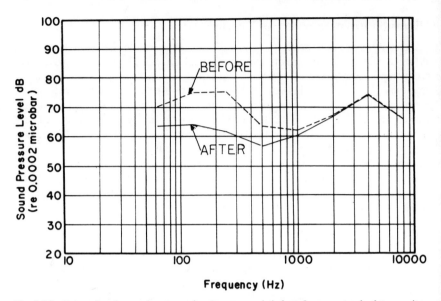

Fig. 7-38. Octave-band sound-pressure levels measured 4 feet from a standard typewriter supported on a typical office table before and after placing 4 neoprene pads (50 durometer) under the machine. [Ref. 21.]

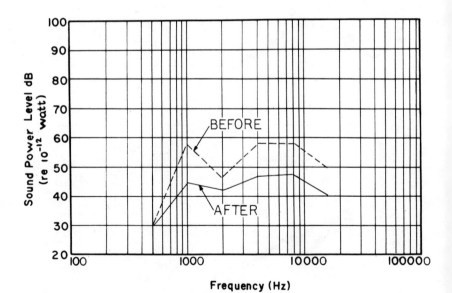

Fig. 7-39. Octave-band sound-power levels of an electrical transformer before and after vibration isolation treatment. [Ref. 25.]

Ch. 7 VIBRATION ISOLATION FOR NOISE REDUCTION

practice which involves vibration isolation for the purpose of obtaining noise reduction. The total reduction in the sound level radiated from the typewriter and its support is achieved by reducing the vibrational energy flow from the machine to its supporting structure, where reradiation may easily take place. This structure is often a light flexible table which, once excited, can efficiently radiate noise into the surrounding space.

Figure 7-38 shows the reduction in the sound-pressure level (measured at four feet from the machine) obtained by installing four neoprene pads (50 durometer) under the metal feet of a standard typewriter supported by a typical office table. Clearly, an improvement in noise level (3 to 12 dB) is achieved only at low frequencies below 1000 Hz with no noise reduction provided above 1000 Hz. This is probably due to the fact that higher order modes are not well excited in the table and radiate much less than the modes set up in the typewriter structure itself.

Other Vibration Isolation Applications

Vibration isolators have been used in a number of other widely varying applications. Mohr (Ref. 23) discusses noise control for outboard motors, rotary lawnmowers, and low-speed three-wheeled transportation vehicles using acoustical enclosures and vibration isolators. Zaborov, et al. (Ref. 24) report noise reduction for shaker tables of about 10 to 15 dB in the middle frequency range and 25 dB at high frequencies. Reed (Ref. 25) reports that a noise reduction of

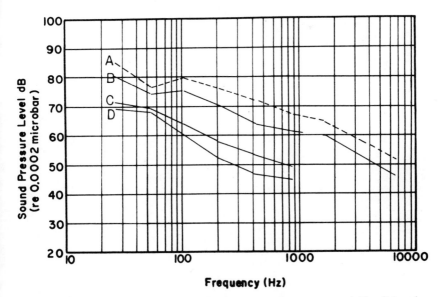

Fig. 7-40. Octave-band sound-pressure levels measured near a portable dishwasher. A—Untreated machine; B—Vibration isolation; C—Vibration isolation, damping, and absorption. [Ref. 26.]

328 VIBRATION ISOLATION FOR NOISE REDUCTION Ch. 7

approximately 10 dB can be achieved in some aircraft electrical components (such as transformers) by vibration isolation, as illustrated in Fig. 7-39. Finally, an example of a combined approach using damping, absorption, and vibration isolation on a portable dishwasher has been reported by Yin (Ref. 26); Fig. 7-40 shows the effect of the different methods in reducing the dishwasher noise level.

8

Machine Element Noise

JAMES E. SHAHAN
and
GEORGE KAMPERMAN

Introduction

Machinery noise and vibration control is a subject which requires a thorough understanding of the dynamics of the basic components that together make up a machine. These components, or machine elements, include gears, cams, linkage mechanisms, bearings, brakes, clutches, chains and sprockets, etc. This chapter does not discuss all possible elements directly, but emphasizes those components which are usually of most importance in machine noise problems: gears, bearings, cams, slider-crank mechanisms, and roller-chain drive systems.

The intent of this chapter is to provide a brief introduction into the dynamics of various machine elements and to outline the fundamental causes of noise and vibration in each of the mechanisms. An understanding of these basics, together with his basic knowledge of engineering principles, is the best equipment available to the engineer when dealing with the design or specifications of new equipment or in attacking an existing machinery noise or vibration problem. Case histories have been avoided because most machinery noise problems are unique to a particular machine, and the number of machines using the various elements is almost limitless.

GEARS

Types of Gears

Gears are the mechanisms that transmit motion in machinery, usually between rotating shafts. They can be used to transmit large amounts of mechanical power as in driving marine propeller shafts or in maintaining precise relative motion between small parts as in a delicate watch mechanism. From a machinery noise standpoint, however, it is primarily power transmission systems that are critical.

Basic types of gears include: involute spur gears, cycloidal gears, parallel helical gears, herringbone gears, crossed helical gears, worm gears, bevel gears, and hypoid gears. These classifications are drawn from the basic form of the outline of the tooth cross-section, the shaft arrangement for which the gears are applicable, the type of rolling surface from which the gears are derived, and the general appearance of the gears.

Involute spur gears transmit rotary motion between parallel shafts and are cylindrical in shape. The teeth have an "involute" profile, are straight, and are parallel to the axis of the gear shaft (Fig. 8-1). An involute tooth profile has an important characteristic in that the center distance between a pair of mating involute gears can be varied without changing the angular velocity ratio. They are also easy to manufacture.

Cycloidal gears were used before the involute tooth form was introduced and

Courtesy of Brad Foote Gear Works, Inc.
Fig. 8-1. Spur gears.

Courtesy of Illinois Gear Div., Wallace-Murray Corp.
Fig. 8-2. Helical gears.

Courtesy of Pennsylvania Gear Corp.
Fig. 8-3. Double helical gears.

are characterized by teeth with cycloidal profiles. These gears are no longer used for power transmission because of their many disadvantages in comparison to the involute tooth profile. Their primary uses today are for watches, clocks, and other applications where great speed reductions are required but where mechanical power transmission is not significant. These applications require low numbers of pinion teeth, but only where interference between the pinion and the mating gear teeth is not tolerable. This is the primary advantage of cycloidal teeth—they do not encounter interference as do involute gears. Their main disadvantage is that they will transmit a constant angular velocity motion only for the theoretically correct center spacing. They are expensive to manufacture.

Helical gears utilize an involute tooth profile with the teeth cut at an angle to the gear shaft. (See Fig. 8-2) The gear shafts are nonintersecting. The gears are *parallel* helical gears if the shafts are parallel and *"crossed* helical gears" when the shafts are at an angle.

Herringbone gears or *double-helical gears* (Fig. 8-3) consist essentially of two

Courtesy of Illinois Gear Div., Wallace-Murray Corp.
Fig. 8-4. Worm gear.

sets of parallel helical gears. The importance of this arrangement is that the axial thrust forces, inherent in a single set of parallel helical gears, are balanced.

Worm gears (Fig. 8-4) are very similar to crossed helical gears in their operation. The *worm*, or *pinion*, gear has a small number of teeth which completely wrap around the shaft to form threads. These gears provide high-speed reduction ratios and can transmit high tooth loads. One of the disadvantages, from a mechanical efficiency standpoint, is the high sliding velocity between the teeth.

Bevel gears transmit power between rotating shafts whose axes intersect (Fig. 8-5). The tooth profile used is the involute form. The gears are *straight-toothed* bevel gears when the teeth are straight and *spiral*-bevel gears when the teeth are curved. Spiral bevel gears are to straight-bevel gears as helical gears are to spur gears.

Courtesy of Illinois Gear Div., Wallace-Murray Corp.
Fig. 8-5. Straight-toothed bevel gears.

When the gear shafts are nonintersecting and nonparallel, *hypoid* gears are used. The tooth action in hypoid gears is similar in nature to that in a worm gear, consisting of a combination of rolling and sliding along the lines of contact on the teeth.

Involute Gear Nomenclature

Because of the extensive use of the involute in gears used for power transmission, the nomenclature used, and the tooth action, should be fully understood. Most of the terminology involved can be shown in terms of the involute spur gear. (See Fig. 8-6.)

The *pitch surface* of a gear is an imaginary surface upon which most calculations are based. The pitch surfaces of mating gears are tangent to each other. In spur gears, the pitch surfaces are right circular cylinders.

The *pitch diameter* is the diameter of the pitch surface cylinder.

The *diametral pitch* is the number of teeth on a gear divided by the pitch diameter.

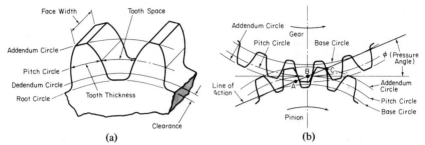

Fig. 8-6. Terminology used in involute gearing. (a) Single involute gear; (b) Meshing involute gears.

The *circular pitch* is the distance between corresponding points on adjacent teeth, measured along the pitch circle.

The *addendum* is the radial distance from the pitch circle to the outside circle of the gear.

The *dedendum* is the radial distance from the pitch circle to the root of the gear teeth.

The *tooth thickness* is the thickness of a single tooth, measured along the pitch circle of a gear.

The *base circle* is the circle which forms the basis for generating the involute tooth form on an involute spur gear. A fixed point on a taut string, unwound from a cylinder of radius equal to the base circle radius, would form the involute profile form of the teeth on a gear (See Fig. 8-7a). With the diameters of the base circles being constant for two mating gears, properties of the involute curve are such that the path traced out by the contact point of two mating teeth (the "line of action") will lie along the common tangent to the two base circles, as indicated in Fig. 8-7b. The pitch circles will always be tangent to each other at the intersection of the line of action and the line connecting the centers of two mating involute gears. Due to this property of involute gear tooth action, the

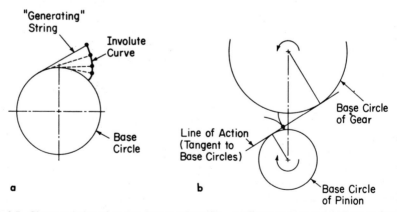

Fig. 8-7. Characteristics of an involute tooth profile. (a) Generating an involute curve from the base circle. (b) Relationship of path of tooth contact relative to involute gear base circles.

angular-velocity ratio between the gears will remain constant (on an averaged time basis) irrespective of the center-to-center spacing of the mating gears.

The *base pitch* is the distance between corresponding points on adjacent teeth on a gear, measured along the base circle.

Other definitions and nomenclature for gears are shown in Fig. 8-6b, where two mating gears are shown:

The *line of action* is the line along which the points of contact between two mating teeth trace a path. For involute gearing, the contact points will lie along the common tangent line of the two base circles.

The *pressure angle* is the angle between the line of action and a line perpendicular to the line of centers between two gears.

The *contact ratio* is the length of the path of contact along the line of action divided by the base pitch, and represents the time-averaged number of teeth which are in contact at any instant.

The *backlash* is the difference between the width of a tooth space and the width of an engaging gear tooth, measured along the pitch circle.

The *clearance* of a mating set of gears is the difference between the dedendum of a gear and the addendum of the mating gear.

Involute Gear Action and Sources of Noise and Vibration

The meshing of two "perfect" gears, as mentioned above, is shown in Fig. 8-6b where the lower, smaller gear (the pinion), is driving the upper, larger gear (the gear). The line of contact will be between Points A and C. As the point of contact progresses along the line of action toward Point B, there is a diminishing relative sliding velocity between the two teeth. At Point B, the point of tangency of the two pitch circles, there is a reversal in the direction of the relative sliding velocity. This reversal results in a force pulse, the magnitude and duration of which depend on the forces transmitted between the gears, the coefficient of friction between the teeth, and the value of the relative sliding velocity. The relative sliding velocity is directly proportional to the rotating speed of the gears and increases with increasing distances from Point B to the tooth contact point.

This force pulse is referred to as *pitch line impulse*. If the gears were perfect, even under operating conditions, this would theoretically be the only origin of gear noise (assuming that the prime mover and driven machinery did not create fluctuating torque on the shafts). The gear noise and vibration frequency spectrum would have components only at the tooth contact frequency

$$f_c = (N_p n_p / 60) \text{ Hz} \tag{8-1}$$

and its higher harmonics. The magnitude of the higher harmonics would depend on the shape of the force-versus-time character of the impulse force. N_p = number of pinion teeth and n_p = pinion speed in rpm.

Materials always deform to some extent when pressure is applied to them. Gear teeth will always deform in operation so that no gears will ever be "perfect." This deformation gives rise to the second basic source of gear noise and vibration known as *engagement impulse*. This is an impulsive force which is directed along the common tangent to the two base circles (the line of action), and is caused by imperfect engagement of gear teeth as they come into contact.

The effects of most gear design and operational parameters on noise and vibration can be related to their effects on the pitch line impulse and the engagement impulse. The primary factors which influence the magnitudes of these forces include the type of gear, pressure angle, contact ratio, face width, gear alignment, surface finish, gear pitch, the accuracy of the tooth profile compared to the ideal involute form, the speed of the gear, and the load being transmitted between the gears.

Noise and Gear Type

In general, sliding action between meshing gear teeth results in a quieter operation. Different types of gears, because of their geometrical characteristics, will result in different types of tooth action. Spur gears, for example, have line contact across the full face width at the instant of impact, as shown in Fig. 8-8a. This results in a large value of engagement impulse. Hence, spur gears in operation can be expected to be relatively noisy, particularly at high values of pitch-line velocity.

Helical gears also have line contact. However, the contact starts at a point on the tooth of the gear and progresses down across the tooth, as shown in Fig. 8-8b. This tooth action results in less impact loading on the gear teeth and, because of the relatively greater sliding motion, instead of line impact, helical gears operate more quietly than spur gears. This difference can be expected to be of the order of from 3 to 10 decibels for equivalent load and speed operating conditions.

Straight-toothed and spiral bevel gears bear about the same relationship to each other as do spur and helical gears. Their noise and vibration characteristics are about the same as their spur and helical-gear counterparts.

Hypoid gears and worm gears are the two most quiet types considered here, due to the relatively large amount of sliding contact in the tooth action. In hypoid gears, the tooth action is a combination of rolling and sliding. Hypoid

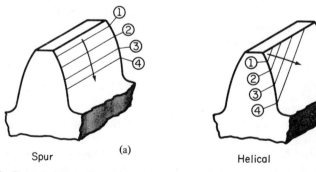

Fig. 8-8. Characteristics of tooth contact between involute gears. (a) Spur gears—line contact across full face width of tooth; (b) Helical gears—progressive sliding line contact across tooth face.

gears are used extensively in automotive differentials where the connecting shafts can be offset to allow a lower body.

Pressure Angle and Gear Noise

Lower pressure angles usually result in lower noise levels. Figure 8-9 shows the schematic action of two involute gears in mesh. Neglecting frictional forces between teeth (acting normal to the line of action), the force transmitted between the gears is directed along the line of action. At the pitch point (point of tangency of pitch circles) F_n is the "useful" force supplying the working torque to the driven gear and, neglecting friction, is determined by the power requirements

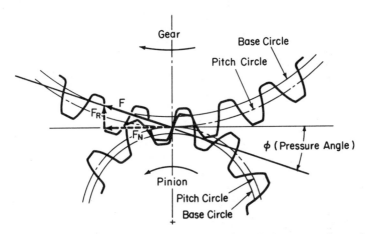

Fig. 8-9. Forces between meshing involute gears depend on the pressure angle, ϕ. F is the resultant force between the teeth in contact at the pitch point neglecting friction. F_n is the component of F transmitting "useful" power between the teeth in contact at the pitch point (neglecting friction).

of the machine connected to the output shaft of the gear assembly. The relation between the resulting total force F along the line of action, and F_n, is given by:

$$F_n = F \cos \phi \qquad (8\text{-}2)$$

This relation indicates that increasing pressure angles cause an increase in the normal forces required to maintain F_n constant. This results in higher noise and vibration levels with increasing pressure angle for given power requirements.

For involute tooth action, the operating pressure angle is determined not only by the design of the gear teeth but also by the actual center distance of the meshing gears. When any pair of involute gears is operating at the theoretical center distance, the operating pressure angle will be equal to the theoretical pressure angle. Manufacturing tolerances permit slight variations in the actual center distance, which may be increased by wear, misalignment, or temperature changes. Within limits, a deviation from the theoretical center distance will not affect the smooth running of involute gears; however, it will cause the pressure angle to change. In the same manner, any other change in the center distance will cause the pressure angle to change. As shown in Fig. 8-10, the pressure angle will increase when the center distance between a pair of gears increases. This causes a change in the forces transmitted between the gear teeth and, hence, a change in the noise and vibration output of the gear set.

Parameters which result in a periodic change in gear center distances during operation will create periodic changes in the pressure angle and tend to modulate the noise and vibration spectrum from a gear set. Eccentricity of a gear on its shaft will vary the center distances at a frequency equal to the rotational rate of

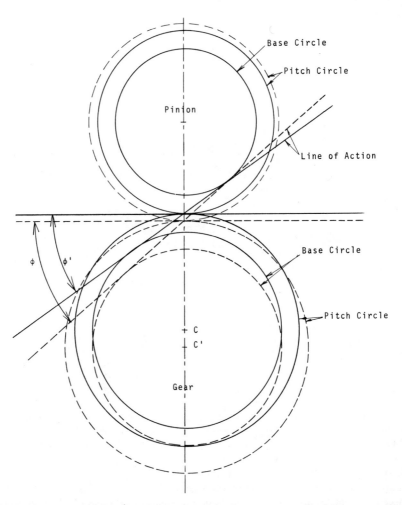

Fig. 8-10. The effect of changing the center distance of two meshing involute gears is to change the operating pressure angle. Solid lines in the figure indicate operation when the gear center is located at Point C with pressure angle ϕ. Dashed lines indicate operation with the gear center shifted to C' and resulting pressure angle ϕ'.

the gear. The resulting noise spectrum would tend to have the characteristics shown in Fig. 8-11. The modulation amplitude will depend on the amount of eccentricity. Other factors which can result in modulation of the tooth forces and noise output are a fluctuating torque due to characteristics of the prime mover or the driven machine and resonances in the gear shafts or bearing support structures. Fluctuating torques cause radial gear forces (and hence shaft deflection) to vary periodically, resulting in center distance variations. This is minimized by stiff shafts.

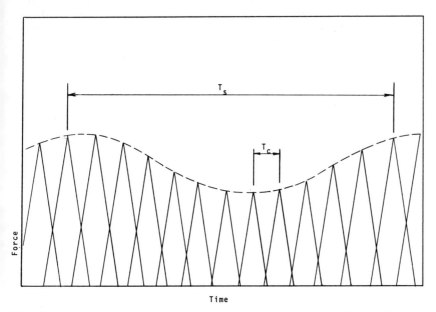

Fig. 8-11. Schematic of the variation in successive tooth contact impulse forces with modulation at the shaft rotational frequency. T_S = the time required for one shaft revolution. T_C = the time between contact of adjacent gear or pinion teeth. Modulation of this type could be caused by eccentricity of a gear on a shaft.

Contact Ratio and Gear Noise

If there were no deflection of gear teeth as they transmitted load, engagement impulse would be negligible for "perfect" gears. However, real gear teeth will always deflect to some extent, resulting in imperfect involute action at the instant of meshing and unmeshing. Imperfect engagement results in engagement shock directed along the line of action of the mating gears. Increasing the average number of teeth in contact at any instant of time distributes the loading over more teeth and results in lower tooth deflections. This improves the engagement and disengagement of meshing teeth thereby lowering noise levels from the gears. For necessarily low contact ratios, a contact ratio of exactly 2 is considered optimum.

Tooth Face Width and Gear Noise

The effect of tooth width on the noise output of gears depends on the relationship between the precision of the gears and the load-induced tooth deflection. Low-quality gears with low tolerances on the pitch and tooth spacing are not affected strongly by higher specific loads (load-per-unit tooth width) of reasonable values. The load-induced effects on the accuracy of the tooth-meshing process will be of secondary importance to the inaccurate meshing caused by loose tolerances. High-precision gears, on the other hand, will experience significant noise-level increases with increasing specific loads since inaccurate meshing between the teeth will be caused primarily by tooth deformation under loading. For equivalent specific tooth loading, high-precision gears will operate more quietly than those with loose tolerances.

Figure 8-12 shows the effect of specific tooth loads on various AGMA (American Gear Manufacturer's Association) gear classes [1].[1] As the gear class (and

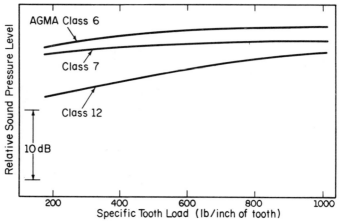

Fig. 8-12. Relative sound pressure levels emitted by various quality gears for different tooth loading. From Ref. [1].

quality) increases, the importance of load-induced tooth deformation increases. AGMA Class 6 and Class 7 gears have tolerances loose enough so that tooth deformation plays a secondary role in tooth meshing as well as in noise generation. AGMA Class 12, a higher-quality gear, has a stronger dependence on tooth loading but operates at a lower absolute noise level.

Alignment and Gear Noise

When gears are not in perfect alignment, large localized tooth deformations will occur in gear teeth at the leading edge of contact causing inaccurate involute tooth action and greater engagement shock. Tooth alignment effects on gear

---

[1] Brackets [ ] indicate numbered references found at the end of the book.

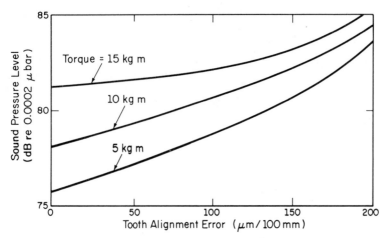

Fig. 8-13. Effect of angular tooth alignment error on sound pressure levels from gears for different transmitted loads. $\mu m = 10^{-6}$ meters; $mm = 10^{-3}$ meters. From Ref. [2].

noise are shown in Fig. 8-13, for different degrees of loading [2]. For small alignment error, the loading plays a significant role in determining the noise output of the gears since load-induced tooth deflection is large relative to the alignment error effect. With large alignment errors, the loading becomes less important since the engagement process is so strongly affected by the angular misalignment.

Tooth Surface Finish and Gear Noise

The coefficient of friction between mating gear teeth can have a significant effect on the noise generated by a set of gears. In general, the lower the coefficient of friction is, the quieter the gears will be. In addition to lubrication, the surface finish of mating gears controls the friction between gear teeth. Higher friction means greater power requirements for a given load, resulting in greater forces between teeth and hence more noise. The noise level curve is almost linear when decibels are plotted against tooth surface finish, with approximately a 4-decibel difference between a smooth tooth finish (smooth ground, fine ground) and a relatively rough tooth finish (milled, heat-treated).

Fundamental effects of the friction between gear teeth can be discussed with reference to Fig. 8-14. The figure illustrates pinion and gear involute teeth in contact at two different instants in time. Frictional forces are tangent to the tooth profiles at the point of tooth contact and, in involute gears, will be perpendicular to the line of action. The solid lines represent the gears in contact at one instant of time (t) and the dashed lines represent the gears in contact at a slightly later instant in time ($t + \Delta t$). F_1 and F_2 are the friction forces acting on pinion teeth 1 and 2, respectively, at time t, and F_1' and F_2' represent these forces at ($t + \Delta t$). The figure illustrates the position of these forces and the perpendicular distance from the center of the pinion shaft at both instants of time.

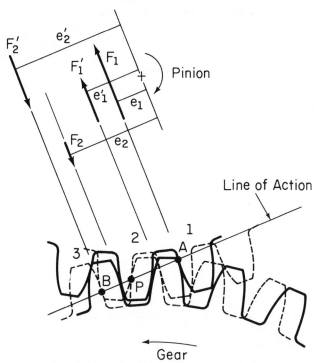

Fig. 8-14. Variation in magnitude and position of involute tooth forces for different contact points along the line of action. Solid tooth outlines represent the pinion and gear position at an arbitrary time, t, and dashed tooth outlines at a short time later $(t + \Delta t)$. Forces F' and positions e' relate to time $(t + \Delta t)$, and forces F and e relate to time, t. Variations in the magnitudes and positions of tooth forces result in periodically varying torques and radial gear forces.

A summation of moments M about the center of the pinion, due to these forces at times t and $(t + \Delta t)$, yields

$$M(t) = F_1 e_1 - F_2 e_2 \tag{8-3}$$

$$M(t + \Delta t) = F'_1 e'_1 - F'_2 e'_2 \tag{8-4}$$

The quantities F and e vary with time as the gears rotate, creating fluctuating torques and radial forces on the gears. The magnitude of F depends on the friction coefficient between the gear teeth and the load being transmitted. The coefficient of friction depends to some extent on the relative velocity between the tooth surfaces. The relative velocity is a maximum at the beginning and end of contact, reversing direction at the point of tangency of the pitch circles of the gears. The higher the coefficient of friction is between the gear teeth, the greater are the exciting forces and moments and, hence, the more the noise and vibration.

The magnitudes and positions of the fluctuating forces caused by tooth friction will depend also on the pressure angle and on the contact ratio. The more teeth that are in contact along the line of action, the smoother will be the tooth action, and the quieter the gears will operate. More teeth in contact also results in reducing the eccentricity of the frictional forces, since the forces of the teeth will act closer to the base circles of the gears. Because of the properties of the involute, a force acting tangent to the tooth at the base circle, has its line of action passing through the center of the gear and would not result in a torque about the gear axis, irrespective of the magnitude of the friction forces.

Gear Pitch and Gear Noise

The effect of different pitch gears has been discussed indirectly, in terms of other parameters related to the pitch of a gear. The number of teeth on a gear is inversely proportional to the pitch for the same gear diameter. Hence, finer pitch gears imply higher contact ratios and tooth forces acting more closely to the base circle of the gears. Both of these effects are expected to result in lower gear noise.

Pitch errors, the difference between ideal tooth spacing and that actually achieved in production, can have a significant influence on gear noise; the effect will depend on the load-induced tooth deflection. In low-precision gears the pitch errors may be great enough so that, except at high specific tooth loads where tooth deflection controls the accuracy of meshing, these influences will be great. In high-precision gears, the pitch errors will be less influential, with load-induced deformations usually governing the meshing process. (See Fig. 8-12).

Tooth Profile Accuracy and Gear Noise

The accuracy of the tooth profile (the variation in the actual profile from the ideal involute), as illustrated in Fig. 8-15, has a strong influence on the tooth

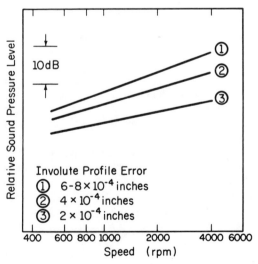

Fig. 8-15. Variation in the relative sound pressure level radiated from gears at different speeds and for various tooth profile errors. From Ref. [1].

meshing process and the noise output of gears [1]. In one instance, under the same operating conditions, a gear tooth profile which deviated from the ideal by about 5×10^{-6} meters (average) was on the order of 10 decibels more quiet than one with a profile error between 15 to 20×10^{-6} meters [2].

Effects of Load and Speed on Gear Noise

Variation in tooth-loading influences gear noise in two ways: (1) Loads high enough to cause sufficient tooth deformation to dominate the accuracy of gear meshing causes the noise to increase due to engagement impulse. In addition, increasing the load increases frictional forces with a corresponding increase in the noise output due to pitch line impulse. (2) At low speeds and low loads the noise output of gears increases by about 3 decibels per doubling of load. This variation probably corresponds to loads low enough so that manufacturing inaccuracies govern the meshing process. At high speeds, greater than about 4000 rpm, gear noise varies approximately with the square of the tooth load, and a doubling of tooth load results in a 6-decibel increase of noise level. For high, specific tooth loading, the sound-power output can be expected to vary with the square of the tooth load, even at low speeds. This corresponds to loads sufficiently high so that the meshing process is controlled by tooth deflection and will depend on the quality of the gears.

The sound power output of gears at high speeds or at high specific tooth loads varies with the square of the tooth load and with the square of the operating speed. The useful mechanical power transmitted is proportional to the product of pitch-line velocity and the tooth force component normal to the line of gear centers. Thus, the acoustic power radiated by a set of gears has a direct relationship with the transmitted mechanical power, increasing at a rate of 6 decibels per doubling of transmitted power.

Other Sources of Gear Noise

In high-speed gears, the expulsion of air from between meshing teeth (air pocketing) can generate air velocities great enough to create acoustic shock waves. The coupling of this acoustic energy to the gears and gear housing (for housed gear sets) is usually not great enough to cause significant noise problems. High-speed gearing also necessarily has a rather rigid housing which acts as an effective acoustic barrier to the airborne shock waves. Air-pocketing effects can be reduced by increasing the backlash of a set of gears to reduce the escape velocities from between meshing teeth.

A more serious problem occurs when the fluids ejected from between the teeth are nearly incompressible lubricants. With the void spaces between teeth filled with lubricant which has difficulty escaping, the gears will experience shock excitation. The maximum amount of lubricant which can be supplied to a gear set without oil-pocketing problems is given approximately by [3]

$$V_{\max} = C \times U_n \times F_t \times 1/231 \text{ in.}^3/\text{min} \tag{8-5}$$

where C = the radial distance between the dedendum circle and the root circle in inches, U_n = pitch line velocity, in inches per minute, and F_t = total face width of the gear, in inches.

Gear Noise and Gear Quality

The primary causes of noise and vibration in gears are fundamentally related to the quality of manufacturing and the tolerances used. Based on the results of an extensive German study of gears used in industry [2], a relationship between the quality of gears and the noise characteristics was formulated (Fig. 8-16). The

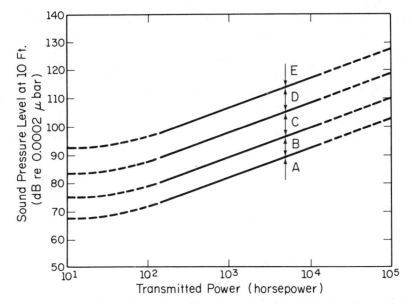

Fig. 8-16. Method of classifying the quality of gears based on their noise characteristics. (See text for definition of gear classes, A through E.) The classification is useful for gross estimations of the noise from different quality gear sets. From Ref. [2].

following classifications were established:

Class A: These requirements cannot reliably be achieved even through highest quality production techniques; additional noise control must be used to ensure these characteristics.

Class B: These characteristics are achieved through the result of very high manufacturing accuracy and control.

Class C: This is the result of high manufacturing quality.

Class D: This is the result of normal manufacturing quality.

Class E: The noise levels of gears having this characteristic can be easily reduced by improving the quality of manufacturing.

This classification system is useful for gross estimates of gear noise to evaluate potential noise problems and, in some cases, to avoid purchasing more expensive higher quality gear sets when a cheaper set would suffice.

Frequency Characteristics of Gear Noise

The dynamic forces responsible for gear noise are both broadband (due to friction and fluid flow) and discrete tone in nature. Discrete components arise due to impulses occurring at the tooth contact frequency. The impulsive nature of gear-tooth contact generates forces at higher order harmonics of the tooth contact frequency, their magnitudes depending on the precise shape of the tooth force versus time. Generally, the more peaked the force (i.e., the shorter its rise time) the greater are the magnitudes of higher harmonics. Add to this the fact that design or operating parameters, such as gear eccentricity or lateral shaft resonances, will cause the tooth contact forces to vary periodically and modulate the forces, thus it is difficult to predict with any degree of confidence the most important forcing frequencies and their relative magnitudes for any given gear set.

Before finally ending up as radiated airborne noise, the forced vibrations are modified by the frequency response of the gear bodies, shafts, and gear housing. It is almost certain that many resonances of these components will be excited by the impulsive and frictional forces acting within the gear assembly. These resonant modes are responsible for the major part of the airborne noise radiated from housed gear sets. The noise spectrum of gear units will then be dominated by:

(1) discrete components corresponding to particularly strong forcing frequencies (e.g., tooth contact frequency)
(2) resonant vibration of efficiently radiating structures. The strongest noise components will occur where a forcing frequency lies close to a structural resonance of a large machine number.

With the exception of the quality of gear design and manufacture, resonances within the gear train are primarily responsible for high noise levels due to amplification. An ordering of the relative importance of these resonances, from experience, is:

1. torsional resonance in the drive train
2. resonance of gear housing surfaces
3. resonance of gear hubs and rings
4. gear tooth resonance.

The most serious resonance problem is torsional resonance in the drive train. This results in high amplitude angular oscillations of meshing gears and severe interference with tooth meshing. This problem is particularly troublesome in large gear sets where shaft lengths are long and torsional resonance frequencies are low.

Noise Control Procedures

It will frequently be found that even the highest quality gear manufacturing processes will not produce gears having an acceptable noise level (Fig. 8-16). Moreover, gears of a quality high enough to meet acceptable noise requirements are expensive to manufacture. For these reasons, it is often less expensive to utilize lower quality gears and apply other noise control procedures. These procedures can be applied quite effectively to reduce gear noise; they will be described in the following paragraphs.

Damping is effective in reducing the resonant response of structures. In gear noise, the airborne noise spectra generally contain many resonant mode vibration-induced sound frequencies, and damping is beneficial. Damping material may be in the form of viscoelastic compounds which can either be spread or sprayed on the surface to be damped (free layer damping), or in the form of a layer which must be held rigidly against the surface to be damped by another metal surface (constrained layer damping). The free layer damping materials are useful mostly for light and thin machine members.

The airborne noise emanating from a housed gear unit is radiated directly from the housing surfaces. Increasing the bending stiffness of these housings will usually result in less vibration amplitude and hence, less radiated noise.

Gear housings are excited by forces transmitted through the bearings which support the gear shafting. Different types of bearings can aid in isolating these forces from the housing and hence, reduce noise levels. In general, the highest damping effect occurs in tapered roller bearings, where noise levels 4 to 5 decibels lower than with ball bearings have consistently been observed [2]. This also will depend somewhat on the type of loading between the shaft and the bearing. In machines where no thrust forces are required of the bearing, tapered roller bearings have been observed to transmit 4 to 5 decibels higher forces than straight roller bearings which are incapable of carrying a thrust load.

BEARINGS

Bearing Types

Bearings generally may be classified into two categories: *rolling-contact* bearings and *sliding-contact* bearings. Sliding-contact bearings, according to their function, may be *journal* bearings, *thrust* bearings, or *guide* bearings. Journal bearings are cylindrical in shape and carry a rotating shaft. The fluid lubrication in journal bearings (most important from a noise standpoint) depends on the ability of the lubricant to adhere to the bearing surfaces. Thrust bearings function to prevent motion of a machine element along its length. Guide bearings serve to guide a machine element along its length, usually without rotation of the element. Materials commonly used for sliding bearings include primarily alloys of lead, tin, and copper. Generally, high lead content infers better lubricating properties while high tin content infers higher bearing strength and hardness.

Rolling-contact bearings are classified according to the general geometry of the rolling elements and to the types of loads they are designed to carry. *Ball* bearings have spherical roller elements and may be classified either as *radial* ball bearings or *angular* ball bearings. Radial bearings are intended to carry forces in the radial direction only, with no axial load. Angular ball bearings are capable of carrying both a radial and an axial (thrust) load.

Roller bearings are a type of rolling-contact bearing in which the elements make "line" contact with the bearing races in contrast to the "point" contact of ball bearings. The rolling elements may be either straight, tapered, barrel-shaped, or spherical. Needle bearings are essentially roller bearings where the rolling elements are straight and cylindrical in shape and have a minimum length-to-diameter ratio of at least four.

Sources of Bearing Noise

As a general rule, bearings with sliding contact are more quiet in operation than are rolling-contact bearings. The main cause of any significant noise produced by sliding bearings is due to frictional effects between the bearing surfaces and the shaft—or journal—in the case of journal bearings. Friction of sufficient magnitude to cause noise and vibration problems is usually due to inadequate or improper lubrication of the bearing.

The lubrication of a journal bearing is accomplished either (1) Through the use of pressurized forced lubrication, or (2) By depending on the basic theory of the converging film in order to support the loads on a thin layer of lubricant. Journal bearings using forced lubrication seldom show any significant amount of frictional excitation. Under certain conditions, however, bearings depending on free film lubrication can experience improper lubrication which allows metal-to-metal contact and a subsequent "screeching" noise caused by "stick-slip" motion between the journal and the bearing surface.

The lubrication of a journal bearing using a converging film is shown in Fig. 8-17 for three different running conditions of the journal; i.e., at rest, starting up, and running under full hydrodynamic lubrication. During the stopped and starting conditions, there is metal-to-metal contact between the bearing and the journal surfaces. As the journal increases in speed, it carries with it a film of

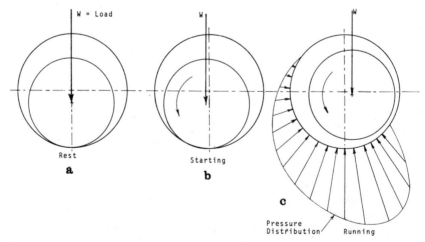

Fig. 8-17. Various positions of a journal (shaft) during starting and running conditions. The journal is in contact with bearing surfaces during stopped and starting conditions. Under running condition, for full hydrodynamic lubrication, the journal will "float" on a thin film of lubricant.

lubricant attached to its surface into the region of contact between the journal and bearing. At a certain speed of the journal, the film of lubricant starts to "float" the journal. Under full hydrodynamic lubrication (thick-film lubrication) conditions, as shown in Fig. 8-17c, the shaft is freely floating on the film. The pressure distribution under running conditions is also shown schematically in Fig. 8-17c.

Since the major noise source in journal bearings is due to inadequate lubrication, it is important to maintain thick-film bearing conditions during operation of the machinery. The "bearing modulus"

$$B = ZN/P \qquad (8\text{-}6)$$

can be related to the conditions necessary to maintain thick-film lubrication. Z = lubricant viscosity, in centipoise, N = journal speed, in rpm, and P = mean pressure on the bearing due to the load, in pounds per square inch.

$$P = \frac{W}{LD} \qquad (8\text{-}7)$$

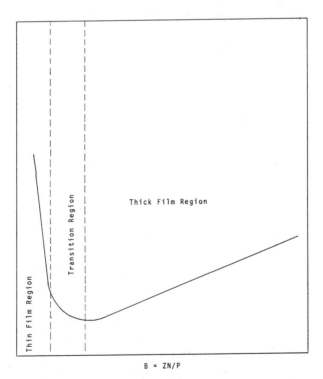

Fig. 8-18. Relationship between the coefficient of friction f and the bearing modulus B for a journal bearing. If the bearing modulus is sufficiently small, there will be a breakdown or rupture of the lubricant film allowing metal-to-metal contact between the journal and bearing. Z = Lubricant Viscosity, in centipoises; N = Journal Speed, in rpm; and P = the Mean Pressure on the bearing, in pounds per square inch.

where W = bearing loads, in pounds, L = length of the journal bearing, in inches, and D = the journal diameter, in inches. The average pressure is determined assuming the projected area of the bearing is uniformly loaded with the mean pressure, P. Figure 8-18 illustrates the basic manner in which the coefficient of friction f varies with the bearing modulus. The slope and intercept of the straight line in the thick-film region depends primarily on the clearance ratio C/D and the ratio L/D, where C = diametral clearance between the journal and the bearing. The equation of this line is given approximately by [4]

$$f = (473/10^{10})(BD/C) + K \qquad (8\text{-}8)$$

K is a function of L/D and is obtained from Fig. 8-19. As Fig. 8-18 illustrates, it is necessary to maintain the bearing modulus at large enough value to prevent a breakdown, or rupture, of the lubricant film. If the breakdown occurs, stick-slip excitation of machine parts at their resonant frequencies will occur. Table 8-1 illustrates the operating ranges of journal bearings in typical applications [4].

Even under conditions at which thick-film lubrication exists at normal operating speeds, it is possible to experience film breakdown when the bearing is subjected to oscillatory conditions or when the direction of rotation is reversed.

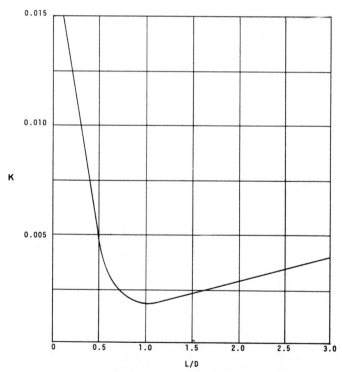

Fig. 8-19. Value of K for use in Equation (6-8) to estimate the coefficient of friction in a journal bearing. L = the Length of the journal bearing, in inches; and D = the journal Diameter, in inches.

Table 8-1. Typical Operating Conditions
for Journal Bearing Applications

EQUIPMENT	BEARING	LUBRICANT*	
		Z	ZN/P
Automobile and Aircraft Engines	Main Crankpin Wrist Pin	7 8 8	15 10
Gas and Oil Engines	Main Crankpin Wrist Pin	20 40 65	20 10
Marine Engines	Main Crankpin Wrist Pin	30 40 50	20 15
Stationary Steam Engines	Main Crankpin Wrist Pin	15–60 30–80 25–60	20 6
Reciprocating Pumps and Compressors	Main Crankpin Wrist Pin	30 50 80	30 20
Steam Turbines	Main	2–16	100
Rotary Motors and Pumps	Shaft	25	200

* Z = lubricant viscosity, in centipoise
N = journal (shaft) speed, in rpm
P = mean pressure on the bearing due to the load. (See Ref. [4].)

Under these circumstances, the bearing modulus passes through zero and film breakdown will occur. Under some conditions of journal loading and motion, it is possible to maintain stable lubrication by grooving the bearing surfaces to distribute the lubricant more effectively [5].

Improperly selected lubricants for the operating conditions can also result in film breakdown at high temperatures. Figure 8-20 illustrates the effect of temperature on the viscosity of various lubricating oils [4]. When the temperature of the lubricant exceeds a certain value, the bearing modulus can become small enough to result in film rupture. This value will, of course, depend on the loading and the operating speed of the bearing.

The decrease in lubricant viscosity with increasing temperatures will also result in reduced vibration-damping capacity of journal bearings; a significant advantage which journal bearings can have over rolling-contact bearings. The damping can be increased by using a more viscous lubricant; however, the frictional power consumption, due to shearing action of the lubricant, will also increase and the amount of damping gained, in many cases will not warrant the loss in mechanical efficiency. This can be determined on an empirical basis for a specific machine.

MACHINE ELEMENT NOISE

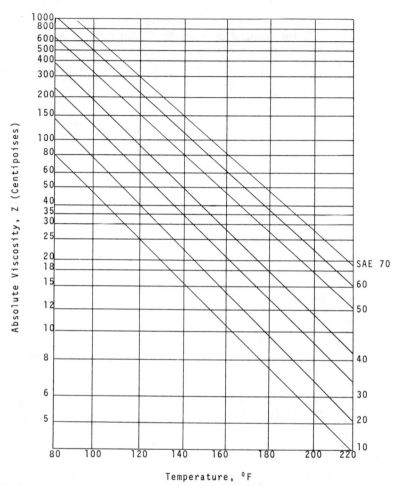

Fig. 8-20. Dependence of the viscosity of various lubricating oils on operating temperature.

Another source of vibration and noise in journal bearings is the phenomenon known as *oil whirl*. This condition can occur in horizontal or vertical bearings in a lightly loaded application with either self-lubricated, or externally pressurized, oiling systems. Oil whirl is identifiable by a frequency of vibration at approximately one-half the shaft rotational frequency. This half-frequency excitation is the precession of the shaft in the bearing induced by the oil surrounding the shaft. From fluid mechanics we know that the oil film next to the shaft in the fluid boundary layer rotates at the shaft speed and the oil film next to the bearing is stationary because the bearing does not move. The average oil velocity is then one-half the shaft speed, which is the frequency at which the shaft precesses within the bearing clearance. With bearing clearances as much as 0.007 inch, there is ample latitude for shaft vibration.

Ch. 8 MACHINE ELEMENT NOISE 353

If a restoring force exists due to the load on the shaft, the effect of the oil whirl can be overcome and no vibration results. For lightly loaded bearings the restoring force may not overcome the forces in the oil and the shaft can precess within the bearing at approximately one-half the shaft rotational frequency. For horizontal shafts, gravity forces may be enough to overcome the oil forces. For vertical shafts and lightly loaded bearings, this phenomenon is most common because restoring forces are minimal and the unbalance in the shaft and rotor can initiate motion in the bearing.

Oil whirl can be eliminated by altering bearing operating conditions or changing the bearing design. A change in oil viscosity or in oil supply pressure are easily implemented solutions. Oil pressure has a marked effect on oil whirl, as illustrated in Fig. 8-21. For the example given in this illustration, the vibration

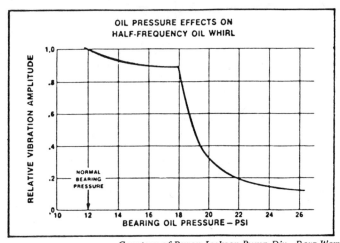

Courtesy of Byron Jackson Pump Div., Borg-Warner Corp.
Fig. 8-21. Illustration of the effects of bearing oil pressure on the relative vibration amplitude of oil-whirl induced motion.

amplitude can be reduced by a factor of 10 to 1 if bearing pressure is increased from 12 psi to 26 psi. The higher oil pressure changes the boundary layer thickness reducing shaft precession at the half-frequency. Pressure changes from 12 psi to 18 psi had little effect, as did changes beyond 22 psi.

In some instances it is not possible to operate at increased pressures because of leakage at the bearing. In these situations an oil dam may be added, as illustrated in Fig. 8-22. Curve A is the amplitude of the shaft vibration in the original bearing. Curve B is a first attempt to use a dam which was not successful. Curve C is for a properly located oil dam, which results in locally slowing down the oil increases, the pressure in the oil resulting in a force opposing the oil whirl pressure force.

Rolling-Contact Bearings

Rolling-contact bearings are generally more noisy than sliding-contact bearings even though they are mechanically more efficient. The major noise components

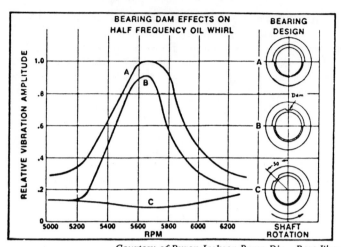

Courtesy of Byron Jackson Pump Div., Borg-Warner Corp.
Fig. 8-22. Reduction of relative vibration amplitude by the use of an oil dam to control motion induced by oil-whirl. Curve A is the original bearing, Curve B an improperly located oil dam, and Curve C a properly located oil dam.

which arise in rolling bearings are due primarily to geometric inaccuracies either machined into the bearing elements or resulting from severe operating conditions. Inaccuracies affecting bearing noise include rolling element or race eccentricities, surface waviness, ball sphericity or roller cylindricity, and groove wobble. These conditions result in vibration of the bearing elements, the bearing support, or connected structure resulting in sound radiation. All of these parameters are dependent upon the quality of manufacturing and the tolerances used.

Because the noise from good quality bearings is generally well below the total noise radiated from machinery, considerable emphasis is usually placed on using the frequency spectra of noise and vibration which could be expected from bearings in order to locate defects. Given the geometry of a bearing, it is possible to relate the operation of the bearing to the frequency spectrum of the noise and vibration. Table 8-2 gives the expected frequencies of a rolling bearing (Fig. 8-23). Measurement instrumentation and an illustration of a measured narrow-band spectrum are illustrated in Chapter 2, Fig. 2-13. A narrow-band analysis of the vibration, or noise signal at a bearing makes it possible in many instances to locate defects in a bearing under operating conditions by correlating the spectrum with the predicted frequencies from the bearing, as computed from the table. If relatively large vibration amplitudes or large acoustic amplitudes exist at one or more of the frequencies given in Fig. 8-23, compared to the amplitudes at other frequencies, it might be suspected that the defect corresponding to this frequency exists in the bearing element.

There are several complications which can develop in comparing measured and calculated frequency spectra to either locate bearing defects or in attempting to isolate bearing noise components from the total machinery noise. Since the nature of bearing excitation is generally impulsive in nature, the dynamic forces from the bearing can excite many resonances of the bearing parts and of the

Table 8-2. Approximate Discrete Frequencies Expected from Roller Bearings.*

FREQUENCIES, in hertz	RELATIONSHIP TO OPERATION
$f = N/60$	Shaft rotational speed
$f = (N/120)(1 - [d/D]\cos\phi)$	Rotational speed of ball cage when outer race is stationary
$f = (N/120)(1 + [d/D]\cos\phi)$	Rotational speed of ball cage when inner race is stationary
$f = (N/120)(D/d)(1 - [d^2/D^2]\cos^2\phi)$	Rotational frequency of a roller element
$f = (\eta N/120)(1 - [d/D]\cos\phi)$	Frequency of contact between a fixed point on a stationary outer race and a rolling element
$f = (\eta N/120)(1 + [d/D]\cos\phi)$	Frequency of contact between a fixed point on a stationary inner race with a rolling element
$f = (N/60)(D/d)(1 - [d^2/D^2]\cos^2\phi)$	Contact frequency between a fixed point on a rolling element with the inner and outer races
$f = (N/60)[1 - \frac{1}{2}(1 - [d/D]\cos\phi)]$	Frequency of relative rotation between the cage and rotating inner race with stationary outer race
$f = (N/60)[1 - \frac{1}{2}(1 + [d/D]\cos\phi)]$	Frequency of relative rotation between the cage and rotating outer race with stationary inner race
$f = (\eta N/60)[1 - \frac{1}{2}(1 - [d/D]\cos\phi)]$	Frequency at which a rolling element contacts a fixed point on a rotating inner race with fixed outer race
$f = (\eta N/60)[1 - \frac{1}{2}(1 + [d/D]\cos\phi)]$	Frequency at which a rolling element contacts a fixed point on a rotating outer race with fixed inner race

*f = frequency, in hertz; N = shaft speed, in rpm; d = roller diameter; D = pitch diameter of bearing; η = number of rolling elements; ϕ = angle of contact between rolling element and raceway, in degrees ($\phi = 0°$ for a simple radial ball bearing).

machine structure. The excited resonances will appear in the narrow-band spectrum of the total noise and often make it difficult to relate to bearing-generated forces.

There will also be, in general, modulation of the frequency components of the bearing noise and vibration. One source of modulation can be due to bearing elements moving in and out of the "loaded zone" of the bearing. Roll-

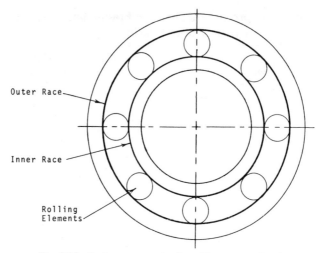

Fig. 8-23. Basic components of a rolling-contact bearing.

ing elements may experience times when they are either not loaded at all or ball bearing elements may rotate to such a position that a spot defect on an element will not contact either the inner or outer races during the shaft rotation. These factors also complicate identification of bearing noise contribution.

The exact nature of a defect or irregularity in rolling bearings can have a significant effect on the character of the noise produced. Very sharp defects such as cracked races or flaking of the surfaces of a rolling element will generate several higher order harmonics of the basic contact frequency, as calculated from Table 8-2. It is almost certain that some of these higher order harmonics will lie close to machine structural resonances and will result in high noise levels at these frequencies. More gradual irregularities such as out-of-round rollers or races will demonstrate the fundamental occurrence frequency in its noise and vibration spectrum with less emphasis of higher harmonics.

Noise Control of Bearings

In sliding type bearings, the primary objective in ensuring low noise and vibration levels is to maintain proper lubrication between the surfaces. This implies designing the bearing to achieve a proper balance between the bearing pressure, lubricant viscosity, and operating speeds to maintain thick-film lubrication and prevent film rupture. When the operation requires oscillatory motion between journals and bearing surfaces, the film will break down. In these circumstances, pressurized lubrication may be used to eliminate the stick-slip excitation characteristic of improperly lubricated sliding bearings. Under certain types of motion and loading conditions, it is possible to achieve sufficient free film lubrication by grooving the bearing surfaces.

The major source of noise and vibration in rolling-contact bearings is impulsive in nature and is due to irregularities or defects in the bearing components. The best noise control for rolling-contact bearings is to select high-quality bearings manufactured under tight tolerances, and to maintain them properly. In most machinery containing gears, cams, etc., the radiated noise due to forces originating from bearings is usually more than 10 decibels below the total noise levels radiated from the machine. It would normally be uneconomical to spend time and effort on more expensive bearings unless the major noise and vibration sources were first reduced.

In critical applications where bearings are not secondary to other machinery noise sources (e.g., some direct- or belt-driven machines, such as fans and typewriters at idle), selecting the proper type of rolling-contact bearing can have an appreciable effect on the resulting noise and vibration. Bearings inherently tend to dampen and isolate dynamic machinery forces from efficiently radiating machine structures. The highest damping effect is usually achieved by tapered-roller bearings [2]. A slight axial load under operation or preloading the rolling elements generally reduces deflections in the bearing elements and results in slightly lower noise levels. Preloading may slightly reduce the life of the bearing.

When the shaft supported by the bearings is allowed some degree of flexibility at the bearings (as would not be the case for precision machine tools), noise levels radiated by a machine may sometimes be reduced by installing a resilient material, such as rubber, between the bearing and the machine structure. The noise reduction realized will depend on the relative contribution of the machine forces exciting the structure through bearing contact points.

Frequently, the resonant response of bearing races, themselves, will be a major source of bearing noise. Damping these resonant vibrations can be accomplished to some degree by providing a slip-fit between the outer bearing surface and the machine structure.

CAMS

Cam Classification and Nomenclature

A cam is a mechanism for imparting a specified motion to another machine member (the cam "follower") via direct contact between the cam and its follower. The follower motions possible are almost only limited by the imagination of the designer of the mechanism, and are used extensively in machinery.

The *plate cam* (also called *disk*, *radial*, or *flat* cam) is used most extensively. This cam is made from a flat piece of cam stock with the follower contact made at the outside perimeter of the cam. (See Fig. 8-24.)

The *face* cam (also called *face-groove*, *plate-groove*, *box*, or *closed-track* cam) is also formed from a flat piece of cam stock, but with the follower contact being made in a groove (track) cut into the face of the cam. (See Fig. 8-25.)

The *double-profile* cam (also called *conjugate-follower*, *complementary*, or *double-disk* cam) essentially consists of two plate cams on a common shaft. (See Fig. 8-26.) This cam is useful when a single follower must have different motion characteristics on the upstroke and downstroke, or when two separate followers must have a constant phase relationship to a common rotating shaft.

The above types of cams are applicable when the follower motion is perpendicular to the cam shaft. The *ribbed-barrel* cam (also called *double-end* cam) is applicable when the follower is to move parallel to the cam-shaft. (See

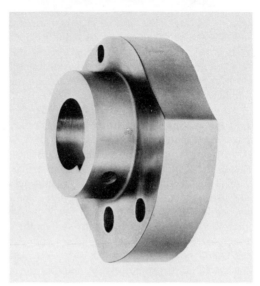

Courtesy of Ferguson Machine Company
Fig. 8-24. Plate cam.

Fig. 8-27.) The *grooved-barrel* cam (also called *plain-cylinder*, or *drum* cam) is also used for applications where follower motion and cam shaft are parallel. (See Fig. 8-28.)

Cam followers are classified according to the type of motion they execute

Courtesy of Ferguson Machine Company
Fig. 8-25. Face cam.

Courtesy of Ferguson Machine Company
Fig. 8-26. Double profile cam.

(translation or oscillation), the relationship between the cam center-of-rotation and follower path (radial or off-set), and by the character of the surface which makes contact with the cam (*flat-face*, *spherical-face*, and *roller-face* followers are common types).

The basic nomenclature for plate and barrel cams is illustrated in Fig. 8-29 (A & B). The following definitions are given for the relevant parameters:

t = time in seconds
S = linear stroke of translating follower, in inches

Courtesy of Ferguson Machine Company
Fig. 8-27. Ribbed barrel cam.

360 MACHINE ELEMENT NOISE Ch. 8

Courtesy of Commercial Cam Div., Emerson Electric Company
Fig. 8-28. Grooved barrel cam.

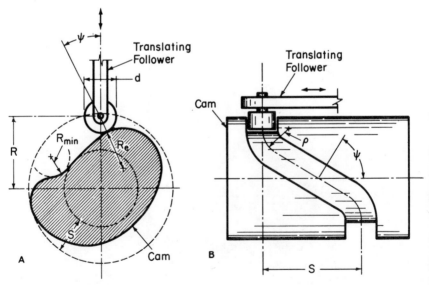

Fig. 8-29. Nomenclature used in cam application. (See text for terminology.) (a) Simple plate cam with translating roller follower; (b) Barrel cam with translating roller follower.

α = amount of rotation of cam to move follower distance S, in degrees
τ = time, in seconds, required to rotate cam α degrees
N = angular speed of cam, in rpm
R_e = radius of cam, in inches, at point where pressure angle is measured
Ψ = pressure angle = the angle between the instantaneous direction of the follower and the normal to the cam surface at the point of contact between the cam and follower, in degrees
ρ = instantaneous radius of curvature of follower trajectory, in inches
d = diameter of cam follower, in inches
R = radial distance between center of cam and center of follower, in inches
R_{min} = minimum radius of curvature of cam, in inches
r = radius of cam roller follower, in inches.

Causes of Cam-induced Noise and Vibration

The basic function of a cam mechanism is to cause a follower system to undergo a prescribed motion which may, in part, be selected by the designer. This very principle results in cams being potentially noisy mechanisms. Figure 8-30 shows, schematically, an example of a cam/follower system contained in a machine structure. The follower system will have mass, stiffness, and damping characteristics which depend on the specific application of the mechanism. As the follower executes its prescribed motion, forces will be generated between the

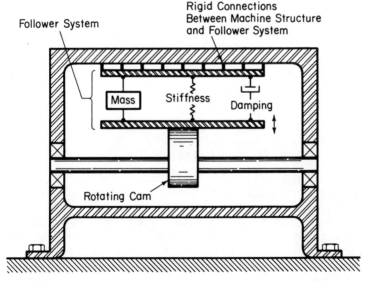

Fig. 8-30. Schematic cam application. Forces generated by cam action depend on the dynamic characteristics of the follower system. Dynamic forces may be transmitted to efficiently radiating machine surfaces through bearings and other connections between the follower system and structure.

cam and the follower system. One component of the resulting force is proportional to the instantaneous *acceleration* of the follower system mass, one component is proportional to the instantaneous *velocity* of the follower system (i.e., a shock-absorbing mechanism), and one force component is proportional to the instantaneous *displacement* of the follower system (i.e., an air-compression device). These three dynamic forces generate vibrations which cause both the cam body and the follower system to oscillate. The vibration is also transmitted to the machine structure via shaft bearings and connections between the structure and the follower system. Usually, cam surfaces are small enough so that they will not radiate an appreciable amount of acoustic energy. However, the follower system (and especially the machine structure) can have large surfaces which radiate sound very efficiently.

The relative importance of the displacement, velocity, and acceleration force components depends on many factors. Of primary importance, are the mass, stiffness, and velocity-dependent damping characteristics of the follower system, the speed of the cam shaft, and the characteristic motion executed by the follower. At low speeds, the displacement-dependent forces may be prominent— as determined by the stiffness of the follower system and the rise, or maximum displacement, of the follower. The velocity-dependent forces will increase with cam-shaft speed and may be dominant at low-to-medium shaft speeds. Acceleration-dependent forces, determined by the mass of the follower system, increase with the square of the shaft speed and are almost always the dominant forces contributing to radiated noise in high-speed machinery.

The manner in which the displacement, velocity, and acceleration-dependent forces vary with the cam-shaft speed can sometimes be used in machinery to determine the dominant causes of cam-induced machine vibration by monitoring the frequency of vibration corresponding to the shaft rpm and noting its change in magnitude as the machine speed is varied. Displacement-dependent forces will remain essentially constant; velocity-dependent forces will increase by a factor of two for each doubling of shaft speed; and acceleration-dependent forces will cause vibration magnitudes to increase by a factor of four for each doubling of shaft speed. This results because:

$$\text{displacement} = X \sin(2\pi ft) \qquad (8\text{-}9)$$
$$\text{velocity} = X 2\pi f \cos(2\pi ft) \qquad (8\text{-}10)$$
$$\text{acceleration} = -X(2\pi f)^2 \cos(2\pi ft) \qquad (8\text{-}11)$$

where f is the rotational frequency, Hz (cycles per second). The magnitude of displacement is independent of f. Velocity is directly proportional to f; and acceleration is proportional to frequency squared.

This procedure is useful when the frequency response of the machine structure is nearly flat over the range of frequencies covered by the highest and lowest shaft speeds. It is therefore necessary that no strong resonances in the machine lie close to the shaft speed (undesirable in any event). If much larger changes in vibration levels occur with changes in shaft speed than are expected, it can usually be concluded that a resonance exists in the machine near the shaft speed. This resonance may be in the machine structure, the cam shaft, or in the follower system itself (which has mass and stiffness and, therefore, resonant char-

Ch. 8 MACHINE ELEMENT NOISE 363

acteristics). One word of caution—determining the type of force component which dominates the vibration at the fundamental cam-shaft speed does not necessarily provide information on dominant forces at higher frequencies. However, for the purposes of vibration isolation of machinery, one is usually interested in the lowest frequency generated. Attenuating this frequency usually leads to greater attenuation at higher frequencies. The exceptions are resonant mode vibrations of machine members.

In addition to the dynamic forces caused by displacement, velocity, and acceleration of the follower system, another, and much more important, source of noise in high-speed cam applications is impact (or shock) excitation. If the follower depends only on gravity and on elastic (spring) forces to keep it in contact with the face of the cam, there is a definite limit to the speed at which the follower will maintain contact with the cam. This limiting speed will depend on the type of motion the follower executes (i.e., on the shape of the cam contour and the mass and stiffness of the follower system). This speed can be estimated knowing these parameters. Above this "speed limit," the cam follower will separate from the cam surface at the high point of its rise and, upon returning, will impact on the cam surface. This impact excitation will cause the machine structure to vibrate at its resonant frequencies and will result in high noise levels.

There are obvious limits to the spring and gravity forces which can be used to maintain contact between a cam and its follower. At high speeds, other means must be used. Figure 8-31 illustrates one alternate method, consisting of a roller follower in conjunction with a face cam. This approach ensures that the follower will undergo its approximate prescribed motion. It does not, however, completely eliminate the impact noise problem at high cam speeds. If the roller is to spin in the cam track, it is necessary to provide clearance between the cam and the roller because of the opposingly directed rotation between the roller and one side of the cam track. This clearance is called "backlash." (See Fig. 8-31.) As the follower changes its direction of translation at the top of its rise and at the bottom of its return, the roller will cross the cam track, resulting in a "cross-

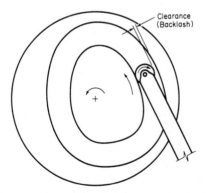

Fig. 8-31. Simple face cam with a roller follower. In order for the roller to turn freely, clearance or backlash, must be provided. This will cause impacts when the follower moves across the track.

over shock" pulse. The greater the amount of backlash, the greater will be the impact on the cam and the higher will be the noise levels radiated from the machine. It is therefore necessary to maintain close tolerances in this type of cam for low-noise, high-speed applications. Preloading of the follower is also useful in maintaining contact.

One method for reducing the effect of cross-over shock is to use an arrangement such as that shown in Fig. 8-32. Utilizing a double-roller follower this method allows simultaneous contact between the follower and the inner and outer cam-track surfaces. Even with this arrangement, however, shock excitation is still possible. The follower and cam materials cannot be infinitely rigid and will deflect to some extent. If the acceleration of the follower changes too rapidly, the materials themselves will deform and result in a shock pulse. The magnitude of the shock forces can be up to three times those which would be predicted assuming that the system did not deform.

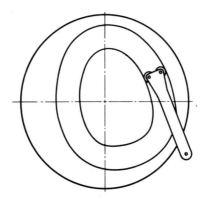

Fig. 8-32. Face cam with a double roller follower. This design permits the elimination of required clearances for the rollers to turn freely within the track of the cam.

Impact excitation will also occur in cam mechanisms if the cam and its follower are improperly proportioned. Figure 8-33 illustrates a roller follower in contact with a cam; if the minimum cam radius R_{min} is less than the roller radius, a severe impact will occur. If R_{min} equals r, there will be a sudden change in direction of the follower at that point and, in medium- and high-speed machinery, would result in shock excitation due to flexure of the system. In high-speed machines, the minimum cam radius should be greater than in the follower radius, as shown in Fig. 8-33c. The minimum radius of curvature of the cam for any given follower will depend on the specific motion the follower is to execute. A minimum value commonly recommended is one and one-half times the follower radius. It may be desirable to increase this value in high-speed applications where noise presents a problem.

High noise levels due to impact excitation in cam mechanisms will occur if the cam contour includes flat portions. Some cams are made by forming the cam contour with circular arcs of different radii connected by their common tangents.

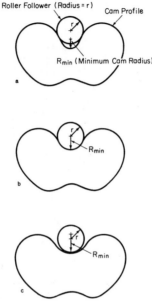

Fig. 8-33. The effect of the relationship between the follower radius and the minimum cam profile radius: (a) Roller radius $r <$ minimum cam radius R_{min}. The roller will impact on the cam at all speeds; (b) Roller radius $r =$ minimum cam radius R_{min}. The cam will suffer serious impacts except at extremely low operating speeds; (c) Roller Radius $>$ minimum cam radius R_{min} (desirable relationship). R_{min} should be at least $1.5r$.

These cams are to be avoided in high-speed applications where it is desirable to have a smoothly varying acceleration of the follower. This cannot be obtained with a cam where the radius of curvature suddenly changes from one value to another.

The force on a cam follower is not directed entirely along the line of motion of the follower. The angle between the follower path and the normal to the cam surface at the point of contact (i.e., the instantaneous pressure angle) determines the relation between the useful force directed parallel to the follower motion and the force component perpendicular to the line of motion. Figure 8-34 illustrates the situation for a translating roller follower on a cam. A force analysis for the situation depicted yields [6]:

$$F_n \sin \Psi = \frac{P}{\cot \Psi - (2\mu l_1/l_2 + \mu - \mu^2 d/l_2)} \qquad (8\text{-}12)$$

$F_n \sin \Psi$ represents an "undesirable" force, from a noise standpoint, since it will result in dynamic forces (varying in magnitude and direction) as the cam rotates. The pressure angle Ψ should be made as small as practicable for low-noise high-speed applications. The maximum pressure angle recommended is 30° for plate or face cams and 45° for barrel cams. The following recommenda-

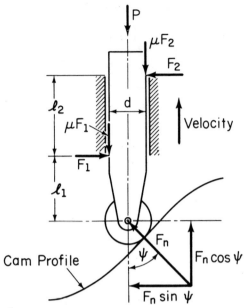

Fig. 8-34. Force analysis of a translating roller follower. The magnitude and direction of the resultant force F_n depends on the instantaneous pressure angle ψ. Only the force component parallel to the direction of motion of the follower is "useful." The lateral component results in undesirable dynamic forces on the follower. Equation (8-12) gives the relationship between the parameters.

tions can be made for minimizing the noise and vibration due to transverse forces on the follower:

1. Make the maximum pressure angle as low as practicable.
2. Make the surfaces of the follower and its support as smooth as possible and of low friction materials, to minimize the coefficient of friction.
3. Use a large guidance ratio (l_2/l_1).
4. Make the diameter of the follower shaft large enough to prevent bending of the shaft.

Basic Cam Follower Motions

Cam mechanisms are usually designed or selected so that a given machine member will be in a certain place at a certain time. There are almost an unlimited number of methods of satisfying this requirement. The noise and vibration characteristics of a cam mechanism critically depend on the precise path taken by the follower in moving from one point to another and this factor should be considered in selecting or designing the cam motion. Several basic motions have been developed and are used extensively. These motions are generally classified in terms of the shape of the displacement pattern that the follower executes as the cam makes a single revolution. The displacement diagram for a follower is a plot of the follower displacement versus the angle of the shaft rotation. Since the shaft rotation is generally constant in machinery the displacement may also be plotted as a function of time. When this is the case, the velocity and acceleration characteristics may be obtained directly from the displacement curve by differentiating the displacement-versus-time curve (if it is given in analytical form), by graphical differentiation (if the displacement curve is given), or by numerical differentiation (if the displacement curve is given in tabular form). The relationship between the displacement, velocity, and acceleration is given by:

displacement = $y(t)$ (8-13)
velocity = $dy(t)/dt$ (8-14)
acceleration = $d^2y(t)/dt^2$ (8-15)

Another important parameter of follower motions is the third time derivative of the displacement curve. This quantity (called "jerk") represents the rate-of-change of follower acceleration and is proportional to the rate-of-change of force between the cam and its follower. Large values of jerk indicate rapidly changing forces between the cam and its follower and lead to shock forces and high noise and vibration levels. The jerk is given by:

jerk = $d^3y(t)/dt^3$ (8-16)

Follower motions which have continuous jerk curves with low maximum values are necessary in high-speed machines for low noise and vibration levels.

Basic follower motions and their characteristic kinematic curves are illustrated in Table 8-3 [6]. The analytical expressions for these curves may be found in textbooks on kinematics of machine elements [6], [7], [8]. The curves shown indicate only the portion of the follower motion experienced from the beginning to the end of a rise. In most instances, the cam system will have a "dwell" period in which the follower remains stationary. The dwell can then be followed by a "return" of the follower to its original position. Figure 8-35 illustrates a simple displacement diagram consisting of a dwell-rise−dwell-return follower action to clarify the terminology.

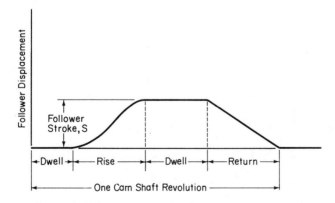

Fig. 8-35. Simple displacement diagram for a cam follower.

Table 8-3. Dynamic Characteristics of Cam Follower Motions.*

TYPE OF FOLLOWER MOTION	ACCELERATION CURVE	VELOCITY FACTOR, C_v	ACCELERATION FACTOR, C_a	JERK FACTOR	PERFORMANCE AT HIGH SPEEDS
Constant Velocity		1.00	∞	...	Poor
Parabolic		2.00	4.00	$\infty(3\times)$	Good
Simple Harmonic		1.57	4.93	$\infty(2\times)$	Good
Cycloidal		2.00	6.28	61	Excellent
Double Harmonic		2.00	5.5/9.9	...	...
Cubic Curve No. 1		3.00	12.00	∞	Poor
Cubic Curve No. 2		1.50	6.00	$\infty(2\times)$	Poor
Cubic Curve No. 3		2.00	8.00	32	Poor
3-4 Polynomial		2.00	6.00	48	Excellent
3-4-5 Polynomial		1.88	5.77	60	Excellent
4-5-6-7 Polynomial		2.19	7.52	52.5	Good to Excellent
Trapezoidal Acceleration		2.00	5.33	42.7	Good to Excellent

Modified Trapezoidal Acceleration	⌀	2.00	4.89	61.4	Excellent
Modified Sinusoidal Acceleration	⌀	1.76	5.53	69.3	Good to Excellent
Circular Cam Profile	⌀	...	...	...	Best of All
Circular Arc Profile	-----	...	...	...	Good (Small Cams)
Circular Arc and Straight-Line Profile	⌀	...	...	...	Good (Small Cams)
Modified Cycloidal	⌀	...	5.89 to ∞	...	Good to Excellent

*The velocity factor C_v and the acceleration factor C_a give, respectively, the maximum velocity and the maximum acceleration attained by the follower for a rise of 1-inch occurring in 1-second. In Equations (8-14) and (8-15), this corresponds to $S = 1$ and $[360(N/60)/\alpha] = 1$. (See Ref. [6]).

Cam follower motions (displacement, velocity, acceleration, and jerk) are kinematic quantities determined only by the shape of the mechanism and its speed of operation. The dynamic forces which result depend on the mass, stiffness, etc., of the machine elements. It is possible, therefore, to directly compare dynamic characteristics of different follower motions without regard to the machine in which they are utilized (assuming that they are to be placed in the same machine structure).

Important kinematic quantities directly related to the dynamic forces generated by cam action are the maximum velocity and maximum acceleration of the follower. These values can be determined from

$$\left[\frac{dy(t)}{dt}\right]_{max} = C_v S [360(N/60)/\alpha] \text{ in./sec} \tag{8-17}$$

$$\left[\frac{d^2y(t)}{dt^2}\right]_{max} = C_a S [360(N/60)/\alpha]^2 \text{ in./sec}^2 \tag{8-18}$$

C_v and C_a are called the velocity coefficient and the acceleration coefficient, respectively. S is the total excursion of the cam follower, as illustrated in Fig. 8-29. These equations show that velocity dependent forces are inversely proportional to the indexing period α (in degrees) and that acceleration dependent forces are inversely proportional to the square of the indexing period. If the number of degrees of cam-shaft rotation used to make the follower execute its motion is halved, velocity dependent forces will increase two-fold, while acceleration dependent forces will increase four-fold. If the frequency response of the structure were uniform in terms of sound radiation, this would correspond to a 3-decibel increase in velocity dependent noise levels and a 6-decibel increase in acceleration dependent noise levels. The velocity coefficient C_v and the acceleration coefficient C_a are included in Table 8-3 for the different basic follower motions described there. It is obviously desirable, from a noise standpoint, to use the maximum index period possible.

Table 8-3 also gives a "jerk factor" for the different follower motions. It is desirable to have a smooth and continuous jerk curve with a low peak value. The jerk factor gives the peak value of jerk for each of the cam follower motions. Theoretically, an infinite value of jerk corresponds to an instantaneous change in acceleration and, hence, in force. Since any machine member has mass, this is impossible. The result is a deformation of the components and shock excitation of the machine structure. This will always yield high noise and vibration levels and will be compounded, by machine members with a high mass, to stiffness ratio and by backlash.

Comparison of Noise from Basic Follower Motions

The airborne machine noise generated by a cam mechanism cannot be determined solely from the known follower motion, even in the event that the mass, stiffness, and velocity-dependent characteristics are given so that dynamic forces can be estimated. It is necessary to also know the sound radiation response of the machine structure to the dynamic forces transmitted from the cam and follower through bearings and follower guides.

If the general frequency response of the machine structure to dynamic forces at the bearing and guide excitation points can be determined by experimental methods, it is possible then to analyze the cam/machine system in terms of the harmonic content of the characteristic cam motions to compare the noise and vibration which would result through the use of a specific cam. Any periodic function (which would always be the case for a cam operating at a constant speed) can be written in terms of a series of sinusoidal terms consisting of harmonics of the fundamental (cam-shaft speed). The amplitude of each of the harmonics will depend on the shape of the periodic function. Mathematically, this can be represented by the Fourier series for the arbitrary periodic function with period $= T$.

The Fourier series allows a periodic function to be represented by a summation of sine or cosine terms of the form,

$$y(t) = A_0 + A_1 \cos\left(\frac{2\pi t}{T}\right) + A_2 \cos\left(\frac{4\pi t}{T}\right) + \cdots$$

$$+ B_1 \sin\left(\frac{2\pi t}{T}\right) + B_2 \sin\left(\frac{4\pi t}{T}\right) + \cdots \quad (8\text{-}19)$$

where:

$y(t) =$ the function to be represented by a series of sine and cosine terms
$t =$ time, in seconds
$T =$ arbitrary time interval for integration (usually one period)
$c =$ an arbitrary constant to account for a shift in integration interval
$A_n, B_n =$ coefficients to be found by integration

which is identical to Equation (8-20) below, expanded. The concept, which is beyond the scope of this book to prove, is that any arbitrary periodic function can be represented as an infinite series of sine and cosine terms at different frequencies and amplitudes; this is illustrated by Fig. 1-3, Chapter 1. Physically this can be interpreted to mean that any arbitrary periodic force, motion, or acoustic pressure has energy in the frequency domain at specific frequencies given by the arguments of the terms of the series. This is illustrated in Fig. 1-3, where the major contributions of some acoustic pressure signals are predominant at specific frequencies and little response occurs at other frequencies.

$$y(t) = \sum_{n=1}^{n=\infty} \left(A_n \cos\frac{2n\pi t}{T} + B_n \sin\frac{2n\pi t}{T}\right) \quad (8\text{-}20)$$

$$A_n = \frac{2}{T} \int_c^{c+T} y(t) \cos\frac{2n\pi t}{T} \, dt \quad (8\text{-}21)$$

$$B_n = \frac{2}{T}\int_c^{c+T} y(t)\sin\frac{2n\pi t}{T}\,dt \tag{8-22}$$

where $n = 0, 1, 2, \ldots$ and c = an arbitrary constant. The Fourier series representation of a specific follower motion provides information regarding the prominent forcing frequencies and their relative magnitudes. This information, in conjunction with the frequency response of the machine structure, can be used to compare different cam actions for the quietest operation.

Figures 8-36, 8-37 and 8-38 show the Fourier coefficients for velocity and acceleration (with unity rise) of parabolic, 3-4-5 polynomial, and cycloidal follower motions, respectively. The unity rise occurs over 320° (i.e., $\alpha = 320°$) of cam-shaft rotation followed by a 40° dwell period. The Fourier coefficients were computed numerically on a minicomputer using 90 equally spaced points

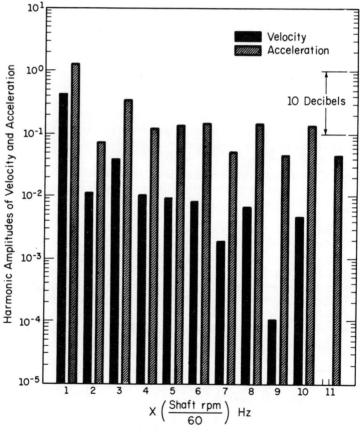

Fig. 8-36. Fourier coefficients of velocity and acceleration for parabolic follower motion. The coefficients are for unity rise ($S = 1$) with the rise occurring over 320° of cam shaft rotation (i.e., 40° dwell period).

Ch. 8 MACHINE ELEMENT NOISE 373

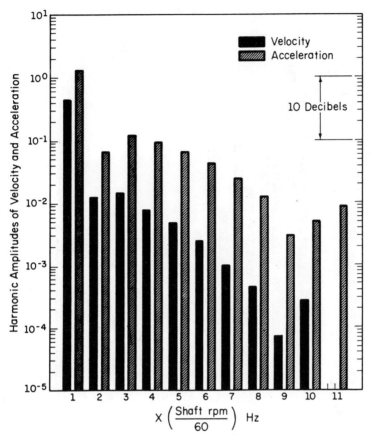

Fig. 8-37. Fourier coefficients of velocity and acceleration for 3-4-5 polynomial follower motion. The coefficients are for unity rise ($S = 1$) with the rise occurring over 320° of cam shaft rotation (i.e., 40° dwell period).

on each curve. The coefficients thus are not exact values but are of sufficient accuracy for a gross comparison of the noise aspects of the different cam motions when applied to the same machine. Numerical procedures using more points on the curves would yield more precise values. Exact values could be obtained by mathematically integrating the equations to obtain the coefficients.

In high-speed cam applications, it is usually the acceleration forces which dominate the noise and vibration. The relative results of using parabolic, 3-4-5 polynomial, and cycloidal cam motions to attain the same rise height in the same period of time (when the machine is operating at the same speed) can be determined from Fig. 8-39. This figure is based on the data in Figs. 8-36, 8-37, and 8-38, and clearly shows that the noise and vibration characteristics of a specific follower motion cannot be forseen based only on the maximum velocity or maximum acceleration values attained by the follower. The noise and vibration

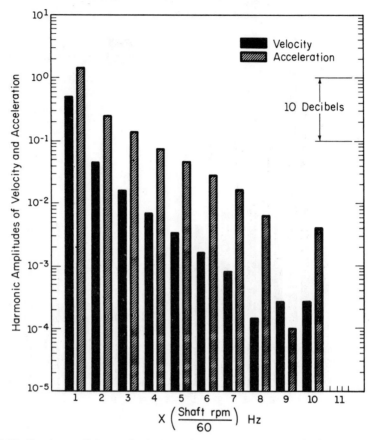

Fig. 8-38. Fourier coefficients of velocity and acceleration for cycloidal follower motion. The coefficients are for unity rise ($S = 1$) with the rise occurring over 320° of cam shaft rotation (i.e., 40° dwell period).

levels which would result at the lower harmonics from the three cam motions would be highest for cycloidal motion. However, at the higher harmonics, where the human ear is more sensitive, the cycloidal follower motion would be expected to be on the order of 10 to 20 decibels quieter than parabolic motion even though Table 8-3 shows that cycloidal motion has an acceleration factor of 6.28 compared to an acceleration factor of 4.00 for parabolic motion.

Selection and Design of Cam Mechanisms for Minimum Noise and Vibration

The design of cams is usually based on kinematic criteria; i.e., causing a machine member to undergo a prescribed motion in order to perform a desired operation. The resulting dynamic forces, along with the accompanying noise and vibration, will depend on the characteristics of the follower system mating

Ch. 8   MACHINE ELEMENT NOISE   375

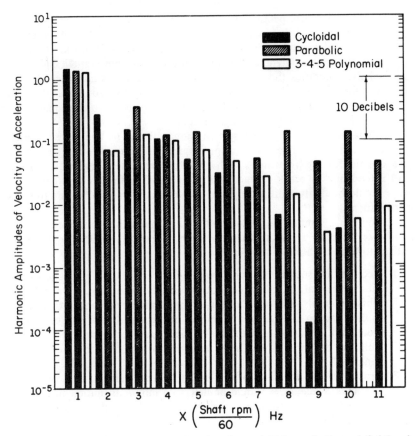

Fig. 8-39. Fourier components of acceleration for cycloidal, parabolic, and 3-4-5 polynomial follower motions. The coefficients are for unity rise ($S = 1$) and for the rise occurring in 320° of shaft rotation ($\alpha = 320°$). The values indicated in the figure are approximate due to the numerical procedure of the calculation. Even though there is not a significant difference in the peak acceleration values for the different motions, the noise and vibration characteristics in the higher harmonics will be greatly different.

with the cam. At low speeds, cam profiles are less critical, and the dominant forces may be due to forces proportional to displacement or velocity. Severe noise and vibration problems with cam mechanisms usually occur in high-speed machinery. Here cam profiles are critical both because of shock excitation (impact from cross-over or follower leaving the cam profile, and shock resulting from machine member deformations caused by jerk) and the harmonic frequency content of the specific follower motion. Cam motions having high-order harmonic frequencies of appreciable amplitude should be avoided. The harmonics may fall within the range of frequencies where the machine structure has its maximum sound radiation response.

Where cam speeds and/or masses are high enough, a cam designed without regard to elastic machine member deformations will result in unacceptable noise and vibration levels irrespective of the manufacturing quality. A cam designed by the "polydyne cam design method" can be used to take these deformations into account and minimize dynamic backlash and cross-over shock. This technique basically uses a polynomial representation of the approximate desired motion and then modifies the equation to take deformations into account. The major drawback is that the cam is designed for a single speed of operation and, because the deforming acceleration forces are proportional to the square of the cam speed, is quite sensitive to changes in speed.

The dynamics of the follower system should always be considered in a cam application. The fundamental resonant frequency of the follower system should be above the forcing frequencies so that the follower, due to its inertia, does not have a tendency to break contact with the cam. The follower system components should be lightweight and rigid to raise resonant frequencies and to reduce inertial forces.

The following summary of important points should be considered when designing or selecting a cam mechanism for high-speed operation or when very critical noise or vibration criteria apply:

1. Acceleration should be smooth, with a finite slope to minimize jerk. Maximum acceleration values do not necessarily mean maximum noise and vibration levels—this will depend on the harmonic content of the follower motion and on the frequency response of the machine structure.
2. Dominant forces at high machine speeds are due to accelerating masses. Follower systems should be made light to minimize the inertial forces generated.
3. Cross-over shock is due to backlash and elastic deformations of machine members. Tolerances should be held to a minimum and members should have high stiffness-to-mass ratios to reduce this effect. Critical situations for very high speeds require consideration of elastic deformation in designing the cam profile—this is specifically for machines which operate at nearly constant speeds.
4. Acceleration is inversely proportional to the square of the time required to displace the follower. Utilize the maximum time available to execute the motion, omitting unnecessarily long dwell periods.
5. Make cams and follower strokes as small as possible. Oversized cams result in unnecessary excessive acceleration and forces.
6. Maximum pressure angles should be 30° for plate cams and 45° for barrel cams. These angles should be less for critical situations.
7. The minimum cam radius of curvature should be one and one-half times the follower radius. This should be made greater for very high speeds.
8. Use large guidance ratios for follower systems to prevent bending of the follower shaft.
9. Lubricate properly and use low-friction materials to minimize friction forces between cam and follower. This becomes more and more important as pressure angles increase.
10. Cam surfaces should not have surface irregularities to cause unwanted accelerations (the magnitude of which are proportional to the square of the distance from the cam-shaft center).

11. Cams are inherently unbalanced mechanisms and should be balanced, preferably with the cam shaft in its own bearings.
12. Follower systems have both inertia and stiffness, the two necessary parameters to yield resonances. Consider the resonance frequencies of the follower system and do not operate near these frequencies. Forcing frequencies should be below the fundamental resonance of the follower system.

SLIDER-CRANK MECHANISMS

The "slider-crank mechanism" is used to transform rotary motion into reciprocating motion or, as in the case of internal combustion engines, to transform the reciprocating motion of the piston into rotary motion of the crankshaft. The nomenclature is given in Fig. 8-40. When the line of motion of Point A passes through the center of rotation of the crank (Fig. 8-40a), the mechanism is called a "centrical slider-crank." If the line of motion does not pass through the center of rotation (Fig. 8-40b), the mechanism is termed an "offset slider-crank."

A simple kinematic analysis of the slider-crank mechanism can provide insight into the parameters that affect noise and vibration characteristics. As an example, consider the horizontal displacement of Point A, relative to the crank center for the centrical slider-crank shown in Fig. 8-40. The displacement X is given by

$$X = R \cos \theta + [L^2 - R^2 \sin^2 \theta]^{1/2} \qquad (8\text{-}23)$$

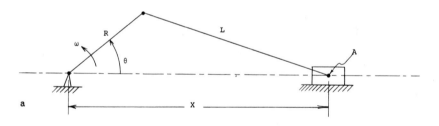

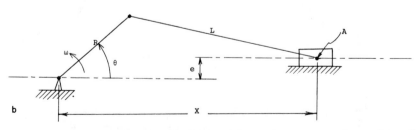

Fig. 8-40. Nomenclature used for describing the slider crank mechanism. R = crank length; L = connecting rod length; ω is the angular velocity of the crank; X = linear distance from center of crank rotation to the pinned joint; and e = the perpendicular distance between the line of motion of Point A to the crank center; (a) Centrical slider crank ($e = 0$); (b) Offset slider crank.

From the standpoint of noise and vibration, the acceleration of Point A is of more importance and is obtained by differentiating Equation (8-23) twice with respect to time. The result is

$$\ddot{X} = \omega^2 R \left[-\cos\theta - \frac{(R/L)^3 \sin^2 2\theta}{[1-(R/L)^2 \sin^2\theta]^{3/2}} - \frac{(R/L)\cos 2\theta}{[1-(R/L)^2 \sin^2\theta]^{1/2}} \right] \quad (8\text{-}24)$$

By dividing the acceleration by $\omega^2 R$, a normalized value results which is independent of the radian frequency, ω. This is convenient because the normalized acceleration is the same for all slider-crank mechanisms. The specific acceleration can be quickly found by multiplying by $\omega^2 R$.

The normalized acceleration ($\ddot{X}/\omega^2 R$) of Point A for the centrical slider-crank is shown plotted (See Fig. 8-41) for various R/L values as a function of the crank angle θ. It is seen that increasing values of R/L result in larger acceleration values and, hence, greater inertia forces and more noise and vibration. In addition, increasing R/L ratios result in more sudden changes in acceleration from positive to negative values and can give rise to impulsive type excitation caused by deflection of machine members. Impulses due to reversal in the sign of the acceleration occurs twice during each revolution of the crank, at a position of about 70 to 80 degrees and at about 280 to 290 degrees of crank rotation (from the top dead center crank position where $\theta = 0°$ and $X = R + L$).

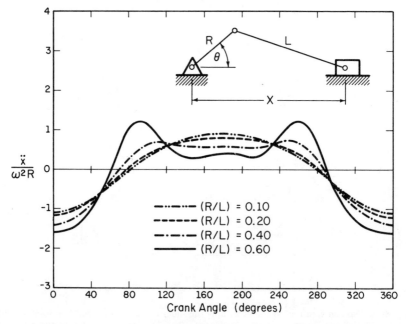

Fig. 8-41. Normalized linear acceleration ($\ddot{X}/\omega^2$) for the centrical slider crank for various R/L ratios as a function of the crank angle θ. Decreasing R/L values result in smoother operation with less sudden changes in accelerations and, hence, quieter operation.

A source of noise in reciprocating compressor and internal combustion engines known as "piston-slap" is inherent in the slider-crank mechanism. Piston-slap is a complicated phenomenon which is due to the combined effects of pressure and inertial forces acting on the system to cause the piston to rotate and to move laterally inside the cylinder. This action causes the piston to impact on the cylinder walls and can be a major source of noise, particularly on large machines. A detailed treatment of piston-slap dynamics has been reported in the literature [9]. Because of the influence of pressure forces on piston-slap, major piston impacts may not occur near 80 and 280 degrees of shaft rotation (from top dead center) where the linear acceleration of the piston changes direction. In internal combustion engines, the interaction of pressure and inertial forces will usually cause the major impacts near the top and bottom dead-center positions.

CHAIN DRIVES

Chain drives are used frequently in transmitting mechanical power between rotating shafts. The dynamics of a chain meshing with a toothed sprocket is similar in nature to meshing gear teeth. There is a basic inability in both cases to transmit kinematically smooth motion, giving rise to impacts and dynamic variations in the forces transmitted between mating chain-drive elements.

The Roller Chain Drive

The "roller chain drive" is the most widely used type of chain drive in power transmission and will be emphasized in this section. The basic construction and terminology of roller chains are illustrated in Fig. 8-42. The chain consists of a series of alternate "pin links" and "roller links." The pins are free to rotate inside the bushings of the roller link and the rollers are free to rotate on the outside of the bushing. A pin link has two pins press-fitted with two pin-link plates while a roller link has two bushings press-fitted with two roller-link plates—rollers are outside each bushing.

The roller chain links mesh with the teeth on a sprocket to provide a positive transmission system. Figure 8-43 illustrates the basic terms used in connection with a roller chain drive sprocket.

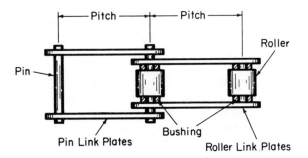

Fig. 8-42. Basic construction and terminology of roller chains.

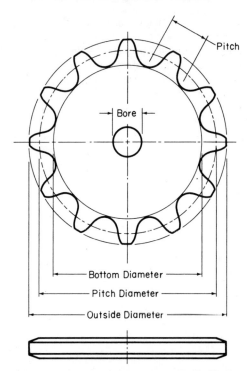

Fig. 8-43. Basic terminology and construction of roller chain sprockets.

Noise and Vibration Characteristics of Roller Chain Drives

From the standpoint of noise and vibration, the most important aspect of chain drives is the impact forces which occur between meshing chain links and sprocket teeth. This results in dominant noise and vibration components at harmonics of the contact frequency of the links and sprocket teeth and at the prominent resonant frequencies of the sprocket and other machine surfaces excited through bearing contact points.

The overall magnitude of the noise radiated from a chain drive will depend on the amount of energy transferred due to the impact of the links and the teeth. Good correlation has been reported between roller damage (breakage) and the energy of impact between sprocket teeth and chain links [10], calculated using

$$\text{impact energy} = \tfrac{1}{2} [W/g] Q_r^2 \qquad (8\text{-}25)$$

where: W = average weight of one chain link, g = gravitational constant, and Q_r = relative velocity between a mating link and tooth at the instant of impact, in a direction normal to the tooth profile at the point of impact (Fig. 8-44). Figure 8-45 shows the impact energy calculated for a special $5/8$-inch-pitch roller chain [10]. It is reasonable to assume that the energy which ends up in structural vibration and radiated noise would be (approximately) proportional to the impact energy. This implies that radiated sound energy from chain-drive sys-

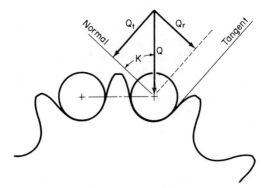

Fig. 8-44. Relative velocities between a roller chain link and a sprocket tooth at the instant of impact. Q = the resultant relative velocity; Q_r = the component of relative velocity normal to the tooth profile; and Q_t = the component of relative velocity tangent to the tooth profile. K depends on the form of the tooth profile. For the tooth form given by the American Standards Association, in American Standard B29.1, the angle K is given as $(55°-240°/N)$, where N is the number of sprocket teeth. Q_r may be used to calculate the impact energy which appears to correlate well with roller damage [10] and should, therefore, be a satisfactory guide to the noise characteristics of a chain guide.

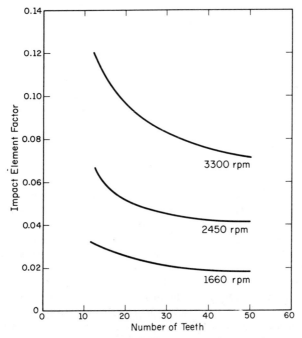

Fig. 8-45. Energy of impact (impact element factor) calculated by Equation (8-25) for a special ⅝-inch-pitch roller chain. Assuming radiated noise energy from the chain drive is in direct proportion to the impact element factor, chain drive noise levels increase about 6 decibels per doubling of sprocket speed for a given number of teeth. From Ref. [10].

tems increases with the square of the rotational speed of the sprockets (increase 6 decibels per doubling of speed). Sprockets with the same pitch decrease in noise radiated as the number of teeth increases.

In addition to impact excitation, the kinematics of the chain-drive system inherently results in fluctuating motions and forces. Figure 8-46 illustrates schematically two possible arrangements for a driving and a driven sprocket. In Fig. 8-46a, the chain span is an even multiple of the chain half-pitch while in Fig. 8-46b the span is an odd multiple of half-chain pitches. The fractional vari-

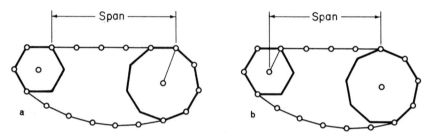

Fig. 8-46. Chain span; (a) An even multiple of the chain half-pitch; and (b) An odd multiple of the chain half-pitch.

ation in angular velocity ratio of the driven sprocket to the driver sprocket [10] may be expressed as:

$$\frac{(n_2/n_1)_{max} - (n_2/n_1)_{min}}{(n_2/n_1)_{min}} = \left[\frac{\cos(180°/N_2)}{\cos(180°/N_1)}\right] - 1 \qquad (8\text{-}26)$$

when the chain span is an even multiple of the half-chain pitch and by

$$\frac{(n_2/n_1)_{max} - (n_2/n_1)_{min}}{(n_2/n_1)_{min}} = \left[\frac{1}{\cos(180°/N_2)\cos(180°/N_1)}\right] - 1 \qquad (8\text{-}27)$$

when the chain span is an odd multiple of the half-chain pitch.

The relationships given in Equations (8-26) and (8-27), above, are shown plotted in Fig. 8-47 for various numbers of teeth on the driving and the driven sprockets. It is apparent that, from the standpoint of smoother and quieter operation, it is desirable to have the span an exact even multiple of the chain pitch. This is especially true for transmission systems having low numbers of teeth. The result is a more uniform angular velocity ratio between the driven and the driving sprockets.

Chain elongation caused by wear between pins and bushings results in the pitch of pin links increasing while the pitch of roller links remains the same. This causes the pressure angle between mating chain links and sprocket teeth to decrease. The force between the first tooth in mesh with the driving strand and the corresponding link, is inversely proportional to the quantity $\sin[(360°/N) + \phi]$ where: N = number of sprocket teeth and ϕ = pressure angle. The force in the first tooth, which initiates in an impulsive manner, therefore increases with decreasing pressure angle and, hence, wear resulting in elongation of the chain

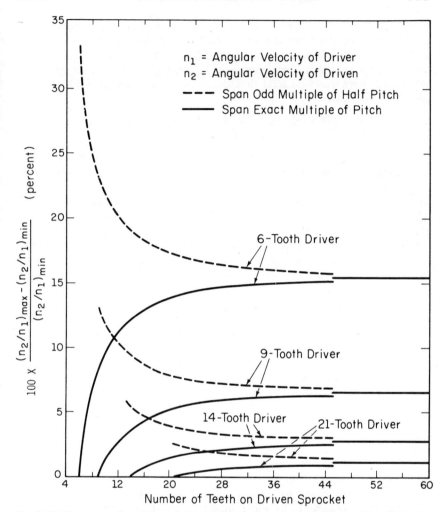

Fig. 8-47. Percent variation in angular velocity between two sprockets of a roller chain drive. The results are calculated from Equations (8-26) and (8-27). Chain spans with an exact multiple of the chain pitch theoretically provide smoother and quieter operation, particularly for small numbers of teeth on the sprockets. From Ref. [10].

will generally result in the chain-drive noise increasing somewhat. This increase is expected to be relatively minor for normal chain wear, probably on the order of less than 2 decibels.

Noise Control of Chain Drive Systems

The radiated sound due to chain-generated noise comes from both forced and resonant vibrations of sprockets, chain guards, and other machine members coupled through bearings. High-quality chains with accurately matched pitches between chain and sprockets will result in lower noise levels. Fine-pitch systems with larger numbers of teeth run considerably more quietly than course-pitch systems because of reduced impact forces between links and teeth. Chain spans as near as possible to an even multiple of half-chain pitches provide the most uniform angular velocity ratio between the driving and the driven sprockets. Damping of large sprockets will help reduce vibrations and radiated sound due to sprocket resonances.

IDENTIFYING SOURCES OF MACHINE NOISE AND VIBRATION

Thus far in this chapter, basic aspects of the dynamic force-generation characteristics of individual machine elements have been discussed. The effects on the noise and vibration characteristics of relevant design and operating parameters have been both qualitatively and quantitatively (as much as possible) outlined. This information should be useful in the specification and design phases of putting together a "quiet" machine.

It has been stressed throughout, however, that the noise and vibration resulting from a machine utilizing any of the basic components (gears, cams, etc.) depends critically on the noise and vibration transfer properties of the particular machine, in addition to the dynamic forcing characteristics of the particular mechanism. This fact makes it virtually impossible in most cases to design a machine having specific noise characteristics, even if the dynamics of all the individual components are known. The transfer characteristics of the machine structure would also be required before any prediction of total machine noise caused by an individual component could be made with any degree of confidence. This information is seldom available. Knowledge of component dynamics makes possible a gross comparison of various component designs when applied to the same machine.

The noise and vibration frequency spectrum at any point on or around rotating mechanical equipment is usually dominated by many discrete frequency components. These components will correspond (1) to forcing frequencies related to operation of the various machine elements, and (2) to resonant mode vibration of machine members.

A method fundamental to successfully solving a machine noise problem is to "define" the problem (1) By identifying forced and resonant mode frequency components, and (2) By identifying those machine elements (sources) which are the primary cause of the important frequency components. Identification of the primary sources of the noise or vibration is second in importance only to establishing the criteria upon which to judge the success of noise-control procedures.

The following paragraphs suggest general techniques of source identification in addition to those already outlined in connection with the various machine elements discussed. However, because most machinery noise problems are unique, application of these guidelines should never be considered as an adequate substitute for a combination of engineering judgment and a basic understanding of the machine elements and their dynamic characteristics.

It is frequently possible to use machine speed variation (when obtainable) in conjunction with narrow-band (e.g., tenth-octave or constant bandwidth frequency analyzers) frequency analysis to identify possible machine noise sources.

When the speed of a machine changes, several things can happen to prominent discrete frequency noise components. Some frequency components ("dynamic" components) will maintain the same harmonic relationship with one another (or with the speed of the prime mover) and normally will increase in amplitude with increasing speeds. Other discrete noise components ("stationary" components) will remain stationary in the frequency domain and may undergo large changes in amplitude as the machine speed is varied. The changes in amplitudes of these stationary components may or may not be in the same direction as the machine speed (i.e., an increase in speed may result in a decrease in amplitude of the frequency component).

The "stationary" discrete noise components will normally be associated with resonant response of machine members. In order to reduce these frequency components, it is necessary to determine the machine members which are resonating. This can usually be accomplished by

(1) Measuring the average acceleration levels of surfaces of suspect machine members under actual operation
(2) Impacting the machine members and performing a frequency analysis of either the free surface vibration or the nearfield sound pressure
(3) Calculating natural frequencies of suspected members when they are of simple geometry (i.e., flat rectangular plates).

Noise radiated due to resonant vibration of a member can be reduced by damping the member, altering the resonant frequencies of the member by design, isolating the member from its source of excitation, or a combination of these procedures. Also, resonant frequencies in machinery are usually excited by impulsive excitation from various machine elements. Reducing or eliminating the impact forces and altering their temporal characteristics can effectively reduce noise from resonant mode excitation.

The "dynamic" discrete noise components radiated from a machine structure are associated with forced vibration of machine members. In many cases, the source of these noise components can be identified by correlating the measured noise frequency spectrum with the expected frequency components from the machine (i.e., tooth-contact frequency or shaft rotational frequencies). Identification and control of noise due to these forced frequency components have been discussed in prior sections of this chapter.

The cause of most noise problems arising in machinery due to the operation of machine elements is due to either actual impacts between parts or shock-type excitation caused by sudden changes in accelerations and deformation of members. Often it is useful in attempting to define noise and vibration sources in machinery to try to correlate the relative phase relationships between impulsive peaks in the noise field creating the problem and the shock-type forces inherent in the machine components. This technique could be used, for example, in determining the relative contribution of shock pulses from a cam mechanism to the vibration at the camshaft bearings by monitoring the overall vibration levels at the bearing and relating impulsive peaks to camshaft angle. If peak impulses occur at the angles expected due to cam action (taking into account vibration signal transit times), it could be concluded that the forces due to the cam mechanism serve as the primary excitation to the vibration at the bearing.

9

Fan and Flow System Noise

J. BARRIE GRAHAM
and
L. L. FAULKNER

Introduction

In the noise analysis of fans, blowers, and air-distributing systems to be discussed in this chapter, the current state of the art of fan noise measurement and analysis, the characteristic noise spectra of a number of typical fan designs, as well as some guidelines for estimating fan noise levels will be treated first. The purpose is to provide air-handling system design engineers with an explanation of the background of fan noise-measuring procedures to furnish typical values for estimating the noise a fan will radiate to a system and to select attenuation facilities for these systems. The two types of fans covered are the centrifugal and axial fans normally used with duct systems.

Duct systems, treated in the latter part of this chapter, are used for transmitting air from the fan or blower to wherever it is required for purposes of central station air-conditioning, an industrial ventilating system, or for industrial process applications such as process cooling, furnace air-supply, and other applications. Although the major source of noise in such systems may be the fan or blower, the flow of air through the duct elements also can be a significant source of sound energy. Procedures for predicting the sound-power levels of duct-system components such as elbows, turning vanes, flow control valves, etc., are given. Methods of reducing the noise level in these systems are also included.

FAN NOISE

Fan Noise Test Standard

From the viewpoint of the ventilation system design engineer, the purpose of establishing a standard fan noise-measuring procedure is:

Ch. 9 FAN AND FLOW SYSTEM NOISE

1. To provide a method of direct comparison of the noise levels of various fans
2. To provide fan noise data for use in the engineering design of air-moving systems.

Fan noise-measurement techniques are specified by the Air Moving and Conditioning Association (AMCA) Standard 300-67, *Standard Test Code for Sound Rating Air Moving Devices*.

This test standard calls for fan noise to be reported in terms of sound-power levels in eight-octave bands and for the laboratory noise measurements to be taken in a reverberant room using a calibrated reference sound source and the substitution technique. The standard does not cover field measurements. Fan noise is reported in terms of sound-power levels re 10^{-12} watt in eight octave bands with center frequencies of 63, 125, 250, 500, 1000, 2000, 4000, and 8000 Hz.

At the present time AMCA Standard 300-67 is intended to apply to the following types of fan equipment: (1) Central station air-conditioning and heating and ventilating units, (2) Centrifugal fans, (3) Industrial, axial, and propeller fans, (4) Power roof- and wall-ventilators, and (5) Steam and hot-water-unit heaters.

SYMBOLS
L_w = Sound-power level re 10^{-12} watt, dB
K_w = Specific sound-power level re 10^{-12} watt, dB
L_p = Sound-pressure level re 2×10^{-5} N/m², dB
R = Room constant, sq ft
rpm = Revolutions per minute
B_f = Blade frequency
BFI = Blade frequency increment
Q = Flow rate, cubic feet per minute (cfm)
P = Pressure, inches of water gage

SUBSCRIPTS
w = Power level
p = Pressure level
r = Reference sound source
s = Fan selected

Sound-Power Level

It is essential to understand the difference between sound-power level and sound-pressure level, which is treated in Chapter 1. Figure 9-1 shows both of these in forms familiar to most persons.

Figure 9-1 shows a typical situation for many conventional ventilating systems. In this case, the *sound-power* level generated by the fan results in a *sound-pressure* level in the fan room that can be objectionable, depending on whether or not the fan room is normally occupied. In an adjacent space, the fan noise is not disturbing since the intervening wall provides sufficient attenuation and there is no opening in the ductwork. Also, the noise level requirement in this shop area is not severe. In the office farther from the equipment room, the in-

388 FAN AND FLOW SYSTEM NOISE Ch. 9

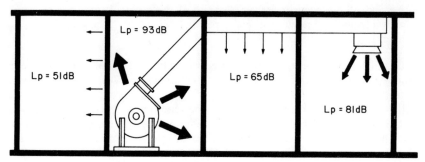

Fig. 9-1. Illustration of an air-distribution system and associated sound-pressure levels.

tervening walls have provided sufficient attenuation, but the fan noise is transmitted through the ductwork and radiated from the discharge grille. An occupant in this space finds the sound-pressure level too high for good working conditions.

The fan noise can also be transmitted through the louvers of the fresh-air inlet to persons outside the building. Anyone close to the louvers may find the sound pressure objectionable, but since noise levels decay rapidly with distance, those persons farther away will not be aware of the fan.

For this reason the question "Will this fan meet specified noise criteria?" cannot be answered by the fan manufacturer. The answer depends on where the listener is located and on the building construction. Fan noise from the same source will be described in various ways by different listeners, since human ears respond only to the sound pressure in their particular environment. The essential factor to be observed is that the sound power radiated by a fan is constant, but the resulting sound pressure is influenced by the environment. The fan manufacturer can give accurate data on the sound power generated by the fan. It is then up to the system designer or the acoustic consultant to calculate the sound-pressure level that will result in the particular environment of the job under consideration.

Fan Noise Measurement

Sound-power levels cannot be measured directly but must be calculated from sound-pressure-level readings.

If a fan is operated in a room where the surfaces are so constructed as to be highly sound reflective, the room is filled with a reasonably uniform sound field, and the sound-pressure level in the room is a function of the sound power generated by the fan and the acoustic characteristics of the room. This relationship between sound pressure and sound power in a semireverberant space is the basis for the current method of determining the sound-power level of fans.

Reverberant Test Room

A semireverberant test room is built with as little absorption as possible. In such a room sound reverberates from surface to surface and the sound-pressure level produced is a function of the sound-power level radiated by the sound

Ch. 9   FAN AND FLOW SYSTEM NOISE

source and of the acoustical characteristics of the room. If the characteristics of such a room are known, the sound-power level can be calculated from measurements of the sound-pressure level:

$$L_w = L_p + 10 \log_{10} R + 0.5 \text{ decibels} \qquad (9\text{-}1)$$

where:

L_w = sound-power level in decibels
  (reference = 10^{-12} watts)
L_p = sound pressure level in decibels
  (reference = 2×10^{-5} N/meter^2)
R = room constant, in feet^2

Determination of the room constant (R) presents some difficult problems as described in Chapter 5, page 183; therefore, a modification of this concept is specified by AMCA Bulletin 300-67. The AMCA procedure calls for the use of a calibrated reference sound source to be operated in the test laboratory under the identical acoustical conditions that the test fan will be operated.

The reference sound source used by the fan industry is a modified centrifugal fan designed to generate broad-band sound without objectionable single frequency components. A description of the ILG sound source is given in Chapter 2, page 94. The reference sound sources are calibrated for acoustic power output by an independent acoustic laboratory. Since the acoustical conditions are identical for both reference sound source and fan tests, the room constant, R, has the same value for both situations and the applicable equations can be combined into the following equation:

$$L_{wt} = L_{wr} - L_{pr} + L_{pt} \text{ decibels} \qquad (9\text{-}2)$$

where:

L_{wt} = sound-power level of the fan tested, in decibels (reference power = 10^{-12} watts)
L_{wr} = sound-power level, in decibels, of the reference source, in decibels. See Chapter 2, Table 2-4.
L_{pr} = sound-pressure level, in decibels, in test room due to reference source (reference pressure = 2×10^{-5} N/meter^2)
L_{pt} = sound-pressure level, in decibels, in the test room due to the fan (reference pressure = 2×10^{-5} N/meter^2)

While the room constant R does not enter directly into the calculations, it is, nevertheless, an important part of the procedure, since the room must be sufficiently reverberant to provide a reasonably uniform sound field so that a reliable determination of the average sound-pressure level may be made.

The test room is qualified according to AMCA Standard 300-67 by using the reference sound source as a noise generator. Sound-pressure-level readings are taken at some arbitrary distance from the reference sound source. A second set of readings is taken at twice this distance from the sound source. A drop of less than 3 dB between the two sets of readings indicates that the room has reverberation characteristics suitable for this type of measurement.

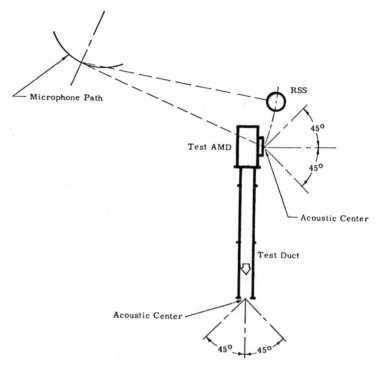

Fig. 9-2. Test arrangement for determining sound-power level of a ducted fan. (From AMCA Standard 300-67.)

Figure 9-2 shows one of the test-room arrangements specified by the AMCA Code.

Since the sound field throughout the room is not completely uniform, a statistical sample of the pressure level is obtained by moving the microphone through a specified distance, as shown in Fig. 9-2.

The distance from the fan to the microphone is determined by the size of the test room. The microphone should be in the reverberant acoustic field, which normally is taken at a location greater than one acoustic wavelength away from the fan. The microphone should not be closer than 6 feet to any reflecting surface, including the fan or duct surfaces.

The present method of measurement gives the total noise generated by the fan and radiated through the inlet and outlet; it does not attempt to separate these two values. Some noise is radiated directly through the housing, but the present method does not evaluate this noise component.

End-Reflection Correction

When noise is radiated from the open end of a tube the amount of low-frequency noise radiated from this open end is a function of the diameter of the tube. The smaller the diameter, the smaller the amount of low-frequency noise radiated into the surrounding space.

Fans are usually tested with test ducts terminating in a nozzle or an orifice plate, and the open inlet and/or outlet of the fan is similar to a duct termination. Both conditions form an open-ended tube; the low frequencies, however, are not radiated efficiently from such terminations. Under the conditions of AMCA Bulletin 300-67, it is assumed that the sound energy reflected back into the fanduct combination is completely dissipated internally and does not appear as a part of the direct sound from the fan. To account for the loss of these low-frequency components a procedure is given in the standard which calls for correction factors (as in Fig. 9-3) to be added to the low frequency part of the spectrum. Thus, the end-reflection correction is *added* to the fan noise data to indicate the total noise generated. Based on this same reasoning, these low frequencies cannot be radiated through small outlet grille openings to the surrounding space. Therefore, in the acoustical design of the system, these low-frequency corrections should be *subtracted* from the sound-power level of the fan to give the correct values for noise radiation from the outlets.

Ratings Program

Obviously, the fan manufacturer cannot possibly test all sizes of fans at all possible speeds, therefore a Ratings Program has been devised to limit the number of tests yet give good data on the noise generated by the fan. The conditions of this program are defined in AMCA Standard 311-67–*Certified Sound Ratings Program*. Reported noise data must be based on a product sample or a production model which must be of the same design and materials that will be sold.

Size and Speed Change

Under the terms of the Certified Ratings Program it is permissible to test one size of a particular fan design and use that sound-power-level data for other sizes of the *same* design. Also, within limits, it is permissible to use test data taken at one speed to calculate the sound-power levels throughout a range of speeds.

AMCA Bulletin 300-67 Limitations

Probably the greatest limitation of AMCA Standard 300-67, as presently written, is the lack of data on pure tone components. The current method of measuring in octave bands tends to conceal the pure tones present in the fan noise spectrum. Most fan noise spectra contain these pure tone components and design procedures for attenuation should take them into account.

Fan Design and Noise Characteristics

The primary purpose of a fan in any air-moving system is to move a given quantity of air against a given pressure differential, as efficiently as possible, and it must do this at a reasonable first cost. Some fans will have secondary requirements, such as an ability to handle dust-laden air, a resistance to abrasion, a construction suitable for production techniques, or be so constructed that they easily can be repaired in the field. Only after these requirements have been

FAN AND FLOW SYSTEM NOISE

TYPE	DESIGN	SPECIFIC SOUND-POWER LEVEL, K_w — CENTER FREQUENCY – Hz — dB re 10^{-12} watt and 1 cfm at 1 inch ftp								BFI	APPLICATIONS
		63	125	250	500	1000	2000	4000	8000		
CENTRIFUGAL FANS											
AIRFOIL		35	35	34	32	31	26	18	10	3	Highest efficiency of all centrifugal fan design contains 10 to 16 blades of airfoil shape. Used for general heating, ventilating, and air-conditioning systems, usually applied to central station units where the horsepower saving will be significant. Can be used on low, medium, and high-pressure systems and will operate satisfactorily in parallel. Is also used in large sizes, for clean-air industrial applications where power savings will be significant. Can be used on industrial exhaust systems, where the air-cleaning system is of high efficiency.
BACKWARD INCLINED BACKWARD CURVED		35	35	34	32	31	26	18	10	3	Efficiency is only slightly less than that of the airfoil fan. Contains 10 to 16 blades. Used for the same general applications as the airfoil fan. Can be used in industrial applications where the gas is essentially clean, but does not meet the standards required for airfoil fan selection.
RADIAL		48	45	43	43	38	33	30	29	5–8	Simplest of all centrifugal fans—relatively low efficiency, usually has 6 to 10 blades; includes both radial blades (R), and modified radial blades (M). Used primarily for industrial exhaust, including dirty gas fans and recirculating gas fans. This design also used for high-pressure industrial applications.
FORWARD CURVED		40	38	38	34	28	24	21	15	2	Efficiency less than the airfoil and backwardly curved fans, this fan is usually fabricated of lightweight and low-cost construction. It may have from 24 to 64 blades. This design will be the smallest of the centrifugal fan types and operates at the lowest speed. Used primarily in low-pressure heating, ventilating, and air-conditioning applications, such as: domestic furnaces, small central station units, and packaged air-conditioning equipment.

FAN AND FLOW SYSTEM NOISE

Type		Octave band levels	Blades	Description
AXIAL FANS — VANEAXIAL	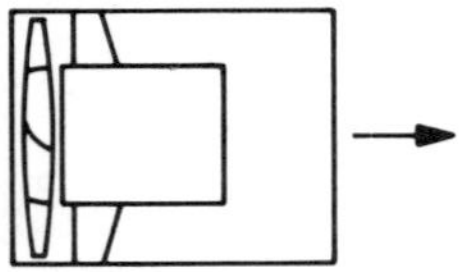 (diagram)	42 39 41 42 40 37 35 25	6-8	high pressure capability. Blades may be fixed or adjustable and the hub diameter is usually greater than 50 per cent of the fan tip diameter. There may be from 3 to 16 blades. This fan design has guide vanes downstream from the wheel which permits good air flow pattern on the discharge side of the fan. Used for general heating, ventilating, and air-conditioning applications in low, medium, and high-pressure systems. May also be used in industrial applications such as: drying ovens, paint spray booths, and fume exhaust systems.
AXIAL FANS — TUBEAXIAL	(diagram)	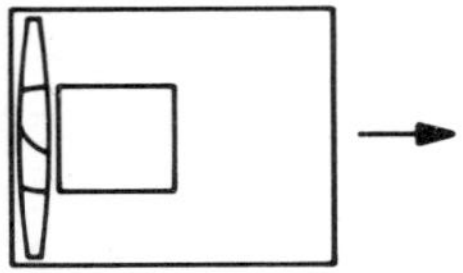 44 42 46 44 42 40 37 30	6-8	This fan is more efficient than the propeller fan design and can develop a more useful pressure capability. The number of blades may vary from 4 to 8 and the hub is usually about 50 per cent of the fan tip diameter. The blades may be of airfoil or single thickness cross-section. The fan is built without downstream guide vanes. Used in low- and medium-pressure ducted heating, ventilating, and air-conditioning applications where the poor air flow pattern downstream from the fan is not detrimental. This fan is also used in some industrial applications such as: drying ovens, paint spray booths and fume exhaust systems.
AXIAL FANS — PROPELLER	(diagram)	 51 48 49 47 45 45 43 31	5-7	Low efficiency wheels are usually of inexpensive construction and are limited to very-low-pressure applications. Usually contains 2 to 8 blades of single thickness construction attached to a relatively small hub. The housing is a simple circular ring or orifice plate. This fan is used for low pressure, high-volume air-moving applications such as air circulation within a space or as exhaust fans in a wall or roof.
AXIAL FANS — TUBULAR CENTRIFUGAL	(diagram)	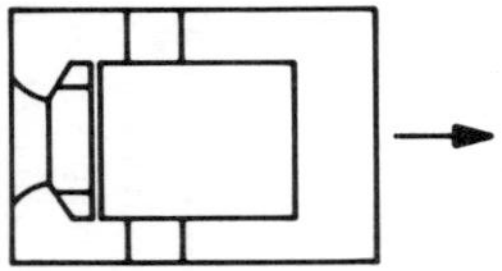 46 43 43 38 37 32 28 25	4-6	This fan usually has a wheel similar to the airfoil or backwardly inclined wheel, described above, which is built into an axial flow type housing. This results in lower efficiencies than the centrifugal fans of similar wheel design. The air is discharged radially from the wheel and must change direction by 90 degrees to flow through the guide vane section. Used primarily for low-pressure return-air systems in heating, ventilating, and air-conditioning applications.

Fig. 9-3. Acoustic properties of various fan types.

satisfied is the fan evaluated from a noise standpoint. It must be understood that every fan generates some amount of noise that is proportional to the volume flow rate, the pressure developed, and the type of fan; the system design engineer must accept this noise as a part of his engineering design problem. It is impossible to design a large, high-speed, high-pressure fan that generates only low sound levels. Fan noise is just as much an integral part of the fan performance as is the horsepower requirement. The horsepower requirements for a given fan cannot arbitrarily be established since the horsepower is determined by the actual operating requirements. Fan noise is a function of these same requirements and noise levels cannot be set at arbitrary levels, but must be based on actual operating needs.

No single type of fan will solve all fan problems. If this were the case, obviously, only that type would be offered by fan manufacturers. However, there are many different types to satisfy the many different fan applications and, for a truly adequate engineering analysis of each system, all applicable fan types should be considered. For example, the radial fan shown in Fig. 9-3, with straight radial blades, is probably one of the noisiest fans in common use, since it has not only a high noise level, but also a predominant blade-frequency tone that can be extremely objectionable. Consideration of noise alone would eliminate this fan from use, but it is a very common industrial fan used in cases where erosive material is to be handled.

The significance of the foregoing comments cannot be overemphasized, because some previous literature has suggested that quiet fans are available for all possible fan applications. Fans are seldom, if ever, designed exclusively on an acoustical basis. The fan is designed for the required duty, and if it is the best fan for that job the fan noise will be an integral part of that design and will be the minimum for that specific application.

There is no magic to fan noise, and minor alterations to a *good* fan design will have no beneficial effects on the noise levels.

There are no abrupt changes in the noise characteristics of fans as the design of the fan is altered slightly to produce a small change in fan performance. The fan noise characteristics also change gradually. For example, it has been explained that the radial-blade fan is probably one of the worst fans from a noise standpoint. By a change in blade shape, for example, in the radial blade with forward curved heel as shown in Fig. 9-3, the fan noise and especially the blade frequency components are lowered. However, it must be emphasized that this is a matter of gradual improvement and the noise level does not drop abruptly, and although the blade frequency component may be lowered, it is *not* eliminated.

Fan Noise Data

In Fig. 9-3 information is given on the principal types of fans used in commercial and industrial installations. The fan noise is listed in terms of specific sound-power levels.

Specific sound-power level is defined as the sound-power level generated by a fan operating at a capacity of 1 cfm and a pressure of 1 inch of water. By reducing all fan noise data to this common base, the specific sound-power-level concept allows direct comparison of the octave-band levels of various fans and serves as a basis for a convenient method of estimating the noise levels of fans at actual operating conditions.

The specific sound-power levels shown in Fig. 9-3 represent the noise generated by the fan when it is operating at an efficient point on the performance curve. The data represent the results of tests on fans obtained from a number of sources, and are representative of commercially available fans which follow the principles of good design.

Fans generate a tone at blade frequency and the strength of this tone depends, in part, on the type of fan. In order to adjust the blade frequency, the power level should be increased in the octave band into which the blade frequency falls. The amount of increase to be added to this band is listed in Fig. 9-3, for each fan, as "Blade Frequency Increment."

Blade frequency (B_f) is determined by:

$$B_f = \frac{\text{rpm} \times \text{No. of blades}}{60} \qquad (9\text{-}3)$$

The number of blades and the fan rpm can be obtained from the catalog being used for fan selection to meet specific flow requirements.

Estimating Fan Noise

The specific sound-power levels (K_w) listed in Fig. 9-3 provide the basis for a method for estimating the sound-power levels of fans under actual operating conditions. The correction is a function of the flow rate (Q), the pressure rise across the fan (P), and a blade frequency increment (BFI).

$$L_w = K_w + (10 \log_{10} Q + 20 \log_{10} P) + \text{BFI} \quad \text{dB} \qquad (9\text{-}4)$$

(Solution for the term $(10 \log_{10} Q + 20 \log_{10} P)$ in Equation (9-4) may be obtained from Fig. 9-4.) Other minor corrections are added as explained below.

To estimate the operating sound-power levels, L_w, it is necessary to determine:

1. Fan type
2. Flow rate—Q, cubic feet per minute

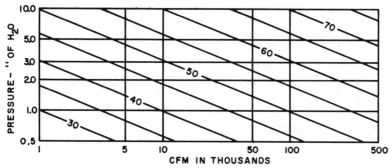

Fig. 9-4. Correction, in decibels, to be added to sound-power levels listed in Fig. 9-3 to determine total sound-power level at actual operating conditions.

3. Total pressure—P, inches of water
4. Fan speed—rpm
5. Number of blades in the fan.

By obtaining the specific sound-power level from Fig. 9-3 for the type of fan selected, and applying the correction factor for the operating cfm and pressure, and applying the blade frequency increment, the sound-power level may be estimated as in the following example:

Example No. 1

Estimate the sound-power level radiated from the inlet of a backward-curved-blade fan operating at 50,000 cfm and 3.1″ H_2O. The fan has 16 blades and will operate at 500 rpm to meet the operating requirements.

Step 1. The specific sound-power level, K_w, of the backward-curved-blade type fan is determined from Fig. 9-3 and listed in row 1 of Table 9-1.

Step 2. Using Equation (9-4), calculate the correction factor $10\log_{10} Q + 20\log_{10} P$, to be added for 50,000 cfm and 3.1″ H_2O.

$$L_w = K_w + (10\log_{10} Q + 20\log_{10} P) + \text{BFI} \quad \text{dB}$$

Computing the bracketed term:

$$(10\log_{10} \text{cfm} + 20\log_{10} P) = (10\log_{10} 50{,}000 + 20\log_{10} 3.1)$$
$$= 47 + 10 \quad \text{(values rounded)}$$
$$= 57 \text{ dB}$$

(Or obtain from Fig. 9-4)

Thus, 57 dB is listed for every octave band in row 2 of Table 9-1.

Step 3. Catalog information tells us this fan has 16 blades and will operate at 500 rpm to meet the operating requirements.
Using Equation (9-3):

$$B_f = \frac{500 \times 16}{60} = 133 \text{ Hz}$$

133 Hz falls in the 125 Hz octave band as shown in Table 2-2, page 71. Figure 9-3 lists a blade frequency increment, BFI, of 3 dB for the backward-curved-blade fan so this value is listed for the 125 Hz octave band and zeros are listed for all other octave bands as indicated in row 3 of Table 9-1.

Step 4. The total sound power developed by the fan is predicted by Equation (9-4). The resulting values for each octave-band center frequency is the sum of rows 1, 2, and 3 in Table 9-1. This includes the sound generated by both the inlet and outlet of the fan.

Of the total amount of sound power generated, it can be assumed that one-half will radiate from the fan inlet and one-half will radiate from the fan outlet. Thus, the inlet and outlet radiated sound power will be equal and will be one-half of the total generated, respectively.

Since, in this problem, we are interested only in the sound-power level

Table 9-1. Estimation of the Sound-Power Level of a Fan Inlet for Example No. 1

Row No.	Calculations	Octave-Band Center Frequency, Hz							
		63	125	250	500	1000	2000	4000	8000
1	Specific sound-power level, dB. (From Fig. 9-3.)	35	35	34	32	31	26	18	10
2	Correction factor, $10 \log_{10} Q + 20 \log_{10} P$. [From Equation (9-4).]	57	57	57	57	57	57	57	57
3	Blade Frequency Increment, BFI. (From Fig. 9-3.)	0	3	0	0	0	0	0	0
4	Correction in dB to convert total sound power to inlet alone.	-3	-3	-3	-3	-3	-3	-3	-3
5	Sound-power level, dB, for fan inlet (or outlet).	89	92	88	86	85	80	72	64

radiated from the inlet, we must correct the value calculated by Equation (9-4). This is done by referring to Table 9-1, and subtracting 3 dB from the sum of rows 1, 2, and 3, as shown in rows 4 and 5. The sound-power level at the inlet for each octave-band center frequency is given in row 5.

The reason for subtracting 3 dB in row 4 is illustrated in Chapter 2, Fig. 2-3. Whenever two equal decibel values are added, the result is always 3 dB greater than the values being added. Conversely, to find the value of the sound-power level that corresponds to one-half of the total sound power, we must *subtract* 3 dB from the total sound-power level.

The sound-power levels calculated in this example can now be used to calculate the sound-pressure level in the space in which the fan is located. To predict the sound-pressure level in the space it is necessary to know the acoustical characteristics of the space in addition to those of the fan. The procedure for these calculations is given in Chapter 12.

The sound-power level calculated in this example is also that of the outlet. Thus this procedure can be used to calculate the outlet sound-power level, i.e., the sound-power level that is radiated into the duct system.

Another useful application of Equation (9-4) and the procedure shown in Example No. 1, is to determine the noise reduction that will result when the fan or blower operating conditions are changed. Noise reduction of fans or blowers is often accomplished by reducing the fan or blower speed to the minimum required to accomplish the necessary air delivery. Of course, noise reduction can also be accomplished by the addition of attenuators on the inlet or outlet of the fan or blower.

Equation (9-4) and the method of Example No. 1 can be used to estimate the sound-power level of a fan, using the information provided in Fig. 9-3. In some instances an oversized fan is selected and deliberately operated at a lower speed to provide an acceptable sound level.

Fan Noise Specifications

As mentioned above, the noise of fans is reported in terms of sound-power levels. An understanding of the sound-power-level concept is essential to the understanding and engineering of air-handling systems. Although the fan noise is reported in terms of sound-power levels, the noise criteria for systems is expressed in terms of sound-pressure levels. Since the fan manufacturer has no knowledge of the type of acoustical environment in which the fan will be located, it is impossible for the fan manufacturer to predict the sound-pressure levels under operating conditions in the field.

The design engineer, or acoustical consultant, must use the sound-power levels as furnished by the fan manufacturer or estimated by Equation (9-4) and, applying the principles found elsewhere in this manual, he must calculate the resulting sound-pressure level in the particular environment of interest. This is illustrated in Chapter 5, page 184.

The Occupational Safety and Health Act (OSHA) of 1970, establishes maximum limits of exposure to noise for employees of industrial and commercial establishments. It is essential to recognize that the OSHA requirements apply to the acoustical environment in which *people* are working and do not establish maximum noise levels for fans. It is incorrect to specify the noise requirements of a fan by stating that "The sound power of the unit shall comply with the OSHA requirements." Obviously, the fan manufacturer cannot relate to this requirement because he does not know where the employees will be located, nor what the acoustical environment of the fan and the employees will be.

OSHA compliance is determined by measuring sound-pressure levels at the employees' work station and *not* by measurements at some arbitrary distance from the fan.

Specification Format

Design engineers may write specifications covering the noise generated by a fan in either of two general forms:

1. *Specification Form No. 1*
   A request for the sound-power level generated by the fan when it is operating at the specified conditions.
2. *Specification Form No. 2*
   A specified upper limit of the sound-power level that will be permitted to be generated by the fan under the operating conditions.

Specification Form No. 1

In this form the design engineer is requesting the total sound-power level generated by the fan which will permit him to calculate the resulting sound-pressure level in his particular fan installation. If these pressure levels are too high for his installation he must select appropriate sound attenuation facilities to be added to the system. A specification aimed at this type of requirement may be written as follows:

"NOISE GENERATED BY THE FAN WHEN OPERATING AT THE SPECIFIED VOLUME FLOW-RATE AND PRESSURE SHALL BE DE-

TERMINED ACCORDING TO THE CONDITIONS OF AMCA[1] STANDARD 300-67, *TEST CODE FOR SOUND RATING*, AND SHALL BE REPORTED IN TERMS OF SOUND-POWER LEVELS RE 10^{-12} WATT IN EIGHT OCTAVE BANDS."

Specification Form No. 2

The second method of specification is used in those cases where the design engineer has calculated the acoustical properties of his system and has determined the maximum sound-power level that can be permitted on this particular job. The design engineer then specifies the maximum sound-power limits for the fan and it is up to the fan manufacturer to provide the necessary attenuation on his fan to meet these levels. In this case the fan and the attenuation facilities are considered to be one package. Obviously, the addition of the attenuation facilities adds to the cost of the fan. Specifications based on this approach may be written as follows:

"NOISE GENERATED BY THE FAN WHEN OPERATING AT THE SPECIFIED VOLUME FLOW-RATE AND PRESSURE SHALL BE DETERMINED ACCORDING TO THE CONDITIONS OF AMCA STANDARD 300-67, *TEST CODE FOR SOUND RATING*, AND SHALL BE REPORTED IN TERMS OF SOUND-POWER LEVELS RE 10^{-12} WATT IN EIGHT OCTAVE BANDS AND SHALL NOT EXCEED THE LIMITS SHOWN IN TABLE 9-2."

Table 9-2. Sample of Maximum Sound-Power-Level Table for Specification of a Fan

Octave-Band Center Frequency, Hz	63	125	250	500	1000	2000	4000	8000
Maximum Sound-Power Levels	X	X	X	X	X	X	X	X

In Table 9-2 the required maximum sound-power level is inserted in place of the x's. The procedure for determining the maximum sound-power level is given in Example No. 1, Chapter 12.

Fan Noise Attenuation

A procedure to estimate the sound-power level that will be radiated by various types of centrifugal and axial flow fans has been outlined in previous sections of this chapter. These values represent reasonable power levels for the various types of fans and it is unlikely that significantly lower levels will be generated from the basic fan configuration. If lower power levels are required, it is necessary to add

---

[1] Air Moving and Conditioning Association, Inc.; 205 West Touhy Ave.; Park Ridge, Illinois 60068. This standard applies primarily to fan and blower manufacturers.

400 FAN AND FLOW SYSTEM NOISE Ch. 9

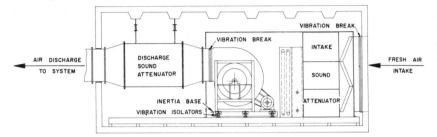

Fig. 9-5. Typical installation of a centrifugal fan with commercial attenuators.

attenuation to the basic fan. Attenuation can be added as separate units in the field, or may be added to the fan as an integral part of the fan assembly. In this latter case, the entire assembly is furnished by the fan manufacturer.

Although specific details of fan noise attenuation will vary, the principles are demonstrated in Figs. 9-5 and 9-6. In Fig. 9-5 a central station centrifugal fan is shown with attenuation on both the inlet and the outlet. The discharge of the fan has been fitted with an attenuator to reduce the amount of noise radiated to the supply ductwork. This attenuator is sized by calculating the noise level that would result in the room with the most critical criteria. The amount by which the calculated level exceeds the allowable level is the amount of noise that the fan attenuator must remove. Each attenuator must be based on the actual requirements for that particular fan installation. In this way the noise level radiated by the discharge of the fan can be brought down to any reasonable level.

Figure 9-5 also shows noise attenuation on the fan intake system. This attenuator is selected in precisely the same way as the one mentioned in connection with the fan discharge. However, in this case, outdoor noise criteria must be used and it is quite likely that this level would be established by applying criteria either at the property line of the building or at the nearest building or residence.

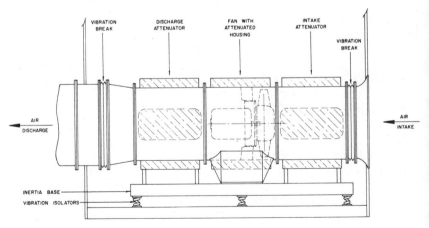

Fig. 9-6. Typical installation of an axial-flow fan with commercial attenuators.

Ch. 9 FAN AND FLOW SYSTEM NOISE 401

The entire fan assembly is supported on a rigid framework which is isolated from the structural floor of the building by spring-mounted vibration isolators. This is important from an acoustical standpoint because vibration from the fan assembly can be transmitted to the building and create acoustical problems at some distance from the fan by initiating resonant vibrations within the structure. These vibration mounts should be carefully selected and should be based on the *rpm* of the fan for their isolation efficiency.

It should also be noted that the fan is equipped with a vibration isolator in the discharge section between the fan and the continuing ductwork. This vibration isolator in the ductwork prevents the transmission of vibration from the fan casing to the attached ductwork. Here again, unless the vibration isolator is used, the vibration from the fan can trigger resonant vibrations in the ductwork which can create an acoustical problem.

Figure 9-6 demonstrates the same principles of fan noise attenuation for an axial flow fan. In this case the attenuators are attached directly to both the upstream and downstream sides of the fan, and the fan with noise attenuators is treated as a unit. It should also be noted that this unit is separated from both the inlet and discharge ductwork by vibration isolators, and that the entire unit is supported on a rigid frame which is isolated from the building structure by spring-mounted vibration mounts.

AIR FLOW SYSTEMS

An analysis of the sound level in a ducted system requires some additional acoustic analysis. As has been mentioned, the major noise source in air-distribution systems is the fan or blower, and generally, the duct system will tend to attenuate (reduce) the fan sound-level. There are instances, however, where air flow through elbows, dampers, branch take-offs, mixing units, plenums, sound traps, and other duct elements produce sound levels greater than the fan unit. For industrial systems and process air-handling systems, high air-flow velocities interacting with duct elements can be a major noise-generating mechanism.

As indicated in Fig. 9-1, all sound paths between the fan unit and the exposed location must be considered. The following factors are important in the resulting noise level:

1. Sound transmission through walls, floors, and ceilings
2. Sound from other ducts in the space that are not for supplying air to the space
3. Sound conducted along the duct
4. Sound transmission through the duct walls.

Sound transmission through walls, floors, and ceilings, item 1, above, was covered in Chapter 5, "Acoustic Absorption and Transmission-Loss Materials"; and an example problem using a fan in a shop area was presented on page 183. The transmission loss of the walls, floor, and ceiling must be sufficient to prevent sound transmission by this path. It was also shown by Example No. 2, in Chapter 5, that the amount of acoustic absorption materials in the source and receiving room can effect the resulting sound level. If air ducts penetrate the wall between a source (possibly the fan) and adjacent rooms, the prevention of sound transmission from the source room to the receiving room may require treatment of the duct walls to prevent sound transmission through the duct walls, i.e., into the duct from the source room and out of the duct into the receiving room. The installation of sound attenuators in the duct may be necessary to prevent sound transmission from one space to another through the duct-system path.

Sound transmitted along the duct can be in the form of mechanical vibration of the duct system. However, the use of good mechanical design practices should minimize this situation. Commercially available flexible connections between fan units and the ducting system are recommended as well as proper isolation of the duct from the building structure by means of isolation hangers.

In-Duct Sound

The airborne sound inside the duct system involves the following factors:

1. Sound power generation of the fan or blower
2. Distribution of sound in the ducted system
3. Attenuation provided by the duct (acoustically treated or untreated) for sound propagating inside the duct
4. Attenuation provided by elbows
5. Attenuation provided by commercial silencers and attenuators
6. Generation of sound due to air flow in fittings, grills, elbows, etc.

The sound-power generation by the fan unit was considered earlier in this chapter. Items 2 through 6 will be treated in order, and finally, a typical industrial process air-distribution system will be discussed.

Distribution of Sound Power at Branch Take-Offs

At a tee-section, or other branch take-off, the sound power upstream of the take-off propagates into each of the duct elements at the take-off location. The proportion of sound power which transmits from a main air-supply duct into a branch take-off is approximately in the same ratio as the duct area ratios at the branch location. As illustrated in Chapter 2, sound levels increase (or decrease) by logarithmic ratios; therefore, if a main duct has airborne sound with a level of 63 dB and this duct divides into two equal area ducts, the sound level will be 60 dB in each. In other words, cutting the sound power in half will result in a 3 dB reduction. The decibel reduction for other area ratios at branch take-offs is given in Table 9-3.

Table 9-3. Sound-Power Level Division at Branch Take-Offs*

Area of continuing duct, in percent, of the total area of all ducts after branch take-off	5	10	15	20	30	40	50	80
Decibels to be subtracted from power level before take-off in order to get power level in continuing duct	13	10	8	7	5	4	3	1

*Reprinted by permission from *ASHRAE Guide and Data Book*, 1973, Table 16.

Ch. 9 FAN AND FLOW SYSTEM NOISE

For multiple outlets, the cfm (cubic feet per minute) at each outlet will be approximately in proportion to the area ratio; therefore, the values in Table 9-3 can be converted to percentage of cfm of each outlet to the total fan cfm, as given in Table 9-4.

Table 9-4. Allotment of Fan Sound Power to Each Air Outlet*

cfm in Percent of Total Fan cfm	1/5	1/2	1	2	5	10	20	50
Decibels to be subtracted from the upstream sound-power level to get fan sound-power level per outlet	27	23	20	17	13	10	7	3

*Reprinted by permission from *ASHRAE Guide and Data Book*, 1967, Table 12.

Attenuation of Untreated Duct

Untreated sheet-metal duct walls will absorb sound energy or transmit sound through the duct walls. For round ducts, the sound attenuation with or without thermal insulation is 0.03 decibels per linear foot for frequencies below 1000 Hz and rises to 0.1 decibel per linear foot at 8000 Hz. Attenuation values for rectangular sheet-metal ducts, round elbows, and square elbows are given in Tables 9-5, 9-6, and 9-7.

Duct-Lining Attenuation

Direct lining in ducts may be used for both thermal insulation and sound attenuation. The sound attenuation can be estimated from Equation (9-4).

$$\text{Lined-Duct Attenuation} = 12.6 \, L \, \alpha^{1.4} \left(\frac{P}{S}\right) \text{ dB} \qquad (9\text{-}4)$$

where:

L = length of the lined duct, in feet.
P = perimeter of duct inside the lining, in inches.

Table 9-5. Approximate Natural Attenuation in Bare, Rectangular Sheet-Metal Ducts*†

Duct	Size, in.	Octave-Band Center Frequency, Hz			
		63	125	250	Above 125
		Attenuation, dB/ft			
Small	6 × 6	0.2	0.2	0.15	0.1
Medium	24 × 24	0.2	0.2	0.1	0.05
Large	72 × 72	0.1	0.1	0.05	0.01

*If duct is covered with thermal insulating material, attenuation will be approximately twice the listed values.
†Reprinted by permission from *ASHRAE Guide and Data Book*, 1973, Table 17.

Table 9-6. Approximate Attenuation of Round Elbows*

Diameter or dimensions, in.	Octave-Band Center Frequency, Hz							
	63	125	250	500	1000	2000	4000	8000
	Attenuation, dB							
5 to 10	0	0	0	0	1	2	3	3
11 to 20	0	0	0	1	2	3	3	3
21 to 40	0	0	1	2	3	3	3	3
41 to 80	0	1	2	3	3	3	3	3

*Reprinted by permission from *ASHRAE Guide and Data Book*, 1973, Table 18.

Table 9-7. Attenuation* of Square Elbows without Turning Vanes,** in Decibels (dB)†

	Octave-Band Center Frequency, Hz							
	63	125	250	500	1000	2000	4000	8000
	Attenuation, dB							
(A) No Lining								
5" Duct width (D)	...	...	...	1	5	7	5	3
10" Duct width	...	...	1	5	7	5	3	3
20" Duct width	...	1	5	7	5	3	3	3
40" Duct width	1	5	7	5	3	3	3	3
(B) Lining* Ahead of Elbow								
5" Duct width	...	...	...	1	5	8	6	8
10" Duct width	...	...	1	5	8	6	8	11
20" Duct width	...	1	5	8	6	8	11	11
40" Duct width	1	5	8	6	8	11	11	11
(C) Lining* After Elbow								
5" Duct width	...	...	...	1	6	11	10	10
10" Duct width	...	...	1	6	11	10	10	10
20" Duct width	...	1	6	11	10	10	10	10
40" Duct width	1	6	11	10	10	10	10	10
(D) Lining* Ahead of and After Bend								
5" Duct width	...	...	...	1	6	12	14	16
10" Duct width	...	...	1	6	12	14	16	18
20" Duct width	...	1	6	12	14	16	18	18
40" Duct width	1	6	12	14	16	18	18	18

*Based on lining extending for a distance of at least two duct widths "D" and lining thickness of 10 percent of duct width "D." (See Fig. 9-7). For thinner lining, lined length must be proportionally longer.

**For square elbows with short turning vanes, use average between Tables 9-5 and 9-6.

†Reprinted by Permission from *ASHRAE Guide and Data Book*, 1973, Table 19.

S = cross-sectional area of the duct inside the lining, in square inches.
α = the acoustic absorption coefficient of the lining material (a function of frequency).

The value of $\alpha^{1.4}$ can be found from Table 9-8.

Table 9-8. Values of $\alpha^{1.4}$ for Equation (9-4)

Absorption Coefficient	0.10	0.15	0.20	0.25	0.30	0.35	0.40	0.45	0.5	0.6	0.7	0.8	0.9	1.0
$\alpha^{1.4}$	.040	.070	.105	.144	.185	.230	.277	.327	.379	.489	.607	.732	.863	1.00

The limitations of Equation (9-4) are:
1. Smallest duct dimension should not exceed 18 in., nor be less than 6 in.
2. Ratio of duct width to height should not exceed 2/1.
3. The absorption coefficient α should be representative for the entire octave band.
4. Air flow velocities should not exceed 4000 fpm.
5. Equation (9-4) does not allow for line-of-sight propagation of sound which limits high frequency attenuation. In a straight 12 in. duct, for instance, the attenuation in the 8000 Hz octave band will be only about 10 dB for any lining length over 3 ft. The attenuation in the next lower octave band (4000 Hz), will be about midway between 10 dB and the value calculated from Equation (9-4). The frequency above which the 10 dB limit applies is inversely proportional to the shortest dimension of the duct.

In selecting the duct lining material, the following factors should be considered:
1. For the absorption of frequencies below 500 Hz, the lining material should be 2- to 12-inches thick. Thin materials, particularly when mounted on hard solid surfaces, will absorb only the high frequencies.
2. It has been shown that increased absorption at frequencies below 700 Hz may be realized by using a perforated facing[2] in which the area of the perforations is from 3 to 10 percent of the surface area. Such facings, however, decrease sound absorption at higher frequencies.
3. Any air space behind the lining material will have considerable effect. Absorption coefficients should be based on the particular mounting method intended to be used.

Sound-absorbing materials suitable for use in air ducts are available in the form of blankets, and semirigid boards. Specifications and absorption coefficients for most materials can be obtained from the Acoustical and Insulating Materials Association (AIMA), 205 West Touhy Avenue, Park Ridge, Illinois 60068.

Figure 9-7 illustrates the correct application of an acoustic lining material in a

---
[2] See Chapter 5, Fig. 5-12, for an example of a perforated facing.

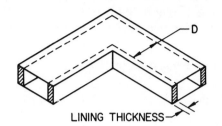

NOTE: LINING ON SIDES ONLY IS EFFECTIVE FOR ELBOW ATTENUATION

Fig. 9-7. Illustration of duct width, D, for Table 9-7.

square or rectangular elbow; the attenuation that can be obtained with this arrangement is given in Table 9-7.

Example No. 2.

Determine the acoustic attenuation of a 24 by 36-in duct, 20 feet long. The lining material is 1-inch glass fiber with absorption coefficients as given in row 1 of Table 9-9. These values must be obtained from the supplier of the acoustic material.

Solution: The perimeter of the duct inside the lining is,

$$P = 2(22 + 34) = 112 \text{ inches}$$

The cross-sectional area inside the lining is,

$$S = 22 \times 34 = 748 \text{ sq in.}$$

For the 20-foot-long duct, the attenuation is computed:

$$\text{Attenuation} = 12.6 \, \alpha^{1.4} \, (P/S) \text{ dB/ft}$$

$$\text{Attenuation} = 12.6 \times 20 \times \alpha^{1.4} \, (112/748) = 37.7 \alpha^{1.4} \text{ dB}.$$

Table 9-9. Attenuation of a Lined Duct, Example No. 2

Row No.	Calculations	Octave-Band Center Frequency, Hz							
		63	125	250	500	1000	2000	4000	8000
1	Absorption Coefficient, α, of duct liner material	.08	.11	.34	.70	.81	.86	.85	.89
2	$\alpha^{1.4} =$	.029	.045	.220	.607	.745	.810	.797	.849
3	Attenuation, dB = $37.7\alpha^{1.4}$	1.1	1.7	8.3	22.9	28.1	30.6	30.1	32.0

Note: The final value for 4000 Hz would be 20 and for 8000 Hz would be 10; per item 5, page 405.

The values for $\alpha^{1.4}$ are given in row 2 of Table 9-9 and the attenuation values, in decibels, are given in row 3. In most calculations these values would be rounded to the nearest decibel.

Sound Attenuation of Plenums

When large values of sound attenuation are required, a sound absorption plenum will often be advantageous. The geometry of a sound-absorbing plenum is shown in Fig. 9-8.

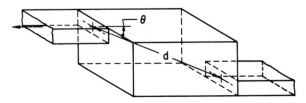

Fig. 9-8. Diagram of a sound-absorbing plenum.

A relation given in Equation (9-5) has been developed for approximating the acoustic attenuation of a plenum:

$$\text{Plenum Attenuation} = 10 \log_{10}\left[\frac{1}{S_E\left(\dfrac{\cos\theta}{2\pi d^2} + \dfrac{1-\alpha}{\alpha S_w}\right)}\right] \text{ dB} \quad (9\text{-}5)$$

where:

α = sound absorption coefficient of the lining material (frequency dependent)
S_E = plenum exit area, square feet
S_w = plenum wall surface area, square feet
d = distance between entrance and exit, feet (see Fig. 8-7)
θ = the angle d makes with the normal to the entrance opening, degrees (see Fig. 7-8).

For sound frequencies sufficiently high so that the wavelength of sound given by:

$$\text{wavelength of sound} = \frac{1128}{\text{frequency, Hz}} \text{ feet} \quad (9\text{-}6)$$

is less than any of the plenum dimensions (height, width, or length), Equation (9-5) is accurate within a few decibels. At lower frequencies, when the wavelength of sound becomes greater than the plenum dimensions, Equation 5 is conservative and the actual attenuation exceeds the calculated value by 5 to 10 decibels.

Example No. 3

Determine the attenuation of a plenum as shown in Fig. 9-8. The exit duct is 10 inches in diameter, the surface area of the plenum is 42 square feet, the distance d is 2.44 feet, and the angle θ is 15 degrees. Compare the attenuation of a bare sheet-metal plenum with the plenum covered with one inch of glass fiber.

Solution: The attenuation, in octave bands, is computed from Equation (9-5) with the following constants:

$$S_E = \pi \left(\frac{5}{12}\right)^2 = .55 \text{ ft}^2$$

$$d^2 = 5.95 \text{ ft}^2$$

$$\cos \theta = .966$$

$$S_w = 42 \text{ ft}^2$$

The values for the absorption coefficient, α, of sheet-metal panels are listed on row 1 of Table 9-10. These values are from Table 5-1, of Chapter 5, taking minimum values for the absorption of panels. The calculated values for $1 - \alpha$ are given in row 2; the values of $\dfrac{1-\alpha}{\alpha S_w}$ are given in row 4. The value for the argument,

$$\frac{\cos \theta}{2\pi d^2} + \frac{1-\alpha}{\alpha S_w}$$

from the log quantity of Equation (9-5) is given in row 5. The computed values for the attenuation of the plenum as computed from Equation (9-5) are given in row 6. These are the values for the attenuation, in decibels, for the various octave bands. For the 4000 and 8000 Hz center-frequency octave bands the resulting attenuation (row 6) is calculated to be -1.17 dB, which would indicate that sound is generated. Since this violates the definition of attenuation of a plenum, as given by Equation (9-5), which does not include a term for noise generation, noise cannot be generated and the true attenuation is zero. Thus, a zero value for the attenuation is recorded in row 6, for the 4000 and 8000 Hz frequencies.

To determine the attenuation of the plenum with the inside surfaces treated with one inch of glass fiber, the calculations are repeated with the absorption coefficients for the glass-fiber treatment as given in row 7 of Table 9-10. Those values must be obtained from the supplier of the acoustic absorption material. Values greater than $\alpha = 1.00$ should not be used for this calculation because the complete surface area is to be treated and values of α greater than 1.00 are obtained from a sample in a test room. The usual procedure is to use a value of 1.00 for any absorption coefficients for materials which are reported greater than 1.00. This is the case for the value of α, for the 8000-Hz octave band in this example.

The steps shown in rows 8, 9, 10, and 11 are identical to the computation for the sheet-metal plenum without acoustic treatment. The final values for attenua-

FAN AND FLOW SYSTEM NOISE

Table 9-10. Computation of Sound Attenuation for a Plenum. For Example No. 3

Row No.	Calculations	\multicolumn{8}{c}{Octave-Band Center Frequency, Hz}
		63	125	250	500	1000	2000	4000	8000
1	Absorption Coefficient, α, for Sheet-metal Panel.	.05	.05	.04	.03	.03	.02	.01	.01
2	Calculation of $1 - \alpha$	.95	.95	.96	.97	.97	.98	.99	.99
3	Calculation of $\dfrac{1-\alpha}{\alpha}$	19.00	19.00	24.00	32.33	32.33	49.0	99.0	99.0
4	Calculation of $\dfrac{1-\alpha}{\alpha S_W}$	.452	.452	.571	.769	.769	1.167	2.357	2.357
5	$\dfrac{\cos\theta}{2\pi d^2} + \dfrac{1-\alpha}{\alpha S_W}$	.478	.478	.597	.795	.795	1.193	2.383	2.383
6	Attenuation, dB Equation (9-5).	5.8	5.8	4.8	3.6	3.6	1.8	0 (–1.17)	0 (–1.17)
7	Absorption Coefficient, α for 1-inch Glass Fiber.	.05	.06	.20	.65	.90	.95	.98	1.00
8	Calculation of $1 - \alpha$	.95	.94	.80	.35	.10	.05	.02	.00
9	Calculation of $\dfrac{1-\alpha}{\alpha}$	19.00	15.70	4.00	.54	.11	.053	.020	.00
10	Calculation of $\dfrac{1-\alpha}{\alpha S_W}$	.452	.374	.095	.013	.0026	.0013	.00048	.00
11	$\dfrac{\cos\theta}{2\pi d^2} + \dfrac{1-\alpha}{\alpha S_W}$	.478	.400	.121	.039	.028	.0271	.0263	.0258
12	Attenuation, dB Equation (9-5).	5.8	6.6	11.8	16.7	18.0	18.3	18.4	18.5

tion are shown in row 12 which are octave-band values of the plenum attenuation, in decibels. By comparison of rows 6 and 12, the addition of the sound treatment is most effective in octave bands above 250 Hz. The attenuation of the plenum was not changed in the 63-Hz and 125-Hz octave bands. To increase attenuation at the lower octave bands, a material which has higher values for the absorption coefficient at low frequency must be used, or angle θ must be made larger.

Duct Lining and Elbows

An elbow in a lined duct can improve the sound attenuation. If the wavelength, as calculated by Equation (9-6), is less than the width of the duct (Fig. 9-7, dimension D), the sound attenuation is improved as compared to a straight duct of equivalent length. On the other hand, if the wavelength is greater than the duct width, the sound attenuation is not improved. The sound attenuation of an elbow involves the following:

1. Only the lining on the *sides* of the duct is effective in attenuating the sound level, as shown in Fig. 9-7.
2. The attenuation of the elbow should be added to the attenuation calculated for lengths of lined duct without elbows as calculated by Equation (9-4).
3. For best results, the sides of the duct should be lined, both before *and* after the elbow, for a length of at least two duct widths. This length is based on a lining thickness of at least 10 percent of the duct width.
4. If the duct is lined only before or after the elbow, there is still some gain in attenuation, rows (B) and (C) of Table 9-7.
5. The attenuation which can be credited to the duct lining is the difference between the total attenuation in rows (B), (C), or (D), of Table 9-7, and the attenuation of the unlined elbow row (A).
6. The listed increases in elbow attenuation are obtained only with duct linings, not with sound traps or lined plenums.

Open-End Reflection Loss

The sudden expansion of the fluid at the open end of a duct into a room or work space will cause some of the sound passing through the duct to reflect back into the system at the opening plane of the duct. Because of this phenomenon, only a portion of the sound energy is radiated from an open duct; the remaining energy is reflected. This effect is most pronounced at low frequencies, but is negligible at high frequencies. In the duct system analysis this result can be treated as an end reflection loss as given as a function of duct size and frequency in Table 9-11.

Air-Flow Noise

The fan is the major noise source in an air supply system, but as mentioned earlier in this chapter, air-flow interaction with other elements in the system can be of significance. Turbulence in the flow is of major importance and in extreme

Table 9-11. Duct End Reflection Loss*†

Duct Dia, inches	Duct Size, sq in.	Octave-Band Center Frequency, Hz							
		63	125	250	500	1000	2000	4000	8000
		Reflection Loss, dB							
5	25	17	12	8	4	1	0	0	0
10	100	12	8	4	1	0	0	0	0
20	400	8	4	1	0	0	0	0	0
40	1600	4	1	0	0	0	0	0	0
80	6400	1	0	0	0	0	0	0	0

*Applies to ducts terminating flush with wall or ceiling and several duct diameters from other room surfaces. If closer to other surfaces, use entry for next larger duct.
†Reprinted by permission from *ASHRAE Guide and Data Book*, 1973, Table 20.

Ch. 9 FAN AND FLOW SYSTEM NOISE

cases of flow separation, considerable sound power can be produced throughout the system. Estimates can be made for the sound-power generation of typical duct system elements from the following relation:

$$L_W = F + G + H \quad \text{decibels} \tag{9-7}$$

where:

L_W = octave-band sound-power level, in decibels, re 10^{-12} watts.
F = *spectrum function* determined from flow characteristics, in decibels.
G = *velocity function* which accounts for the flow velocity through the duct element, in decibels.
H = *correction function* for the octave band of interest.

The values of the spectrum function, F, are determined from a nondimensional flow parameter called the *Strouhal Number*, computed from,

$$S_t = \frac{5fD}{V} \tag{9-8}$$

where:

f = the octave-band center frequency, in Hertz.
D = the duct diameter, in inches, or for rectangular ducts, $D = \sqrt{(4/\pi)(\text{area})}$
V = average air-flow velocity in the duct, in feet per minute.

The value of the *spectrum function*, F, in Equation (9-7) can be determined for elbows and branch take-offs from Figs. 9-9, 9-10, and 9-11. In all of these figures, the *Strouhal Number*, S_t, must be computed from Equation (9-8) for entry to the figures.

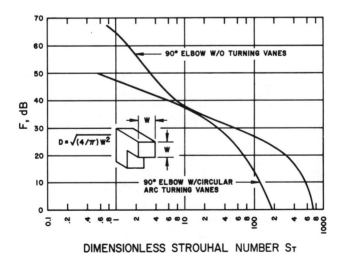

Fig. 9-9. Spectrum function, F, for square cross-section, 90-degree elbows for use in Equation (9-7). Reprinted by permission of *ASHRAE Guide and Data Book*, 1973, Figure 10.

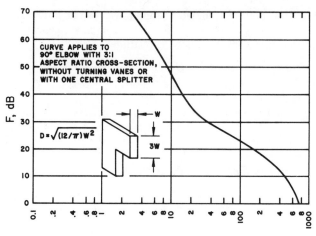

Fig. 9-10. Spectrum function, F, for rectangular cross-section, 90-degree elbows with 3 to 1 aspect ratio for use in Equation (9-7). Reprinted by permission of *ASHRAE Guide and Data Book*, 1973, Figure 9.

Values for the *velocity function*, G, can be determined from Figs. 9-12 and 9-13 for elbows and branch take-offs as a function of the average flow velocity in the duct element.

The values for the octave band with *Correction Function*, H, are obtained from Table 9-12. The values of H are not a function of the duct element, but rather only depend on the octave band of interest.

The effect of flow-generated sound power for 90-degree elbows is easily seen in Fig. 9-12. For a doubling of the average flow velocity in the duct element, the sound power increases approximately 12 dB.

Table 9-12. Octave-Band Width Correction Function, H, for Equation (9-7)*

Octave-Band Center Frequency, hertz	H, dB
63	16
125	19
250	22
500	25
1000	28
2000	31
4000	34
8000	37

*Reprinted by permission from *ASHRAE Guide and Data Book*, 1973, Table 13.

Ch. 9 FAN AND FLOW SYSTEM NOISE 413

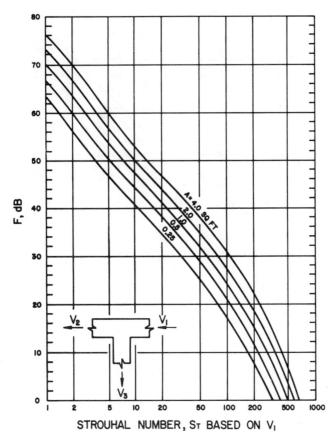

Fig. 9-11. Spectrum function, F, for 90-degree branch take-offs for use in Equation (9-7). Reprinted by permission of *ASHRAE Guide and Data Book*, 1973, Figure 13.

Two possibilities exist for controlling noise due to this effect:

1. Size the duct elements to minimize flow velocity and therefore obtain minimum flow noise source generation, or
2. Utilize high-velocity systems with additional silencers to attenuate flow noise. There is a limitation with the second method; the air flow through silencers will also generate noise and thus reduce the attenuation which can be achieved. This will be treated in the next section.

Example No. 4

A 36 by 24-inch rectangular 90-degree elbow without turning vanes handles a flow rate of 8000 cfm air. Determine the sound power generated by the flow in this elbow.

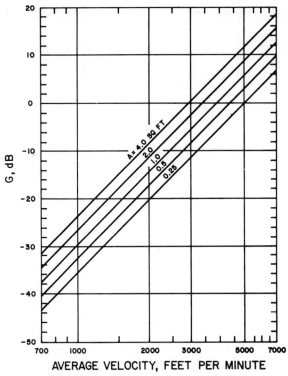

Fig. 9-12. Velocity function, G, for 90-degree elbows for use in Equation (9-7). Reprinted by permission of *ASHRAE Guide and Data Book*, 1973, Figure 12.

Step 1. The equivalent duct diameter is computed from:

$$D = \left(\frac{4}{\pi} \times 24 \times 36\right)^{1/2} = 33.2 \text{ in.}$$

The average velocity in the duct is:

$$V_{\text{avg}} = \frac{Q}{A}$$

$$V_{\text{avg}} = \frac{8000}{2 \times 3} = 1333 \text{ fpm.}$$

Using Equation (9-8), the Strouhal Number is computed for each octave band. Equation (9-8) becomes:

$$S_t = \frac{5D}{V} f = 0.125 f$$

These values are computed and listed in separate rows in Table 9-13.

FAN AND FLOW SYSTEM NOISE

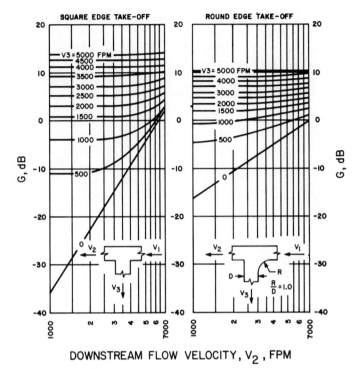

Fig. 9-13. Velocity function, G, for 90-degree branch take-offs for use in Equation (9-7). Reprinted by permission of *ASHRAE Guide and Data Book*, 1973, Figure 14.

Table 9-13. Sound-Power Level Generated by a 90° Elbow, From Example No. 4

Row No.	Calculations	Octave-Band Center Frequency, Hz							
		63	125	250	500	1000	2000	4000	8000
1	S_f [Equation (9-8)]	7.8	15.5	31	62	125	250	500	1000
2	Spectrum Function, F, from Fig. 9-9	48	35	33	30	26	20	10	0
3	Velocity Function, G, from Fig. 9-12	-17	-17	-17	-17	-17	-17	-17	-17
4	Octave-Band Width Function, H. Table 9-10	16	19	22	25	28	31	34	37
5	Sound-Power Levels, dB, $F + G + H$. Sum of rows 2, 3, and 4	47	37	38	38	37	34	27	20

416 FAN AND FLOW SYSTEM NOISE Ch. 9

Step 2. The spectrum function, F, is obtained from Fig. 9-9 for square cross-section, 90-degree elbows without turning vanes. This condition is closer to the actual situation than Fig. 9-10. The values for the spectrum function are given in row 2 of Table 9-13.

Step 3. The values for the velocity function, G, are obtained from Fig. 9-12 for an average velocity of 1333 fpm and a cross-sectional area of 6 square feet. These values are given in row 3 of Table 9-13.

Step 4. The values for the octave-bandwidth function, H, are obtained from Table 9-10 and listed in row 4 of Table 9-13.

Step 5. The sound-power levels for the various octave bands are obtained by summing $F + G + H$, which is the sum of rows 2, 3, and 4 of Table 9-13. The result is given in row 5. The values are the sound-power levels, in decibels, generated by the air flow in the elbow.

Example No. 5

A 36 by 24-inch air duct has a square-edged $90°$ branch take-off. The branch take-off is 18 by 12 inches. If the upstream main-duct flow rate is 8000 cfm and the branch flow rate is 1000 cfm, determine the octave-band sound-power level downstream of the take-off in the main duct and downstream of the branch take-off.

Solution:

Step 1. Compute the main air-duct equivalent diameter:

$$D = \sqrt{\frac{4}{\pi} \times 36 \times 24} = 33.2 \text{ in.}$$

Step 2. Compute the Strouhal Number for each octave band. First, the upstream average velocity is computed from:

$$V_{avg} = \frac{8000}{3 \times 2} = 1333 \text{ fpm}$$

The Strouhal Number is computed for each octave-band center frequency, from Equation (9-8),

$$S_t = \frac{5fD}{V} = \frac{(5 \times 33.2)}{1333} = 0.125 f$$

These values are recorded in row 1 of Table 9-14.

Step 3. Determine the spectrum function, F, from Fig. 9-11, for the *branch take-off*, using the area $A = 12 \times 18/144 = 1.5$ square feet. The values are recorded on row 2 of Table 9-14.

Step 4. Determine the velocity function, G, from the left-hand plot in Fig. 9-13. The downstream main-duct velocity is $7000/6 = 1167$ fpm and the branch take-off velocity is $1000/1.5 = 667$ fpm. The value for G is found to be -8 dB and is recorded in row 3 of Table 9-14.

Step 5. Values for the octave-band-width correction function, H, are obtained from Table 9-12 and are recorded in row 4 of Table 9-14.

FAN AND FLOW SYSTEM NOISE

Table 9-14. Sound Power Level Generated by a Branch Take-off. For Example No. 5

Row No.	Calculations	Octave-Band Center Frequency, Hz							
		63	125	250	500	1000	2000	4000	8000
1	Strouhal Number, S_t, Equation (9-8).	7.8	15.5	31	62	125	250	500	1000
2	Spectrum Function, F, for Branch, Fig. 9-11.	51	45	38	31	23	13	2	–
3	Velocity Function, G, Fig. 9-13.	–8	–8	–8	–8	–8	–8	–8	–8
4	Octave-Band Width Correction Function, H, Table 9-12.	16	19	22	25	28	31	34	37
5	Sound-Power Level, L_W dB, for the Branch. Sum values in rows 2, 3, and 4.	59	56	52	48	43	36	28	29
6	Spectrum Function, F, for Main Duct, Fig. 9-11.	57	51	44	37	30	20	9	–
7	Sound-Power Level, L_W dB, for the Main Duct. Sum values in rows 3, 4, and 6.	65	62	58	54	50	43	35	29

Step 6. The octave-band sound power levels L_W dB are obtained by the sum of $F + G + H$; row 2, row 3, and row 4 of Table 9-14. These values are given in row 5 of the table.

Step 7. The values for the spectrum function, F, for the main duct are obtained from Fig. 9-11 using $A = 6$ square feet. The curves in the figure are extrapolated to give values for 6 square feet by drawing a line parallel to the 4-square-feet line and a distance one-half of the spacing of the lines in the figure. The values for F are given in row 6 of Table 9-14.

Step 8. The values for the velocity function, G, and the correction function, H, are the same as given in rows 3 and 4, respectively, in Table 9-14.

Step 9. The sound-power levels, L_W dB, are obtained by the sum of F (main duct) $+ G + H$ which are the sum of row 3, 4, and 6. The resulting sound-power level generated in the main duct is given in row 7.

The values in rows 5 and 7 of Table 9-14 are the octave-band sound-power levels, in decibels, generated by the air flow in the take-off and main branch, respectively.

418 FAN AND FLOW SYSTEM NOISE Ch. 9

Flow Noise Generation of Silencers

Duct silencers are intended to attenuate the sound which has been generated upstream of the silencer unit. In order to obtain maximum sound attenuation in a silencer, the flow is disturbed by baffles, liners, splitters, etc. Air flow over the sound attenuation surfaces can act as a flow noise generator; therefore a silencer can itself be a generator of sound power. The self-generated sound power of a silencer with a center body can be predicted from Equation (9-9).

$$L_W = -145 + 55 \log_{10} V + 10 \log_{10} A - 45 \log_{10} \frac{P}{100} - 20 \log_{10} \frac{460 + T}{530} \quad (9\text{-}9)$$

where:

L_W = Octave-band sound-power level re 10^{-12} watts.
V = Flow velocity, in feet per minute.
A = Cross-sectional area, in square feet.
P = Percentage of open cross-sectional area in the silencer's open area divided by the total cross-sectional area of the silencer.
T = Temperature of the air (or gas), in °F.

Equation (9-9) predicts the same sound-power level for all octave bands, which agrees with experiments. This equation may be treated as a method of predicting the maximum sound-power level of silencers without regard to flow streamlining the unit. Streamlining the baffles in the silencer may also reduce the self-generated sound power by as much as 10 dB.

When selecting silencers one should determine if the source attenuation provided by the supplier includes the self noise of the silencer. If the silencer was tested under no-flow conditions, or if the tests were made at a flow velocity different from the application, Equation (9-9) may be used to estimate the effects of flow on the silencer's acoustic performance.

Example No. 6

Determine the sound-power level generated by a rectangular duct silencer in a 36- by 36-inch duct with a flow of 8600 cfm of 75 °F air. The silencer has an open area of 80 percent.

Solution: Compute L_W from Equation (9-9).

Step 1. Determine the flow velocity in the silencer.

$$V = 8600/(3 \times 3) = 956 \text{ fpm}$$

Calculate the term $55 \log_{10} V$ from Equation (9-9):

$$55 \log_{10} (956) = 55(2.98) = 164 \text{ dB}$$

Step 2. Compute $10 \log_{10} A$, where $A = 9$ sq ft:

$$10 \log_{10} (9) = 10 (.95) = 9.5 \text{ dB}$$

Step 3. Compute the term $-45 \log_{10} \frac{P}{100}$, where $P = 80$ percent:

$$-45 \log_{10} \left(\frac{80}{100}\right) = -45 \log_{10}(.8)$$
$$= -45(-.097)$$
$$= 4.4 \text{ dB}$$

Step 4. Compute the term $-20 \log_{10} \left(\frac{460 + T}{530}\right)$, where $T = 75\ °F$:

$$-20 \log_{10} \left(\frac{460 + 75}{530}\right) = -20 \log_{10}(1.01)$$
$$= -.08 \text{ dB}$$

Step 5. Compute the sound-power level generated by the air flow in the silencer from the above calculated values.

$$L_W = -145 + 164 + 9.5 + 4.4 - .08 \text{ dB}$$
$$L_W = 33 \text{ dB}$$

The value of $L_W = 33$ dB is the sound-power level generated by the air flow through the silencer for each octave band.

Sound Generation of Obstructions in the Flow

Obstructions such as rods, rings, strips, valves, dampers, etc., in an air-supply system generate sound due to interaction with the flow. Fluctuating forces due to the turbulence in the flow and the vortex shedding action at the trailing edge of an obstruction have been studied as noise-generation mechanisms. Only a limited number of flow obstructions, however, have been investigated, but they have indicated that the sound-power generated is proportional to the pressure drop across the obstruction raised to some power.

Acoustic Power Generated by A Flow Obstruction:

$$\sim \text{(Pressure Drop)}^n$$

where: n is 3 to 4 for many types of flow obstructions. For n equal to 3, the sound-power level is reduced approximately 10 dB if the pressure drop across the obstruction is reduced by a factor of 2, and the power level reduces approximately 30 dB if the pressure drop is reduced by a factor of 10. The design procedure should be to minimize the pressure drop for flow obstructions to minimize the sound generated by such devices.

Conditions for the estimation of sound generation by flow obstructions, apply

in general, and will be applicable to all of the geometries to be treated in this section:
1. The flow obstruction is small compared to the cross-sectional dimensions of the ducting system.
2. The walls of the ducting system have low values of acoustic absorption coefficient and are thus reflective in nature to sound incident upon them.
3. The flow obstruction is located upstream from the exit of the duct system of a distance equal to, or greater than, three duct diameters (for rectangular cross-section use three times the square root of the cross-sectional area). Obstructions near the exit do not produce as much sound power due to a reduction in flow-duct interaction downstream of the obstruction, subtract 10 dB from the overall sound-power level if the flow obstruction is at the exit plane of the duct system.

Rod or Bar in the Flow

The overall sound power in watts generated by a uniform rod or bar oriented in a flow, as shown in Fig. 9-14, is given by:

$$W = \frac{2.5 \times 10^{-4} \Delta P^3 D^2}{\rho^2 c^3} \text{ watts} \quad (9\text{-}10)$$

where:

D = duct diameter, meters
ΔP = total pressure drop across the rod, N/meter^2
ρ = density, Kg/meter^3
c = sound velocity, meter/sec
T_{abs} = absolute temperature, degrees Kelvin (273.16 + degrees Celsius)

and

$$c = 20.04 \sqrt{T_{abs}}$$

If the rod is sufficiently near the exit such that the pressure downstream of the rod location is nearly equal to atmospheric pressure, the pressure drop can be estimated from:

$$\Delta P = \tfrac{1}{2} \rho V^2 \quad (9\text{-}11)$$

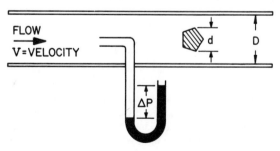

Fig. 9-14. Illustration of a rod obstruction in a flow.

Ch. 9 FAN AND FLOW SYSTEM NOISE 421

with V the upstream velocity in meters per second. The overall sound-power level can be computed from:

$$L_W = 10 \log_{10} \left(\frac{W}{10^{-12}}\right) \text{ dB} \tag{9-12}$$

$$= 10 \log_{10} \left(\frac{2.5 \times 10^8 \Delta P^3 D^2}{\rho^2 c^3}\right) \text{ dB} \tag{9-13}$$

$$= 10 \log_{10} \left(\frac{3.125 \times 10^7 \rho V^6 D^2}{c^3}\right) \text{ dB} \tag{9-14}$$

The above equations predict the overall sound-power level which is a summation over frequency. The frequency dependence of the sound-power level is illustrated in Fig. 9-15, with the peak frequency denoted by f_p. The peak frequency can be estimated for noncircular rods from:

$$f_p = \frac{\beta V}{d} \tag{9-15}$$

where:

f_p = peak frequency for sound-power level, Hz
V = upstream flow velocity, meters per second
d = projected height of rod normal to the flow, as indicated in Fig. 9-14, meter
β = frequency factor related to the flow velocity as given in Fig. 9-16. For circular rods, $\beta = 0.20$.

The factor β is a parameter dependent on the flow velocity upstream of the obstruction. An experimentally verified relation for β as a function of flow velocity is given in Fig. 9-16.

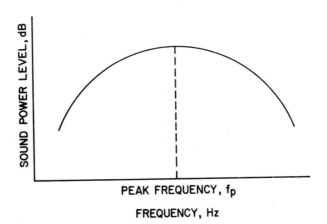

Fig. 9-15. Variation of sound-power level with frequency.

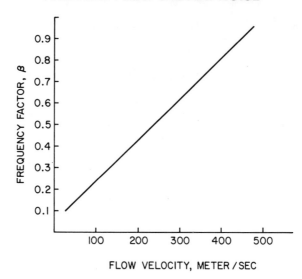

Fig. 9-16. Frequency factor, β, as a function of flow velocity for use in Equation (9-15).

From the overall sound-power level from Equation (9-14) and from the peak frequency as given by Equation (9-15), the relation for sound-power level as a function of frequency can be determined by the use of Table 9-15. The sound-power level at fractions or multiples of the peak frequency can be obtained by subtracting the appropriate value in Table 9-15.

Table 9-15. Frequency Corrections to Overall Sound-Power Level for Rods, Rings, and Strip Obstructions

Frequency Related to Peak Frequency, f_p	Correction to Overall Sound-Power Level, dB
$\frac{1}{16} f_p$	-23
$\frac{1}{8} f_p$	-16
$\frac{1}{4} f_p$	-11
$\frac{1}{2} f_p$	-6
$1 f_p$	-4
$2 f_p$	-7
$4 f_p$	-12
$8 f_p$	-18
$16 f_p$	-23

Ch. 9 FAN AND FLOW SYSTEM NOISE

Example No. 7

A $\frac{1}{2}$-inch-diameter retainer rod is located in an 18-inch-diameter flow passage, as shown in Fig. 9-14. The flow velocity upstream of the rod is 95 ft/sec and the fluid is air at 130 °F. Determine the octave-band sound-power level generated by the air flow over the rod.

Solution: The overall sound-power level can be determined from Equation (9-14) using the following values:

$$T_{abs} = 273.16 + \tfrac{5}{9}(130 - 32) = 327.60\ °K$$

$$c = 20.04\ \sqrt{T_{abs}} = 20.04\ \sqrt{327.60} = 363\ \text{meters/sec}$$

ρ of air at 130 °F = 0.068 lb/ft^3 = 1.09 Kg/m^3

$$V = 95\ \text{ft/sec} = 28.96\ \text{m/sec}$$

$$D = 18\ \text{inches} = 0.46\ \text{meter}$$

$$L_W\ \text{overall} = 10\ \log_{10}\left(\frac{3.125 \cdot 10^7 \rho V^6 D^2}{c^3}\right)$$

$$= 10\ \log_{10}\left(\frac{3.125 \cdot 10^7 \cdot 1.09 \cdot 28.96^6 \cdot .46^2}{363^3}\right) = 79\ \text{dB}$$

This value is listed for all octave bands in row 1 of Table 9-16. The peak frequency is computed from Equation (9-15) where:

$$d = \tfrac{1}{2}\ \text{in.} = .01225\ \text{meter}$$

$$f_p = \frac{\beta V}{d} = \frac{0.20\ (28.96)}{.01225} = 473\ \text{Hz}$$

The next step is to determine the band width within which the peak frequency occurs by noting within which octave-band width the calculated peak frequency occurs in Table 2-2, Chapter 2. In this case the calculated peak frequency of 473 Hz occurs within the 500 Hz octave band.

Table 9-16. Frequency Corrections to the Overall Sound-Power Level for Example No. 7

Row No.	Calculations	Octave-Band Center Frequency, Hz							
		63	125	250	500	1000	2000	4000	8000
1	Overall Sound-Power Level, dB, from Equation (9-14)	79	79	79	79	79	79	79	79
2	Frequency corrections, dB, from Table 9-15	−16	−11	−6	−4	−7	−12	−18	−23
3	Octave-Band Sound-Power Level, dB	63	68	73	75	72	67	61	56

In Table 9-15, the frequency related to the peak frequency is 500 Hz; for this frequency the correction factor is then −4dB. The 2 × 473 Hz frequency will occur within the 1000 Hz octave band (−7dB correction) and the 4 × 273 Hz frequency will occur within the 2000 Hz octave band (−12dB correction). The corrections for the other octave bands are determined in a like manner from Table 9-15 and the resulting values are shown in row 2, Table 9-16.

The octave-band sound-power levels are given in row 3 of Table 9-16 and are obtained by summing values in rows 1 and 2.

Ring Obstruction

The overall sound-power level generated by a ring obstruction in the flow can also be estimated by Equation (9-14). In this case, the velocity, V, in meters per

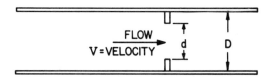

Fig. 9-17. Illustration of a ring obstruction in a flow.

second, is the velocity through the ring as shown in Fig. 9-17. The peak frequency, f_p, for a ring in the duct flow is computed from:

$$f_p = \frac{V}{\frac{1}{2}(D-d)} \tag{9-16}$$

where:

f_p = peak frequency, Hz
V = velocity through the ring, meters/sec
D = duct diameter, meters
d = ring inner-diameter, meters.

The frequency distribution of the sound-power level can be estimated from Table 9-15.

Example No. 8

A ring having a diameter, d = 17 in., is built into a duct having a diameter, D = 18 in. The average flow velocity through the ring has been determined to be 95 feet per second. Determine the octave-band sound-power level produced by the flow through the ring, which is air at 130 °F.

Solution: The overall sound-power level is computed from Equation (9-14) and will be a calculation identical to Example No. 7 when the sound-power level of a rod was computed with identical flow parameters.

$$L_W \text{ overall} = 79 \text{ dB}$$

Table 9-17. Frequency Corrections to the Overall Sound-Power Level, for Example No. 8

Row No.	Description	Octave-Band Center Frequency, Hz							
		63	125	250	500	1000	2000	4000	8000
1	Sound-Power Level, dB [Equation (9-14)]	79	79	79	79	79	79	79	79
2	Frequency Correction, dB (Table 9-15)	−30	−23	−16	−11	−6	−4	−7	−12
3	Octave-Band Sound-Power Level, dB	49	56	63	68	73	75	72	67

The peak frequency, f_p, is computed from Equation (9-16), where $D = .457$ meter and $d = .432$ meter.

$$f_p = \frac{28.96}{\frac{1}{2}(.025)} = 2317 \text{ Hz}$$

The frequency corrections are given in Table 9-17 to obtain the octave-band sound-pressure levels. The procedure used to determine the corrections to the overall sound-power level, as mentioned above, is identical to the procedure given in Example No. 7. In this case the peak frequency occurs within the 2000 Hz octave band and the corrections are shown in row 2, Table 9-17.

The sound-power level of a strip obstruction in the flow as shown in Fig. 9-18 can also be estimated by the use of Equation (9-14), where the velocity is the value as shown in Fig. 9-18. The peak frequency is computed from:

$$f_p = \frac{.8DV}{l \sin \alpha} \qquad (9\text{-}17)$$

where:

V = the average flow velocity in the opening between the duct and the strip, meter/sec
l = width of the strip, meters
α = angle the strip makes with the axis of the duct, radians.

The frequency distribution can be estimated by using Table 9-15.

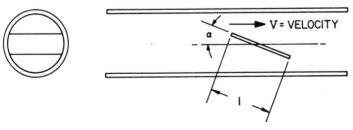

Fig. 9-18. Strip obstruction in a flow.

Flow Control Valve Noise

Damper or butterfly type valves are typically employed to control the flow in a process air system, as shown in Fig. 9-19, a and b. In the partially open position, these devices do not meet the criterion established at the beginning of this section which was small flow obstructions compared to the duct dimensions. For rather low pressure ratios across the valves (approximately 3) the flow can become choked and the resulting velocity through the valve construction can become sonic. The condition of choked flow through the valve separates the noise generation mechanisms into two classes; below choked flow conditions and choked flow conditions. The sound-power generation, in watts, has been related

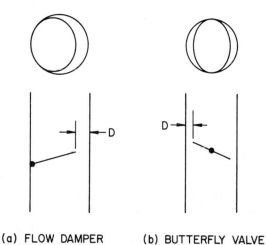

(a) FLOW DAMPER   (b) BUTTERFLY VALVE
Fig. 9-19. Flow control valves.

to the mass flow rate through the valve and the speed of sound in the flow medium by:

For pressure ratio less than 3

$$W = 3 \times 10^{-5} \dot{m} c^2 \text{ watts} \tag{9-18}$$

For pressure ratio of 3, or greater (choked flow)

$$W = 12 \times 10^{-4} \dot{m} c^2 \text{ watts} \tag{9-19}$$

where:

W = overall sound power, watts

$\dot{m} = \dfrac{dm}{dt}$ = mass flow rate through the valve, kg/sec

  = ρV, where ρ is density (kg/meter^3) and V is the average flow velocity in the duct (meter/sec).[3]

c = speed of sound in the flow media, meter/sec.

---

[3] The term $\dot{m}$ is used to denote dm/dt, or the first derivative of mass with respect to time.

The overall sound-power level, in decibels, can be computed from:

$$L_W = 10 \log_{10} \left(\frac{W}{10^{-12}}\right) \text{ decibels} \qquad (9\text{-}20)$$

Therefore, the sound-power levels become:
For a pressure ratio less than 3

$$L_W = 10 \log_{10} (3 \times 10^7 \; \dot{m}c^2) \text{ decibels} \qquad (9\text{-}21)$$

For a pressure ratio of 3 or greater

$$L_W = 10 \log_{10} (6 \times 10^8 \; \dot{m}c^2) \text{ decibels} \qquad (9\text{-}22)$$

The frequency distribution of the sound-power level can be obtained for the unchoked flow case (pressure ratio ≤ 3) from the peak frequency, f_p,

$$f_p = \frac{c}{5D} \text{ Hz} \qquad (9\text{-}23)$$

where:

f_p = the frequency at which the largest sound-power level exists, Hz.
c = speed of sound in the flow, meters/sec.
D = distance shown in Fig. 9-19, meters.

The sound-power level at other frequencies which are multiples or fractions of the peak frequency can be obtained from Table 9-18.
For the choked flow condition, where the pressure ratio across the valve is greater than 3, the peak frequency, f_p, is dependent upon the pressure ratio.

Table 9-18. Frequency Corrections to Overall Sound-Power Level for Air Valves

Frequency Related to Peak Frequency, f_p	Correction to Overall Sound-Power Level, dB	
	Pressure Ratio $<$ 3	Pressure Ratio $\geq$ 3
$\frac{1}{16} f_p$	-32	-40
$\frac{1}{8} f_p$	-24	-30
$\frac{1}{4} f_p$	-15	-19
$\frac{1}{2} f_p$	-7	-10
$1 f_p$	-4	-7
$2 f_p$	-6	-10
$4 f_p$	-10	-16
$8 f_p$	-14	-25
$16 f_p$	-18	-34

The peak frequency is given by,

$$f_p = \frac{S_t c}{D} \qquad (9\text{-}24)$$

where:

f_p = peak frequency, Hz
c = speed of sound, meters/sec
D = flow characteristic dimension from Fig. 9-19, meters
S_t = Strouhal Number from Fig. 9-20.

For the choked flow case, the peak frequency is related to the peak Strouhal Number as given by Fig. 9-20. The frequency distribution for the choked flow case is obtained for fractions or multiples of f_p from Table 9-18.

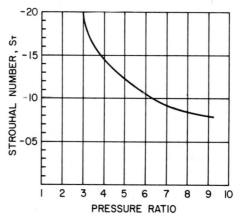

Fig. 9-20. Strouhal Number, S_t, as a function of pressure ratio for use in Equation (9-24).

Example No. 9

A butterfly air-valve is located in an 18-inch-diameter duct as shown in Fig. 9-19b. 7000 cfm of 130 °F air flows through the valve with the opening dimension, D, equal to 0.1 meter and the pressure drop across the valve, Δp, is equal to 3. Calculate the octave-band sound-power level generated by the flow interaction with the valve.

Solution: For a pressure ratio of 3, or greater, the overall sound-power level is computed from Equation (9-22). The mass flow rate, $\dot{m}$, is

$$\dot{m} = \rho V = 0.068 \text{ lb/ft}^3 \times 7000 \text{ ft}^3/\text{min} = 476 \text{ lb/min}$$

$$= 216 \text{ Kg/min}$$

$$= 3.6 \text{ Kg/sec}$$

c = speed of sound in air at 130 °F = $49.03 \sqrt{T_{abs}}$ = $49.03 \sqrt{130 + 460}$ ft/sec
    = 1191 ft/sec = 363 m/sec

FAN AND FLOW SYSTEM NOISE

Table 9-19. Frequency Corrections to the Overall Sound-Power Level for Example No. 9

Row No.	Description	Octave-Band Center Frequency, Hz							
		63	125	250	500	1000	2000	4000	8000
1	Overall Sound-Power Level, dB Computed from Equation (9-22)	145	145	145	145	145	145	145	145
2	Frequency Corrections, dB (from Table 9-18)	−40	−30	−19	−10	−7	−10	−16	−25
3	Octave-Band Sound-Power Level, dB	105	115	126	135	138	135	129	120

The overall sound-power level is computed from:

$$L_W \text{ overall} = 10 \log_{10} (6 \times 10^8 \, \dot{m}c^2)$$
$$= 10 \log_{10} (6 \times 10^8 \cdot 3.6 \cdot 363^2) = 145 \text{ dB}$$

This value is entered for every frequency band in row 1 of Table 9-19.

The peak frequency for the sound-power-level frequency is computed from Equation (9-23),

$$f_p = \frac{S_t c}{D} = \frac{363}{5 \times 0.1} = 726 \text{ Hz}$$

The frequency corrections for the octave-band sound-power levels are obtained from Table 9-18. The value of −7 dB Hz, is entered in the 1000 Hz octave band. This is because the bandwidth of the 1000 Hz octave band is from 710 Hz to 1420 Hz, as given by Table 2-2 of Chapter 2. The value for $2f_p$ would be at a frequency of 1452 Hz which is −10 dB. This value is entered in the 2000 Hz octave band because the bandwidth is from 1420 Hz to 2,840 Hz. The other corrections are entered in row 2, as shown in Table 9-19.

Noise of Air Jets

Air jets are commonly used in process systems for exhaust ports, ejectors for molded or stamped parts, process cooling, drying, etc. The sound generation of the free jet without flow interaction with solid objects can be estimated from the mechanical power in the jet flow. The mechanical power is computed from

$$W_{\text{mech}} = \tfrac{1}{2} \dot{m} V^2 \text{ watts} \qquad (9\text{-}25)$$

where:

W_{mech} = mechanical power of jet, watts

$\dot{m} = \dfrac{dm}{dt}$ = mass flow rate of the jet, Kg/sec

V = exit velocity of the jet, meter/sec.

The overall sound-power level of jet noise can be estimated from Equation (9-26), which is obtained from Equation (9-25) and accounting for the reference

FAN AND FLOW SYSTEM NOISE

Table 9-20. Jet Noise Efficiency Constant as a Function of Mach Number

Jet Mach Number $M = \dfrac{V}{c} = \dfrac{\text{exit velocity}}{\text{sound speed}}$	Efficiency Constant, ξ
$M < 0.3$	$8 M^3 \times 10^{-5}$
$0.4 < M < 2$	$M^5 \times 10^{-4}$
$M > 2$	2×10^{-3}

power of 10^{-12} watt and a radiation efficiency, ξ. The acoustic power is given by ξW_{mech} and the sound-power level by $10 \log_{10} (\xi W_{\text{mech}}/10^{-12})$.

$$L_W = 10 \log_{10} (5 \dot{m} V^2 \xi \times 10^{11}) \text{ decibels} \tag{9-26}$$

where:

L_W = overall sound-power level, decibels
$\dot{m}$ = mass flow rate of the jet, Kg/sec
V = exit velocity of the jet, meter/sec
ξ = efficiency constant dependent upon the jet exit mach number as given by Table 9-20

The sound power radiating from a free jet is not uniform over a hemisphere surrounding the jet. Due to the nature of the sound-generation mechanism, the sound-pressure level is greatest in a direction from 30° to 45° from the jet axis, as shown in Fig. 9-21.

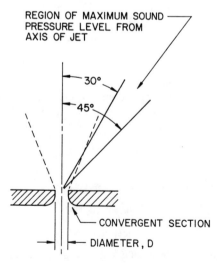

Fig. 9-21. Sectional view of a free jet.

FAN AND FLOW SYSTEM NOISE

The frequency distribution for jet sound-power level can be found from the peak frequency f_p, given by Equation (9-27).

$$f_p = \frac{V}{5D} \qquad (9\text{-}27)$$

where:

f_p = peak frequency for a free jet sound-power level, Hz
V = exit velocity of the jet, meters/sec
D = jet diameter, meters.

The sound-power level as a function of frequency can be determined from Table 9-21, where the decibel corrections are given for various multiples and fractions of the peak frequency, f_p.

Table 9-21. Frequency Corrections to Overall Sound-Power Level for Jet Noise

Frequency Related to Peak Frequency, f_p	Correction to Overall Power Level, dB
$\frac{1}{16} f_p$	-33
$\frac{1}{8} f_p$	-24
$\frac{1}{4} f_p$	-16
$\frac{1}{2} f_p$	-8
$1 f_p$	-4
$2 f_p$	-6
$4 f_p$	-10
$8 f_p$	-13
$16 f_p$	-18

The sound power generated by a free jet is given by Equation (9-25) with an efficiency constant, ξ, added: $\xi \dot{m} V^2 / 2$. It can be seen from Table 9-20 with the mach number given by V/c, that the sound power is related to the exit jet velocity as shown below:

$$M < 0.3 \qquad W_{acoustic} \sim V^5$$
$$0.4 < M < 2 \qquad W_{acoustic} \sim V^7$$
$$M > 2 \qquad W_{acoustic} \sim V^2$$

For mach numbers less than 2, the most effective noise-reduction parameter is the exit velocity, V, since in all cases the sound-power dependence on mass flow is to the first power. For a jet with $M < 0.3$ and $\dot{m}$ constant, a reduction in exit velocity by a factor of 2, will result in a 15 dB reduction in sound-power level;

for $0.4 < M < 2$ a reduction in exit velocity will result in a 20 dB reduction in sound-power level. Multiple jets then, may be a method of reducing the exit velocity while maintaining the same flow rate.

Example No. 10

A one-inch-diameter jet exhausts into ambient room air (70 °F) at sonic ($M = 1$) velocity at the exit. The mass flow-rate is 0.4 lb/sec. Determine the octave-band sound-power level for the air jet.

Solution: The overall sound-power level is calculated from Equation (9-26) with $\xi = M^5 \times 10^{-4}$ as obtained from Table 9-20 for $M = 1$. The overall sound-power level is calculated below:

where:

$\dot{m} = 0.4$ lb/sec = 0.18 Kg/sec
V = exit velocity = speed of sound at 70 °F = $49.03\sqrt{T_{abs}}$
$\quad = 49.03 \sqrt{460 + 70}$ ft/sec = 1129 ft/sec = 344 m/sec
M = exit mach number = 1.0
$L_W = 10 \log_{10} (5 \dot{m} V^2 M^5 \cdot 10^7)$ dB
$\quad = 10 \log_{10} (5 \cdot 0.18 \cdot 344^2 \cdot 1 \cdot 10^7) = 140$ dB

This value is listed for every octave band of Table 9-22.

The peak frequency is computed using Equation (9-27)

$$f_p = \frac{344}{5(.0254)} = 2709 \text{ Hz}$$

This peak frequency is within the 2000 Hz octave band as determined by Table 2-2, Chapter 2. The correction factors to the overall sound-power level of the jet are obtained from Table 9-14. The computations for obtaining the octave-band sound-power levels are illustrated in Table 9-22.

Flow Diffusers

The usual method for reducing flow velocity in a process air-supply system is to utilize a diffuser which expands the flow cross-sectional area gradually until the desired velocity is achieved. Noise generated by diffusers typically results

Table 9-22. Frequency Corrections to the Overall Sound-Power Level for Example No. 10

Row No.	Calculations	Octave-Band Center Frequency, Hz						
		125	250	500	1000	2000	4000	8000
1	Overall Sound-Power Level, dB, from Equation (9-26)	140	140	140	140	140	140	140
2	Frequency Corrections, dB, from (Table 9-21)	-33	-24	-16	-8	-4	-6	-10
3	Octave-Band Sound-Power Level, dB, for the Jet Noise	107	116	124	132	136	134	130

Ch. 9 FAN AND FLOW SYSTEM NOISE 433

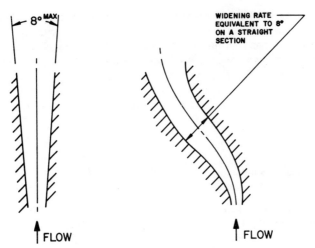

Fig. 9-22. Examples of flow diffusers.

when flow separation occurs in the diffuser section. To prevent flow separation, the maximum included angle of the diffuser section should not exceed 8 degrees, as shown in Fig. 9-22. In many instances a straight diffuser section is not possible and a curved diffuser section is required as is also seen in this figure. The curved diffuser cross-section can be developed from an equivalent straight section and obtaining an equivalent widening rate equal to 8 degrees included angle for the straight section.

Mathematical relations to predict the noise of diffusers without separation have not yet been developed. Quantitative data are usually obtained by taking measurements on existing setups, an example of which is listed in Reference [15] at the end of the book.

Sound Transmission Through Duct Walls

In many industrial applications, sound energy generated inside a duct or piping system by some type of flow-obstruction is transmitted through the walls of the duct system. For this condition, the sound transmission-loss of the duct system must be considered. For rectangular cross-sections the flat surfaces are treated as flat plates and the transmission-loss can be determined from Equation (5-12) which is repeated here.

$$TL = 20 \log_{10} W + 20 \log_{10} f - 33 \text{ dB} \qquad (9\text{-}28)$$

where:

  TL = transmission loss, in decibels
  W = weight, in pounds per square foot, of the panel
  f = frequency, in Hz.

The transmission loss is computed for the octave-band center frequencies and used to determine the reduction in sound-power level from inside the duct to the outside of the duct.

FAN AND FLOW SYSTEM NOISE

Round cross-section ducts behave differently from rectangular cross-sections because of added stiffness due to curvature at low frequency. At high frequencies a round duct has transmission loss much like a rectangular flat plate. The frequency which separates the stiffness regions from the plate-like region is called the ring frequency, f_r, given by:

$$f_r = \frac{c_l}{\pi D} \text{ Hz} \quad (9\text{-}29)$$

where:

f_r = the ring frequency, in Hz
c_l = speed of sound for the material from which the duct is made, m/sec
D = nominal duct diameter, meters

In the frequency region below the ring frequency, the transmission loss is found by first finding the nominal transmission-loss from

$$\text{TL}_{\text{nominal}} = 10 \log_{10}\left(\frac{t}{D}\right) + 50 \text{ dB} \quad (9\text{-}30)$$

where:

$\text{TL}_{\text{nominal}}$ = the nominal transmission-loss below the ring frequency, dB
t = wall-thickness of the duct
D = nominal duct-diameter

t and D must be in the same units.

The nominal transmission-loss is corrected for frequencies below the ring frequency by the values in Table 9-23.

Table 9-23. Corrections to the Nominal Transmission-Loss of Equation (9-30) for Frequencies which are Less Than or Equal To the Ring Frequency, f_r

Frequency, f	$\frac{1}{64} f_r$	$\frac{1}{32} f_r$	$\frac{1}{16} f_r$	$\frac{1}{8} f_r$	$\frac{1}{4} f_r$	$\frac{1}{2} f_r$	$1 f_r$
Correction, dB	-7	-6	-5	-4	-3	-2	-4

For frequencies above the ring frequency, the transmission loss is computed from Equation (9-28).

Example No. 11

Compute the A-scale sound-pressure level 6 feet from the sheet steel duct section containing the butterfly air valve of Example No. 9. The duct is 18 inches in diameter and is of standard 15-gage sheet metal which is .0673-inch thick.

Solution: The octave-band sound-power levels for the sound generated by the air flow through the valve must be determined and then the duct attenuation computed to determine the sound power outside the duct.

Ch. 9 FAN AND FLOW SYSTEM NOISE 435

Step 1. Determine the sound-power level of the sound generated by the air flow through the air valve. This was computed in Example 9 for this problem. The values are repeated in row 1 of Table 9-24.

Step 2. Compute the transmission loss of the duct. The ring frequency for the duct is computed from Equation (9-29). The duct diameter is 18 inches or 0.4572 meter. The speed of sound, c_l, for sheet steel is 5100 meters per second.

$$f_r = \frac{c_l}{\pi D} = \frac{5100}{\pi \times 0.4572} \text{ Hz} = 3550 \text{ Hz}$$

The nominal transmission-loss of the duct is calculated for the ring frequency and frequencies below the ring frequency from Equation (9-30).

$$\frac{t}{D} = \frac{.0673}{18} = .0037.$$

where:

$$\text{TL}_{\text{nominal}} = 10 \log_{10}\left(\frac{t}{D}\right) + 50$$
$$= 10 \log_{10}(.0034) + 50 \text{ dB}$$
$$= -24.3 + 50$$
$$= 25.7 \text{ dB}$$

This value is entered in row 2 for all octave bands of 4000 Hz or less. Note from Table 2-2 of Chapter 2, that the ring frequency of 3550 Hz is within the bandwidth of the octave band centered at 4000 Hz. The values for the correction factor for the nominal transmission loss are obtained from Table 9-23 and entered in row 3 of Table 9-24 for all octave bands except 8000 Hz.

The transmission loss for the duct is computed by adding the values in rows 2 and 3 for all octave bands except the octave band with center frequency of 8000 Hz. This octave band has a frequency above the ring frequency and the transmission loss must be computed from Equation (9-28). A standard 15-gage sheet metal is 2.8125 pounds per square foot.

$$\text{TL} = 20 \log_{10} W + 20 \log_{10} f - 33 \text{ dB}$$
$$= 20 \log_{10}(2.8125) + 20 \log_{10}(8000) - 33$$
$$= 9.0 + 78.1 - 33$$
$$= 54.1 \text{ dB}$$

This value is listed in row 4 for the 8000-Hz octave band.

Step 3. The equivalent sound-power level outside the duct is obtained by subtracting the duct attenuation from row 4 from the sound-power level inside the duct, as given in row 1. The resulting values are given in row 5.

Step 4. Compute the octave-band sound-pressure level at a distance of 6 feet from the duct. For this computation Equation (4-1) of chapter 4, is utilized. The acoustic properties of the room in which the noise source

Table 9-24. Computation of the A-scale Sound-Pressure Level for Example No. 11

Row No.	Calculations	Octave-Band Center Frequency, Hz							
		63	125	250	500	1000	2000	4000	8000
1	Sound-Power Level of the Air Valve, dB, Example No. 9	105	115	126	135	138	135	129	120
2	Nominal Transmission Loss, dB Equation (9-30)	25.7	25.7	25.7	25.7	25.7	25.7	25.7	—
3	Correction Factors from Table 9-23	−7	−6	−5	−4	−3	−2	−4	—
4	Transmission-loss, dB, for the duct	18.7	19.7	20.7	21.7	22.7	23.7	21.7	54.1
5	Equivalent Sound-Power level L_W, outside duct, dB. Row 1 minus row 4	86.3	95.3	105.3	113.3	115.3	111.3	107.3	65.9
6	Correction to L_W, dB $L_p = L_{wi} - 16.2$	−16.2	−16.2	−16.2	−16.2	−16.2	−16.2	−16.2	−16.2

7	Sound-Pressure Level, dB. Sum rows 5 and 6.	70.1	79.1	89.1	97.1	99.1	95.1	91.1	49.7
8	A-scale Correction Factors. Table 2-3, Chapter 2	−26.2	−16.1	−8.6	−3.2	0	+1.2	+1.0	−1.1
9	A-weighted Sound-Pressure Levels, dB. Sum rows 7 and 8.	43.9	63.0	80.5	93.9	99.1	96.3	92.1	48.6
10	Combining Weighted Octave-Band Levels To Obtain the A-scale Sound-Pressure Level. Fig 2-3, Chapter 2.	will not contribute significantly to the A-scale level			100.2		96.3 → 101.7 → 102.1 dBA	92.1	

is located must be known. If the reflected sound is neglected or can be shown to be minimal, a simplified method of estimating the sound-pressure level results. This assumption can be made if no large reflecting surfaces are near the source and the observation point is close to the source. For this example and an observation point 6 feet from the source, it will be assumed that the direct sound energy from the source is predominant and the reflected sound will be neglected. This assumption cannot be made in general. With the above assumption, Equation (4-1) becomes

$$L_{pi} = L_{wi} + 10 \log \left[\frac{Q_{\theta i}}{4\pi r^2} \right]$$

where:

L_{pi} = octave-band sound-pressure level to be computed
L_{wi} = octave-band sound-power level of the source
$Q_{\theta i}$ = directivity of the source, = 1.0 for source radiating uniformly in all directions without reflecting surfaces nearby
r = distance from source to observation point, meters.

For r = 6 feet = 1.83 meters

$$L_{pi} = L_{wi} - 16.2 \text{ dB}$$

−16.2 dB is entered for all octave bands in row 6. The sound-pressure level for each octave band is obtained by summing the values in rows 5 and 6 with the results given in row 7.

Step 5. Compute the A-weighted sound-pressure level 6 feet from the duct. The octave-band sound-pressure levels are corrected for the A-weighting by use of the factors in Table 2-3, Chapter 2. The A-scale correction factors from Table 2-3 are given in row 8 of Table 9-24. The A-weighted sound-pressure level is obtained by summing the values in rows 7 and 8 with the resulting values as shown in row 9.

The A-scale sound-pressure level is obtained by combining the values in row 9, two at a time, until all octave bands are included. Figure 1-14, in Chapter 1, can be utilized for this calculation. Note that if two decibel values differ by 20 dB, or more, the sum of the two values will not be significantly greater than the larger of the two values; therefore, in this example only the values in the 500, 1000, 2000, and 4000-Hz octave bands need to be combined to determine the A-scale level, since all other values are 20 dB lower than the sum of these four octave bands. The resulting value in row 10 is rounded to 102 dBA for the A-weighted level at the 6-foot distance from the duct with a flow value. To reduce this level to below 90 dBA, some noise-control treatment is required. A duct wrapping material may be suitable in this case if the treatment will provide a minimum of 15 dB transmission-loss in the 500, 1000, 2000, and 4000-Hz octave bands. Duct wrapping materials and the decibel transmission-loss should be obtained from suppliers of acoustical materials.

10

Combustion and Furnace Noise

ABBOTT A. PUTNAM

Introduction

Historically, the main interest in noise produced by combustion systems, was in the high-frequency screech and low-frequency pulsations that occasionally emanated from such systems, and because these noises could be of an intolerable level, could change the combustor performance, or even cause physical deterioration and damage of the furnace equipment. However, both because of the increased realization of the detrimental effects of noise on human work output, and the increased attention of various legislative bodies to problems of industrial noise, the combustion industry is now concerned about both combustion-driven oscillations and the industrial noise counterpart of the comforting "roaring" fireplace of old.

There are many sources of noise in addition to the combustion process in an industrial combustion system. Considering the problem in detail, it is apparent that many acoustic interactions and amplifications are also possible. Figure 10-1 is a schematic diagram indicating the major elements of the industrial combustion system:

1. The general operating environment (the room in which the furnace is to be located)
2. The furnace
3. The air- and fuel-handling system
4. The burner.

Only noise generated in the burner will be discussed here, in detail, because the other noise sources have been covered in various chapters of this book. Burner noise will be discussed in the following pages with an attempt to distinguish combustion and flame noise from the concurrent noises of ancillary

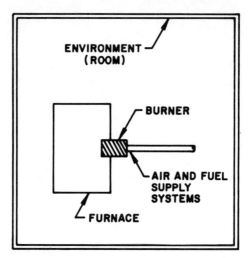

Fig. 10-1. Major elements in industrial combustion systems.

equipment. Engineering approaches to the suppression or elimination of burner noise also will be given. At the outset it must be said that the combustion process is complex and only imperfectly understood at present. Partly for this reason, the analysis of sound generation is more qualitative than quantitative; thus, the ability to predict noise levels in combustion processes by analytical means or calculation, is limited. Also, combustion noise reduction is based more on experience or empirical relations than on analytical relationships.

The difficulty in predicting noise levels of industrial burners or even in obtaining an understanding of the combustion acoustics associated with these burners is that often many interrelated factors affect the noise output. These factors include: flame size, combustion intensity, turbulence, firing rate, flow velocity, air/fuel ratio, and structural effects.

It is difficult to conduct an experiment in which all but one of the above variables can be held constant while obtaining accurate measurements of a dependent variable of interest. For this reason, extreme discretion must be used in interpreting observations and assigning a cause to the observed effect. Firing rate and flow velocity are two of the most common interrelated causes that make interpretation of results difficult. These phenomena will be discussed in this chapter. An example of furnace construction to reduce the noise level is given in Case Study 4, Chapter 12.

CLASSIFICATION OF COMBUSTION-INDUCED NOISE

Before discussing the four classifications of combustion noise, namely: 1. Combustion-driven oscillations, 2. Combustion roar, 3. Unstable combustion noise; and 4. Combustion amplification of periodic flow phenomena. Some comment might be made about the parameter of the combustion process that is involved in the production of noise. The kinetics of a laminar flame are well understood in a gross sense, and for identifying such a flame it is often sufficient only to know the laminar burning velocity of the mixture and possibly, a characteristic flame thickness. For turbulent flames, such as encountered in most industrial applications, the situation is not as clear. There is not universal agreement that the kinetic processes remain unchanged. On the other hand, the turbulence itself is the controlling factor in the gross rate of reaction, and the details of the kinetics are of secondary importance. Thus, even the location of the combustion pattern is not simple for a turbulent flame. However, for either a laminar or turbulent flame, the important parameter is the volume expansion of the gases as a result of the heat release during the combustion process (because heat-release rate is the significant output parameter in combustion equipment, attention is usually focused on this variable rather than on the related volume expansion ratio). From a thermodynamic point of view, this is easy to understand. Acoustic energy is produced when the volume expansion rate acting against the local pressure (relative to the average pressure), is positive. This is in the same sense as in an automobile engine, wherein chemical energy is converted to mechanical energy by the expansion of the combustion products against the force of the piston. Summarizing, the volume expansion ratio of a combustible mixture during combustion is the common mixture parameter entering into all four types of combustion noise listed above, which will now be discussed.

Combustion-driven oscillations, often termed pulsating combustion in larger pieces of equipment and associated low frequencies, involve a feedback cycle that converts chemical energy to oscillatory energy in the gas flow to the combustion region. The noise spectrum ordinarily involves one specific frequency and its harmonics. Figure 10-2 shows such a spectrum.[1] Typical efficiencies[2] observed for conversion of chemical energy to noise, for combustion driven oscillations are 10^{-4}, although purposely produced pulsations in pulse combustors can be considerably higher, [Ref. 1.2]. This can be expressed by the following:

$$\text{acoustic power} = \frac{\text{chemical energy}}{\text{time}} \times \text{efficiency}$$

or

$$W_{\text{acoustic}} = W_{\text{chemical}} \times \eta \qquad (10\text{-}1)$$

For example, consider a 30,000 Btu/hr burner. The chemical power is given by:

$$W_{\text{chemical}} = \frac{30{,}000 \text{ Btu}}{\text{hr}} \times \frac{1055 \text{ joule}}{\text{Btu}} \times \frac{1 \text{ hr}}{3600 \text{ sec}} = 8.8 \times 10^3 \text{ watts}$$

then

$$W_{\text{acoustic}} = (8.8 \times 10^3)\, 10^{-4} = .88 \text{ watts}$$

---

[1] The sharpness of the peaks as recorded involves not only the physical characteristics of the pulsation but the averaging characteristics of the recording system. See Chap. 2, Fig. 2-11.
[2] Efficiency is the ratio of acoustic power output to thermal power input from combustion.

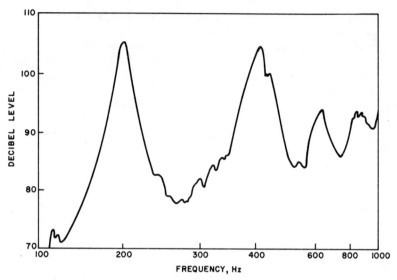

Fig. 10-2. Two pulse combustors of Muller design, 4-in. apart, connected at combustion chambers by 24-in.-long tube, and firing at 200,000 Btu/hr each.

The sound power level is given by Equation (1-28).

$$L_W = 10 \log_{10}\left(\frac{W_{\text{acoustic}}}{10^{-12}}\right) \text{ dB}$$

$$= 10 \log_{10}\left(\frac{.88}{10^{-12}}\right) \text{ dB}$$

$$= 119 \text{ dB} \quad (1\text{-}28)$$

For a one million Btu per hour burner

$$W_{\text{chemical}} = 10^6 \,\frac{\text{Btu}}{\text{hr}} \times \frac{1055 \text{ joule}}{\text{Btu}} \times \frac{1 \text{ hr}}{3600 \text{ sec}} = 2.9 \times 10^5 \text{ watts}$$

and

$$W_{\text{acoustic}} = 2.9 \times 10^5 \times 10^{-4} = 29 \text{ watts}$$

$$L_W = 10 \log_{10}\left(\frac{29}{10^{-12}}\right) \text{ dB}$$

$$= 135 \text{ dB}$$

As a result of this high conversion efficiency of chemical energy to sound, the noise produced can be unacceptable and cause physical destruction of equipment: also, large amounts of heat release can occur. Thus, combustion-driven oscillations have always received the largest share of attention, both by industry, in respect to residential and industrial combustors, and by government agencies, in respect to rockets, ram jets, afterburners, and such.

Combustion roar has no specific frequency but is composed of a broad-band frequency spectrum in its basic form. Figure 10-3 shows such a spectrum. Typical observed efficiencies of conversion to combustion roar from chemical energy input of a burner are in the range 10^{-8} to 10^{-6}. Little attention has been given to this form of combustion-produced noise until recently, when it was realized that it might be a significant contributor to environmental noise.

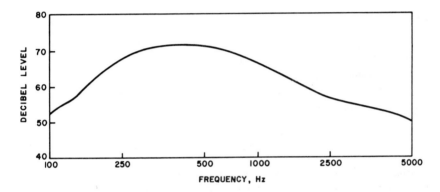

Fig. 10-3. Combustion-roar amplitude frequency spectrum at 40,000 Btu/hr, using two impinging natural gas jets.

The third class of noise, *unstable combustion noise*, is related to incipient flashback, blowoff, and other random or semirandom gross movements of the flame and associated velocity or pressure pulsations. However, it does not involve an acoustic feedback cycle; if it did, a combustion-driven oscillation would result. The amplitude of noise output is above that of the combustion roar, which might be considered as a baseline noise level, and ordinarily below that of a combustion-driven oscillation. The spectrum resembles that of combustion roar.

The fourth class is related to the *combustion amplification of periodic flow noise*. Sources in this class, to date, can be subdivided into (a) external noise sources, (b) Strouhal-related or vortex shedding phenomena, and (c) swirl-burning precession noise. The external noise sources include jet noise from sudden contractions in the flow system, flow noise, and purposely imposed high frequencies. Strouhal-related phenomena are: (a) those phenomena in which vortex shedding, often axially symmetric in combustion systems, occur and have a significant effect on the rate of heat output, or (b) those phenomena, like jet noise, whose peak frequency has the characteristics of Strouhal frequency. The

Strouhal number, which is a dimensionless quantity, is given by the following relation:

$$S_t = \frac{fD}{U} \qquad (10\text{-}2)$$

where f is the frequency in Hz, D is a characteristic dimension in meters, and U is an average velocity in meter per second. The Strouhal number is constant or varies mildly with varying Reynolds number and swirl number. The Reynolds number is obtained from the relation:

$$R_e = \frac{UD}{\nu} \qquad (10\text{-}3)$$

where:

U = Average velocity, meter/sec
D = Characteristic dimension, meter
ν = Kinematic viscosity of the fluid

The swirl number is defined for cases where the fuel and air have an angular velocity, typically in swirl burners, and is defined by:

$$N_S = \frac{\text{angular momentum} \times \text{radius}}{\text{linear momentum}} \qquad (10\text{-}4)$$

Swirl-burning noise is associated with swirl burners and, under certain conditions, the combustion process can amplify the associated frequencies and produce undesirable sound levels.

COMBUSTION-DRIVEN OSCILLATIONS

While a moderate amount of work has been sponsored by industry on combustion-driven oscillations in residential, commercial, and industrial combustion systems, the most extensive work has been sponsored by government agencies in connection with the presence of combustion-driven oscillations in rockets, ramjets, and afterburners; but principally, rockets. However, many portions of this government-sponsored work are directly applicable to the industrial area. As the use of oxygen and of higher combustion pressures occurs in industry, even more of this information will be pertinent.

As mentioned above, combustion-driven oscillations involve a feedback cycle, where a change in the acoustic pressure and/or velocity causes a change, directly or indirectly, in the heat release rate. In turn, the change in heat release supplies the energy to maintain the acoustic oscillations. There is a wide variety of possible feedback mechanisms. Four of these mechanisms prevalent in the industrial combustion systems will be discussed following a brief treatment of Lord Rayleigh's criterion for occurrence of oscillations. Some comments will also be made on the various modes of oscillation, interaction with room effects, and on potential suppression methods. Further details may be found in Refs. 3 and 4.

Because of our emphasis on prevention and cure, positive uses of combustion-driven oscillations will not be covered. However, they have been rather thoroughly reviewed in the literature. [Refs. 2, 7].

Rayleigh Criterion

In industrial combustors a simple and, in most instances, adequate[3] criterion of whether combustion-driven oscillations will occur was advanced in the 19th Century by Lord Rayleigh [Ref. 8]. Rayleigh has described this criterion in detail:

If heat is periodically communicated to, and abstracted from, a mass of air vibrating (for example) in a cylinder bounded by a piston, the effect produced will depend upon the phase of the vibration at which the transfer of heat takes place. If heat is given to the air at the moment of greatest condensation (maximum pressure), or taken from it at the moment of greatest rarefaction (minimum pressure), the vibration is discouraged. When the transfer of heat takes place at the moment of greatest condensation or of greatest rarefaction, the pitch is not affected.

If the air be at its normal density at the moment when the transfer of heat takes place, the vibration is neither encouraged nor discouraged, but the pitch is altered. Thus the pitch is raised if heat be communicated to the air a quarter period before the phase of greatest condensation.[4]

In general both kinds of effects are produced by a periodic transfer of heat. The pitch is altered, and the vibrations are either encouraged or discouraged. But there is no effect of the second kind if the air concerned be at a loop, i.e., a place where the density does not vary, nor if the communication of heat be the same at any stage of rarefaction as at the corresponding stage of condensation.

A mathematical formulation of this criterion of driving is obtained by an integration over a cycle given by:

$$\oint hp \, dt > 0$$

where h is the rate of heat release, p is the oscillating component of the pressure, and t is the time. This is a simple and convenient way of taking a first look at the driving of an oscillation by heat addition.

Singing Flames

When a fuel-supply line is inserted sufficiently far into a vertical combustion tube, as in Fig. 10-4, the diffusion flame may "sing" and emit a periodic sound. High-speed movies of the flame show the movement of annular undulations down the surface of the flame; the number of undulations is a function principally of fD^2/D_v, where D_v is the molecular diffusivity. This is the "singing

---

[3] In high-velocity systems, there can be a significant conversion of flow energy to driving energy (Ref. 6).

[4] It is well known that a periodic function can be divided into two components, one component in phase with an arbitrary cycle of the same frequency and one component a one-quarter cycle out-of-phase.

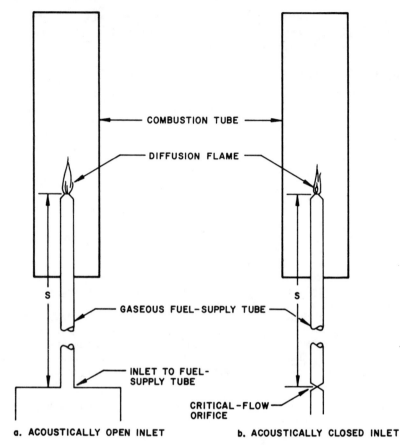

Fig. 10-4. Singing-flame apparatus.

flame" first reported about 1800. The mechanism of driving is as follows. A periodic change in pressure in the region of the flame at a natural frequency of the combustion tube causes a periodic change in the fuel supply rate. This causes a periodic change in the flame shape, and thus, in the heat release rate. If the periodic change in heat release rate is phased properly, relative to the pressure change; that is, if Rayleigh's criterion is satisfied, the oscillation will continue. The lag time[5] of the periodic heat release rate behind the periodic fuel supply rate is significant in whether the oscillation occurs or not. This time lag, τ, can be computed for simple shapes of laminar flames. For a circular fuel port the time lag can be determined from:

$$\tau = \frac{\delta}{V_p} + \frac{a}{3F} \text{ seconds} \qquad (10\text{-}5)$$

---

[5] An alternate interpretation of time lag is the time between the fuel entering the combustion chamber and the time it is burned.

Ch. 10   COMBUSTION AND FURNACE NOISE

where:

τ = Time lag, in seconds
V_p = Average velocity of fuel-air mixture at the face of the port, ft/sec
a = Radius of the port, or one-half the slot width, ft
F = Flame front velocity or burning velocity of the fuel-air mixture, ft/sec
δ = Thickness of the dark space or unilluminated fuel between the port and the luminous cone, inch

The thickness of the dark space, δ, should be the minimum value specified in combustion literature for the fuel used. Some values are given in Table 10-1.

Table 10-1. Properties of Some Fuels

Fuel	Minimum Dark Space, δ, inch	Burning Velocity, F, 65% stoichiometric, ft/sec
Natural Gas	0.039	0.4
Manufactured Gas	0.029	1.3
Propane	0.032	1.1

For wedge-shaped flames over slot-type openings the lag time is given by:

$$\tau = \frac{\delta}{V_p} + \frac{a}{2F} \text{ seconds} \qquad (10\text{-}6)$$

It is important to note that the above line of reasoning holds not only for the combustion system shown in Fig. 10-4, but for other similar diffusion flame systems, both with laminar flames and with turbulent flames. Thus, rules for predicting the occurrence of oscillations in the simple singing-flame experiments may be extended to a more general class of "singing flames."

Table 10-2 represents such a set of rules. Whether or not a singing flame occurs depends upon the time lag, τ, the frequency of oscillation, f, and the geo-

Table 10-2. Conditions for Occurrence of a Singing Flame (for n = 0, 1, 2, ... and m = 0, 1, 2, ...)

Condition	Value of Time Lag, τ	
	$0 < f\tau - m < \frac{1}{2}$	$\frac{1}{2} < f\tau - m < 1$
Acoustically open inlet to fuel supply line of length, S (Fig. 10-4a)	$0 < \frac{2S}{\lambda} - n < \frac{1}{2}$	$\frac{1}{2} < \frac{2S}{\lambda} - n < 1$
Acoustically closed inlet to fuel supply line of length, S (Fig. 10-4b)	$\frac{1}{2} < \frac{2S}{\lambda} - n < 1$	$0 < \frac{2S}{\lambda} - n < \frac{1}{2}$

metric conditions of the supply tube, namely, the supply-line length, S, and the wavelength[6], λ. The inequalities shown in Table 10-2 give the conditions at which a singing flame can occur, where m and n are integers. The integer m relates to the pressure nodes in the combustion tube and n relates to the pressure nodes in the fuel supply tube. For example, the most easily excited condition corresponds to the fundamental nodes or frequencies which occur when m and n are both equal to zero. If, in this case, the product of the frequency, f, and the time lag, τ, is greater than zero and less than $1/2$, and if the quotient of two times the supply-line length, S, divided by the wavelength, λ, is greater than $1/2$ and less than one, the burner with the acoustically closed fuel inlet can have a singing flame. For the same conditions, Table 10-2 will indicate that the acoustically open fuel-supply tube will not have a singing flame. Thus, Table 10-2 can be used as a guide to predict the occurrence of a singing flame. It should be pointed out that other values of m and n are possible, although these conditions are not excited as readily; moreover, m does not have to be equal to n.

As an example of the tie-in between Rayleigh's criterion and Table 10-2, one can consider Fig. 10-4b, with a length, S, less than $1/4$ wavelength. Then, the pressure in the fuel-supply line is entirely in phase with that in the combustion chamber. According to Bernoulli's equation when the pressure is increasing in the fuel-supply line, the gas flow velocity *relative to the average flow rate* must be negative. Thus, the gas flow velocity must lag the pressure. For Rayleigh's criterion to be satisfied, a component of the rate of heat release must be in phase with the pressure. Therefore, the heat release rate must lag the velocity between one-half and one period of oscillation. Therefore, for the oscillation to be driven with $0 < 2S/\lambda < 1/2$, $1/2 < f\tau < 1$ for an acoustically closed inlet. This agrees with Table 10-2.

It is obvious that the system can be moved out of the oscillating region by changes in the fuel-supply length, the fuel-supply inlet condition, the time lag (for instance, by changing the burner diameter), or even by changing the natural frequency of the entire system. However, most combustion systems have more than one acoustic mode that can be involved in a combustion-driven oscillation, and one oscillation may be eliminated only to have another occur as indicated by $n = 0, 1, 2, \ldots$, etc., for various modes. Also, according to the second half of Rayleigh's criterion, usually not mentioned, the component of the periodic heat-release rate that is not in proper phase for aid in driving an oscillation will tend to shift the system frequency toward one more conducive to driving.

---

[6] $\text{wavelength}, \lambda = \dfrac{\text{speed of sound}}{\text{frequency}} = \dfrac{c}{f}$

where:

$c = \sqrt{kgRT}$ ft/sec
k = ratio of specific heats for the gas mixture, dimensionless
R = gas constant = $53.3 \dfrac{\text{ft}}{\text{sec-}^\circ\text{R gr mole}}$
T = absolute temperature, $^\circ\text{R} = 460 + T_{\text{Fahrenheit}}$
g = gravitational acceleration, ft/sec
f = frequency, Hz

Ch. 10 COMBUSTION AND FURNACE NOISE 449

Many investigators attribute the driving of oscillations in such large installations as blast-furnace stoves shown in Fig. 10-5 to the "singing-flame" mode of feedback. In this type of system, both the fuel and air-supply systems are ducted; and thus, with this assumption of mechanism for driving, both supply systems must be considered in the light of Table 10-2. In many types of flames, the time lags for the fuel and air are different and must be considered separately. Because of practical problems associated with a full-scale furnace, a $1/25$-scale model of a system was constructed. In the model, the lengths of the fuel-supply line and the air-supply line lengths were varied and the acoustic modes studied. Figure 10-6 summarizes the results of the study. When both the fuel and air supply lines are of a length such that a singing region would be predicted by the "singing-flame" theory for short time lags, severe pulsations were observed with conditions as in Fig. 10-6a. When both air and fuel-supply-line lengths are such

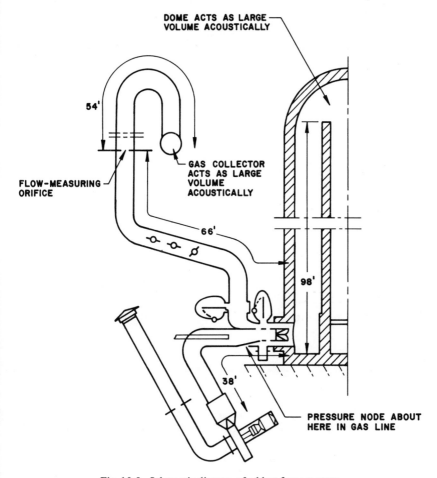

Fig. 10-5. Schematic diagram of a blast-furnace stove.

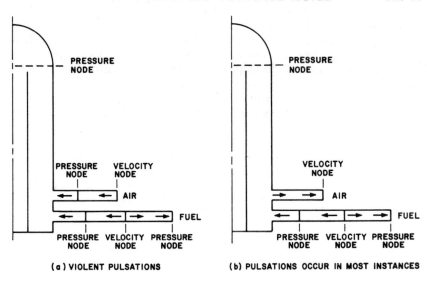
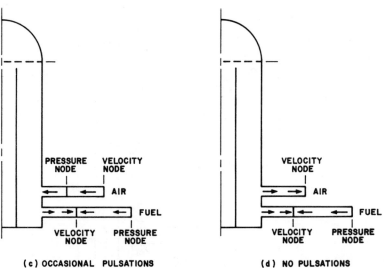

Fig. 10-6. Schematic diagram of a scale model blast-furnace stove. Arrows indicate direction of the acoustic velocity when the pressure is increasing at the bottom of the well.

that no pulsations are predicted by the "singing flame" theory no pulsations were observed with the conditions of Fig. 10-6d. With a fuel-line length incorrect for pulsations and the air-supply line of a length for pulsations, as in Fig. 10-6b, oscillations occurred for most operating conditions of pressure and flow. In the case of Fig. 10-6c the fuel-line length was correct for pulsations but the air-line length was incorrect and oscillations occurred only occasionally.

Ch. 10   COMBUSTION AND FURNACE NOISE

To reduce the noise associated with cases a, b, and c of Fig. 10-6, the length of fuel- and air-supply lines should be shortened so that a velocity node rather than a pressure node exists immediately upstream of the combustion chamber as is shown in Fig. 10-6d. This may be accomplished by actually shortening the lengths or by installing plenum chambers in the supply lines.

As the amount of information on large combustion systems has increased, anomalies have appeared that indicate the existence of an alternate feedback mechanism in some instances. These large combustors are characterized by multiple fuel or multiple air jets, not necessarily parallel to the axis of the closed combustion system. In many situations, such as blast furnace stoves, the stoichiometric air/fuel ratio is approximately 1:1 rather than 10:1, or more, of hydrocarbon fuels with air. Thus, one needs to give equal weight to both fuel and air for driving, and to compute both time lags. While these time lag computations (properly weighted for frequency effects) are not easy to make on a basis of the presently available knowledge of mixing and combustion in the complex combustion chamber, approximate computations give lags that appear in line with physical observations of large flames but are too long as compared to the short time lag ($f\tau > \frac{1}{2}$) necessary to explain observed results on the basis of Table 10-2.

The second observed discrepancy in some large combustors was the observation, both visually and by ionization probes, that the flame was flashing or jumping back and forth along the combustion chamber in phase with the oscillation, when combustion-driven oscillations were present. This does not agree with the flame concept for the "singing flame" wherein the flame is held at the fuel (or air) inlet. However, it can be shown that a short "time lag" is associated with the flame flashing back and forth between two positions in the combustion chamber as a result of periodic blow-off of the upstream flame because of a high total velocity followed by a periodic flashback from downstream due to a low total velocity. Thus, Table 10-2, assuming a short time lag, can be used for predicting oscillation regions for flashing flames, as well as for singing flames.

When fans are used to supply the air, a question arises as to the proper treatment of the inlet and where the fan is located. Early reports in the literature indicated that in some instances fans should be treated as open ends of simple acoustic systems, and in other instances, as simple closed ends. More advanced treatments indicate that the proper assumption depends on the point of operation on the fan characteristic curves. But experiments and analytical treatments are needed to determine the prediction rule for the equivalent length for the fan scroll plus air inlet duct.

Fuel-Oil Combustors

The most common example of the fuel-oil combustor is the residential heating unit shown in Fig. 10-7. And at least one feedback mechanism whereby oscillations may be driven in fuel-oil-fired combustors seems to be common for residential through industrial-size combustors using either gun-type or two-fluid atomizing nozzles. Based on observations with high-speed movies (which are not too difficult to obtain since even in residential units 40 Hz is a high frequency) and simultaneous acoustic pressure records, the mechanism appears to be as

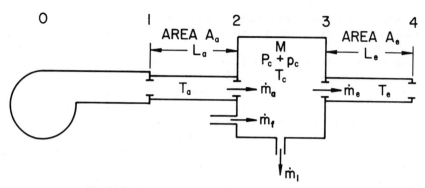

Fig. 10-7. Schematic diagram of an oil-fired combustor.

follows: If there is an oscillation in the combustion chamber, a higher rate of air is supplied to the combustor when the pressure in the combustion chamber is below average (while the fuel-supply rate remains constant). When the pressure in the combustion chamber is above average, the rate of air supply is below average and there may even be flow reversal at the edge of the blast tube. As a result, the smaller droplets of fuel are burned periodically rather than at a steady rate. In fact, the reverse flow effect has been observed to carry flame remnants back into the blast tube, from where they are shot out along the axis to ignite a fresh mixture of small droplets and air. The rapid expansion of the flame through the new mixture supplies the periodic energy necessary for driving.

Several methods of eliminating this specific type of oscillation have been tried with various degrees of success. Since, for a specific fuel-nozzle spray angle, the oscillations only occur over a finite range of mixture ratios, specific to the particular nozzle, changing the spray angle may help. However, the change in spray angle may also cause flame deterioration. Several flame-stabilizing devices, such as grids, have also been used, but they may also coke up. Perforations or openings around the blast tube inlet have been used very successfully, but care must be taken not to unbalance the overall air supply. More drastic changes include a complete change of fan to move the operating point on the characteristic curve of the fan and gain a damping effect.

Several other possible modes of driving have been reported in various oil-fired combustors [Ref. 4], but sufficient detail has not been given to prove the feedback mode. One system, using air atomization of the fuel, appeared to have feedback in the atomizing air line—much in the spirit of the singing flame. Some definite cases of flashing flames have been observed, but no measurements were made. Some compact, high-intensity burners have apparently shown spinning modes of oscillation, just as in rocket engines. The investigator encountering these types of driving have little guidance for suppression from the available literature.

Laboratory studies of residential and commercial-sized fuel-oil heating units have been conducted. The conventional heating unit is illustrated in Fig. 10-8a. The difficulty in experimenting with this system is twofold: the air-flow rate must be computed from a measurement of the product composition, and the

Ch. 10 COMBUSTION AND FURNACE NOISE 453

control of air-flow rate also changes the acoustic properties of the system. In Fig. 10-8b, the fan housing is closed off and the air is supplied through a critical flow orifice so that there is no acoustic feedback into the air-supply duct. The difficulty with this situation is that the maximum acoustic-pressure amplitude for the fundamental mode occurs at the flow orifice, which then causes the acoustic velocity to lead the acoustic pressure by $3/4$ cycle as compared to a lead of $1/4$ cycle in an actual system such as Fig. 10-8a. Thus, the acoustic system is changed from the conventional system to be investigated.

Figure 10-8c shows the best system for studying the performance of an oil-fired heating unit. It combines a controllable and measurable air supply with a

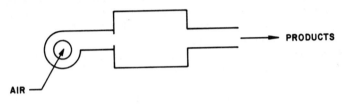

(a) CONVENTIONAL COMBUSTOR WITH FAN

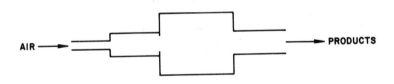

(b) COMPRESSED AIR SUPPLIED TO COMBUSTOR

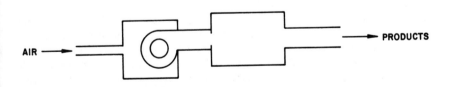

(c) COMPRESSED AIR SUPPLIED THROUGH
LARGE PLENUM-SURROUNDING FAN

Fig. 10-8. Schematic diagrams of systems used to study pulsations in oil-fired heating units.

simulation of the actual conditions for a heating unit if the pressure in the plenum is balanced with atmospheric pressure, and the plenum volume is considerably larger than the combustion chamber volume, so that the overall acoustic frequencies are not changed.

Tunnel Burners

For the purpose of this discussion, tunnel burners are considered to be high-velocity burners that are long in comparison to their diameter. They may be gas- or oil-fired, metal or ceramic lined, and, essentially, open or closed at the upstream end. Industrial units normally have grids, obstacles, spreaders, and enlargements to hold the flame, which may be a premix, nozzle-mix, or delayed-mix type. A premix burner type is characterized by the fact that the fuel and air are mixed in proper proportions before entering the combustion region. In a nozzle-mix system, fuel and air are separately introduced in proper ratios by nozzles at the combustion location. A delayed-mix situation occurs when fuel is supplied by a nozzle and the air for combustion is supplied at locations in the combustion region. Because of this variety of design, there appears to be a variety of feedback mechanisms. Some of these will now be discussed. Shown in Fig. 10-9a is a premixed tunnel burner with a variable tube length used to investigate acoustic resonances. The results for a typical set of data for a particular fuel is given in Fig. 10-9b. The solid line gives the burning oscillations due to a time lag required for the fuel fed into the tube to ignite. This oscillation is due to the fact that the fuel requires a certain time interval to enter the combustion tube, but burns rapidly when it ignites. This results in a series of explosions at a constant repetitive rate. The dashed line in Fig. 10-9b is the resonant frequency of the tube for various lengths. When the firing rate oscillations coincide with the tube resonances (when the solid and dashed lines cross) the acoustic intensity reaches a maximum as shown by the peak in the acoustic intensity curve.

Many of the combustors have a multiple-orifice grid holding a premixed flame. The premixed situation with a nonturbulent flame has been quite extensively discussed in the literature. [Refs. 3, 4]. Compare a combustion-driven oscillation with a premixed flame and the singing diffusion flame; in both instances the potential of the combustion chamber to drive the supply line at approximately the natural frequencies of the combustion chamber is considered, but for the premixed flame a combustible mixture is in the supply line. The time lag involved is still based upon the time from the supply of the combustible mixture to the average burning time. Approximately, the time lag is given by Equations (10-5) or (10-6). More accurate predictions of driving mechanisms can be made by considering the acoustics of the combustion system in more detail, and the driving force (in the Rayleigh sense) as the sum of the increments along the length of the flame, for each frequency. However, for practical burner shapes, the time lag for a given burner face can better be obtained at the present state of the art by carefully conducted laboratory tests in which the furnace frequency and the mixture supply-line length are used as controllable variables.

For turbulent premixed flames, the case in many tunnel burners, the same rules hold but the proper computations of time lag is still a problem needing

Ch. 10 COMBUSTION AND FURNACE NOISE 455

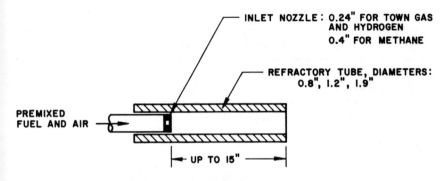

(a) EXPERIMENTAL REFRACTORY BURNER

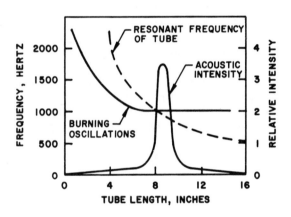

(b) VARIATION OF FREQUENCY AND INTENSITY
WITH LENGTH OF 1.2"-DIAMETER TUBE, FOR AIR/TOWN GAS
RATIO OF 5.25, AT 0.4 CFS

Fig. 10-9. Schematic diagram of a tunnel burner and data obtained by Howland and Simmonds.

study. Thus, the experimental approach to evaluation of time lag is even more important for the turbulent premixed flame.

Studies with fuel-air rockets have also revealed a scheme of driving that appears to be related to a natural frequency of the combustible mixture (the chemical reaction-rate per unit volume or reciprocal chemical time lag). This may be related to the combustion roar phenomenon discussed later.

It would appear that some of the problems in tunnel burners are related to flashing flames, similar to those mentioned above, in connection with the blast-furnace stoves. The flashing flame in the case of the premixed flame is different in detail, however, from the flashing diffusion flame. The variables of mixture ratio, burning velocity, and chemically controlled reaction-rate per unit volume within the combustion chamber are removed from the problem. This makes for an easier analysis of the phenomenon, but leaves more possibilities of temporary holding regions in a cycle of oscillation to be checked.

Finally, there is a possibility that some of the oscillations in tunnel burners are interrelated with vortex shedding phenomena.

Vortex Shedding

The sequence of smoke rings emanating from the famous cigarette advertising sign on Times Square in New York City was a good example of vortex shedding that is occasionally involved in combustion-driven oscillations. The main difference is that the smoke rings were forced at a controlled frequency while those from the end of a burner or a flame spreader tend to shed at a natural frequency. The natural frequency is usually presented in the dimensionless form of a Strouhal number, discussed earlier. This phenomenon is associated with an imbalance between the transverse component of the drag forces on a recirculation or shear region and the transverse acceleration of the flow streams.

Vortex shedding can form an essential part of a feedback cycle. For instance, for single-port residential gas-fired furnaces, it has been shown that there is a tendency for vorticles to form off the end of the burner and move upward around the flame spread. When the natural vortex-shedding frequency is near the natural acoustic frequency of the furnace, then a sequence of events can occur in which the acoustic oscillation triggers the vortex shedding, the periodic change in flame surface area, because of the vortex, causes a periodic heat release, and the periodic heat release maintains the acoustical oscillation. Dorresteijn [Refs. 4, 9] has proved the existence of a similar phenomenon for a crude-oil heater.

The most successful preventive measure reported for this type of combustion-driven oscillation has been to move the flame front away from the most active vortex shedding region.

SUPPRESSION OF COMBUSTION-DRIVEN OSCILLATIONS

While there are many approaches to the suppression of combustion-driven oscillations, there is no one universal solution or even one best solution, in most cases. Each case must be analyzed both in reference to the possible and probable feedback cycles that might be involved, and in reference to the impact of any possible cures on performance, safety, initial cost and maintenance cost.

If the feedback system is known, suppression of the amplifying effect of any or all the steps in the feedback cycle will help. That is, the ratio of output amplitude to input amplitude in each step should be reduced. When a fuel-supply line or air-supply line is involved, the most effective cure is often a change in length between the upstream end as defined by a high-pressure-drop valve, a large plenum, or a fan inlet, and the exit at the burner. In some instances, the burner can also be moved away from the region of highest acoustic pressure

Ch. 10 COMBUSTION AND FURNACE NOISE 457

amplitude in a furnace. If a flashing flame—as a result of a periodicity in velocity—is involved, the flame might be stabilized by use of a flame holding region with a greater pressure drop and thus greater, potential flame-holding ability, by adding a pilot flame, or by changing the mixture ratio in the critical region to improve the flame-holding ability. Alternately, one might decrease the flame-holding ability of the entire combustion region to the downstream holding region of the flashing flame.

An alternative approach to changing the amplification is to shift response times between outputs and inputs. A common method is to change the time lag between fuel (or combustible mixture) injection and burning. As noted previously, this can be done by changing the size of the individual flame elements in a burner face. For premix burners with secondary air added to produce a diffusion flame to complete combustion, a simple and common method of changing time lag is to change the primary ratio. In some industrial burners, recirculated hot products of combustion can be used to change the time lag, and also decrease the amplification effects.

A second approach is to relieve the pressure in the region of maximum pressure-amplitude by an orifice or port to the surroundings. The use of this method in oil-fired units has been mentioned above. However, this is not always a practical solution for various reasons. An alternative is to return part of a pressure signal out-of-phase with the acoustic-pressure oscillation in the combustion system. This can be done by the use of quarter-wave tubes or a Helmholtz resonator. This solution is often quite satisfactory for experimental units, but may not be satisfactory over the normal range of operating conditions of units used in industry.

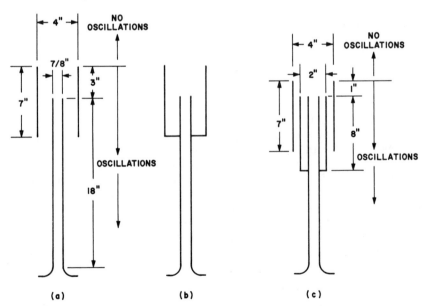

Fig. 10-10. Three types of exit shields to reduce combustion-driven oscillations.

Damping and sound-absorbing materials may also be used; alone, or in combination with other approaches. In the case of frequencies below about 200–400 Hz, this is not usually a practical approach, however, since low-frequency absorption is difficult to obtain with acoustic materials.

Baffles at the entrance, exit, and in the interior of the combustion system have been used successfully on various occasions, but the reason for the success has not always been clear. Two specific examples are porous duct sections and restrictive orifices at the exit of tunnel burners.

Shown in Fig. 10-10a is the placement of a concentric tube over the outlet of a combustion duct. When the top of the outer tube extended less than 3 inches above the duct exit, there was no effect on acoustic oscillations. However, for extensions of 3 inches or more, the oscillations were abruptly suppressed. These oscillations were re-established by blocking the annular space at the lower end, as shown in Fig. 10-10b. For two concentric tubes over the combustion duct, as shown in Fig. 10-10c, oscillations were suppressed when the outer tube extended one inch above the inner tube and duct. Observations have been made on industrial combustors that a collar similar to Fig. 10-10a when placed around the burner eliminated combustion-driven oscillations.

Room Effects

A brief comment should be made about the consequences of various comparative values of the natural frequencies of furnace and enclosure (or room). There will be little tendency of a room to oscillate in response to the furnace frequency if no natural frequency of the room of enclosure is close to the furnace frequency. But if a frequency of room and the furnace frequency are close together, and the furnace is near a pressure antinode in the enclosure, the installation will be unacceptable. Solutions of this problem have included the removal of walls, the erection of false walls, and the movement and even changing of the combustion system.

COMBUSTION ROAR

The relatively new subject of combustion roar has received constantly increasing attention in the last few years.[7] The following comments are based on the application of turbulent flames of industrially significant size. There is some disagreement with certain commonly cited conclusions based on experiments on small flames and considerations of such flames as single monopole sources.[8] However, it is believed that the presentation herein is both technically and intuitively acceptable. Furthermore, the author has found the approach forms a useful basis for analyzing experimental data.

In spite of many years of study of turbulent flames, the understanding of neither turbulent premixed nor turbulent diffusion flames is completely acceptable. On a macroscale, it appears that the intensity of turbulence controls

---

[7] References 10 and 11 contain lengthy reference lists which identify current workers in the area.

[8] A monopole source can be pictured as a sphere, periodically expanding and contracting radially.

Ch. 10 COMBUSTION AND FURNACE NOISE 459

the gross reaction rate. On the other hand, the character of the combustion processes, the sizes of burning cells,[9] appear to be functions of the scale of the turbulence and the burning velocity of the fuel as well as the intensity of turbulence. Finally, if the turbulence is too intense, it would appear that the flame is quenched. Whether these burning cells should be considered as having a continuous connected flame front, as for small, mildly turbulent flames, or as having a homogeneous burning process occurring within them, is still a matter of argument. Nevertheless, the model for considering noise production is either of the cases above. As will be shown, the actual size of the cells and the number thereof, in a flame, cancel out of the noise production equations, and thus, the conceptual picture need not be sized for the discussion.[10]

From the acoustic point of view, then, the array of turbulent burning cells in a turbulent flame of commercially significant size is considered to be an array of monopole sources. [Refs. 12, 13]. The sound power, W, produced by N monopole sources is given by:

$$W = \frac{\rho N}{4\pi c} \left[\frac{d}{dt}(E-1)q \right]^2 \qquad (10\text{-}7)$$

where E is the volume expansion ratio of the burned to the unburned gas, ρ is the density, c is the velocity of sound, and q is the volume rate of consumption of the combustible gas in each burning cell. We assume that the time of consumption of a burning cell or element is proportional to d_f/u', where d_f is the element thickness and u' is the intensity of turbulence.[11] Thus, we can replace d/dt by u'/d_f. We also assume that $qN = Q$, where Q is the volume firing rate of the flame. Making these substitutions

$$W = \left(\frac{\rho c q Q}{4\pi} \right) \left(\frac{E-1}{d_f} \right)^2 \left(\frac{u'}{c} \right)^2 \qquad (10\text{-}8)$$

With some further manipulation and the assumption that the burning velocity, F, equals q/d_f^2, the efficiency of conversion of chemical energy to acoustic energy, η, is given by

$$\eta = \left(\frac{\gamma - 1}{4\pi} \right) (E-1) \left(\frac{F}{c} \right) \left(\frac{u'}{c} \right)^2 \qquad (10\text{-}9)$$

where γ is the ratio of specific heats, c_p/c_v, or K.

We note that the efficiency is a function of the product of a term depending on the fuel and mixture ratio, and a term depending on the turbulence level in the combustion region.

Further considerations of this model, discussed in Ref. 13, justify a proportionality of the peak frequency of the combustion-roar spectrum (see Fig. 10-2)

---

[9] In combustion processes combustion occurs in many localized regions instead of homogeneously throughout the flame. These regions are called burning cells.

[10] In the case discussed subsequently where a high-frequency signal can result in the reduction of combustion roar, the size may be significant.

[11] The intensity of turbulence is usually about 5 to 20 percent of the characteristic flow velocity in a burner.

to the chemically limited reaction-rate per unit volume of the combustible mixture. For the usual hydrocarbon fuels in air, this does not vary by a large amount.

General Conclusion from Combustion-Roar Theory

Several conclusions may be drawn from the above theory. These will be applied to thrust-controlled flames.[12] For buoyancy-controlled flame, the intensity of turbulence assumption that is made in this discussion would have to be modified. This would result in slightly altered conclusions which would be dependent on the shape of the flame.

Change in Firing Rate of Single Burner.—For a single burner with a fixed fuel and fuel/air ratio, we can assume that $u' \approx U \approx Q$, where U is a typical flow velocity and Q is the heat release rate. If W is the noise power output,[13] then $W \approx Q^3$ and L_W (decibels) $\approx 30 \log_{10} Q$.

Effect of Change in Burner Size.—If a homologous series of burners is considered, each firing at its rated capacity and pressure drop (ΔP), then (neglecting Reynolds number effects) u'/c is a constant. It follows that the efficiency of conversion is a constant. Thus, $W \approx Q$ and $L_W \approx 10 \log_{10} Q$. Increases in Reynolds number tend to decrease the efficiency of noise production.

It also follows that a change in the number of burners for a given application, if from the same series and all fired at rated capacity, will not change the combustion-roar output. Figure 10-11 shows a correlation of three sizes of premix burners, singly and in pairs, on this basis.

It can be concluded, as a consequence of this observation, that four 1-million-Btuh burners will produce the same amount of noise as one 4-million-Btuh burner if both systems are operated at the same back pressure. Thus, a designer given the choice of the number of burners to use in the furnace—either one large one or several smaller ones—should base his selection on the desired heat distribution within a furnace rather than on a noise-output consideration. This does not take into account a possible flame-size effect discussed below.

Effect of Intensity of Combustion.—The intensity of combustion, I, is proportional to U/L where L is the modular size[14] associated with a particular combustor. With some manipulation, one can determine that $W \approx Q^{5/3} I^{4/3}$, or $L_W \approx 16.7 \log_{10} Q = 13.3 \log_{10} I$. Thus, if the intensity of combustion is maintained throughout a series of designs, say, by increasing the number of modules (rather than scaling up all the burner dimensions) and maintaining the characteristic velocity, the noise output will increase with somewhat less than the

---

[12] Thrust-controlled flames are those in which the level of turbulent mixing in the flame is controlled by the thrust forces of the fluid jets (fuel and air for diffusion flames, and mixture for premix flames). Buoyancy-controlled flames are those in which the level of turbulent mixing in the flame is controlled by the buoyancy forces imparted by the hot gases. Most commercial combustion systems have thrust-controlled flames, in the sense outlined. Free-burning fires, as from an accidental spill, and ignition of a combustible are the common examples of a diffusion flame.

[13] $L_p(\text{dB}) = L_W(\text{dB}) - 20 \log_{10} R$ (feet) $- 0.5$ where L_p is the sound-pressure level at a point R feet from the source, assuming spherical expansion of the sound from the source.

[14] Many burners can be considered as arrays of small identical modules, square, hexagonal, pie-shaped, etc. The intensity of combustion may then be related to the module size and shape, and not the overall size of the array.

Ch. 10 COMBUSTION AND FURNACE NOISE 461

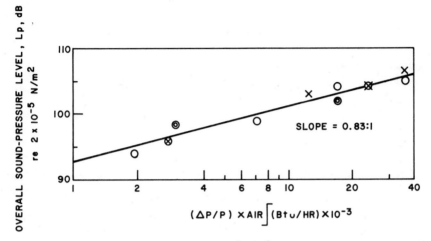

Fig. 10-11. Overall decibel level (L_p ref. 2×10^{-5} N/m^2) at 3 feet from premix burner(s) as a product of pressure drop across the burner and Btu/hr based on available air. Different symbols are for different burner sizes. Two symbols at the same point indicates two burners fired together at a 20-inch spacing.

square of the combustion energy input. It might be noted some series of burner designs are constructed on this basis. This could explain the results reported by Bitterlich [Ref. 14] and Seebold [Ref. 15].

Effect of Flame Size. — The effect of flame size is not well understood and may involve more than one phenomenon. First of all, there is an overall effect from change in intensity, if the firing rate is held constant. But second, there is a change in combustion-roar frequency spectrum shape; as the flame gets larger a gradual cut-off on the high-frequency side moves to long wavelength or lower frequencies [Ref. 10]. Because of the response of the human ear, the effect can become quite pronounced for large flames. This could make the use of a single large burner preferable to two smaller burners.

Effect of Fuel Composition. — In a case of a premixed flame or a nozzle-mix burner in which all combustion takes place within the burner, the expansion ratio and the burning velocity are known. One therefore expects that if the overall flow rate is held constant, the efficiency of noise output will peak near stoichiometric[15] and fall off on both sides. Since the heat release rate is limited on the fuel-rich side by the amount of air present, and on the excess-air side by the amount of fuel present, a further peaking will be found in plotting dB output against air/fuel ratio. Figure 10-12 shows the total effect; the peaking at the stoichiometric ratio of about 10, is noted.

For the case of constant air-flow rate, increasing the fuel rate (decreasing the air/fuel ratio) toward the stoichiometric value increases the noise output. Increasing the amount of fuel past the stoichiometric value has little or no effect

---

[15] The stoichiometric mixture of fuel and air is one in which the exact amount of oxygen needed for complete combustion of the fuel is provided.

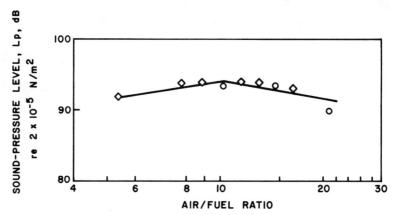

Fig. 10-12. Decibel level at peak frequency of a nozzle-mix burner, with data at constant natural-gas flow rate (◇), and constant air-flow rate (○), corrected to constant, total flow rate.

on the noise output (the flow velocity does not change appreciably) because there will be insufficient air for complete combustion.

Figure 10-12 is the noise output corrected to constant rate flow as a function of air/fuel ratio. Decibel levels were corrected using the V^2 relation. From this curve it becomes apparent that the following relations hold:

$$\text{Sound-pressure-level output} \approx V^2 Q$$

where:

V = average flow velocity
$Q = Q_{\text{fuel}}$ for air/fuel > stoichiometric
$Q = Q_{\text{air}}$ for air/fuel < stoichiometric

For nozzle-mix and diffusion flames, as compared to premixed flames, one expects the burning to take place over a wide range of mixture ratios and to involve more local recirculation of hot products, thus decreasing the effective value of E, the volume expansion ratio. As a result of this effect and of change in the effective value of burning velocity, the noise output should be reduced and the noise spectrum flattened and shifted to lower frequencies.

Effect of Fuel Type.—Most hydrocarbon fuels do not vary greatly in burning velocity at stoichiometric, or in expansion ratio. However, for high percentages of ethylene, hydrogen, and some other fuels with high burning velocity, the efficiency of noise output is seen from Equation (10-9) to be increased accordingly.

Effect of Environment

While the basic noise-frequency spectrum is shaped much like a jet noise-frequency spectrum, components of the spectrum are amplified at the natural frequencies of the surrounding environment. For instance, a burner tile around the flame can amplify the combustion roar at the natural axis modes of the tile,

Ch. 10 COMBUSTION AND FURNACE NOISE 463

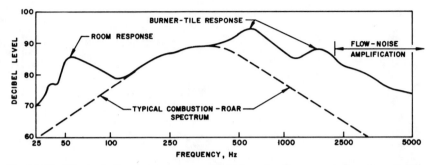

Fig. 10-13. Effect of other noise sources on typical combustion-roar spectrum for a nozzle-mix burner.

as in Fig. 10-13. The same holds true for the acoustic modes of the furnace. In Ref. 10 it is shown that the amplification factor can be computed by conventional acoustic methods by comparing the output of a monopole source within the tile to the output of the same source in free space.

Another environmental effect is that of noise from valves or orifices in the fuel- or air-supply lines. These noises are amplified by the flame, although at the same time the combustion-roar spectrum may decrease somewhat, as discussed later.

Application of the Combustion-Roar Theory

It is not convenient or even possible, in most instances, to specify the intensity of turbulence in a burner. However, for a series of similar burners one might assume that $(u'/c)^2 \approx (U/c)^2 \approx \Delta P/P$, where ΔP is the pressure drop across the air inlet into the burner (or fuel and air, if premixed), and P is the absolute pressure.[16] In case the air is inspirated by fuel jets, the pressure drop across the fuel jets divided by the square of the air-to-fuel ratio, can be used. The data for the efficiency of noise output of a series of similar burners is expected to vary with $(\Delta P/P)^\alpha$, where α is close to unity.

Figure 10-14 shows data presented on this basis. It can be seen that the correlation is quite acceptable for specifying noise output.

Often, the sound-power level, in decibels, is preferred. In this case, the sound-pressure level can be plotted against $\log_{10} (\Delta P/P)^\alpha Q$. Within the accuracy of the data, $\log (\Delta P/P) Q$ is often equally acceptable even if not absolutely correct, when α is other than unity. Figure 10-11 shows data presented on this basis.

Suppression of Combustion Roar

There are several different possibilities for suppressing combustion roar. One of the most obvious, and often very effective, is the installation of mufflers in the inlet and/or exhaust lines, depending from where the prominent noise

---

[16] Design air-pressures are often about 1/10 atmosphere, mixture pressures about 1/100 atmosphere, and fuel pressures from 1/1000 to 2 atmospheres.

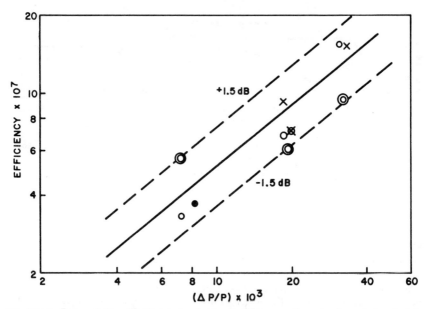

Fig. 10-14. Data of Fig. 10-10 on the basis of efficiency of conversion of chemical to acoustic energy as a function of dimensioned pressure drop.

emanates. Figure 10-15 is a sketch of the installation of an inlet muffler on a gas burner for use in burner banks in oil refineries. Such mufflers quite commonly reduce the dB level by as much as 25.

Figure 10-13 shows that the roar is amplified at the natural frequency of the burner tile and the furnace (room in test). As a result of using a multiple Helmholtz[17] unit built into the tile, the portion of the spectrum of Figure 10-13 above 500 Hz was decreased in total amplitude and amplitude range to where it closely coincided with the dashed curve. Fortunately, the mufflers are of reasonable size for the frequencies most easily detected by the human ear.

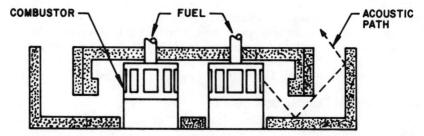

Fig. 10-15. Typical installation of an inlet muffler on a gas burner for use in an oil refinery [Ref. 15].

---

[17] A Helmholtz resonator is a type of furnace-wall construction where a series of hollow volumes in the wall is connected to the interior of the furnace by an open "neck." This resonator will cancel a narrow band or pure-tone frequency to which it is tuned.

There is evidence that the quarter wave tube concept has been successfully used in some units to decrease the dB level at low frequencies.

High frequencies imposed on the flame are known to decrease the combustion roar amplitude. Thus Briffa and Romaine [Ref. 16] have suggested ultrasonic frequencies to decrease the combustion roar.

Since the combustion roar is known to increase about with the square of the intensity of turbulence, the turbulence level should be reduced as far as possible while still maintaining a satisfactory flame from other points of view.

UNSTABLE COMBUSTION NOISE

There are several types of combustion-generated noise that do not fall conveniently into the two classes discussed above. In many instances, these types have not been studied carefully from the noise standpoint, or alternately, it has not been realized that they fall into a separate class. One of these types of combustion-generated noise is that related to unstable combustion.

Flames are often unstable and tend to blow off or flash back. However, a flame may not be sufficiently unstable to result in an undesirable situation under normal operating conditions; in fact, visual inspection of the flame may show it to be satisfactory. Alternatively, as the flame changes toward blow-off, for example, there will be a change in flow pattern and a tendency to re-establish the flame, or a local region thereof, at the expense of some other region. Without a feedback mechanism present in either of these instances, the frequency of such a phenomenon is highly erratic. However, the large, gross movements of the turbulent flame can produce noise from the moving front of ignition of local combustion cells. Increases of 10 dB are not unusual when the mixture ratio, or flow rate, is changed sufficiently to put a flame into a metastable or unstable regime.

Smith and Kilham [Ref. 17] reported the first data on this particular phenomenon in 1963.

In a bunsen-type turbulent propane flame of small size, an annular hydrogen pilot flame was used. As the hydrogen input decreased, the flame suddenly became stable. Before the flame became tenuous and the combustion incomplete, the sound-pressure level increased 3 dB. This corresponded to a doubling of the efficiency of conversion of chemical energy to noise.

In a study of the noise from a premixed, continuous-flame, linetype, industrial burner, Westberg [Ref. 18] observed a sudden jump of 8 dB in sound-pressure level with increase in firing rate near the rated firing rate. This could be interpreted as the flame being partly unstable. The noise-spectrum data before and after the jump show little change in shape. This indicates that the overall sound-generating phenomenon was the same, but the change of the flame into an unstable operating condition resulted in an effective increase in the turbulence level in the combustion region.

Fricker [Ref. 19] discusses the noise produced by a swirl burner with natural gas injected axially at 200 m^3/hr. In this type of burner, the swirl produces a strong recirculation zone. In Type I flames of the two types produced, the gas penetrates the recirculation zone with subsequent burning, but leaves enough behind in the reverse flow region to burn and stabilize the flame. In Type II flames, the central fuel jet is stopped by the recirculation zone and spreads outward to give an intense blue flame. Because the Type II flame was noisy and the Type I, quiet, Fricker hypothesized that

(fuel + heat) + combustion air → quiet mixing controlled ignition
(fuel + combustion air) + heat → noisy "explosion" ignition.

The most noisy flame, however, was produced in the transition zone between the two types of flame, where neither type was stable. The sound level was up about 5 dB in the transition region.

Unpublished data obtained by the author on a 4,000,000 Btu/hr nozzle-mix burner indicated about a 10 dB increase in noise level when the unstable region of combustion was entered. However, from an appearance standpoint, and from a performance standpoint relative to heat output, the burner was still operating satisfactorily.

To summarize, data are not yet available to permit a prediction of the increase in decibel level that can be expected when a burner goes into the zone of metastable or unstable operation. Available data indicate that anything from a three to a ten decibel increase can be expected. They also indicate that the basic phenomenon is still one of combustion roar. The change comes about from an increase in the effective level of turbulence at the point of burning. Thus, one would expect that the basic criterion in reducing the noise in cases where the partially unstable burning has to be accepted would be the same criterion of noise reduction one would use for reducing combustion roar.

COMBUSTION AMPLIFICATION OF PERIODIC FLOW PHENOMENA

There are several variations on the process of noise amplification of periodic flow phenomena by the flame. Because of the lack of understanding of the details of this amplification process, these several variations have been grouped in only three subclasses (a) periodic flow phenomena resulting from an input of a high frequency noise to the combustion system, (b) input of a periodic phenomenon by a Strouhal-type vortex shedding in the flow system, upstream of or at the location of the combustion process, and (c) periodic flow phenomena resulting from the swirl flow in swirl-type combustors.

External Noise Source

Briffa and Romaine [Ref. 16] showed that subjecting a flame to an ultrasonic frequency (30,000 Hz) suppressed the combustion roar as much as 3 dB at 200 Hz and 10 dB in the audible range above 600 Hz. At the same time, the ultrasonic frequency was amplified. However, the presence of two input frequencies resulted in the flame amplifications of not only the beat frequencies but subharmonics thereof.

Giammar and Putnam [Ref. 10] observed that the imposition of jet noise in the mixture flow line upstream of a flame produced a decrease of up to 3 dB in the amplitude of the combustion-roar spectrum, while at the same time the jet noise itself was amplified by the flame by as much as 25 dB at its characteristic frequency. In this case the jet noise was within the audible range, and this made the burner much more noisy. Examination of data on diffusion flames wherein the fuel was supplied by a multiple set of spuds showed the same type of phenomena to be present.

Strouhal-Related Phenomenon

In small turbulent flames, a periodic structure with a frequency spectra having a Strouhal-like characteristic can occur. See Equation (10-2). That is, the peak frequency will have a linear functional relation with the flow velocity. Because the combustion roar may also be present at about the same level of intensity, and in the same frequency range, a clearcut indication of a vortex shedding (Strouhal-like) phenomenon is not always observed.

It is possible that the combustion noise observed by Smith and Kilham [Ref. 17] for turbulent flames from bunsen-like burners falls into this category. While their variation of peak frequency with flow velocity was less rapid than one might expect, the computed Strouhal numbers were of the order of the expected 0.2. Their data on three different tube sizes, over a range of flow rates, when plotted in terms of sound-power level, show a variation of the sound power with the square of the volume flow rate. This agrees with an observation on cold jets of a similar diameter by Giammar and Putnam. [Ref. 10].

Swirl-Burner Precession Noise

Precession of the vortex core in swirl burners was investigated as a possible source of noise by Syred and various co-workers [Ref. 20, 21], using both cold flow and combustion experiments. The air was always swirled in the experimental burners, but the fuel was admitted in various ways. For premixed flames, the mixture of fuel and air entered through the swirl vanes. For diffusion flames, fuel was added: (1) axially, (2) radially near the exit, or (3) tangentially. In the tangential case, the fuel was added through a tube on the axis of the tangential air inlet, terminating at the outer shell wall with the air-inlet termination.

These workers found that the precession noise observed in cold flow was suppressed by the various diffusion flames. However, the flames did produce the characteristic combustion-roar spectrum. With increasing firing rate and closer approach to overall stoichiometric conditions, the processing frequency was finally submerged completely in the roar spectrum. On the other hand, they point out that the premixed natural gas and air excite the precessing frequency, amplifying the noise intensity by as much as a factor of 20 times the level, without combustion. They found that a diffusion flame using town gas with a high percentage of hydrogen tends to act like a premixed flame, because of the high hydrogen diffusion rate.

Effect of Upstream-Generated Noise

The noise contribution of the air-handling system to the combustion system noise is often overlooked. The aerodynamic (flow) noise generated by the flow of fuel gas and combustion air through the various components of the system can be transmitted throughout the entire combustion system and can be amplified by the combustion process as much as 20 dB; it simultaneously can excite standing waves in the burner, the furnace, and the room. Among the components of the system that should be considered as noise sources are (1) valves, meters, and orifices; (2) bends, obstacles, restrictions and bellows in the ducting; and (3) fans.

Aerodynamic noise is associated with the level at which turbulence is generated in the flow stream. Naturally, any obstacle or obstruction that disturbs the smooth flow contours creates noise. This aerodynamic noise usually occurs at higher frequencies (over 1000 Hz) than combustion roar and is more objectionable to an observer. (The human ear is most sensitive to noise in the 1000 to 4000 Hz range.) The amplification of upstream noise due to a combusting flame can be estimated from the following relation;

$$\text{Amplification}, A = 20 \log_{10} \left(\frac{N_d T_d}{N_u T_u}\right) \text{ dB} \qquad (10\text{-}10)$$

where:

A = the amplification of upstream sound due to the combustion process, dB
N_d = number of moles of gas beyond the flame front after combustion
N_u = number of moles of gas (fuel and air) upstream before combustion
T_d = temperature after the combustion process, degrees Kelvin (°C + 273)
T_u = Temperature of gases before the combustion process, degrees Kelvin.

Thus, the flame amplification involves the ratio of temperatures after the flame front to the temperature upstream of the combustion process and the ratio of the number of moles of gas in the combustion products to the number of moles in the fuel and air.

The flame temperature, T_d, may be calculated from enthalpies of combustion or empirical data for flame temperature may be used. The mole ratio N_d/N_u depends on the type of fuel and oxidizer employed. For a hydrocarbon fuel and air, the mole ratio may be calculated using the reaction equation. If the hydrocarbon fuel is given by $C_\alpha H_\beta$, a model for amplification can be developed. The amount of air in the combustion process is usually specified by the air-to-fuel ratio, AF, by volume, or by the ratio $AF/AF_{\min}$, where $AF_{\min}$ is the minimum air-to-fuel ratio to support complete combustion of the carbon and hydrogen elements in the fuel. This amount of air is sometimes called "Theoretical Air."

Two cases exist for the hydrogen fuel.

Case 1

The air-to-fuel ratio is greater than the minimum conditions for complete combustion; $AF \geqslant AF_{\min}$

For this case,

$$\frac{N_d}{N_u} = \frac{\beta/4 + AF}{1 + AF} \qquad (10\text{-}11)$$

where:

β = Number of hydrogen atoms in hydrocarbon $(C_\alpha H_\beta)$

Case 2

The air-to-fuel ratio is less than the minimum for complete combustion:
$AF < AF_{min}$ In this case there is more fuel than can be burned by the available oxygen. For this situation,

$$\frac{N_d}{N_u} = \frac{1 + AF + \frac{AF}{AF_{min}}(\beta/4 - 1)}{1 + AF} \qquad (10\text{-}12)$$

For methane, CH_4, as a fuel the number of moles downstream of the flame front is equal to the number of moles upstream of the flame front as seen by taking $\beta = 4$ in Equations (10-11) and (10-12). The amplification of upstream generated noise here is simply the ratio of flame temperature to fuel- and air-supply temperature. Amplification values, in decibels, are shown for a methane-air flame in Table 10-3, both for calculated and measured values. [Ref. 22].

Table 10-3. Acoustic Amplifications of a Methane-Air Flame

Air-to-Fuel Ratio, % Theoretical Air	Temperature of Primary Flow, °C	Flow Temperature, °C	Calculated Flame Amplification (dB)	Measured Flame Amplification (dB)
80	25	1980	17.5	15
90	25	1990	17.6	15
100	25	1955	17.5	15
110	25	1845	17.0	14
120	25	1760	16.7	13
130	25	1665	16.3	13
140	25	1580	15.9	13

Identification of Combustion Noise

Table 10-4 shows the four types of combustion noise delineated in the discussion, and some key ways of identifying each. Combustion-driven oscillations are characterized by relatively high conversion efficiencies of chemical energy to noise, specific frequencies with associated standing and/or traveling waves, and a variety of feedback cycles. Combustion roar is characterized by a broadband spectrum and a lower efficiency of conversion of chemical energy to noise. The intensity of turbulence (and thus the turbulence-inducing pressure drop) has a major effect on noise level. Large noise amplifications at natural frequencies of burner tiles and furnace enclosures can occur. Unstable combustion noise is indicated by flow changes that bring on an incipient flame instability. The sudden significant increase in noise level is related to the increase in turbulence level in the combustion region; thus, the spectral characteristics resemble those of combustion roar. The telltale characteristic of combustion-amplified noise of a periodic flow phenomenon is the frequency of the noise source without combustion. However, it must be noted that noise sources in the vicinity of the flame may have their frequency altered by the flame.

Table 10-4. Classification of Combustion Noise

Acoustic Source	Efficiency, η	Frequency Spectrum	Comments
Combustion-driven oscillations	10^{-4}	Discrete frequencies	Feedback cycle
Combustion roar	10^{-8} to 10^{-6}	Smooth*	Highly dependent on turbulence level
Unstable combustion noise	High end of combustion roar	Smooth*	Close to limit of flame stability
Combustion amplification of periodic flow phenomena	Unknown	Discrete frequencies**	Driven by separate signal source

*There can be amplification at natural frequencies of enclosures.
**Amplification of broad-band frequency sound can occur.

After the general class of combustion source is identified, specific possibilities of control can be investigated. If combustion-driven oscillation occurs, they will ordinarily be of such high intensity that they will have to be eliminated. But even after eliminating the occasional combustion-driven oscillations, the possible unstable flame noise, and the often-confusing jet noise and flow noise, there remains a basic combustion roar associated with a given intensity of combustion and thermal output. At present, one can prevent amplification by configuration effects. In some applications, it might also be possible to decrease the intensity of combustion and thus, the combustion roar. Beyond these approaches one must resort to muffling.

MEASUREMENT OF COMBUSTION-SYSTEM NOISE

Many difficulties are associated with the measurement of combustion noise from a burner in either a laboratory or a field-installation situation. To alleviate some of these difficulties, the following suggestions are made with the aim that sufficient data can be acquired within a minimal amount of time so that noise generated by a given combustor in its environment can be explained unambiguously.

1. A dimensional sketch should be made of the environment, including doors, windows, etc. Floor, wall, and ceiling finishes should be noted. Photographs of the installation are recommended.
2. All instruments should be checked out and calibrated so that absolute values may be recovered from the data. The circuitry used in the study should be sketched and the instrument identified as to type. Specific identification numbers should be recorded.
3. The ambient noise spectra at each working station without the burner in operation should be obtained.

Ch. 10 COMBUSTION AND FURNACE NOISE 471

4. Accurate flow measurements of air and fuel should be made. Barometric pressure, humidity, fuel characteristics, and similar data should also be obtained.
5. The noise of the unit should be recorded with (1) air flow only, (2) if safe, with the fuel flow only, without burning, and (3) air and fuel flow together without burning. The flow rates should match those used in the combustion studies.
6. When possible, data should be obtained at a variety of firing rates and fuel/air ratios.
7. When more than one variation in design is studied, these differences should be carefully noted.
8. Microphone location should be varied to check for near-field (sound-power level is not a simple function of distance from source), or far-field (sound-power level varies inversely with the square of the distance from source), and room (sound-power level varies with mode pattern) noise generation for the frequency of interest.
9. If possible, the response to a white noise source at the burner location should be obtained at the microphone location.
10. A check should be made for noise radiation from flexible tubes, ducts, and connectors. This may not be simple.
11. Strong coupling or response of room modes should be checked. Several microphone traverses (wall-to-wall and floor-to-ceiling) should be made if room modes cannot be easily distinguished. Room modes sometimes can be identified from peak frequencies (on a narrow-band analysis) that can be associated with dimensions of the room. Sometimes, this can be done merely by slowly hand-carrying the microphone across the room while the analyzer is set at a prescribed frequency.
12. In high-noise-level environments, structural response at natural frequencies can occur and can transport signals mechanically or acoustically to the instruments. Sound-analysis equipment should be shielded (possibly in another room), while the microphone should be so mounted that it does not vibrate.
13. Detailed drawings of the burner and installation should be obtained. The unit should be inspected thoroughly, noting details of construction and assembly. The designation of the unit should be recorded.
14. To insure that the above procedures are followed, a standard form or data sheet should be filled out at the time the measurements are taken.

TECHNIQUES FOR DESIGNING A LOW-NOISE-LEVEL COMBUSTION SYSTEM

Based upon the considerations discussed above, specific techniques can be suggested for suppressing the noise of a combustion system. To suppress this noise, noise sources in each of the major elements should be reduced to a minimum. This is because low-level noise generated in one of the system elements may be amplified in another. The following listing summarizes noise-suppression techniques.

Air- and Gas-Handling Systems

1. Install "quiet" valves that are currently on the market and select proper valve body to keep flow velocities to a minimum.
2. Eliminate obstacles, sharp edges, orifices, and other turbulence generators in ducts.
3. Lag acoustical insulation around regulators, valves, ducting, and other flow components that cannot be muffled internally.
4. Minimize the number of turns, elbows, headers, and sudden expansions.
5. Use rigid ducting when noise abatement is of prime concern and flexible ducting when vibration abatement is of prime concern.
6. Install in-line mufflers where possible.

Burner Design

1. Utilize all the available momentum for effective mixing; that is, the mixing zone and combustion region should coincide.
2. Minimize flow velocity through the burner (use the largest possible fuel and/or air jets for desired flame shape).
3. Avoid all unnecessary turbulence generation within the burner.
4. Design burner-tile refractories so that the geometry of the tile does not set up strong natural modes of oscillation.
5. Utilize the burner tile as a muffler when possible, either as an absorptive or reactive type.
6. Consider alternative combustion schemes such as recirculation or two-stage combustion.

Operating Environment

1. Relocate a furnace within a room or a burner within a furnace, so that strongly driven modes of oscillations do not exist.
2. Change the dimensions of furnace to reduce standing waves; extreme discretion must be used, however.
3. Use auxiliary walls to change the natural room frequency.

11

Fluid Piping System Noise

E. E. ALLEN

Introduction

Fluid process and transmission systems are a major source of industrial noise. Elements within the system that contribute to the noise are: control valves, abrupt expansions of high-velocity flow streams, compressors, and pumps. This chapter deals with control-valve noise, a principal source of noise in many fluid systems. Control-valve noise is a result of the turbulence introduced into the flow stream in producing the permanent head loss required to fulfill the basic function of the valve. Specific topics covered are: identification of the source of a valve noise, relation of gross flow parameters to the noise potential of a valve, techniques for prediction of control-valve noise, methods of noise control, and noise analysis of existing fluid transmission systems.

SOURCES OF VALVE NOISE

The major sources of control-valve noise are:
1. Mechanical vibration of valve components
2. Fluid noise
    a. Hydrodynamic noise
    b. Aerodynamic noise

Mechanical Noise

Vibration of valve components is a result of random pressure fluctuations within the valve body and/or fluid impingement upon the movable or flexible parts. The most prevalent source of noise resulting from mechanical vibration is the lateral movement of the valve plug relative to the guide surfaces. Sound produced by this type of vibration will normally have a frequency less than 1500 hertz and is often described as a metallic rattling. The physical damage incurred by the valve plug and/or associated guide surfaces is generally of more concern than the noise emitted.

Early control valves commonly employed valve plugs having skirts into which flow openings were cast or machined. These skirts guided the plug in the valve body ports. Relatively large clearances between the skirts and the body guides made this construction quite susceptible to vibration. An improvement was effected when the guiding was transferred from skirts to posts incorporated into each end of the valve plug. These posts were guided by bushings rigidly held in the bonnet and blind flanges of the body. For difficult valve applications a further improvement was made by increasing the diameter of the posts and reducing clearances as much as possible. Today's more-or-less standard control-valve features *cage* guiding. In this design a cage member containing the flow openings is rigidly retained in the valve body, and the movable valve plug is closely guided on its inside diameter. These three types of construction are illustrated in Fig. 11-1. As a result of the evolution in valve trim design, vibrational problems resulting from lateral movement of the plug have been minimized.

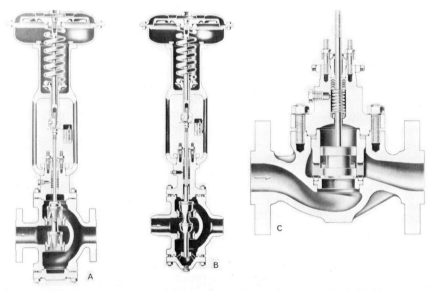

Fig. 11-1. Examples of: (A) Skirt guiding, (B) Post guiding, (C) Cage guiding.

A second source of mechanical vibration noise derives from a valve component resonating at its natural frequency. Resonant vibration of valve components produces a single pitched tone, normally having a frequency between 3000 and 7000 hertz. This type of vibration produces high levels of stress that may ultimately produce fatigue failure of the vibrating part. Valve components susceptible to natural frequency vibration are contoured valve plugs with hollow skirts and flexible members, such as the metal seal ring of a ball valve.

In summary, noise that is a by-product of the mechanical vibration of valve components is:

1. Not predictable

Ch. 11 FLUID PIPING SYSTEM NOISE 475

2. Of secondary concern relative to the physical damage that can occur
3. Even deemed beneficial in that it warns of existing conditions which could produce valve failure
4. For the most part, eliminated by improved valve design.

Hydrodynamic Noise

Control valves handling liquid flow streams can be substantial sources of noise. The flow noise produced is referred to as *hydrodynamic noise* and may be categorized with respect to the specific flow classification or characteristic from whence it is generated. Liquid flow can be divided into three general classifications:

1. Non-cavitating
2. Cavitating
3. Flashing.

Non-cavitating liquid flow generally results in very low ambient noise levels. It is generally accepted that the mechanism from which the noise is generated is a function of the turbulent velocity fluctuations of the fluid stream, which are usually referred to as the *Reynolds stresses* or *turbulent momentum flux*. The high intensity turbulence in control valves occurs as a result of the rapid deceleration of the fluid, downstream of the vena contracta, due to the abrupt area change.

The major source of hydrodynamic noise is *cavitation*. This noise is caused by the implosion of vapor bubbles that are formed in the cavitation process. Cavitation occurs in valves controlling liquids when the service conditions are such that the static pressure, downstream of the valve, is greater than the vapor pressure and at some point within the valve the local static pressure, either because of high velocity and/or intense turbulence, is less than or equal to, the liquid vapor pressure.

Figure 11-2 depicts the pressure profile of a cavitating flow stream as a function of distance along the stream. Vapor bubbles are formed in the region of

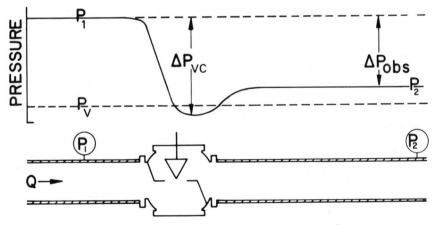

Fig. 11-2. Static pressure along a stream line cavitating flow.

476 FLUID PIPING SYSTEM NOISE Ch. 11

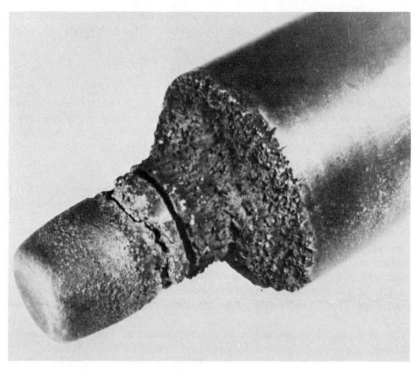

Fig. 11-3. Example of cavitation damage to a valve plug.

minimum static pressure and subsequently are collapsed or imploded as they pass downstream into an area of higher static pressure. Noise produced by cavitation has a broad frequency range and is often described as a rattling sound similar to that which would be heard if gravel were present in the fluid stream.

Cavitation may produce severe damage to the solid boundary surfaces that confine the cavitating fluid. Thus, generally speaking, the noise produced by cavitation is of secondary concern. Physical damage, the result of cavitation erosion, is depicted in Fig. 11-3.

Flashing is a phenomenon that occurs in liquid flow when the differential pressure ΔP, across a restriction is greater than the differential between the absolute static and vapor pressure at the inlet to the restriction, $\Delta P > P_1 - P_v$ where P_1 is the absolute static pressure and P_v the vapor pressure at the inlet. The resulting flow stream is a mixture of the liquid and gas phases of the fluid. Noise resulting from a valve handling a flashing fluid is a result of the deceleration and expansion of the two-phase flow stream.

Test results supported with field experience, indicate that noise levels from non-cavitating liquid applications are quite low and generally would not be considered a noise problem. Figure 11-4 depicts the typical characteristic of hydrodynamic noise as a function of the ratio of differential pressure across the valve

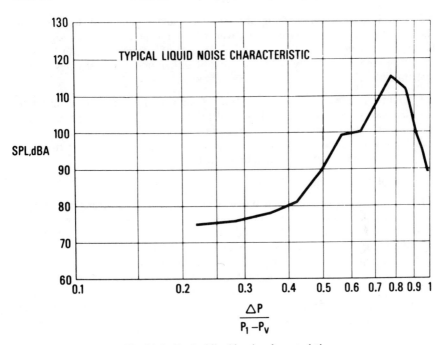

Fig. 11-4. Typical liquid noise characteristic.

(ΔP) to the static pressure at the inlet (P_1 - psia) minus the vapor pressure (P_v - psia).

Aerodynamic Noise

The major source of valve noise is *aerodynamic noise*. This is a noise generated as a by-product of a turbulent gas stream, or a noise produced without the interaction of the fluid with vibrating boundaries or other external energy sources.

Aerodynamic noise is a result of the Reynolds stresses or shear forces created in the flow stream as a result of deceleration, expansion, or impingement. The principal area of noise generation in a control valve is the recovery region immediately downstream of the vena contracta where the flow field, characterized by intense turbulence and mixing, is of chaotic quality with phase being completely random and discontinuous.

VALVE NOISE PREDICTION

Aerodynamic Noise

Most of the theories, to date, on aerodynamic noise have been developed with special reference to the noise of free jets as opposed to a gas stream restrained by solid walls. A standard reference on this topic is Lighthill's paper [1].

Several hypotheses describing the acoustic power generated by gas flow through a control valve have been advanced on the postulates: (a) Lighthill's theory is applicable; and (b) the turbulent level of a flow stream can be accurately defined analytically. The validity of either or both postulates is very questionable, because:

1. Lighthill's theory is based on a free jet and would not appear applicable to the vastly more complex flow field produced by high subsonic and sonic gas streams contained within the solid boundaries of a control valve.
2. Very little, either qualitatively or quantitatively, is known concerning the turbulence level of a bounded fluid stream.

Empirical Advantage

Development of an analytical model which accurately describes the precise mechanism of aerodynamic noise generation by control valves is very unlikely, in view of the complexity of the problem and the lack of knowledge concerning absolute values of the relevant parameters. It would seem obvious, therefore, that an accurate technique for predicting valve noise must be based on empiricism. Such an approach requires the noise characteristic for each valve style and adjacent piping configuration to be established by actual test. Obviously, it is neither practical nor feasible to determine noise levels for all valve applications by laboratory tests. Dimensional analysis, a very useful and powerful tool in formulating problems which defy analytical solution and must be solved experimentally, would be the logical tool for correlation and extrapolation of the test data.

In fluid mechanics, the relevant parameters in any physical situation may be combined into independent, dimensionless groups that characterize the flow. Two flow streams are considered similar, if they are geometrically similar and if the dimensionless groups have constant values for both flows. This technique of expressing a physical relationship between quantities by utilizing dimensionless ratios of relevant parameters, is the field of mathematics known as *dimensional analysis*.

Dimensionless groups (or as they are called, numbers) may be obtained directly either from the differential equations that describe the flow field or by simply combining the relevant variables into as many independent dimensionless numbers as possible. An advantage of the latter system is that one does not need to know the governing equations or the laws for complex problems; however, one must identify all the variables.

The variables to be considered for a dimensional analysis approach to the problem of predicting aerodynamic noise generated by compressible fluid flow through control valves are p_s, ΔP, V, ρ, D, μ, c, c_p, c_v and λ. The independent dimensionless numbers formed from the above list are:

1. Prandtl number, $(P_r = c_p \, \mu/\lambda)$
2. Reynolds number, $(R_e = \rho V D/\mu)$
3. Mach number, $(M = V/c)$
4. Specific heat ratio, $(k = c_p/c_v)$
5. Ratio of sound pressure to differential pressure across valve, $(p/\Delta P)$.

Ch. 11 FLUID PIPING SYSTEM NOISE

Relative to the generation of aerodynamic noise, the significance of each of the dimensionless numbers listed above is briefed below.

1. The Prandtl number, P_r, is nearly the same for all gases and can be considered insignificant in the analysis of processes where heat transfer does not play an important role. Generation of aerodynamic noise is considered an adiabatic process and thus P_r can be dropped from consideration.
2. Reynolds number, R_e, is a measure of the ratio of inertia forces to viscous forces. Thus, in the study of compressible fluids, the role of the Reynolds number as an important parameter is limited to areas concerning low velocities or the boundary layer. For control-valve noise levels of concern, the principal noise-generation mechanism is located in the area of high subsonic or supersonic velocity, thus the Reynolds number would not be considered a relevant parameter.
3. From the preceding argument it is obvious that the mach number, M, should be considered the most important factor. It can be shown that M can be presented as a function of ΔP and P_1 by $\phi(\Delta P/P_1)$.
4. Not enough experimental data have been obtained to accurately define the influence that the specific heat ratio, k, has on noise generation; however, it has a relatively weak influence on the flow stream characteristic.

Based on the dimensional analysis arguments presented above the acoustic power generated by compressible flow through a control valve obeys the following relationship:

$$W \propto C_g^2 (\Delta P)^2 \, \phi(\Delta P/P_1) \qquad (11\text{-}1)$$

where C_g is the gas sizing coefficient, which can be found from the following three equations:

For Gas Flow

$$C_g = \frac{Q}{P_1} \sqrt{\frac{GT}{520}} \; \frac{1}{\sin\left(\frac{3417}{C_1}\sqrt{\frac{\Delta P}{P_1}}\right)_{\text{degrees}}} \qquad (11\text{-}2)$$

For Steam and Vapor Flow

$$C_g = \frac{Q_s}{1.06\sqrt{d_1 P_1}} \; \frac{1}{\sin\left(\frac{3417}{C_1}\sqrt{\frac{\Delta P}{P_1}}\right)_{\text{degrees}}} \qquad (11\text{-}3)$$

For Liquid Flow the liquid sizing coefficient, C_V, is given by:

$$C_V = Q_L \sqrt{\frac{G_L}{\Delta P}} \qquad (11\text{-}4)$$

where:

C_g = Gas sizing coefficient
C_V = Liquid sizing coefficient

$C_1 = \dfrac{C_g}{C_V}$ [From Table 11-1]

Q = Gas flow rate, scfh
Q_s = Steam flow rate, lb per hr
Q_L = Liquid flow rate, gpm
G = Gas specific gravity [From Table 11-2]
G_L = Liquid specific gravity (water = 1.0)
P_1 = Valve inlet pressure, psia
ΔP = Pressure drop across valve, psi
T = Absolute temperature of the gas at the valve inlet, deg Rankine
d_1 = Density of steam or vapor, lb per cu ft.

If the sine term in Equations (11-2) and (11-3) equals or exceeds 90 degrees, critical flow is indicated and the angle must be limited to 90 degrees for subse-

Table 11-1. Representative C_1 Values for Various Valve Configurations

Body Design	Body Style	Description	Flow Direction	Representative C_1
"A"	Globe	Single Port	Open	35.0
"A"	Globe	Double Port	Normal	35.0
"AA"	Angle	Specs. B & D	Open	35.5
"AA"	Angle	Specs. B & D	Close	30.5
"AA"	Angle	Specs. C & E	Close	17.5
"BF"	Globe	Single Port	Open	32.0
"D"	Globe	Any Valve Plug	Open	30.0
"D"	Globe	Micro-Form	Close	21.5
"D"	Globe	Micro-Flute	Close	24.0
"DA"	Angle	Micro-Form	Close	18.5
"DA"	Angle	Micro-Flute	Close	22.5
"DBQ"	Globe	Single Port	Open	33.0
"DBAQ"	Angle	Single Port	Open	34.5
"GS"	Globe	Single Port	Open	35.0
"IC"	Globe	Single Port	Open	33.5
"U"	150 & 300 lb	Vee-Ball Valve	...	22.0
V25	Ball	Hi-Ball	...	20.0
461	Angle	Single Port	Close	18.0

Note: The representative C_1 value for Fisher valve bodies not listed above is 35.0. C_1 values for Continental Equipment Company butterfly valves are shown in their Technical Data Bulletin 141.05.

Table 11-2. Values for G for Various Gases

Gas	G
Acetylene	0.90
Air	1.00
Butane	2.00
Ethane	1.03
Helium	0.14
Hydrogen	0.07
Methane	0.55
0.6 G Natural Gas	0.60
Nitrogen	0.97
Oxygen	1.10
Propane	1.52
Propylene	1.45

quent calculations. The rationale used in the development of the above relationships assumed a single port or noise generator. In the following equation an extension is made to the basic relationship to include the effects of multiple ports.

$$W \propto \frac{1}{N} C_g^2 (\Delta P)^2 \phi \left(\frac{\Delta P}{P_1}\right) \qquad (11\text{-}5)$$

where: N = number of ports open to the flow stream.

AERODYNAMIC NOISE PREDICTION TECHNIQUE

Graphical solution of the following equation, based on relationship Equation (11-1) and the arguments above, provides a very expedient and accurate technique for predicting ambient noise levels resulting from the flow of compressible fluids through control valves.

$$SPL = SPL_{\Delta P} + \Delta SPL_{C_g} + \Delta SPL_{\Delta P/P_1} + \Delta SPL_K \qquad (11\text{-}6)$$

where:

SPL^1 = overall noise level in dB at a predetermined point (48 in. downstream of the valve outlet and 29 in. from the pipe surface)
$SPL_{\Delta P}$ = base SPL in dB, determined as a function of ΔP.
ΔSPL_{C_g} = correction in dB for C_g.
$\Delta SPL_{\Delta P/P_1}$ = correction in dB for valve style and pressure ratio.
ΔSPL_K = correction in dB for acoustical treatment; i.e., heavy wall pipe, insulation, inline silencers, etc. See Table 11-3 for correction for pipe wall attenuation.

[1] Editor's Note: Throughout this handbook the symbol L_p is used for sound-pressure level while in this chapter SPL is used, with appropriate subscripts. In the fluid piping and control-valve industry SPL is usually used in company literature instead of the standard symbol, L_p.

Table 11-3. ΔSPL_K Correction for Pipe Wall Attenuation

Nominal Pipe Size, in.	SCH 10	SCH 20	SCH 30	SCH 40	SCH 60	SCH 80	SCH 100	SCH 120	SCH 140	SCH 160	STD	XS	XXS
1				0		-4.5				-9.5	0	-4.5	-15.5
1½				0		-4.5				-9.5	0	-4.5	-15.5
2				0		-5				-11.5	0	-5	-15.5
3				0		-4.5				-10	0	-4.5	-15
4				0		-5		-8.5		-11.5	0	-5	-15
6				0		-6		-9.5		-13	0	-6	-16
8		+3.5	+2.0	0	-3	-6	-8.5	-11	-13	-14.5	0	-6	-14
10		+5.0	+2.5	0	-4.5	-6.5	-9.5	-11.5	-14	-15.5	0	-4.5	
12		+5.5	+1.5	-1	-5.5	-8	-11	-13.5	-15	-17.5	0	-4	
14	+5.5	+2.5	0	-2	-6	-9.5	-12.5	-14.5	-16.5	-18.0	0	-4	
16	+5.5	+2.5	0	-4	-7.5	-11	-13.5	-16	-18.5	-20	0	-4	
18	+5.5	+2.5	-2	-5.5	-9.5	-12.5	-15.5	-17.5	-19.5	-21.5	0	-4	
20	+5.5	0	-4	-6	-10.5	-13.5	-16.5	-19.0	-21.0	-22.5	0	-4	
24	+5.5	0	-5.5	-8	-12.5	-16	-19	-21.5	-23	-25	0	-4	
30	+2.5	-4	-7								0	-4	
36	+2.5	-4	-7								0	-4	
42				-9							0		

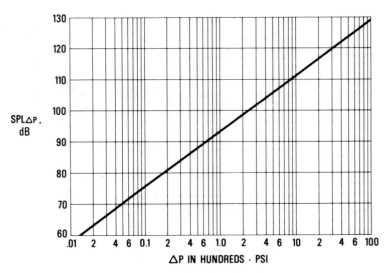

Fig. 11-5. Base *SPL*, all valve styles.

The following example serves as a guide for using the noise-prediction technique. Information presented in Figs. 11-5 through 11-16 facilitates the prediction technique.

The example below is presented to illustrate the procedure for predicting aerodynamic noise:

Given:

4" - 300 lb ANSI, Fisher Design ES (a cage-style globe body) installed in an 8 in. Sch 40 pipeline, required flow, Q = 600,000 scfh; temperature T = 60 °F.

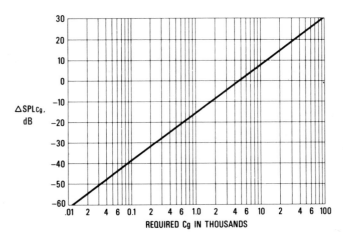

Fig. 11-6. SPL_{CG} correction, all valve styles.

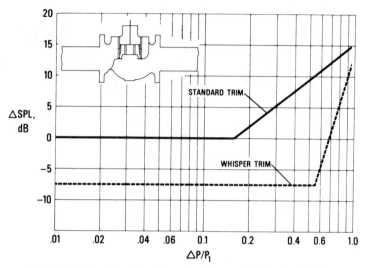

Fig. 11-7. Cage style globe valves—line size equals body size (4–12 in.)

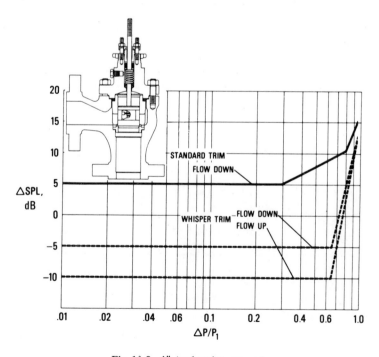

Fig. 11-8. 4" Angle valve—swaged.

Ch. 11 FLUID PIPING SYSTEM NOISE 485

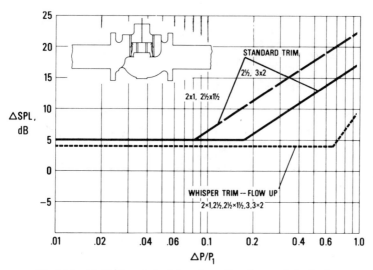

Fig. 11-9. 2″–3″ Cage style globe valves—line size equals body size.

$P_1 = 615$ psia
$\Delta P = 348$ psi

$$C_g = \frac{Q}{P_1}\sqrt{\frac{GT}{520}}\,\frac{1}{\sin\left(\frac{3417}{C_1}\sqrt{\frac{\Delta P}{P_1}}\right)_{\text{degrees}}} \quad \text{[Equation (11-2)]}$$

$C_1 = 35.0$ [Footnote to Table 11-1]

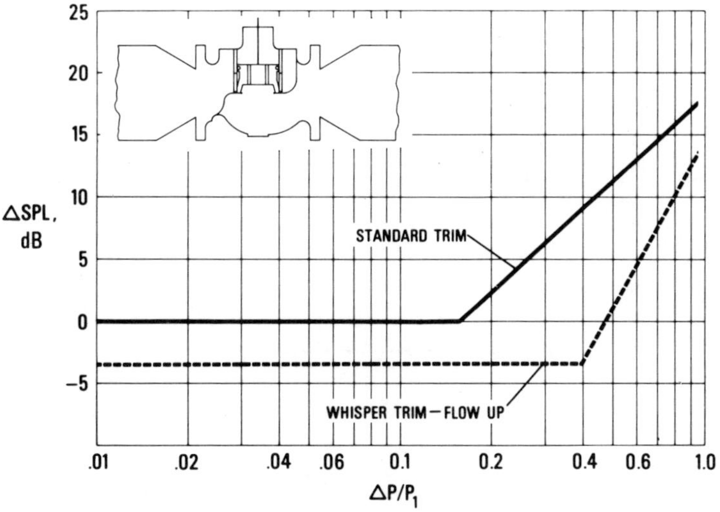

Fig. 11-10. 3″–6″ Cage style globe valves—swaged.

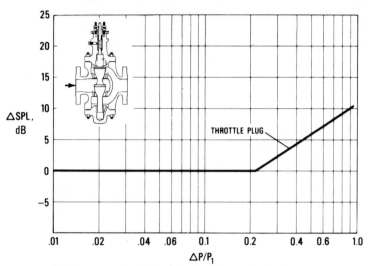

Fig. 11-11. Top and bottom guided globe valves—all line to body size ratios (all valve sizes).

$$\therefore C_g = \frac{600,000}{615} \sqrt{\frac{1.0 \times 520}{520}} \frac{1}{\sin\left(\frac{3417}{35}\sqrt{\frac{348}{615}}\right)_{\text{degrees}}}$$

$$= 1018$$

$$\frac{\Delta P}{P_1} = .566$$

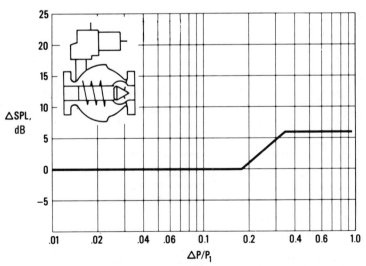

Fig. 11-12. Sliding sleeve regulators—all line to body size ratios (all valve sizes).

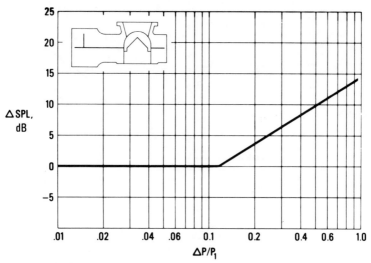

Fig. 11-13. V-Notch ball valves—all line to body size ratios (all sizes).

$\Delta SPL_{\Delta P} = 103.5$ [Fig. 11-5]
$\Delta SPL_{C_g} = -5$ dB [Fig. 11-6]
$\Delta SPL_{\Delta P/P_1} = +9.5$ dB [Fig. 11-10]
$\Delta SPL_K = 0$ db [Table 11-2]
$SPL = [103.5 + (-5) + 9.5 + 0]$ dB
$SPL = 109$ dB

As a possible alternate valve selected to minimize noise, we will repeat the above problem using an 8 in. × 4 in. Design EW (a cage-style globe body)

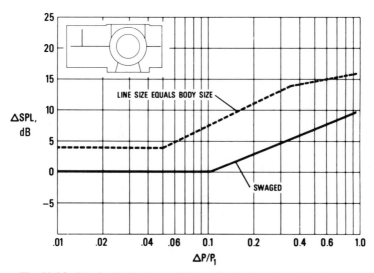

Fig. 11-14. Standard ball valves—all line to body size ratios (all valve sizes).

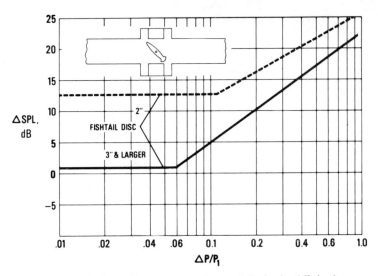

Fig. 11-15. Butterfly valves—line size equals body size (all sizes).

with Whisper Trim®. Note: the only change in the calculation will be the $\Delta SPL_{\Delta P/P_1}$ value.

$\Delta SPL_{\Delta P/P_1} = -7.5$ [Fig. 11-7]
$SPL = [103.5 + (-5) + (-7.5) + 0]$ dB   [Equation (11-6)]
$SPL = 91$ dB

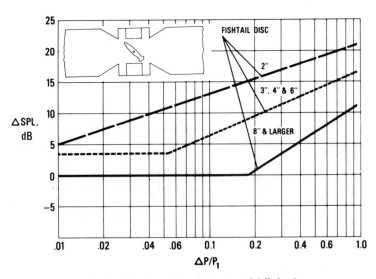

Fig. 11-16. Butterfly valves—swaged (all sizes).

If we were to use Schedule 80 pipe from Table 11-2,

$\Delta SPL_K = -6$ dB
$SPL = (91 - 6)$ dB
$SPL = 85$ dB

Hydrodynamic Noise Prediction

As with aerodynamic noise, a purely mathematical approach to the prediction of hydrodynamic noise would be questionable due to approximations and assumptions that must be made during the course of derivation. However, by considering a dimensional analysis approach, some insight into the governing parameters can be achieved. The variables considered as possibly pertinent to hydrodynamic noise generated by a control valve are:

$$p, P_1, P_2, P_v, V, \rho, D, E, \sigma, \nu$$

The above variables can be combined to form independent dimensionless numbers:

1. Reynolds number $\quad (R_e = VD/\nu)$
2. Cauchy number $\quad (C_u = \rho V^2/E)$
3. Weber number $\quad \left(W_e = \dfrac{\rho D V^2}{\sigma}\right)$
4. Euler number $\quad \left(E_u = \dfrac{\rho V^2}{2(P_1 - P_v)}\right)$
5. Ratio of the rms pressure fluctuation to the pressure drop. $\quad (p/\Delta P)$

Not all of the above dimensionless numbers are pertinent in describing hydrodynamic noise generation and scrutinization of each will result in eliminating those of little importance.

1. The Reynolds number, R_e, is the ratio of inertia forces to viscous forces. For the most part, the viscosity effect on the flow is unimportant for fully developed turbulent flow. Since turbulent flow exists in all control valves of interest from a noise standpoint, Reynolds number will be dropped from consideration.
2. The Cauchy number, C_u, is the ratio of inertia force to the compressibility force of the fluid. The Cauchy number becomes very important for numbers approaching 1.0 and relatively unimportant for small numbers less than 0.1. Since the Cauchy number for most liquid flow applications is less than 0.1, the Cauchy number can be ignored. However, some error may result as flow conditions approach flashing and downstream velocities become quite high.
3. The Weber number, W_e, equivalent to the ratio of inertia force to the surface tension force, becomes important when considering small gaseous cavitation and this should play a minor role in influencing the noise gener-

ated in control valves since the inertia forces far exceed the surface tension forces.

4. The Euler number, E_u, is the ratio of inertia forces to the pressure forces affecting the flow stream. Since pressure conditions are of major importance when describing the occurrence of cavitation, the Euler number, as expressed above, is probably the most important of the dimensionless numbers discussed thus far and will be of major interest.

Based on the dimensional analysis arguments briefed above the acoustic power generated by incompressible flow through a control valve obeys the following relationship:

$$W \propto C_V^2 \, \phi \left(\Delta P, \frac{\Delta P}{P_1 - P_v} \right) \qquad (11\text{-}7)$$

where:

W = Acoustic power, watts
C_V = Liquid sizing coefficient [see Equation (11-4)]
P_1 = Valve inlet pressure, psia
P_v = Vapor pressure at upstream conditions, psia
ΔP = Pressure drop across valve, psi

HYDRODYNAMIC NOISE PREDICTION TECHNIQUE

Based on the relationship seen in Equation (11-6), a technique of the same format used for aerodynamic noise is presented for hydrodynamic noise:

$$SPL = SPL_{\Delta P} + \Delta SPL_{C_V} + \Delta SPL_{\Delta P/(P_1 - P_v)} + \Delta SPL_K \qquad (11\text{-}8)$$

where:

SPL = overall noise level, in dBA, at a predetermined point (48" downstream of valve outlet and 29" from pipe surface)
$SPL_{\Delta P}$ = base SPL in dBA determined as a function of ΔP and $\Delta P/(P_1 - P_v)$
$\Delta SPL_{\Delta C_V}$ = correction in dBA for C_V
$\Delta SPL_{\Delta P/(P_1 - P_v)}$ = correction in dBA for valve style and flow regime
ΔSPL_K = correction in dBA for acoustical path treatment; i.e., heavy-wall pipe, insulation, etc.

The noise levels are based on standard-weight pipe; however, a correction factor, ΔSPL_K, to be used for other schedule of pipe is provided in Table 11-2.

Figure 11-17 is the basic SPL as a function of total pressure drop across the valve. Note that the base $SPL_{\Delta P}$ is also dependent upon the pressure drop ratio $\Delta P/(P_1 - P_v)$. The correction for C_V is shown in Fig. 11-18. Figures 11-19 through 11-23 are the corrections for a few valve styles as a function of the flow regime parameter $\Delta P/(P_1 - P_v)$.

Ch. 11 FLUID PIPING SYSTEM NOISE 491

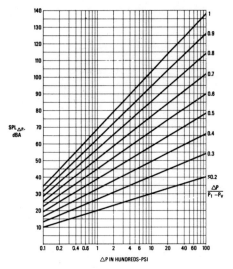

Fig. 11-17. Base $SPL_{\Delta P}$ for all valve styles.

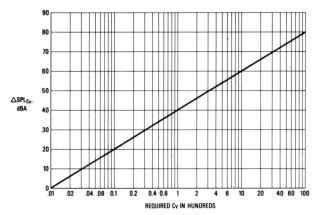

Fig. 11-18. ΔSPL_{C_V} correction for all valve styles.

An example problem is given below:

Given:

4″ - 300 lb ANSI, Fisher Design ED with Cavitrol® Trim installed in a 6″ Schedule 80 pipe, the flow is water at 120 °C, the flow rate, Q_L, is 2000 gpm. The valve inlet pressure is 600 psia, and the pressure drop, ΔP, across the valve, is 400 psi.

Solution: The vapor pressure, P_v, must be obtained from a reference which tabulates the vapor pressure of water at various temperatures. *The Handbook of*

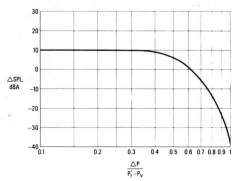

Fig. 11-19. $\Delta SPL_{\Delta P/P_1-P_v}$ correction for standard cage trim in a globe body, flow down.

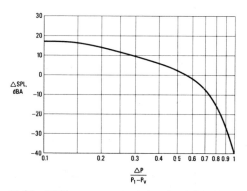

Fig. 11-20. $\Delta SPL_{\Delta P/P_1-P_v}$ correction for globe valve styles.

Chemistry and Physics is one useful reference. From such a source, the vapor pressure of water at 400 psia is found to be 100 psia.

C_V is computed from Equation (11-7).

$$C_V = Q_L \sqrt{\frac{G_L}{\Delta P}} \quad \text{[Equation (11-4)]}$$

$$= 2000 \sqrt{\frac{1.0}{400}}$$

$$= 100$$

$$\frac{\Delta P}{(P_1 - P_v)} = \frac{400}{(600 - 100)} = 0.8$$

$SPL_{\Delta P} = 74$ dBA   [From Fig. 11-17]
$\Delta SPL_{C_V} = 40$ dBA   [From Fig. 11-18]
$\Delta SPL_{\Delta P/(P_1-P_v)} = -22$ dBA   [From Fig. 11-21]
$\Delta SPL_K = -6$ dBA   [From Table 11-3]

Ch. 11   FLUID PIPING SYSTEM NOISE   493

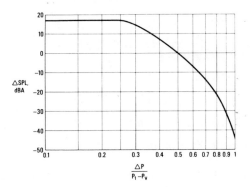

Fig. 11-21. $\Delta SPL_{\Delta P/P_1-P_v}$ correction for Cavitrol® i trim in a globe valve, flow down.

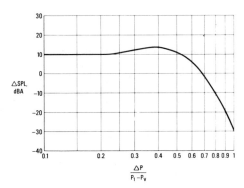

Fig. 11-22. $\Delta SPL_{\Delta P/P_1-P_v}$ correction for 4″ butterfly valve in a swaged setup.

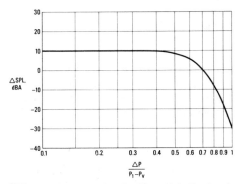

Fig. 11-23. $\Delta SPL_{\Delta P/P_1-P_v}$ correction for V-notch ball valves in swaged setup.

[From Equation (11-8)]

$$SPL = SPL_{\Delta P} + \Delta SPL_{C_V} + \Delta SPL_{\Delta P/(P_1 - P_v)} + \Delta SPL_K$$
$$= 74 + 40 - 22 - 6$$
$$= 86 \text{ dBA}$$

NOISE CONTROL

Noise control employs either one or both of the following basic approaches:

1. Source Treatment—Prevention or attenuation of the acoustic power at the source.
2. Path Treatment—Reduction of noise transmitted from a source to a receiver.

Figure 11-24 illustrates the treatment methods that can be used to reduce noise contained inside of a piping system. In this illustration, the source of the noise can be a pump or the valve shown, or a combination of the pump and valve noise. The source could also be caused by turbulent flow resulting from other

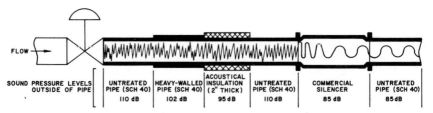

Fig. 11-24. Comparison of path treatment methods used to reduce aerodynamic noise.

elements in the piping system ahead of the valve shown. Valve-generated noise can be reduced by the installation of a quiet valve (see next paragraph); however, a quiet valve is not used for reducing upstream-generated noise. Downstream elements that can be used to reduce the noise are: heavier-walled pipe, acoustical insulation, or commercial silencers. In Fig. 11-24 the effect of these elements is illustrated by the decibel values shown; where the initial noise level at the untreated pipe is 110 dB. The treatment methods are covered in greater detail in the following paragraphs.

Quiet Valves[2]

From the preceding, we know that the parameters which determine the level of noise generated by compressible flow through a control valve for a given application are: the number of ports or restrictions exposed to the flow stream, the total C_g, the differential pressure across the valve, and the ratio of differential pressure to absolute inlet pressure.

---

[2] Comments made in this section are applicable to the design of valves that introduce turbulence into the flow stream in producing the permanent head loss required to fulfill the basic function of the valve. It is conceivable to design a valve that utilizes viscous losses to produce the permanent head loss required. Such an approach would require valve trim with a very high FL/D ratio which becomes impractical from the standpoint of both economics and physical size.

FLUID PIPING SYSTEM NOISE

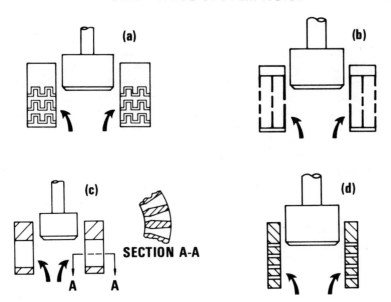

Fig. 11-25. Illustrations of value trim designed for noise attenuation.

Figure 11-25 shows four basic approaches to the design of quiet control valves. Common to all, is the use of a multiplicity of restrictions versus either a single or a few large restrictions.

Approach (a) utilizes a small tortuous path (Dia. $< \frac{1}{32}''$) providing a flow path with a high FL/D ratio designed to maximize the per cent of total pressure drop obtained by viscous stresses induced by the shearing action of the fluid and boundary layer turbulence.

The noise characteristic or noise potential of a regulator increases as a function of the square of the differential pressure (ΔP^2) and the ratio of differential pressure to absolute static pressure at the inlet ($\Delta P/P_1$). Thus, for high pressure ratio applications ($\Delta P/P_1 > 0.7$), an appreciable reduction in noise can be effected by staging the pressure loss through a series of restrictions to produce the total pressure head loss required. Concept (b) in Fig. 11-25 uses multiple series restrictions to limit the $\Delta P/P_1$ ratio across individual restrictions to optimum operating points and provide a favorable velocity distribution in the expansion area. Approaches (a) and (b) can easily be incorporated into cage-style trim fabricated from stacks of discs machined to stage the total pressure drop through a series of circumferential restrictions. It should be pointed out that both (a) and (b) act as strainers and are very susceptible to plugging as a result of either solid particles in the gas stream (dirty gas) or as a result of hydrate formation prior to the last stage.

The third approach, (c), shown in Fig. 11-25 uses a multiplicity of narrow parallel slots specifically designed to both minimize the turbulence level and provide a favorable velocity distribution in the expansion area. This is an economical approach to quiet valve design and can provide substantial noise reduction (15-20 dB) with little or no decrease in total flow capacity.

It can be shown that the acoustic power of a single flow restriction increases as a function of area squared. Changing the area by a factor of 2 results in a corresponding 6 dB change of power level; whereas, the power level is changed only 3 dB when the number of equal noise sources, acting independently, is changed by a factor of 2. Critical to the total noise reduction that can be derived from utilization of many small restrictions versus a single or a few large restrictions, is the proper size and spacing of the restrictions such that the noise generated by jet interaction is not greater than the summation of the noise generated by the jets individually. It has been found that the optimum size and spacing is very sensitive to pressure ratio, $\Delta P/P_1$. The author's company has recently developed the technology to utilize the above approach for the design of quiet valve trim using single-stage pressure reduction. This approach depicted by concept (d) in Fig. 11-25 provides an equivalent or better noise performance than either (a), (b), or (c).

For control-valve applications operating at high-pressure ratios ($\Delta P/P_1 \geq 0.8$) the series restriction approach, splitting the total pressure drop between the control valve and a fixed restriction (diffuser) downstream of the valve can be very effective in minimizing the noise. In order to optimize the effectiveness of a diffuser, it must be designed (special shape and sizing) for each given installation so that the noise levels generated by the valve and diffuser are equal. Figure 11-26 depicts a typical valve-plus-diffuser installation.

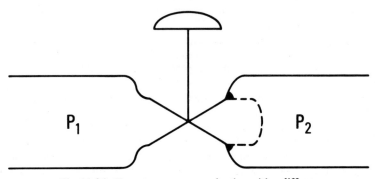

Fig. 11-26. Two-stage pressure reduction with a diffuser.

Pertaining to the design of quiet valves for liquid application, the problem resolves itself to one of designing to eliminate cavitation. Service conditions that will produce cavitation can readily be calculated [3]. The use of staged or series reductions provides a very viable solution to cavitation and hence, hydrodynamic noise.

Path Treatment

A second approach to noise control is that of path treatment. Sound is transmitted via longitudinal waves through the elastic medium or media that separate the source from the receiver. The speed and efficiency of sound transmission is dependent on the properties of the medium through which it is propagated.

Ch. 11 FLUID PIPING SYSTEM NOISE

Path treatment consists of regulating the impedance of the transmission path to reduce the acoustic energy that is communicated to the receiver.

The fluid stream is an excellent noise transmission path. When critical flow exists (fluid velocity at the vena contracta is at least at the sonic level), the vena contracta acts as a barrier to the propagation of sound upstream via the fluid. At subcritical flow, however, valve noise can be propagated in the upstream direction almost as efficiently as it is downstream. The impedance to the transmission of noise upstream at subcritical flow is primarily a function of valve geometry. The valve geometry that provides a direct line of sight through the valve (i.e., ball valves and butterfly valves), offers little resistance to noise propagation. Globe-style valves provide approximately 10 dB attenuation. In any path treatment approach to control-valve noise abatement, consideration must be given to the amplitude of noise radiated by both the upstream and downstream piping.

Dissipation of acoustic energy by use of acoustical absorbent materials is one of the most effective methods of path treatment. Whenever possible the acoustical material should be located in the flow stream either at, or immediately downstream of, the noise source. This approach to abatement of aerodynamic noise is accommodated by inline silencers. Inline silencers effectively dissipate the noise within the fluid stream and attenuate the noise level transmitted to the solid boundaries. Where high mass flow rates and/or high-pressure ratios across the valve exist, inline silencers are often the most realistic and economical approach to noise control. Use of absorption-type inline silencers can provide almost any degree of attenuation desired. However, economic considerations generally limit the insertion loss to approximately 25 dB. Figure 11-27 is a cross-sectional view of a typical inline silencer.

Noise that cannot be eliminated within the boundaries of the flow stream must be eliminated by external treatment or isolation. This approach to the abatement of control-valve noise includes the use of heavy walled piping, acoustical insulation of the exposed solid boundaries of the fluid stream, and use of insulated boxes, rooms, and buildings to isolate the noise source.

In closed systems (not vented to atmosphere) any noise produced in the process becomes airborne only by transmission through the solid boundaries that contain the flow stream. The sound field in the contained flow stream forces

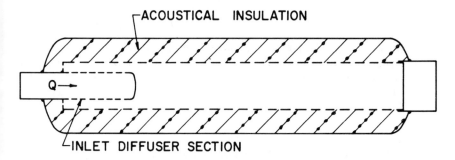

Fig. 11-27. Typical inline silencer.

the solid boundaries to vibrate which, in turn, cause pressure disturbances in the ambient atmosphere that are propagated as sound to the receiver. Because of the relative mass of most valve bodies the primary surface of noise radiation to the atmosphere is the piping adjacent to the valve. An understanding of the relative noise transmission loss as a function of the physical properties of the solid boundaries of the flow stream can be beneficial in noise control of fluid transmission systems.

A detailed analysis of noise transmission loss, beyond the scope of this chapter, is well covered in Chapter 5. However, it should be recognized that the spectral density of the noise radiated by the pipe has been shaped by the transmission-loss characteristic of the pipe and is not that of the noise field within the confined flow stream.

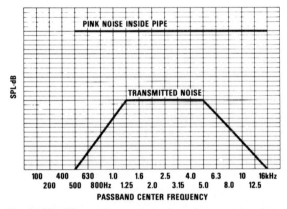

Fig. 11-28. Effects of transmission-loss characteristic of pipe.

Figure 11-28 depicts the general transmission-loss characteristic for commerical piping. The vibrational response of pipe wall to an acoustic field within the pipe is characterized by the following frequencies:

1. Coincident frequency, f_c, is defined as that frequency at which the acoustic wavespeed is equal to the phase velocity of a flexural wave in the pipe wall

$$f_c = \frac{\sqrt{3}\,c}{\pi t c_s} \text{ Hz} \qquad (11\text{-}9)$$

2. Acoustic cutoff frequency, f_{co}, lower limiting frequency where acoustic energy will propagate freely within the pipe, below f_{co} acoustic energy propagates as a plane wave without a radial component to excite the pipe wall

$$f_{co} = \frac{0.586\,c}{d} \text{ Hz} \qquad (11\text{-}10)$$

3. Ring frequency, f_r, is the frequency at which the longitudinal wave in the pipe material equals the pipe circumference.

FLUID PIPING SYSTEM NOISE

$$f_r = \frac{c_s}{\pi d} \tag{11-11}$$

where:

c = velocity of sound of the gas or liquid inside the pipe, inches per sec
c_s = velocity of sound of a wave in the pipe wall material, inches per sec
t = pipe wall thickness, inches
d = mean diameter of pipe, inches.

Above the ring frequency, f_r, the pipe wall tends to behave much like a flat plate, and mass law applies. Below the greater of the coincident frequency, f_c, or cutoff frequency, f_{co}, the vibration response is stiffness controlled. In the frequency range between the coincident and ring frequencies the resonant modes of pipe vibration are classified acoustically fast and are relatively efficient in the transmission of sound.

The relative noise transmission loss as a function of pipe size and schedule has been quantified in Table 11-3. For a comprehensive analysis of pipe transmission-loss, see Reference [4].

Acoustical insulation of the exposed solid boundaries of the fluid stream is an effective means of noise abatement for localized areas. Test results indicate that ambient noise levels can be attenuated as much as 10 dB per inch of insulation thickness.

Path treatment such as heavy walled pipe or external acoustical insulation can be a very economical and effective technique for localized noise abatement. However, it should be pointed out that noise is propagated for long distances via the fluid stream and that the effectiveness of the heavy wall pipe or external insulation terminates where the treatment is terminated. See Chapter 5, page 190 for a treatment of pipe wrapping systems.

Noise Analysis

Control valve noise can be accurately predicted. The noise in the entire system must include noise resulting from other sources, such as abrupt expansion of the fluid in the pipe, compressors, pumps, pipe fittings, etc. Fan noise, blower noise, and noise resulting from obstructions in the flow, such as rods, strips, and rings are treated in Chapter 9. For noise resulting from machinery, such as pumps and compressors, the supplier must be consulted. In some instances actual measurements of the sound radiated to the atmosphere may be required to determine the total contribution made by the piping system to the overall noise level.

The contribution to excessive noise made by machinery that produces pulsations in the flow, such as pumps and compressors, can be reduced by the use of accumulators which reduce the pulsations in the flow. Mechanical vibrations induced by vibration excitation to the piping system can be reduced by the use of flexible couplings in the piping system. Spring-type isolators can be utilized for pipe hangers, to reduce the vibrational energy of the pipe, which may excite other mechanical systems. Isolation system design and isolator selection are discussed in Chapter 7, "Vibration Control for Noise Reduction."

A simple sound survey of a given area will establish compliance or noncompliance to the governing noise criterion, but not necessarily either identify the primary source of noise or quantify the contribution of individual sources. Frequently piping systems are installed in environments where the background noise due to highly reflective surfaces and other sources of noise in the area make it impossible to use a sound survey to measure the contribution a single source makes to the overall ambient noise level.

Vibration measurements of the pipe can be utilized to calculate sound pressure radiated to the atmosphere [4]. The acoustic pressure, at a point in space, r distance from the centerline of the pipe, radiated by a vibrating pipe is related to the pipe surface velocity as follows:

$$p_{rms}^2 = \rho_a^2 c_a^2 v^2 \frac{d}{2r} \xi \tag{11-12}$$

where:

p_{rms} = root mean square (rms) acoustic pressure
ρ_a = density of atmosphere
v = vibrational velocity of pipe wall
c_a = velocity of sound in the atmosphere
d = mean diameter of pipe
r = distance from center of pipe to location where p is required
ξ = acoustic radiation efficiency

English or metric units can be used in Equation (11-12), however, the units used must be consistent. The radiation efficiency, ξ, is defined as the ratio of acoustic power at the source, to the acoustic power transmitted W/W_a. The radiation efficiency has been found to equal unity above the coincident frequency (with the pipe in the atmosphere), f_{ca}, and is directly proportional to the frequency below coincidence.

$$\xi = 1, \quad \text{for } f \geq f_{ca}$$

$$\xi = \frac{f}{f_c}, \quad \text{for } f < f_{ca}$$

The equation for the coincident frequency for the pipe with the atmosphere, f_{ca}, is given below:

$$f_{ca} = \frac{\sqrt{3} \, c_a^2}{\pi t c_s} \tag{11-13}$$

where t is the pipe wall thickness, and c_s is the velocity of sound in the pipe material.

In summary:

$$p^2 = v^2 \rho_a^2 c_a^2 \frac{d}{2r} \frac{f}{f_{ca}} \quad f < f_c \tag{11-14}$$

$$p^2 = v^2 \rho_a^2 c_a^2 \frac{d}{2r} \quad f \geq f_c \tag{11-15}$$

FLUID PIPING SYSTEM NOISE

Acceleration measurements are often made rather than velocity measurements because accelerometers are small and convenient to use.

The following relation,

$$v^2 = \frac{a^2}{\omega^2} \tag{11-16}$$

can be used in Equations (11-14) and (11-15) to yield:

$$p^2 = a^2 \frac{\rho_a^2 c_a^2}{4\pi^2} \frac{d}{2r} \frac{1}{ff_c} \quad f < f_c \tag{11-17}$$

$$p^2 = a^2 \frac{\rho_a^2 c_a^2}{4\pi^2} \frac{d}{2r} \frac{1}{f^2} \quad f \geqslant f_c \tag{11-18}$$

Discussion

The limitations of this approach must be outlined to insure the proper application of the results.

As presented, the theory is appropriate for the shell modes only. This type of response is what would be present if a flat vibrating plate were rolled up into a cylinder. At low frequencies, however, the shell modes are not present and the response of the pipe is due to the entire length of pipe acting as a beam. These beam modes can vibrate with very high amplitudes. However, the efficiency of their coupling to the acoustic field on the outside of the pipe is extremely low. This means a vibration measurement may indicate high energy content at low frequencies with very little contribution to the observed sound-pressure level. Frequencies associated with the lowest shell mode are tabulated for each standard pipe size in Table 11-4 and should be considered as a low frequency cutoff for the direct applicability of the theory. This is generally not restrictive in evaluating control-valve noise or other broadband high-frequency noise.

Table 11-4. 1st Shell Modes, Hz

Pipe Dia.	Pipe Schedule		
	40	80	160
2	3115	4675	8289
4	1301	1940	3363
6	694	1124	2053
8	466	756	1517
10*	338	475	
12*	244	332	
16*	153	208	
24*	67	90	

*Standard and extra strong.

FLUID PIPING SYSTEM NOISE

In order to use the above formulation it is necessary to convert the mean square values to decibels. This can be accomplished using the following definitions:

$$SPL = 10 \log_{10}\left(\frac{p^2}{p_0^2}\right) \quad \text{Sound-pressure level, in dB}$$

$$V_{dB} = 10 \log_{10}\left(\frac{v^2}{v_0^2}\right) \quad \text{Wall velocity level, in dB}$$

$$A_{dB} = 10 \log_{10}\left(\frac{a^2}{a_0^2}\right) \quad \text{Wall acceleration level, in dB}$$

Widely accepted values for the various references in the above definitions are as follows:

$$p_o = .0002 \text{ dynes/cm}^2$$
$$v_o = 10^{-6} \text{ cm/sec}$$
$$a_o = 10^{-3} \text{ cm/sec}^2$$

Using the above definitions Equations (11-4) and (11-5) can be changed to decibel notation:

For $f < f_c$;

$$SPL = 10 \log_{10}\left(\frac{v^2}{10^{-12}}\right) + 10 \log_{10}\left(\frac{d}{2r}\right) + 10 \log_{10}\left(\frac{f}{f_c}\right) - 13.7 \text{ dB} \quad (11\text{-}19)$$

For $f \geq f_c$;

$$SPL = 10 \log_{10}\left(\frac{v^2}{10^{-12}}\right) + 10 \log_{10}\left(\frac{d}{2r}\right) - 13.7 \text{ dB} \quad (11\text{-}20)$$

If the absolute value of the pipe wall vibrational velocity is obtained, as with a vibration meter, then this value may be substituted directly in the first term (v) on the right-hand side of the equations. This velocity must be expressed in cm/sec.

When velocity measurements are taken in decibels referenced to 10^{-6} cm/sec this velocity dB level may be substituted for the entire first term, $10 \log_{10}(v^2/10^{-12})$.

In the same manner as above, Equations (11-17) and (11-18) may be changed to decibel notation for acceleration measurements.

For $f < f_c$;

$$SPL = 10 \log_{10}\left(\frac{a^2}{10^{-6}}\right) + 10 \log_{10}\left(\frac{d}{2r}\right) - 10 \log_{10}(ff_c) + 30.4 \text{ dB} \quad (11\text{-}21)$$

For $f \geq f_c$;

$$SPL = 10 \log_{10}\left(\frac{a^2}{10^{-6}}\right) + 10 \log_{10}\left(\frac{d}{2r}\right) - 20 \log_{10}(f) + 30.4 \text{ dB} \quad (11\text{-}22)$$

Absolute values of the acceleration in cm/sec^2 can be substituted directly for (a) in the first term of Equations (11-21) and (11-22). If acceleration levels are taken in decibels relative to 10^{-3} cm/sec^2 then this level may be substituted for the term $10 \log_{10} (a^2/10^{-6})$.

When vibration levels are taken in decibels relative to a reference value other than presented here, it will require solving for the absolute value using the definitions of A_{dB} or V_{dB} and substituting for (a) or (v) in appropriate equations above.

The equations, as presented for velocity and acceleration, allow corrections as a function of frequency. Ideally, to gain an overall equivalent sound-pressure level from vibration measurements a summation of the corrected levels from each frequency band would be made. This is essential when acceleration is the quantity measured. However, when converting overall wall velocity to overall acoustic pressure, Equation (11-20) will serve as a reasonable approximation, as long as the velocity spectrum is not dominated by low-frequency components.

Measurement Techniques

The accuracy of the conversion method is a function of the accuracy with which the vibration measurements may be taken. Assuming that all equipment is operating properly, the important variable in making a measurement is the attachment of the accelerometer to the pipe wall. The ideal condition is to have a small metal pad welded to the pipe with mechanical means, such as a threaded stud, to attach the accelerometer. Some device for electrically isolating the accelerometer from the pipe should be used such as insulated studs or washers placed between the surfaces. This attachment method will yield valid information over the entire frequency range for which the particular probe is specified. An alternative to this method is to attach pads or studs to the pipe wall using an adhesive. As long as a stiff, thin-layered adhesive is used this can be effective over the specified range of the probe. Different adhesives are necessary depending on the temperature of the application. Magnetic attachments have to be of a special design to give firm bond to a cylindrical surface. Even with a good magnetic attachment the high frequency response is limited. If a magnetic base is used the surface should be clean and free of paint and dirt to ensure maximum contact. Hand-held accelerometers generally are limited to very-low-frequency measurements.

This method cannot be used if the accelerometer is located on a flange or other pipe fitting. Measurements should be taken at a minimum of 2 pipe diameters from a flanged end or pipe turn.

Vibration measurements thus provide a very practical method of determining the noise level radiated by a piping system in an environment that precludes the use of sound-level measurements.

It should be recognized that vibration measurements are not a panacea for all of the problems associated with the noise analysis of a fluid transmission system. Sound and/or vibration surveys can quantify the noise levels radiated by a fluid transmission system; however, because the noise generated within the flow stream is both structure borne and fluid borne it is not necessarily obvious what the primary source of noise generation is, and hence, which sources (if any) are controlling the sound field.

Closure

From the foregoing discussion it should be obvious that a substantial amount of progress in the area of control-valve noise has been achieved in the very recent past. Prior to about 1968, however, the control-valve industry as a whole had done little work on this extremely important area. Perhaps the consensus was that the noise which these devices generated in performing their jobs was a price which had to be paid. More probably, the real reason was simply the absence of a sufficient reason for industry to spend the time and money required for their improvement. Although the control-valve buying public may have, by their occasional complaints of excessive noise, helped to provide the incentive to attack the problem, the major impetus was furnished by the federal government in the form of legislation designed to protect working people from industrial noise pollution.

From strictly noise considerations, there are very few control-valve installations which can be considered truly standard. They will be unique from the standpoint of installation geometry, service conditions, noise attenuation requirements, or some combination of the three. With so many possible installation variables and the numerous pieces of control-valve noise-abatement equipment, it becomes extremely important that knowledgeable persons are consulted in the application of this equipment. For example, several approaches may be taken to the same problem. One approach might produce the most quiet installation but at a prohibitive cost, whereas another approach might meet the required noise specification at a substantial saving. Without the ability to predict noise levels and without the choices of equipment, optimizing a given installation from the noise and cost standpoints would not be possible.

Where do we go from here? Comparing current noise technology with other important control-valve technologies—such as system analysis and valve sizing—indicates that the noise technology is in an emerging status. If this be the case, then dramatic progress can be expected in the immediate future. Studies are in progress which are intended to increase the understanding of the noise-generation mechanisms and to identify parameters not presently being considered. These studies, and others, should generate new and more efficient items of equipment and result in more precise techniques for the prediction of control-valve noise.

NOMENCLATURE

a = Acceleration
c = Velocity of sound
c_a = Velocity of sound in atmosphere
C_g = Gas sizing coefficient
c_s = Velocity of sound in the pipe material
C_u = Cauchy number $\dfrac{\rho V^2}{E}$
c_p = Specific heat at constant pressure
c_v = Specific heat at constant volume
C_V = liquid sizing coefficient

d = Pipe diameter
dBA = Decibel, A-weighting
D = Characteristic diameter
E = Modulus of elasticity
E_u = Euler number $\dfrac{\rho V^2}{2(P_1 - Pv)}$
f = Frequency
f_c = Coincident frequency
f_{ca} = Coincident frequency with atmosphere
f_{co} = Cutoff frequency
f_r = Ring frequency
F = Friction factor
k = Specific heat ratio, c_p/c_v
L = Characteristic length
M = Mach number, V/c
P_1 = Upstream pressure, psia
ΔP = Pressure differential
P_r = Prandtl number, $c_p\, \mu/\lambda$
p = Sound pressure, rms
P_v = Vapor pressure at upstream conditions
r = Radial distance from axial centerline of pipe
R_e = Reynolds number, $\rho\, VD/\mu$
SPL = Sound-pressure level
t = Pipe wall thickness
W = Acoustic power of source
W_a = Acoustic power radiated
W_e = Weber number $\dfrac{\rho D V^2}{\sigma}$
V = Flow velocity inside pipe
v = Vibrational velocity of pipe wall
ν = Kinematic viscosity
λ = Thermal conductivity
μ = Absolute viscosity
ρ = Density
ρ_a = Density of atmosphere
ρ_o = Density of water at 60 °F
σ = Surface tension
ζ = Radiation efficiency
ϕ = Denotes a function operator

12

Industrial Noise Control Studies

L. L. FAULKNER

Introduction

The objective of this chapter is to illustrate how the principles treated in the other chapters of this book are applied for the solution of typical industrial noise control problems. While in the other chapters these principles were applied to isolated examples, that were limited in scope, it is the intention here to demonstrate more fully, by various examples, how these principles can be combined to analyze or solve the typical problems usually encountered in industrial situations. Such problems may be considered as falling into three general classifications:

1. Writing noise specifications for equipment to be purchased
2. Controlling factory space noise
3. Analyzing equipment and noise reduction.

Moreover, an attempt has been made to treat the types of problems that are most frequently encountered so that they may be used as a guide in similar situations. In addition, all of the examples selected for this chapter are actual cases which have been taken from industrial practice.

Case Study 1. *Determining Acoustic Specifications for New Equipment*

A common situation which occurs with all types of industrial machinery is the need to determine an acoustic specification on a new piece of equipment for which bids are to be requested from suppliers. The specification of the maximum *sound power* allowed by the items to be purchased is one method of insuring that desired *sound-pressure* levels will not be exceeded in the work area.

Ch. 12 INDUSTRIAL NOISE CONTROL STUDIES 507

One popular method is that this specification require a maximum A-weighted sound-pressure level at some distance from the device. This method may be acceptable in some instances; however, in critical situations, the recommended specification is the sound-power level in octave bands. For locations near an office, cafeteria, or for outdoor equipment, etc., a level considerably lower than the OSHA level is required in many instances; and computations are necessary to determine the maximum allowable sound-power level.

For example, a new process air fan is to be installed in a shop area and it is desired that 40 feet away from the fan the sound-pressure level is not to exceed the PNC-65 criteria. A specification of a sound-pressure level close to the source thus may be overly restrictive. Moreover, the sound-pressure level is also a function of the acoustic properties of the space in which the source will be located. The desirable method then, is to specify the sound-power level of the unit by octave bands, to meet a given sound criterion for the installed condition and location. This method is the most accurate one; it eliminates overspecification and possibly will reduce the cost of an item by eliminating any nonessential noise control.

For convenience, assume the shop area to be identical to the space described in Example 2 of Chapter 5, on page 183. The acoustic absorption of the existing shop area is assumed to have been determined experimentally. The task is to specify the sound-power level, in octave bands, for a new blower to be placed in the shop to meet the PNC-65 criteria.

The design sound-pressure levels, as described by the PNC-65 criteria, were given in Chapter 4, Table 4-5, and are repeated below, in Table 12-1. The acoustic absorption for the shop area, in sabins, is determined by the method of Example 2, Chapter 5. The values for the absorption are given in row 2 of Table 12-1.

Table 12-1. PNC-65 Criterion and Acoustic Absorption for the Shop Area

Row No.	Description	Octave-Band Center Frequency, Hz							
		63	125	250	500	1000	2000	4000	8000
1	PNC-65 Criteria, in decibels. (From Table 4-5, Chapter 4.)	76	73	70	67	64	61	58	58
2	Room Absorption,[1] A, in sabins. (Table 5-9, Chapter 5, Row 6.)	726^2	1413	891	1120	1778	1778	2239	2872^2

[1] For this example the total room absorption, including the air absorption in the room, is to be used.
[2] These values were not given in Table 5-9, but are experimentally found in the same manner as indicated by Example 2 in Chapter 5.

508 INDUSTRIAL NOISE CONTROL STUDIES Ch. 12

To specify the maximum sound-power levels for the fan, Equation (4-1) is used:

$$L_{pi} = L_{wi} + 10 \log_{10} \left[\frac{Q_{\theta i}}{4\pi r^2} + \frac{4}{A} \right] \quad (4\text{-}1)$$

where:

L_{pi} = octave-band sound-pressure level, dB reference 2×10^{-5} N/meter^2.
L_{wi} = octave-band sound-power level, dB reference 10^{-12} watt.
$Q_{\theta i}$ = directivity of the source. (The value of $Q_{\theta i}$ is given in Chapter 4, Fig. 4-1)
r = distance from source to receiver, meters.
A = room absorption, sabins.

As was shown in Fig. 5-7, when the receiver is situated at a distance of $0.5\sqrt{A}$ feet, or more, away from the source, the sound pressure at this location is not dependent on the term $Q_{\theta i}/4\pi r^2$, thus this term can be neglected; and the receiver is considered to be in the reverberant acoustic field. For this example we are interested in a receiver location 40 feet away from the source. A computation of $0.5\sqrt{A}$ for the largest value of A from Table 12-1 gives:

$$0.5\sqrt{A} = 0.5\sqrt{2872}$$

$$= 27 \text{ feet (rounded)}$$

therefore the receiver location is in the reverberant acoustic field and Equation 4-1 can be utilized in the following form:

$$L_{pi} = L_{wi} + 10 \log_{10} \left[\frac{4}{A} \right]$$

Since we are interested in specifying the sound-power level the above equation is arranged to give:

$$L_{wi} = L_{pi} - 10 \log_{10} \left[\frac{4}{A} \right]$$

The calculation of the sound-power level is illustrated in Table 12-2. Rows 1 and 2 are the sound-pressure level desired for the shop area and the acoustic absorption for the space, as given by Table 12-1. $4/A$ is computed in row 3 and $10 \log_{10} [4/A]$ is computed in row 4; note that these values are negative. The sound-power level in row 5 is computed by summing the values in rows 1 and 4. The summation results because of the negative of a negative value.

In Table 12-2, row 6, the computed sound-power levels are rounded to the nearest whole decibel. In this step some individuals may prefer to round the levels to the next lower whole decibel to allow some margin of safety. The values in row 6 also will specify, for numerous types of mechanical machinery, the maximum allowable sound-power levels by octave bands.

The specific device for this example is a fan; the fan outlet may be ducted and, therefore, will not influence the sound level in the shop area. For this case, a

Ch. 12 INDUSTRIAL NOISE CONTROL STUDIES

Table 12-2. Maximum Sound-Power Levels for Case Study 1

Row No.	Description	Octave-Band Center Frequency, Hz							
		63	125	250	500	1000	2000	4000	8000
1	L_{pi}, decibels (Table 12-1, row 1)	76	73	70	67	64	61	58	58
2	A, sabins (Table 12-1, row 2)	726	1413	891	1122	1778	1778	2239	2872
3	$\dfrac{4}{A}$	.0055	.0028	.0045	.0036	.0022	.0022	.0018	.0014
4	$10 \log_{10} \left[\dfrac{4}{A}\right]$, dB	−22.6	−25.5	−23.5	−24.5	−26.7	−26.7	−27.5	−28.5
5	$L_{Wi} = L_{pi} - 10 \log_{10} \left[\dfrac{4}{A}\right]$, dB	98.6	98.5	93.5	91.5	90.7	87.7	85.5	86.5
6	Sound-Power Level, L_{Wi} (Rounded)	99	99	94	92	91	88	86	87
7	Correction for Total Sound-Power Level if outlet is ducted	+3	+3	+3	+3	+3	+3	+3	+3
8	Total Maximum Sound-Power Level of Fan, dB	102	102	97	95	94	91	89	90

Table 12-3. Typical Specification Sheet Sent to Vendors for Quotation Purposes

	Octave-Band Sound-Power Levels (Reference Power = 10^{-12} watt)			
	Specified Sound-Power Level (dB)	Vendor to Complete:		
Octave-Band Center Frequency (Hz)		Actual (A)	Special Design (B)	Acoustic Treatment (C)
63	102			
125	102			
250	97			
500	95			
1000	94			
2000	91			
4000	89			
8000	90			

Vendor to complete:
Equipment will comply with specified
  Sound-Power Level: _____.
Additional Costs Required:
  For B: _____.
  For C: _____.

correction can be applied to row 6. Since we wish to specify the total sound-power level of the fan, including the inlet and outlet, and since only the inlet radiates sound into the shop area, 3 dB can be added to the values of row 6. This is shown by rows 7 and 8 of Table 12-2. The reason for this was explained in Example 1 of Chapter 9, page 397. The values in row 8, Table 12-2, are the maximum allowable sound-power levels for the fan to meet the PNC-65 criteria in the shop 40 feet away from the fan inlet. If both the inlet and the outlet radiate into the shop area, the corrections of row 7 should not be applied and the maximum sound-power levels would be those in row 6 of Table 12-2.

The acoustic specifications to be sent to the fan supplier could be as shown in Table 12-3 to help in the selection and pricing of a fan which will perform satisfactorily. If special acoustic treatment is required the fan supplier can give proper advice for any given situation.

Case Study 2. *Design of a Control Room in a Noisy Mechanical Equipment Area*

In some instances the most easily implemented and most cost effective noise-control modification is to isolate individuals from the source of excessive noise. This technique is most effective in situations where the machinery operator can be located remotely from the equipment and where control panels are remotely placed. Shown in Fig. 12-1 is a mechanical-equipment room which requires continual monitoring by an operator. The mechanical equipment room has a 24-ft-high ceiling. The decision was made to construct a control room within which the operator can be placed to reduce the noise exposure.

The requirements for the control room are as follows:

1. Inside dimensions shall be 18 × 10 × 7.5 feet in height.
2. Two windows shall be located on the 18-ft side with dimensions of 18 × 27 inches.
3. A door shall be provided on the wall of the control room in line with the equipment-room door in the upper-left-hand corner of Fig. 12-1. The door will be 7 × 3 ft and contain a window 12 × 18 inches, centered 5 feet above the floor level.
4. The A-weighted sound-pressure level must not exceed 85 dB inside the control room.
5. The control room shall be constructed such that telephone conversation can take place within the room.

A noise survey was performed in the mechanical-equipment room with all mechanical machinery operating and the octave-band sound-pressure levels at the three locations as indicated in Fig. 12-1, are given by Table 12-4.

The calculations for the design of the control room are explained in the following steps:

Step 1. The maximum sound-pressure levels outside the proposed control room must be determined. To simplify the calculations and to be conservative, the values for location 2 of Fig. 12-1 will be used as the incident sound-pressure

Ch. 12 INDUSTRIAL NOISE CONTROL STUDIES 511

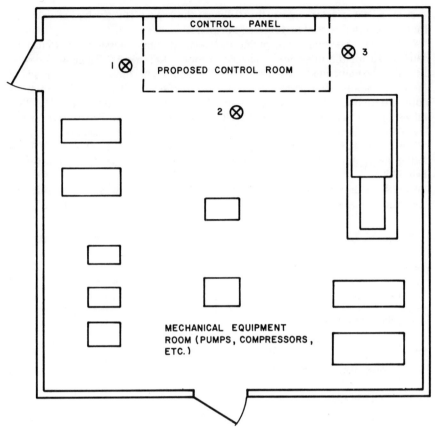

Fig. 12-1. Mechanical equipment room and the location of the proposed control room.

Table 12-4. Sound-Pressure Levels in Mechanical Equipment Room Shown in Fig. 12-1

Location as Indicated in Fig. 12-1	Octave-Band Center Frequency, Hz								
	31.5	63	125	250	500	1000	2000	4000	8000
	Sound-Pressure Levels, dB								
1	86	94	93	90	91	98	99	98	90
2	88	96	96	90	93	98	101	100	95
3	81	78	76	72	68	70	72	71	73

level on all three sides and on the ceiling of the room. These values are given in Table 12-4 and repeated in Table 12-5, row 1.

Step 2. The maximum acceptable octave-band sound-pressure levels must be determined for the interior of the control room. Referring to Table 4-4 on page 132, the recommended criteria for electrical equipment rooms and rooms where speech communication is required is NC-60 or PNC-60. The PNC-60 preferred noise criterion is selected as the design criterion for the control room. The PNC values are chosen, rather than the NC values, because the PNC criterion has lower allowable sound-pressure-level values at high and low frequencies.

The octave-band sound-pressure levels which correspond to the PNC-60 criterion are found in Table 4-5, on page 132. These values are listed in row 2 of Table 12-5.

Step 3. Determine the noise reduction, NR, required between the outside and the inside of the control room. This is not the transmission loss of the walls,

Table 12-5. Sound Transmission-Loss Requirements for Control Room in Case Study 2

Row No.	Description	Octave-Band Center Frequency, Hz								
		31.5	63	125	250	500	1000	2000	4000	8000
1	Maximum Sound-Pressure Level Outside the Control Room, Table 12-4.	88	96	96	90	93	98	101	100	95
2	Desired Sound-Pressure Level Inside the Control Room, PNC-60.	76	73	69	66	63	59	56	53	53
3	Noise Reduction Required, NR. Row 1 minus row 2.	12	23	27	24	30	39	45	47	42
4	Absorption coefficient, α, of the Ceiling System, #7 mounting.	.15	.34	.41	.38	.54	.83	.92	.75	.71
5	Absorption, A, of the ceiling, in sabins, α times ceiling area of 180 ft².	27.0	61.2	73.8	68.4	97.2	149.4	165.6	135.0	127.8
6	$\dfrac{S}{A} = \dfrac{465}{A}$	17.22	7.60	6.30	6.80	4.78	3.11	2.81	3.44	3.64
7	$10 \log_{10}\left(\dfrac{S}{A}\right)$, dB.	12.4	8.8	8.0	8.3	6.8	4.9	4.5	5.4	5.6
8	$TL = NR + 10 \log_{10}\left(\dfrac{S}{A}\right)$, dB. Row 3 plus row 7 (rounded)	24	32	35	32	37	44	50	52	48
9	STC-50 Transmission-Loss Values.	16	25	34	43	50	53	54	54	54
10	Deficiency of the STC-50 Values; row 9 minus row 8.	−8	−7	−1	0	0	0	0	0	0

Ch. 12 INDUSTRIAL NOISE CONTROL STUDIES 513

but is a quantity needed in computation of the required transmission loss. The noise reduction is the difference between the sound-pressure levels outside the control room and the desired sound-pressure levels inside the control room, in octave bands. This calculation is made by subtracting row 2 from row 1 (Table 12-5), with the result as given by row 3.

Step 4. Determine the acoustic absorption in the control room. The transmission loss is a function of the noise reduction, NR, and the acoustic absorption inside the proposed control room. For rooms which do not have extensive acoustic treatments on the surfaces, Equation (5-9), in Chapter 5, can be used to compute the transmission loss, thus:

$$TL = NR + 10 \log_{10}\left(\frac{S}{A}\right) \quad dB$$

where:

TL = Transmission-loss of wall, dB
NR = Difference in sound-pressure level on either side of wall, dB
S = Area through which sound transmits, ft^2
A = Absorption of receiving room, sabins.

The proposed control room would not be considered an extensively treated space. Typically, the walls would be of painted masonry; the floor, concrete; and the other surfaces of low acoustic absorption. Only the ceiling would typically be acoustically treated to prevent the room from being reverberant. With an acoustical ceiling, a conservative solution to the transmission-loss equation will result if the other surface-absorption values are neglected. For this room an acoustical ceiling was chosen, constructed of square acoustical tile suspended with metal supports. This system is known as a #7 mounting. Having selected the ceiling, the supplier was asked to provide octave-band acoustic absorption coefficients for the installed condition. These values are listed in row 4 of Table 12-5.

Step 5. Compute the absorption, in sabins, for the control room. If the total absorption is assumed to be that of the ceiling alone, a conservative calculation will result for the required acoustic transmission-loss of the walls and ceiling. For this reason the absorption of other surfaces and the air absorption in the control room will not be considered. The total absorption is computed by the product of ceiling area and the coefficients of row 4. The resulting absorption, in sabins, for the various octave bands is given in row 5.

Step 6. Compute the correction term $10 \log_{10} (S/A)$ where S is the area through which the sound energy transmits and A is the absorption of the control room, in sabins. The ratio S/A is given in row 6, computed from $S = 465$ ft^2, the area of the three walls and ceiling through which sound transmits, and A, the values of absorption as given by row 5. $10 \log_{10} (S/A)$ is given in row 7.

Step 7. Compute the required transmission-loss in octave bands from:

$$TL = NR + 10 \log_{10}\left(\frac{S}{A}\right)$$

The TL is determined by adding the values in rows 3 and 7, as listed in row 8.

Step 8. Determine the STC rating of the wall and ceiling system. From Table 5-12A of Chapter 5, the STC rating required is determined by the instructions accompanying this table. It was determined that the STC-50 contour would meet the criteria; that the deficiency at any one frequency would not exceed 8 dB; and that the sum of the deficiencies would not exceed 32 dB, as indicated by row 10. Note that the values of the STC-50 contour for 63 and 31.5 Hz was obtained by assuming a linear extrapolation of subtracting 9 dB from each succeedingly higher frequency.

From the requirement of an STC-50 wall and ceiling system the actual construction can be specified. From Table 5-5, page 170, various constructions are given in terms of the STC rating. From this table it can be seen that an 8-inch, dense cement-block wall (not cinder block), has the required STC-50 rating. If the block wall is painted on both sides it will exceed the STC-50 rating. The windows and the door must be selected to also have an STC-50 rating to meet the noise criterion. STC-50 systems can be obtained from suppliers of architectural materials. In installing the door and windows the suppliers instructions must be followed to meet the desired rating. The structural ceiling was specified as 4-inch poured concrete with a density of 144 lb per cu ft.

To insure that the construction will perform as desired, through openings must not exist; all electrical outlets must be mortared and all utilities which penetrate the room must be packed with acoustical packing material. The heating, ventilating, and air-conditioning system for the control room must be so designed that the PNC-60 criterion is not exceeded in the space; also ducting for this system, which passes through the walls or ceiling, must provide STC-50 transmission-loss rating as prescribed. For the small volume of the control room a low-velocity system was utilized with sound-pressure levels at least 10 dB below the PNC-60 criteria of all octave bands.

Case Study 3. *Process Air Supply System*

Air supply systems are common in all industrial plants for heating, ventilating, and air conditioning (HVAC) offices, laboratories, and factory spaces. Process air supply systems are used to supply air for such industrial processes as cooling air, air for combustion, etc. In all cases, the air is supplied by a blower and conveyed through ducting elements.

Because of the frequent occurrence of air supply systems in industry and the need to estimate the noise level encountered in new designs, a very thorough example follows. Although, of necessity, this example is unusually lengthy, it is felt that its inclusion is warranted. The reader who will persevere through the entire example will gain an insight into the method of analysis and the practical assumptions that occasionally must be made for want of more complete data; he will have a very useful guide for predicting the noise level associated with any air supply system.

The problem is illustrated in Fig. 12-2; it is a process air system designed to provide air to cool a die-casting mold located centrally in a large factory building. In this case, the A-weighted sound-pressure level at the operator position must be estimated to determine if noise-control measures are required. The

Ch. 12 INDUSTRIAL NOISE CONTROL STUDIES 515

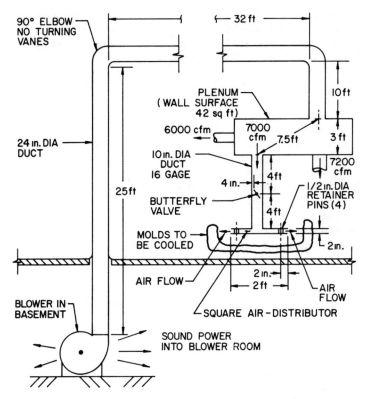

Fig. 12-2. Schematic view of process air supply system to cool die-casting mold.

operator position is 10 feet from the butterfly valve (see Fig. 12-2) and 12 feet from the opening of the square air-distributor. The blower specifications are as follows:

> Six centrifugal radial blades
> Diameter of blades = $45\frac{1}{8}$ inches
> Rotational speed = 1650 rpm
> Static pressure head, P = 32 inches H_2O
> Flow rate, Q = 20,200 cfm air
> Air temperature = 75 °F

The dimensions of the ducting elements are also given in Fig. 12-2. The flow area of the outlet of the square air-distributor is 20 in. × 2 in. and each side of the outlet can be considered to be a duct opening, having an area of 48 sq in.

Solution: The recommended procedure is to calculate the sound power generated and transmitted into the ducting system by the blower; the sound power generated and attenuated by each element of the ducting system is then calculated in the order of its occurrence when starting from the blower. This will allow the sound-power level to be determined at any location in the system, from which the sound-power level near any of the elements can be predicted.

Step 1. Estimate the octave-band sound-power level generated by the blower, using Equation (9-4) which is repeated below:

$$L_W = K_W + (10 \log_{10} Q + 20 \log_{10} P) + \text{BFI dB} \qquad (9\text{-}4)$$

where:

L_W = Sound-power level, dB
K_W = Specific sound-power level, dependent on blower type, dB
Q = Flow rate, cfm
P = Pressure head, inches H_2O
BFI = Blade frequency increment, dependent on blade pass frequency.

The steps in the computation of L_W are given in Table 12-6 and described here as follows:

a. The specific sound-power level, K_W, is obtained from Fig. 9-3 for the radial fan type. These values are listed in Row 1 of Table 12-6.
b. Compute the correction term $(10 \log_{10} Q + 20 \log_{10} P)$.

$$10 \log_{10} Q + 20 \log_{10} P = 10 \log_{10} (20{,}200) + 20 \log_{10} (32)$$

$$= 43.1 + 30.1 \text{ (values rounded)}$$

$$= 73 \text{ dB (rounded)}$$

(This value is listed for every octave band in row 2 of Table 12-6.)

Table 12-6. Calculated Values of Sound-Power Level of Blower for Case Study 3

Row No.	Description	Octave-Band Center Frequency, Hz							
		63	125	250	500	1000	2000	4000	8000
1	Specific sound-power level, dB. From Fig. 9-3.	48	45	43	43	38	33	30	29
2	Correction factor $10 \log_{10} Q + 20 \log_{10} P$. From Equation (9-4).	73	73	73	73	73	73	73	73
3	Blade Frequency Increment, BFI. From Fig. 9-3.	0	8	0	0	0	0	0	0
4	Total sound-power level, dB, of the blower. Includes inlet and outlet. Sum rows 1, 2, and 3.	121	126	116	116	111	106	103	102
5	Correction, in dB, to convert total sound-power level to inlet or outlet alone.	-3	-3	-3	-3	-3	-3	-3	-3
6	Sound-power level, dB, of either inlet or outlet of blower.	118	123	113	113	108	103	100	99

Ch. 12 INDUSTRIAL NOISE CONTROL STUDIES 517

c. The blade frequency increment, BFI, from Fig. 9-3 is given as a maximum value of 8 dB. The octave band within which this blade pass frequency occurs is computed from Equation (9-3):

$$B_f = \frac{\text{rpm} \times \text{No. of blades}}{60} \text{ Hz}$$

$$= \frac{1650 \times 6}{60} = 165 \text{ Hz} \tag{9-3}$$

From Chapter 2, Table 2-2, 165 Hz is within the 125-Hz octave band. The value of 8 dB is listed for the 125-Hz octave band in row 3 of Table 12-6; zeros are listed for all other octave bands.

d. The total sound-power level is obtained by a summation of the values in rows 1, 2, and 3 for each octave band. The resulting sound-power levels are shown by row 4 of Table 12-6.

e. To determine the sound-power level of the inlet alone or the outlet alone, 3 dB must be subtracted from the total sound-power level given by row 4. The reason for this is explained in Example No. 1 of Chapter 9, Step 4. In row 5, –3 dB is listed for every octave-band center frequency and the values in rows 4 and 5 are summed in row 6 to give the inlet or outlet sound-power levels.

The sound-power levels in row 6 of Table 12-6 represent the sound generation of the inlet or the outlet of the blower. If the sound-pressure levels in the blower room are of interest, the method described by Example No. 1 of Chapter 4 may be used. In this example we are interested in the sound power radiated into the duct system and will not consider the blower room.

The sound power levels computed in row 6 of Table 12-6 are also representative of the sound power radiated into the duct system. These values are listed in row 1 of Table 12-7 as the sound-power-level input to the duct system.

Step 2. Determine the attenuation in the 24-inch duct between the blower and the first 90-degree elbow. The octave-band attenuation values, in decibels, are obtained from the information in the section of Chapter 9 entitled "Attenuation of Untreated Duct," page 403. The attenuation is given as,

0.03 dB/linear ft for frequencies below 1000 Hz

and,

rising to 0.1 dB/linear ft at 8000 Hz.

By assuming a linear increase between the 500-Hz octave band to the 8000-Hz octave band, the following attenuation values are obtained:

Frequency, Hz	63	125	250	500	1000	2000	4000	8000
Attenuation, dB/ft	.030	.030	.030	.030	.048	.065	.083	0.100

518 INDUSTRIAL NOISE CONTROL STUDIES Ch. 12

Table 12-7. System Sound-Power Levels to Determine Level at Exit (See Case Study 3)

Row No.	Description	Octave-Band Center Frequency, Hz							
		63	125	250	500	1000	2000	4000	8000
1	Sound-power level of blower outlet, dB. From row 6, Table 12-6.	118	123	113	113	108	103	100	99
2	Attenuation of 25 ft. of round duct, dB.	.8	.8	.8	.8	1.2	1.6	2.1	2.5
3	Attenuation of first 90-degree elbow, dB.	0	0	1.0	2.0	3.0	3.0	3.0	3.0
4	Sound-power level of the blower after the first 90-degree elbow, dB.	117	122	111	110	104	98	95	94
5	Sound-power level produced by the first 90-degree elbow, dB. Table 12-8.	94	87	81	77	78	79	77	76
6	Total sound-power level after the first 90-degree elbow, dB.	117	122	111	110	104	98	95	94
7	Attenuation of 32 ft of round duct, dB.	1.0	1.0	1.0	1.0	1.5	2.1	2.7	3.2
8	Attenuation of second 90-degree elbow, dB.	0	0	1.0	2.0	3.0	3.0	3.0	3.0
9	Sound-power level of blower after second 90-degree elbow, dB.	116	121	109	107	100	93	89	88
10	Sound-power level produced by the second 90-degree elbow, dB. Table 12-8.	94	87	81	77	78	79	77	76
11	Total sound-power level after the second 90-degree elbow, dB.	116	121	109	107	100	93	89	88
12	Attenuation of 10 ft of round duct, dB	.3	.3	.3	.3	.5	.7	.8	1.0
13	Attenuation of plenum, dB. Table 12-9.	6.0	6.0	5.0	3.7	3.7	1.9	0	0
14	Attenuation of 4 ft of round duct, dB.	.1	.1	.1	.1	.2	.3	.3	.4
15	Sound-power level from the blower and system at the butterfly valve, dB.	110	115	104	103	96	90	88	87
16	Sound-power level produced by the butterfly valve, dB. Table 12-10.	114	125	134	137	134	128	119	110
17	Sound-power level after the butterfly valve, dB. Logarithmic combination of row 15 and row 16 (rounded).	115	125	134	137	134	128	119	110
18	Attenuation of 4 ft of round duct, dB.	.1	.1	.1	.1	.2	.3	.3	.4
19	Sound-power level at square corner after butterfly valve, dB. Row 17 minus row 18 (rounded).	115	125	134	137	134	128	119	110
20	Sound-power level produced by the square corner after the butterfly valve, dB. Table 12-11.	117	118	118	114	108	102	99	99
21	Sound-power level produced by the four circular retainer rods, dB. Table 12-12.	65	70	75	77	74	69	63	58
22	Sound-power level at the flow exit of the system, dB. Logarithmic combination of rows 19, 20, and 21.	119	126	134	137	134	128	119	110
23	End reflection loss, dB. Table 9-11.	12	8	4	1	0	0	0	0

Ch. 12   INDUSTRIAL NOISE CONTROL STUDIES   519

Table 12-7 (cont.). System Sound-Power Levels to Determine Level at Exit (See Case Study 3)

Row No.	Description	Octave-Band Center Frequency, Hz							
		63	125	250	500	1000	2000	4000	8000
24	Sound-power level at the flow exit of the system, dB. Row 22 minus row 23.	107	118	130	136	134	128	119	110
25	Sound-power level generated by the flow acting as a jet, dB. Table 12-13.	68	65	60	55	50	45	40	35
26	Total sound-power level at the exit of the system, dB.	107	118	130	136	134	128	119	110

Step 3. Determine the attenuation of the first 90-degree elbow. These values, obtained from Table 9-6, are listed in row 3 of Table 12-7, and represent the attenuation of the sound-power level that was generated upstream of the elbow.

Step 4. Compute the sound-power level produced by the blower immediately after the first 90-degree elbow. This can be obtained by subtracting from row 1 the sum of values in rows 2 and 3. The attenuation values are empirical values presented so that they can be subtracted algebraically from the sound-power levels inside the duct system. The resulting values of sound-power level are given in row 4 of Table 12-7.

At this point the sound-power level which the blower generates and is subsequently attenuated by the 25 ft of duct and the first 90-degree elbow is given by row 4 of Table 12-7. At this location in the system the sound-power level generated by the flow through the 90-degree elbow must be included. The elbow attenuates acoustic energy which was generated upstream; however, it also generates acoustic energy due to flow which propagates downstream of the elbow. The sound power generated by the elbow is computed in the next step.

Step 5. Compute the sound-power level generated by the first 90-degree elbow. This computation is accomplished by utilizing Equation (9-7):

$$L_W = F + G + H \text{ dB} \qquad (9\text{-}7)$$

and is similar to the procedure of Example No. 4 in Chapter 9. For this example the computation is illustrated in Table 12-8.

a. The average velocity in the duct is,

$$V_{avg} = \frac{20{,}200 \text{ cfm}}{\pi \times (1)^2 \text{ ft}^2} = 6430 \text{ fpm}$$

The Strouhal Number is obtained from Equation (9-8), where $V = V_{avg}$ and D = diameter, in inches.

$$S_t = \frac{5D}{V} f = \frac{5 \times 24}{6430} f = .019 f$$

where f is the octave-band center frequency of interest. In row 1 of Table 12-8 are the resulting values for each octave band.

Table 12-8. Sound-Power Levels Generated by the 90-Degree Elbow (See Case Study 3)

Row No.	Description	Octave-Band Center Frequency, Hz							
		63	125	250	500	1000	2000	4000	8000
1	Strouhal Number, S_t, Equation 9-8	1.20	2.4	4.8	9.5	19.0	38.0	76.0	152.0
2	Spectrum Function, F, dB. From Fig. 9-9.	63	53	44	37	35	33	28	24
3	Velocity Function, G, dB. From Fig. 9-12.	15	15	15	15	15	15	15	15
4	Octave Band Width Function, H, dB. Table 9-12.	16	19	22	25	28	31	34	37
5	Sound-power level, dB $F + G + H$. Sum rows 2, 3, and 4.	94	87	81	77	78	79	77	76

b. The spectrum function, F, is obtained from Fig. 9-9 for the various values of Strouhal Number from row 1. The resulting values are given by row 2 of Table 12-8.

c. Determination of the velocity function, G, from Fig. 9-12 is obtained by entering the figure with the flow velocity of 6430 feet per minute and a flow area of three feet as the nearest rounded-off value. The resulting value of 15 dB is entered for every octave band in row 3 of Table 12-8.

d. The correction function, H, is obtained from Table 9-12 for every octave band and given by the values in row 4 of Table 9-27.

e. The octave-band sound-power level generated by the elbow is obtained by summing the values in rows 2, 3, and 4 of Table 12-8. The resulting values are given in row 5.

The sound-power levels generated by the 90-degree elbow and given by row 5 of Table 12-8 are listed in row 5 of Table 12-7.

Step 6. The sound-power level from the blower at the location downstream of the first 90-degree elbow must be combined with the same power level generated by the 90-degree elbow. This is done by combining logarithmically, the values in row 4 with the values in row 5, and listing the result in row 6. Since sound-power levels are decibel quantities, their combination must be treated by logarithmic methods.

From the definition of sound-power level,

$$L_W = 10 \log_{10} \left(\frac{W}{W_{ref}} \right) \text{ dB}$$

the combination of two levels, L_{W1} and L_{W2}, is given by,

$$L_W \text{ (combined)} = 10 \log_{10} \left(\frac{W_1}{W_{ref}} + \frac{W_2}{W_{ref}} \right) \text{ dB}$$

where:

$$\frac{W_1}{W_{ref}} = \log_{10}^{-1} \left(\frac{L_{W1}}{10} \right)$$

Ch. 12 INDUSTRIAL NOISE CONTROL STUDIES

and

$$\frac{W_2}{W_{\text{ref}}} = \log_{10}^{-1}\left(\frac{L_{W2}}{10}\right)$$

Thus,

$$L_W \text{ (combined)} = 10 \log_{10}\left[\log_{10}^{-1}\left(\frac{L_{W1}}{10}\right) + \log_{10}^{-1}\left(\frac{L_{W2}}{10}\right)\right]$$

For the 63-Hz octave band this calculation becomes:

$$L_W \text{ (combined)} = 10 \log_{10}\left[\log_{10}^{-1}\left(\frac{117}{10}\right) + \log_{10}^{-1}\left(\frac{94}{10}\right)\right]$$

$$= 117 \text{ dB (rounded)}$$

This computation is performed for every octave band and the result is given in row 6 of Table 12-7. With a modern electronic engineering calculator this computation can be performed quite easily. Note that the source power level generated by the elbow is at least 10 dB below the blower-generated sound at the location considered. Therefore, the combined level is equal to that due to the blower only and the generation of the elbow can be ignored. This may not be true in general for systems of this type; therefore, the calculation must be completed to determine the relative contribution of various sources.

Step 7. Determine the attenuation of the 24-inch duct for the length of 32 feet between the first and second elbow. The attenuation values in dB/ft are the same as in Step 2, above. Multiplying the values listed in Step 2 by 32 feet results in the attenuation values in row 7 of Table 12-7.

Step 8. Determine the attenuation of the second 90-degree elbow. These values are identical to the first 90-degree elbow as described in Step 3, and are listed in row 8, repeated from the values as given in row 3.

Step 9. Compute the sound-power level after the second 90-degree elbow contributed by the blower and duct elements at a location immediately after the second 90-degree elbow. This is done by subtracting rows 7 and 8 from row 6 in Table 12-7. The resulting values are given in row 9. The sound-power levels in row 8 do not include the sound power generated by the second 90-degree elbow.

Step 10. Calculate the sound-power levels generated by the air flow through the second 90-degree elbow. For this example the sound-power level is identical to the sound generation at the first 90-degree elbow, since the flow parameters are identical. The sound-power levels given in row 5 for the first 90-degree elbow are repeated in row 10 for the second 90-degree elbow.

Step 11. Compute the total sound-power levels after the second 90-degree elbow. This is done by using the identical procedure as described in Step 6, to combine the respective values in rows 9 and 10. Again, for this location, the sound-power levels due to the air-flow noise generation in the elbow are all at least 10 dB less than the sound-power levels in row 9, and therefore, the combined values are identical to the values in row 9, when rounded. The total sound-power level in the air flow after the second elbow are listed in row 11.

522 INDUSTRIAL NOISE CONTROL STUDIES Ch. 12

Step 12. Calculate the attenuation of the 10-foot length of round duct between the elbow and the plenum. The attenuation values in dB/ft were obtained previously, in Step 2, for the 24-inch duct. Multiplying the dB/ft attenuation values of Step 2 by 10 ft gives the values in row 12.

Step 13. Compute the attenuation of the plenum. The equation for the attenuation is given by Equation (9-5) repeated here as follows:

$$\text{Attenuation} = 10 \log_{10} \left[\frac{1}{S_E \left(\frac{\cos \theta}{2\pi d^2} + \frac{1-\alpha}{\alpha S_w} \right)} \right] \text{ dB} \qquad (9\text{-}5)$$

where:

α = acoustic absorption coefficient for the material inside the plenum
S_E = plenum exit area, ft^2
S_w = plenum wall surface area, ft^2
d = distance between entrance and exit, ft
θ = the angle d makes with the normal to the entrance opening.

The above equation predicts the acoustic attenuation for a plenum with a single inlet and a single outlet. For the plenum in this example, there is one inlet and three outlets, as shown by Fig. 12-2A. For the present case of multiple exits there is a proportioning of acoustic energy which transmits to each of the outlets, dependent upon the locations of the outlets and the relative areas of the exit ducts. For a conservative estimation of the plenum attenuation Equation (9-5) can be used as given, although it will show a lower value for the plenum attenuation than will actually result because some of the sound energy propagates to the other ducts. The calculation follows and is tabulated in Table 12-9.

Table 12-9. Calculations of Plenum Attenuation (See Case Study 3)

Row No.	Description	Octave-Band Center Frequency, Hz							
		63	125	250	500	1000	2000	4000	8000
1	Absorption coefficient, α, for the sheet-metal plenum.	.05	.05	.04	.03	.03	.02	.01	.01
2	Calculation of $1 - \alpha$	.95	.95	.96	.97	.97	.98	.99	.99
3	Calculation of $\frac{1-\alpha}{\alpha}$	19.00	19.00	24.00	32.33	32.33	49.00	99.00	99.00
4	Calculation of $\frac{1-\alpha}{\alpha S_w}$	.452	.452	.571	.770	.770	1.167	2.357	2.357
5	Calculation of $\frac{\cos \theta}{2\pi d^2} + \frac{1-\alpha}{\alpha S_w}$	.453	.453	.572	.771	.771	1.168	2.358	2.358
6	Attenuation, dB.	6.0	6.0	5.0	3.7	3.7	1.9	0 (−1.1)	0 (−1.1)

a. The acoustic absorption coefficients for a sheet metal plenum not lined with acoustic material are given in row 1 of Table 12-9. These values are the minimum values from Table 5-1.
b. The term $1 - \alpha/S_w\alpha$ is computed in three steps as outlined in rows 2, 3, and 4 of Table 12-9. The values of $1 - \alpha$ are computed for every octave band from the corresponding values of α in row 1. The values of $(1 - \alpha)/\alpha$ are computed from the values of α and $1 - \alpha$ as given by rows 1 and 2. These values are listed in row 3. The wall surface area, S_w, for the plenum is 42 ft² and the term $(1 - \alpha)/\alpha S_w$ is given in row 4.
c. The term $\cos\theta/2\pi d^2$ is computed from the geometry of the plenum.

$$\cos\theta = \frac{3}{7.5} = 0.4$$

$$d = 7.5 \text{ ft}$$

$$\frac{\cos\theta}{2\pi d^2} = \frac{0.4}{2\pi(7.5)^2} = 0.00113$$

The term $\cos\theta/2\pi d^2 + (1 - \alpha)/\alpha S_w$ is computed by adding 0.00113 to each value in row 4. The sum is given by row 5.

d. Compute the attenuation of the plenum from the values in row 6 and from the exit area S_E given by:

$$S_E = \pi \left(\frac{5}{12}\right)^2 \text{ ft}^2 = .55 \text{ ft}^2$$

The values for attenuation are computed from Equation (9-5) with the values given in row 6. The values of attenuation, in decibels, are listed in row 7 of Table 12-9. These resulting attenuation values are also listed in row 13 of Table 12-7.

Step 14. Compute the attenuation of the 4 ft of round duct between the plenum and the butterfly valve. The attenuation values in dB/ft for the various octave-band center frequencies were given in Step 2; they are independent of the diameter of the duct. Multiplying the values listed in Step 2 by 4 ft results in the attenuation values given by row 14 of Table 12-7.

Step 15. The sound-power level from the blower and the duct system is computed at the butterfly valve. This is done by subtracting the values in rows 12, 13, and 14 of Table 12-7 from the values in row 11. The resulting sound-power level at the butterfly valve which was generated upstream of the valve is given in row 15 of Table 12-7.

Step 16. Compute the sound-power level of the sound generated by the air flow through the butterfly valve. The sound-power level for air flow through a butterfly valve in the choked flow condition was given by Equation (9-22), repeated here:

$$L_W = 10 \log_{10}(6 \times 10^8 \, \dot{m}c^2) \text{ dB} \qquad (9\text{-}22)$$

where:

L_W = overall sound power level, dB
$\dot{m}$ = mass flow rate through the valve, Kg/sec^2
c = speed of sound in the flow media, meter/sec

a. The flow rate through the valve is 7000 cfm which is converted to mass flow rate by,

$$\dot{m} = \rho Q$$

where ρ is the density and Q is the flow rate through the valve. The density of air at 75 °F is .074 lb/ft^3; therefore,

$$\dot{m} = .074 \text{ lb/ft}^3 \times 7000 \text{ ft}^3/\text{min}$$

$$= 518 \text{ lb/min}$$

$$= 518 \times \frac{.4535924}{60}$$

$$= 3.9 \text{ Kg/sec}$$

b. The speed of sound is computed from:

$$c = 49.03 \sqrt{460 + T \, °F} \text{ ft/sec}$$

$$= 49.03 \sqrt{460 + 75}$$

$$= 1134 \text{ ft/sec}$$

$$= 1134 \times .3048$$

$$= 346 \text{ meter/sec}$$

c. The overall sound-power level is computed from Equation (9-22):

$$L_W = 10 \log_{10} (6 \times 10^8 \, \dot{m}c^2)$$

$$= 10 \log_{10} [6 \times 10^8 (3.9) \, 346^2]$$

$$= 144 \text{ dB}$$

This value is listed for every octave-band center frequency in row 1, Table 12-10.

d. The frequency distribution of the air valve sound-power level is determined by Equation (9-23) and Table 9-18:

$$f_p = \frac{c}{5D} \text{ Hz} \qquad (9\text{-}23)$$

where:

f_p = frequency at which the largest sound-power level exists, Hz
c = speed of sound in the flow, meter/sec

Table 12-10. Sound-Power Levels Generated by the Air Valve (See Case Study 3)

Row No.	Description	Octave-Band Center Frequency, Hz							
		63	125	250	500	1000	2000	4000	8000
1	Overall sound-power level, dB. Equation (9-22).	144	144	144	144	144	144	144	144
2	Frequency corrections, dB. Table 9-18.	−30	−19	−10	−7	−10	−16	−25	−34
3	Octave-band sound-power level, dB. Row 1 plus row 2.	114	125	134	137	134	128	119	110

D = flow distance between the duct wall and the air valve, in meters (4 inches in Fig. 12-2, which is .1016 meter)

$$f_p = \frac{346}{5(.1016)} = 681 \text{ Hz}$$

e. The frequency corrections for the octave-band sound-power levels are obtained from Table 9-18 with f_p equal to 681 Hz, as computed above. The value of −7 dB is entered in row 2 of the 500-Hz octave band of Table 12-9 since 681 Hz is within this octave band as given by Table 2-2, Chapter 2. The corrections for the other octave bands are entered in row 2, Table 12-10.

f. The octave-band sound-power levels are obtained from Table 12-10 by subtracting the values in row 2 from those in row 1. The resulting levels are given in row 3.

(The values for the sound-power level produced by the butterfly valve are seen in row 16, Table 12-7.)

Step 17. Compute the sound-power level in the system immediately after the butterfly valve. This calculation is made by logarithmically combining the values for each octave band in rows 15 and 16 in Table 12-7. The procedure is the same as that described in Step 6 of this example. The combined sound-power levels are listed in row 17. Note that except for the 63-Hz octave band the combined levels are identical to the levels produced by the butterfly valve. This is due to the fact that the logarithmic combination of two sound-power levels which differ by 10 dB or more is equal to the larger of the two values.

Step 18. Calculate the attenuation of the 4 ft of duct between the butterfly valve and the square corner in the duct system. This calculation is identical to Step 14, above, and the values in row 14 are repeated in row 18.

Step 19. Compute the sound-power level of the system at the square corner beyond the butterfly valve. This is done by subtracting the values in row 18 from those in row 17, Table 12-7. The resulting values, rounded to the nearest decibel, are found in row 19.

Step 20. Compute the sound-power level generated by the air flow interaction with the square corner. Prediction equations are not available for this exact case; however a method of predicting the sound-power-level generation due to air flow in a square elbow, was given in Chapter 9 in the section entitled "Air Flow Noise." In order to estimate the sound-power level by the 90-degree turn after the butterfly valve, the system will be assumed to be a rectangular 90-degree elbow with area upstream of the elbow equal to the area of the 10-inch-diameter duct and the area downstream, equal to 63 in.^{2}, i.e., the circumference times the height of 2 inches, from Fig. 12-3.

The sound-power level is determined from Equation (9-7).

$$L_W = F + G + H \text{ decibels} \tag{9-7}$$

where:

L_W = octave-band sound-power level, dB
F = *Spectrum Function*, dB
G = *Velocity Function*, dB
H = *Correction Function*, dB

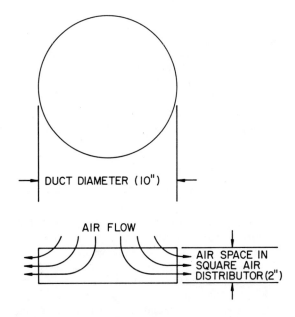

Fig. 12-3. Entrance area to square air-distributor duct. Required to determine the air flow at the entrance to the square air-distributor in order to obtain the sound power generated in this region.

Ch. 12   INDUSTRIAL NOISE CONTROL STUDIES

Table 12-11. Sound-Power Levels Generated by the Square 90-Degree Elbow (See Case Study 3)

Row No.	Description	Octave-Band Center Frequency, Hz							
		63	125	250	500	1000	2000	4000	8000
1	Strouhal Number, S_t. Equation 9-8.	.18	.35	.70	1.40	2.80	5.60	11.20	22.40
2	Spectrum Function, F, dB. Fig. 9-9.	73	71	68	61	52	43	37	34
3	Velocity Function, G, dB. Fig. 9-12.	28	28	28	28	28	28	28	28
4	Correction Function, H, dB. Table 9-12.	16	19	22	25	28	31	34	37
5	Sound-power level, dB, $F + G + H$. Sum rows 2, 3, and 4.	117	118	118	114	108	102	99	99

The calculation is given in Table 12-11 for the various functions and is described as follows:

a. The values of the *Spectrum Function*, F, are obtained from the *Strouhal Number* given by Equation (9-8):

$$S_t = \frac{5fD}{V} \qquad (9\text{-}8)$$

where:

    f = octave-band center-frequency, Hz
    D = duct diameter, in inches, or for rectangular ducts $D = (4/\pi)$ (area)
    V = average flow velocity, in feet per minute

The area used in calculating D is the entrance area to the air distributor from the duct, which is 63 sq in., as shown in Fig. 12-3.

$$D = (4/\pi)(63 \text{ in.}^2) = 9 \text{ inches (rounded)}$$

$$V = \frac{7000 \text{ cfm}}{63 \text{ in.}^2 \times 1 \text{ ft}^2/144 \text{ in.}^2} = 16000 \text{ fpm}$$

$$S_t = \frac{5fD}{16000} = \frac{5(9)}{16000} f$$

$$= .0028 f$$

The *Strouhal Number* is computed for each octave band by multiplying the octave-band center frequency by .0028. The resulting values are listed in Row 1 of Table 12-11. From the values of the *Strouhal Number*, S_t, in row 1, the values of the *Spectrum Function*, F, are obtained from Fig. 9-9. The values obtained are listed in row 2 of the table.

b. The *Velocity Function*, G, is obtained from Fig. 9-12 by entering the graph with 16000 fpm and an equivalent area of 63 in.^{2}, which is equal to 0.44 ft^2. The flow velocity of 16,000 fpm, however, is off the scale of the figure; therefore an extrapolation is made of the logarithmic scale to the

16,000 fpm value. A linear extrapolation of the vertical scale is also made to obtain a value of G equal to 28 dB. Row 3 lists this value of G for every octave-band center frequency.

c. Values for the *Correction Function*, H, are obtained from Table 9-12 and are listed in Table 12-11 on row 4.

d. The sound-power level generated by the flow through the 90-degree square elbow is obtained by summing the values in rows 2, 3, and 4. The resulting octave-band sound-power levels are then given in row 5.

The sound-power levels computed for the 90-degree square elbow given in row 5 are also listed in row 20 of Table 12-7. These are estimates of the sound power generated by the square corner which was modeled as a rectangular elbow.

Step 21. Compute the sound-power level produced by the flow over the ½-inch-diameter retainer rods in the flow. The overall sound-power level generated by one rod is computed from Equation (9-14) repeated here:

$$L_W = 10 \log_{10} \left(\frac{3.125 \times 10^7 \, \rho V^6 D^2}{c^3} \right) \tag{9-14}$$

where:

L_W = overall sound-power level, dB
ρ = density of air in the flow, Kg/meter³
V = flow velocity, meter/sec
D = equivalent duct diameter, meter
c = sound velocity in the flow, meter/sec

The frequency corrections are given by the values in Table 9-15 where the peak frequency, f_p, is computed from Equation (9-15) for circular rods:

$$f_p = \frac{0.2 \, V}{d} \tag{9-15}$$

where d is the diameter of the rod, in meters.

The parameters needed for the calculation are as follows, with the retainers located 2 inches from the outlet of the flow system:

Flow area of retainers = (20 inches × 2 inches) × 4

= 160 in.² = 1.11 ft²

Flow velocity, $V = \dfrac{7000 \text{ cfm}}{1.11 \text{ ft}^2}$ = 6306.3 ft/min

= 105 ft/sec = 32 meter/sec

ρ = 1.19 Kg/meter³ for air at 75 °F

Ch. 12   INDUSTRIAL NOISE CONTROL STUDIES   529

$$\text{Equivalent } D = \left[\frac{4}{\pi} \times 40\right]^{1/2} = 7.1 \text{ in.} = .18 \text{ meter}$$

$$c = 49.03 \sqrt{460 + 75} = 1134 \text{ ft/sec} = 346 \text{ meter/sec}$$

$$d = 0.5 \text{ inch} = .0127 \text{ meter}$$

a. Compute the overall sound-power level from Equation (9-15) repeated above:

$$L_W = 10 \log_{10} \left(\frac{3.125 \times 10^7 \, \rho V^6 D^2}{c^3}\right)$$

$$= 10 \log_{10} \left(\frac{3.125 \times 10^7 \times 1.19 \times 32^6 \times .18^2}{346^3}\right)$$

$$= 75 \text{ dB (rounded)}$$

This value is listed for every frequency in row 1 of Table 12-12.

b. Compute the peak frequency, f_p.

$$f_p = \frac{(0.2)(32)}{.0127} = 504 \text{ Hz}$$

This frequency is in the 500-Hz octave band.

c. Obtain the frequency corrections from Table 9-15 and enter the value of -4 dB for f_p in the 500-Hz octave band. The remaining values are given by row 2 of Table 12-12.

d. Compute the sound-power levels for each octave band by adding the values in rows 1 and 2 of the table. The resulting values are the sound-power levels for one rod in the flow and are given in row 3.

Table 12-12. Sound-Power Levels Generated by Retainer Rods in Square Air Distributor (See Case Study 3)

Row No.	Description	Octave-Band Center Frequency, Hz							
		63	125	250	500	1000	2000	4000	8000
1	Overall sound-power level, dB. Equation (9-14).	75	75	75	75	75	75	75	75
2	Frequency corrections, dB. Table 9-15.	−16	−11	−6	−4	−7	−12	−18	−23
3	Sound-power level of one rod, dB. Row 1 plus row 2.	59	64	69	71	68	63	57	52
4	Correction for four rods, dB.	+6	+6	+6	+6	+6	+6	+6	+6
5	Sound-power level for four rods, dB. Row 3 plus row 4.	65	70	75	77	74	69	63	58

e. Compute the sound-power level for four rods. Since the four rods are identical they will each produce identical acoustic power, in watts, if it is assumed that the flow over each rod is also identical. From the definition of sound-power level the sound-power level of the four identical rods can be determined as follows:

$$L_W = 10 \log_{10} \left(\frac{W}{W_{\text{ref}}}\right) \text{dB}$$

where:

L_W = the sound power level of one rod, dB
W = acoustic power of one rod, watts
W_{ref} = reference power, 10^{-12} watts.

The sound-power level of four rods at equal sound power would be,

$$L_W \text{ (four rods)} = 10 \log_{10} \left(\frac{4W}{W_{\text{ref}}}\right)$$

$$= 10 \log_{10} \left(\frac{W}{W_{\text{ref}}}\right) + 10 \log_{10} (4)$$

$$= L_W \text{ (one rod) dB} + 6 \text{ dB}$$

Therefore, the sound-power level of four identical rods is 6 dB greater than the sound-power level of one of the rods. To convert the values in row 3 of Table 12-12, for one rod, to values for the four rods, 6 dB is added to each octave band as indicated by Row 4. Summing rows 3 and 4 will give the sound-power level generated by the four rods with the result given in row 5 of the table. These values are also listed in row 21 of Table 12-7.

Step 22. Calculate the total sound-power level from the system incident upon the exit opening of the flow system. This computation is made by the logarithmic combination of the values in rows 19, 20, and 21 of Table 12-7. The procedure is the same as described in Step 6 of this example except that three sound-power levels are combined. The following example for the 63-octave-band center frequency will illustrate the method:

$$L_W = 10 \log_{10} \left[\log^{-1} \left(\frac{115}{10}\right) + \log^{-1} \left(\frac{117}{10}\right) + \log^{-1} \left(\frac{65}{10}\right) \right]$$

$$= 119 \text{ dB}$$

The resulting values are given in row 22 of Table 12-7. These values are the octave-band sound-power levels incident upon the opening at the system outlet at the molds.

Step 23. Determine the end reflection loss at the outlet of the system. These values were given in Table 9-11 for various duct sizes, in square inches. If each side of the outlet is considered a duct opening with a 48 sq in. area, the values in row 23 of Table 12-7 are obtained for the end reflection loss. The footnote

Ch. 12 INDUSTRIAL NOISE CONTROL STUDIES 531

to Table 9-11 specifies that if the outlet is near other surfaces the next larger entry should be used. For this reason the values for 100 sq in. are utilized for this example.

Step 24. Compute the sound-power level at the flow exit due to the system upstream of the exit. This is calculated by subtracting the values in row 23 from the values of row 22. The resulting sound-power levels are given in row 24 of Table 12-7.

Step 25. Calculate the sound-power level of the air flow leaving the system and acting as an air jet. The expression which can be used for this computation is given by Equation (9-26) repeated here:

$$L_W = 10 \log_{10} (5 \dot{m} V^2 \xi \times 10^{11}) \text{ decibels} \qquad (9\text{-}26)$$

where:

L_W = overall sound-power level, dB
$\dot{m}$ = mass flow rate, Kg/sec
V = exit velocity of the jet, meter/sec
ξ = efficiency constant dependent on the jet velocity as given by Table 9-20.

Equation (9-26) predicts the sound-power level for a circular jet and does not strictly apply for the slot type of jet in this example. The above equation will be used as a means to estimate the sound-power level generated by the flow at the slot.

a. The average velocity at the exit is computed from the volumetric flow rate of 7000 cfm and the exit area.

$$\text{exit area} = 4 \, (2 \text{ in.} \times 24 \text{ in.}) \times \frac{1 \text{ ft}^2}{144 \text{ in.}^2}$$

$$= 1.33 \text{ ft}^2$$

$$V_{avg} = \frac{7000 \text{ cfm}}{1.33 \text{ ft}^2} \times \frac{1 \text{ min}}{60 \text{ sec}} = 87.7 \text{ ft/sec}$$

$$= 26.7 \text{ meter/sec}$$

$$\text{mach number}, M = \frac{V_{avg}}{\text{velocity of sound}}$$

The velocity of sound was computed in Step 16b, and was found to be 1134 ft/sec; therefore,

$$M = \frac{V_{avg}}{1134 \text{ ft/sec}} = \frac{87.7}{1134} = 0.08$$

From Table 9-20, with the mach number equal to 0.08, the efficiency constant ξ is found to be,

$$\xi = 8 M^3 \times 10^{-5}$$

The mass flow rate for the section of the duct system containing the air valve was previously computed in Step 16a to be:

$$\dot{m} = 3.9 \text{ Kg/sec}$$

From the above values, Equation (9-26) becomes:

$$L_W = 10 \log_{10} [5(3.9)(26.7)^2 (8 \times 0.08^3 \times 10^{-5})(10)^{11}]$$

$$L_W = 78 \text{ dB}$$

This value is listed for every octave band in row 1 of Table 12-13. The overall sound-power level must be corrected by means of Table 9-22 to obtain the octave-band sound-power levels. The peak frequency, f_p, is computed from Equation (9-27) repeated here:

$$f_p = \frac{V}{5D} \text{ Hz} \qquad (9\text{-}27)$$

where:

f_p = peak frequency for jet sound-power level, Hz
V = exit velocity of the jet, ft/sec or meter/sec
D = jet diameter, ft or meter.

The equivalent circular-jet diameter is computed from:

$$D = \left(\frac{4}{\pi} \text{area}\right)^{1/2} = \left(\frac{4}{\pi} 192 \text{ in.}^2\right)^{1/2}$$

$$= 15.6 \text{ in.}$$

$$= .4 \text{ meter}$$

Therefore:

$$f_p = \frac{26.7}{5(.4)} = 13.4 \text{ Hz}$$

Table 12-13. Sound-Power Levels of Flow at System Exit, Acting as a Jet (See Case Study 3)

Row No.	Description	Octave-Band Center Frequency, Hz							
		63	125	250	500	1000	2000	4000	8000
1	Overall sound-power level, dB. Equation (9-26).	78	78	78	78	78	78	78	78
2	Frequency correction, dB. Table 9-22.	−10	−13	−18	−23	−28	−33	−38	−43
3	Octave-band sound-power level, dB. Row 1 plus row 2.	68	65	60	55	50	45	40	35

Ch. 12   INDUSTRIAL NOISE CONTROL STUDIES   533

From Table 2-2, of Chapter 2, 13.4 Hz is within the 16-Hz octave band and four times f_p is 53.6 Hz which is within the 63-Hz octave band. The correction factors for the sound-power level are obtained from Table 9-22, with the correction for $4f_p$ being -10 dB. The value of -10 is listed in row 2 of Table 12-13, for the 63-Hz octave band. The values of -13 and -18 are obtained for $8f_p$ and $16f_p$, respectively, from Table 9-22. Values higher than $16f_p$ do not appear in the table; therefore, values are extrapolated for higher frequencies by subtracting 5 dB from each proceeding octave-band value. This assumption can be made for the higher frequencies (above $16f_p$). In this manner the values in row 2 of the table are obtained.

Step 26. The octave-band sound-power level at the exit of the flow system of Fig. 12-2 is computed by the logarithmic combination of the values in rows 24 and 25, with the resulting values given in row 26 of Table 12-7. Note that the values of sound-power level generated by the jet action at the exit is approximately 80 decibels below the sound-power level generated upstream in the system. Because of this the jet-action sound generation is not a factor in this example, as is indicated by a comparison of rows 24, 25, and 26.

The octave-band sound-power levels at the exit of the flow system are given by row 26 of Table 12-7. However, the sound-power level at any other location in the flow system can be obtained from the appropriate row in the table.

From the system calculations summarized in Table 12-7 the sound-pressure level at the operator location can be determined. The operator location was given as a position 10 feet away from the butterfly valve in the duct and 12 feet away from the flow exit near the ½-inch retainers. The sound-power level after the butterfly valve is given in row 17 of Table 12-7 and the sound-power level at the exit, by row 26. The sound-power level of the butterfly valve is attenuated by the duct wall since the values computed in this table were the levels inside the duct. The calculations for the operator location sound-pressure level are described below and summarized in Table 12-14.

Calculate the sound-power level outside the duct at the butterfly valve. This is done by subtracting the transmission loss of the duct wall from the sound-power level inside the duct.

a. The octave-band sound-power levels inside the duct are obtained from row 17 of Table 12-7 and are listed in row 1 of Table 12-14.
b. The nominal transmission loss of the duct wall is computed from Equation (9-30) for the frequency equal to and below the ring frequency.

$$TL_{nominal} = 10 \log_{10}\left(\frac{t}{D}\right) + 50 \text{ dB} \qquad (9\text{-}30)$$

where:

$TL_{nominal}$ = nominal transmission loss below the ring frequency, dB
t = wall thickness of the duct
D = nominal duct diameter.

Table 12-14. Determining the A-Scale Sound-Pressure Level of the Operator Location (See Case Study 3)

Row No.	Description	Octave-Band Center Frequency, Hz							
		63	125	250	500	1000	2000	4000	8000
1	Sound-power level, dB, inside duct after the butterfly valve. Row 17, Table 12-7.	115	125	134	137	134	128	119	110
2	Nominal transmission-loss of duct wall, dB. Equation (9-28).	28	28	28	28	28	28	28	28
3	Frequency corrections to nominal transmission-loss, dB. Table 9-23.	−8	−7	−6	−5	−4	−3	−2	−4
4	Transmission loss of the duct wall, dB. Row 2 plus row 3.	20	21	22	23	24	25	26	24
5	Equivalent sound-power level outside the duct, dB. Row 1 minus row 4.	95	104	112	114	110	103	93	86
6	Correction to convert L_{Wi} to L_{pi}, dB. Equation (4-1).	−20.7	−20.7	−20.7	−20.7	−20.7	−20.7	−20.7	−20.7
7	Octave-band sound-pressure level of valve at operator location. Row 5 plus row 6.	74	83	91	93	89	82	72	65
8	Sound-power level of exit to flow system, dB. Row 26, Table 12-7.	107	118	130	136	134	128	119	110
9	Correction to convert L_{Wi} to L_{pi}, dB. Equation (4-1).	−22.3	−22.3	−22.3	−22.3	−22.3	−22.3	−22.3	−22.3
10	Octave-band sound-pressure level of exit at operator location, dB. Row 8 plus row 9.	85	96	108	114	112	106	97	88
11	Total sound-pressure level at operator location.	85.3	96.2	108.1	114.0	112.0	106.1	97.0	88.0
12	A-scale weighting factors, dB. Table 2-3.	−26.3	−16.1	−8.6	−3.2	0	+1.2	+1.0	−1.1
13	A-weighted sound-pressure levels, dB. Row 11 plus row 12.	59.0	80.1	99.5	110.8	112.0	107.3	98.0	86.9
14	A-scale sound-pressure level, dB.	115 dB(A)							

The ring frequency, f_r, is given by Equation (9-27):

$$f_r = \frac{c_l}{\pi D} \quad \text{Hz} \tag{9-29}$$

where:

f_r = ring frequency, Hz
c_l = speed of sound in the duct material, meter/sec
D = nominal duct diameter, meter

The thickness of 16-gage sheet metal is .0598 in.; therefore:

$$\text{TL}_{\text{nominal}} = 10 \log_{10} (.00598) + 50 \text{ dB}$$
$$= -22 + 50$$
$$= 28 \text{ dB (rounded)}$$

The ring frequency is computed from Equation (9-29) above, with c_l = 5100 meter/sec for steel and D = 10 in. = 0.254 meter,

$$f_r = \frac{5100}{\pi \times .0254} = 6391 \text{ Hz}$$

The value of 6391 Hz is within the 8000-Hz octave band; therefore, in this example, all octave-band duct transmission-loss values can be computed from Equation (9-30) and 28 dB is listed for every octave band in row 2 of Table 12-14.

c. The corrections to the nominal transmission loss are obtained from Table 9-23, with f_r equal to 6391 Hz and being within the 8000-Hz octave band. The corrections are given in row 3 of Table 12-14. The value for the 63-Hz octave band is obtained by extrapolation of Table 9-23, where one decibel is subtracted for each successively lower octave band.
d. The transmission loss for the 10-inch duct is obtained by summing the values in rows 2 and 3, with the results given in row 4.
e. The equivalent sound-power level outside the duct is computed by subtracting the values in row 4 from those in row 1. The results of this calculation are listed in row 5.
f. The sound-pressure level at a distance of 10 feet, due to the sound-power-level outside duct at the butterfly valve, is computed from Equation (4-1) and repeated here:

$$L_{pi} = L_{Wi} + 10 \log_{10} \left(\frac{Q_{\theta i}}{4\pi r^2} + \frac{4}{A} \right) \text{ dB} \tag{4-1}$$

where:

L_{pi} = octave-band sound-pressure level of the i'th octave band, dB reference 2×10^{-5} N/meter^2

L_{Wi} = octave-band sound-power level of the i'th octave band, dB reference 10^{-12} watt

$Q_{\theta i}$ = directivity of the source ($Q_{\theta i}$ is given in Fig. 4-1)

r = distance from source to receiver, meters

A = room absorption, sabins

If the source is assumed to be located in a large building or factory space such that the reflecting surfaces are far from the source, compared to 10 feet, an estimate of the sound pressure can be obtained with the term $4/A$ neglected. Thus,

$$L_{pi} \simeq L_{Wi} + 10 \log_{10}\left(\frac{Q_{\theta i}}{4\pi r^2}\right) \text{ dB}$$

From Fig. 4-1, the source is not near reflecting surfaces $Q_{\theta i} = 1$, and

$$r = 10 \text{ feet} = 3.048 \text{ meter}$$

$$L_{pi} = L_{Wi} + 10 \log_{10}\left[\frac{1}{4\pi(3.048)^2}\right]$$

$$\simeq L_{Wi} - 20.7 \text{ dB}$$

The above equation applies for every octave band and the sound-pressure level is 20.7 dB less than the sound-power level. The value of -20.7 dB is entered for every octave band in row 6 of Table 12-14. The octave-band sound-pressure level at the operator location is obtained by the sum of values of rows 5 and 6 as given in row 7. These values are the sound-pressure levels at the operator location due to the butterfly valve only, and are not the total sound-pressure level at this location.

g. Calculate the contribution to the operator location sound-pressure level due to the sound power from the exit of the flow system. The sound-power levels were given by row 26 of Table 12-7 and are repeated in row 8 of Table 12-14. The correction to the sound power level is again given by Equation (4-1) in the form from the previous step.

$$L_{pi} = L_{Wi} + 10 \log_{10}\left(\frac{Q_{\theta i}}{4\pi r^2}\right) \text{ dB} \qquad (4\text{-}1)$$

with

$$Q_{\theta i} = 1$$

$$r = 12 \text{ feet} = 3.66 \text{ meters}$$

$$L_{pi} = L_{Wi} + 10 \log_{10}\left[\frac{1}{4\pi(3.66)^2}\right]$$

$$= L_{Wi} - 22.3 \text{ dB}$$

The value of -22.3 dB is listed for every octave band of Table 12-14, in row 9. The contributions of the exit to the operator-location sound-pressure levels are obtained by adding the values in rows 8 and 9 and the result is shown in row 10.

h. Compute the sound-pressure levels at the operator location due to both sources. This calculation is made by combining logarithmically, the values in rows 7 and 10. From the definition of sound-pressure level,

$$L_p = 10 \log_{10} \left(\frac{p}{p_{\text{ref}}}\right)^2 \text{ dB}$$

The combination of two values L_{p1} and L_{p2} is given by:

$$L_{p(\text{combined})} = 10 \log_{10} \left[\left(\frac{p_1}{p_{\text{ref}}}\right)^2 + \left(\frac{p_2}{p_{\text{ref}}}\right)^2\right] \text{ dB}$$

where:

$$\left(\frac{p_1}{p_{\text{ref}}}\right)^2 = \log_{10}^{-1}\left(\frac{L_{p1}}{10}\right)$$

and

$$\left(\frac{p_2}{p_{\text{ref}}}\right)^2 = \log_{10}^{-1}\left(\frac{L_{p2}}{10}\right)$$

Thus,

$$L_p(\text{combined}) = 10 \log_{10} \left[\log_{10}^{-1}\left(\frac{L_{p1}}{10}\right) + \log_{10}^{-1}\left(\frac{L_{p2}}{10}\right)\right] \text{ dB}$$

For the 63-Hz octave band this calculation becomes,

$$L_p(\text{combined}) = 10 \log_{10} \left[\log_{10}^{-1}\left(\frac{74}{10}\right) + \log_{10}^{-1}\left(\frac{85}{10}\right)\right] \text{ dB}$$

$$= 85.3 \text{ dB}$$

This computation is performed for the remaining octave bands, the results are given in row 11.

i. Compute the A-scale sound-pressure level at the operator location. First, the octave band A-weighting factors are obtained from Table 2-3 of Chapter 2, and listed in row 12 of Table 12-14. The A-weighted sound-pressure levels are determined by summing the values in rows 11 and 12, listing the results in row 13.

The A-scale sound-pressure level is computed by the logarithmic combination of all values in row 13 as follows:

$$L_p(\text{A-scale}) = 10 \log_{10} \left[\log_{10}^{-1}\left(\frac{L_{p1}}{10}\right) + \log_{10}^{-1}\left(\frac{L_{p2}}{10}\right) + \cdots \log_{10}^{-1}\left(\frac{L_{pn}}{10}\right)\right]$$

where the values $L_{p1}, L_{p2}, \ldots L_{pn}$ are the octave-band, A-weighted sound-pressure levels. This calculation gives 115 dBA for the sound-pressure level at the operator location.

538 INDUSTRIAL NOISE CONTROL STUDIES Ch. 12

The computed sound level of 115 dBA at the operator location is in excess of the OSHA requirement of 90 dBA for an 8-hour exposure. This means that the system must be redesigned to reduce the sound level at the operator position. Note from Table 12-7 that the butterfly control valve is a major noise source; it is, in fact, the source which dominates the remaining sources in the system. A flow-control valve which produces lower sound levels could be considered from supplies of flow systems. The alternative is to enclose the section of duct containing the flow valve with a duct-wrapping material or build an enclosure to surround the duct. A partial enclosure around the flow exit may also be necessary to meet the desired operator level. Techniques for the design of enclosures to meet specified noise-reduction requirements are described in Chapter 5.

Case Study 4. *Process Furnace Noise Prediction and Reduction*

Shown in Fig. 12-4 is a proposed gas-fired furnace for a continuous process. A conveyor system moves material through the furnace through openings at each end. The bottom of the furnace is open to accommodate the conveyor system

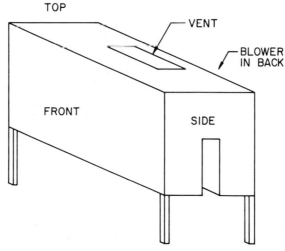

Fig. 12-4. Schematic view of surface areas required to calculate the sound-power level of a process furnace.

and the top has openings for the exhaust gases vented to the factory space. A blower located behind the furnace is to be used to supply the combustion air. The blower is to be supplied with an inlet silencer to control sound radiated directly from the blower to work locations.

The blower specifications are as follows:

> Number of blades = 27 (radial)
> Rotational speed = 3520 rpm
> Static pressure head = 36 inches H_2O
> Flow Rate, Q = 1000 cfm
> Inlet air temperature = 70 °F

Ch. 12 INDUSTRIAL NOISE CONTROL STUDIES 539

Table 12-15. Surface Areas of Process Furnace for Case Study 4

Surface	Area, meter^2
End openings, each	0.088
Bottom opening,	0.176
Exhaust on top	0.064
Total exit areas	0.416
Total inside area (including openings)	14.22
Area covered with refractory	13.80

The surface areas of the furnace interior are as given in Table 12-15; all inside surfaces are covered with fire brick. The temperature at the end openings is 800 °F.

Six identical gas burners are to be used inside the furnace. The acoustic power output, in watts, of one burner is given in Column 2, Table 12-17. These values were supplied by the burner manufacturer as obtained experimentally in an acoustic rating furnace. The burner firing rates are optimal for minimum noise generation at necessary Btu rating.

The steps in predicting the sound level for the system are given as follows:

Step 1. Compute the total acoustic power input to the furnace. The power input is due to the six burners and the combustion blower.

a. The acoustic power produced by one burner is obtained from the burner manufacturer as given in Column 2, Table 12-17. Multiplying these values by six results in the sound power, in watts, produced by the six burners. The resulting values are listed in Column 3 of Table 12-17.

b. The sound-power level of the blower is estimated by Equation (9-4) of Chapter 9, and repeated here:

$$L_W = K_W + (10 \log_{10} Q + 20 \log_{10} P) + \text{BFI dB} \qquad (9\text{-}4)$$

where:

L_W = Octave-band sound-power level, dB
K_W = Specific sound-power level, dB
Q = Flow rate, cfm
P = Total pressure, inches of water
BFI = Blade frequency increment

The procedure for computing the sound-power level, L_W, was given by Example No. 1 of Chapter 9. The calculations are summarized in Table 12-16 for this example without presenting the details of the procedure.

The sound power, in watts, is obtained from the definition of sound-power level, L_W

$$L_W = 10 \log_{10} \left(\frac{W}{10^{-12}} \right) \text{dB}$$

Table 12-16. Sound-Power Levels at Blower Inlet and Outlet (See Case Study 4)

Description	Octave-Band Center Frequency, Hz							
	63	125	250	500	1000	2000	4000	8000
Specific Sound-Power Level (K_W), dB. From Fig. 9-3.	48	45	43	43	38	33	30	29
Correction factor $10 \log_{10} Q + 20 \log_{10} P$, dB.	61	61	61	61	61	61	61	61
Blade frequency increment, BFI From Fig. 9-3.	0	0	0	0	0	8	0	0
Correction to convert total sound power to inlet or outlet alone.	−3	−3	−3	−3	−3	−3	−3	−3
Sound-power level, dB, for inlet or outlet.	106	103	101	101	96	99	88	87

where:

L_W = sound power level, dB
W = acoustic power, watts

Therefore,

$$W = 10^{-12} \log_{10}^{-1} \left(\frac{L_W}{10}\right) \text{ watts}$$

Using this relation, the values in the last row of Table 12-17 were converted to acoustic power, in watts, and listed in Column 4 of Table 12-17.

Table 12-17. Acoustic Power Input to the Furnace (See Case Study 4)

Octave-Band Center Frequency, Hz	Acoustic Power of One Burner,* Watts	Acoustic Power of Six Burners, Watts	Acoustic Power of Blower, Watts	Total Acoustic Power, Watts
63	0.01000	0.0600	0.0398	0.0998
125	0.01180	0.0708	0.0200	0.0908
250	0.01485	0.0891	0.0126	0.1017
500	0.04697	0.2818	0.0126	0.2944
1000	0.0118	0.0708	0.0040	0.0748
2000	0.00075	0.0045	0.0080	0.0125
4000	0.00012	0.0007	0.0006	0.0013
8000	0.00010	0.0006	0.0005	0.0011

*The acoustic power is determined experimentally with a single burner in an acoustic rating furnace. These values are supplied by the burner manufacturer.

Ch. 12   INDUSTRIAL NOISE CONTROL STUDIES   541

The total sound-power input to the furnace, in watts, is given by the values in Column 5 and are obtained by summing values in Columns 3 and 4.

Step 2. Compute the acoustic intensity inside the furnace from Equation (1-36) Chapter 1, repeated here,

$$I = W\left(\frac{4}{R'} + \frac{1}{S_T}\right) \qquad (1\text{-}36)$$

where:

I = acoustic intensity in watt/meter^2
W = sound power, watts
R' = enclosure constant = $S_T \bar{\alpha}/(1 - \bar{\alpha})$, meter^2
S_T = surface area of enclosure, meter^2
$\bar{\alpha}$ = average acoustic absorption coefficient

The average acoustic absorption coefficient is calculated from:

$$\bar{\alpha} = \frac{1}{S_T} \sum_{i=1}^{i=N} \alpha_i S_i$$

with

α_i = acoustic absorption of the i'th surface
S_i = area of the i'th surface, meter^2

a. The acoustic absorption of the furnace fire-brick lining is assumed to be the same as construction brick. The values for the acoustic absorption coefficient are listed in row 1 of Table 12-18. In Chapter 5, Table 5-1 gives these values as the minimum values for brick.
b. The interior surface area covered with brick is 13.80 meters^2, obtained from Table 12-15 and listed in row 2 of Table 12-18.
c. The acoustic absorption coefficient for the openings in the furnace are assumed to be 1.00, as listed in row 3.
d. The total opening area of the furnace, given in row 4, is .42 meter^2.
e. The average absorption coefficient, $\bar{\alpha}$, is given by the equation in row 5. It is computed from the product of values in row 1 and row 2, plus the product of values in row 3 and row 4. This sum is then divided by the total interior area of 14.22 meters^2. The resulting values are given in row 5.
f. The enclosure constant, R', is computed from:

$$R' = S_T \frac{\bar{\alpha}}{1 - \bar{\alpha}}$$

where S_T is the total interior area, 14.22 meters^2 and $\bar{\alpha}$ values are given in row 5 of Table 12-18. The values for the enclosure constant resulting from the calculation are given in row 6.

g. Acoustic intensity inside the furnace is computed from Equation (1-36). The term $(4/R' + 1/S_T)$ is computed from the values of R' in row 6 of

Table 12-18. Sound-Pressure Level at Worker Position
(See Case Study 4)

Row No.	Description	Octave-Band Center Frequency, Hz							
		63	125	250	500	1000	2000	4000	8000
1	Acoustic absorption coefficient of brick, α_i	.01	.01	.01	.01	.02	.02	.02	.02
2	Furnace area covered with brick, meter^2.	13.80	13.80	13.80	13.80	13.80	13.80	13.80	13.80
3	Acoustic absorption coefficient of openings.	1.00	1.00	1.00	1.00	1.00	1.00	1.00	1.00
4	Total open area, meter^2.	.42	.42	.42	.42	.42	.42	.42	.42
5	Average acoustic absorption coefficient, $\bar{\alpha}$ $$\bar{\alpha} = \frac{1}{S_T} \sum_{i=1}^{i=N} \alpha_i S_i$$	.04	.04	.04	.04	.05	.05	.05	.05
6	Enclosure Constant; $R' = S_T \dfrac{\bar{\alpha}}{1 - \bar{\alpha}}$	.59	.59	.59	.59	.75	.75	.75	.75
7	$\dfrac{4}{R'} + \dfrac{1}{S_T}$	6.85	6.85	6.85	6.85	5.40	5.40	5.40	5.40
8	Acoustic power inside furnace, W, watt. (From Table 12-17, Column 5)	.0998	.0908	.1017	.2944	.0748	.0125	.0013	.0011
9	Acoustic intensity inside furnace, watt/meter^2, $I = W\left(\dfrac{4}{R'} + \dfrac{1}{S_T}\right)$; row 7 times row 8.	.684	.622	.697	2.017	.404	.068	.007	.006
10	Effective acoustic power from top, watt. $W_{eff} = \eta IA = (.3)I(.064)$	.0131	.0119	.0134	.0387	.0078	.0013	.0001	.0001
11	Effective acoustic power from each end, watt. $W_{eff} = \eta IA = (.3)I(.088)$	.0181	.0164	.0184	.0532	.0107	.0018	.0002	.0002
12	Effective acoustic power from bottom, watt. $W_{eff} = \eta IA = (1)I(.176)$	.1204	.1095	.1227	.3550	.0711	.0120	.0012	.0011
13	Total effective acoustic power, W_T, watt. Row 10 plus two times row 11, plus row 12.	.1697	.1542	.1729	.5001	.1003	.0169	.0017	.0016
14	Effective sound-power level, dB. $L_{Wi} = 10 \log_{10} (W_T/10^{-12})$	112.3	111.9	112.4	117.0	110.0	102.3	92.3	92.0
15	Sound-pressure level at worker location, dB. $L_{pi} = L_{Wi} - 20 \log_{10} r - 11$	88.6	88.2	88.7	93.3	86.3	78.6	68.6	68.3
16	A-Scale weighting factors, dB. (From Table 2-3, Chapter 2).	−26.2	−16.1	−8.6	−3.2	0.0	+1.2	+1.0	−1.1

Table 12-18 (cont.). Sound-Pressure Level at Worker Position (See Case Study 4)

Row No.	Description	Octave-Band Center Frequency, Hz							
		63	125	250	500	1000	2000	4000	8000
17	A-weighted sound-pressure levels, dB. Row 15 plus row 16.	62.4	72.1	80.1	90.1	86.3	79.8	69.6	67.2
18	Equivalent A-Scale sound-pressure level, dBA.					92			

Table 12-18 and from the total surface area of 14.22 meters². The results of the calculations are listed in row 7. The acoustic power inside the furnace is obtained from Table 12-17, Column 5, as repeated here in row 8 of Table 12-18. The intensity inside the furnace is computed by multiplying the values of row 7 by the values in row 8. As given by Equation (1-36), the resulting acoustic intensity inside the furnace is the product of rows 7 and 8, with the answer as given in row 9.

Step 3. Calculate the sound-pressure level of the worker location.

a. The effective acoustic power at each location is given by $W_{eff} = \eta I A$ where η is the radiation coefficient given in Table 5-16, Chapter 5. I is the acoustic intensity of each opening and A is the area of the opening. The effective power from the vent at the *top* of the furnace is computed in row 10 of Table 12-18 with $\eta = 0.3$, $A = 0.064$ meter², and I taken from row 9.

The effective power from *each end* is calculated in row 11 of Table 12-18 for $\eta = 0.3$, $A = 0.088$ meter², and I from row 9.

The results of the calculation for the effective power at the *bottom* opening of the furnace is given in row 12. In this case η is taken as 1.0 because the bottom of the furnace is directed to the floor which is assumed to be a reflector of sound. The open area at the bottom is 0.176 meter² and I is again given by row 9.

b. The total effective sound power is the summation of rows 10 and 12 and two times row 11 (to account for the openings at each end). The results of this calculation are given in row 13 as W_T, watts.

c. The effective sound-power level is computed from the definition,

$$L_{Wi} = 10 \log_{10} \left(\frac{W_T}{10^{-12}} \right) \text{ dB}$$

and the results of the calculation are given in row 14 of Table 12-18.

d. The relation between the sound-power level at the source and the sound-pressure level at the worker location, is given by Equation (4-1) and repeated here:

$$L_{pi} = L_{Wi} + 10 \log_{10} \left[\frac{Q_{\theta i}}{4 \pi r^2} + \frac{4}{A} \right] \text{ dB} \qquad (4\text{-}1)$$

where:

L_{pi} = sound-pressure level at worker location, dB
L_{Wi} = sound-power level at source, dB
$Q_{\theta i}$ = directivity of the source, dimensionless
r = distance from source to receiver, meter
A = room constant, meter^2

For this example we will assume the furnace is located in a large room and the worker location is close to the source such that the term $4/A$ can be neglected. The directivity, $Q_{\theta i}$, will be taken as 1 since the directionality of the source was accounted for in the values of η in Step 3-d, above. Equation (4-1) becomes:

$$L_{pi} = L_{Wi} + 10 \log_{10} \left[\frac{1}{4\pi r^2} \right]$$

$$= L_{Wi} + 10 \log_{10} [r^{-2}] + 10 \log_{10} \left[\frac{1}{4\pi} \right]$$

$$= L_{Wi} - 20 \log_{10} [r] - 11 \text{ dB}$$

From the furnace geometry given in Fig. 12-4, the distance from the worker location to the opening on top or on the bottom of the furnace is 14 feet and, by geometry, the distance from the worker location to the end openings is 14.3 feet. For estimation purposes, the distance to the center of all openings will be taken as 14 feet or 4.3 meters. Then,

$$L_{pi} = L_{Wi} - 20 \log_{10} [4.3] - 11$$

$$L_{pi} = L_{Wi} - 12.7 - 11$$

$$L_{pi} = L_{Wi} - 23.7 \text{ dB}$$

The values for sound-pressure level are given in row 15 by subtracting 23.7 from the values in row 14.

e. The A-weighting factors from Table 2-3 in Chapter 2, are listed in row 16 of Table 12-18. The A-weighted sound-pressure levels are obtained by summing values in rows 15 and 16 and the results are given in row 17.

f. The equivalent A-weighted sound-pressure level is obtained from the following equation and is shown in row 18:

$$L_p = 10 \log_{10} \left[\log_{10}^{-1} \left(\frac{L_{p1}}{10} \right) + \log_{10}^{-1} \left(\frac{L_{p2}}{10} \right) + \cdots \log_{10}^{-1} \left(\frac{L_{pn}}{10} \right) \right] \text{ dBA}$$

where L_p = A-weighted sound-pressure level, dBA

$L_{p1}, L_{p2} \cdots L_{pn}$ = octave-band sound-pressure levels, dB

The result of this calculation gives a worker-location level of 92 dBA which exceeds the present 90 dBA upper level for an eight-hour exposure.

To reduce the sound-pressure level at the worker location, modifications to the furnace can be considered. By examining rows 10, 11, and 12 of Table 12-18 it is evident that the effective sound power from the bottom of the furnace is greater than the sound power from the top or ends of the furnace. Apparently, if the bottom were completely enclosed, the sound-pressure level at the worker location would decrease.

Step 4. Calculate the sound-pressure level at the worker location when the bottom of the furnace is enclosed. In this case the open area of the furnace will be less and the enclosed bottom opening will be covered with fire brick. The open area will be .24 square meter and the closed area will be 14.04 square meters. Enclosing the bottom of the conveyor, however, does not alter the end openings of the furnace. Thus:

$$S_{openings} = .24 \text{ meter}^2$$

$$S_{fire\ brick} = 14.04 \text{ meters}^2$$

$$S_T = 14.22 \text{ meters}^2 \text{ (Total area)}$$

The calculations are identical to those made in Table 12-18 for the initial conditions; of course, where changes in the furnace design have been made, different values to reflect these changes will have to be used. Table 12-19 provides the order of the calculations and the resulting answers. For the convenience of the reader, the rows in Table 12-19 are numbered to correspond with the equivalent rows in Table 12-18.

The resulting sound-power level with the furnace bottom enclosed is found to be 88 dBA, which is within the presently allowable 90 dBA for an eight-hour exposure. However, there are often other noise sources in the furnace area and a further noise reduction is necessary. This can be accomplished by constructing a muffler, or expansion chamber, at each furnace opening. To achieve the desired noise reduction the length of the expansion chamber is important and must be calculated. The muffler, itself, may be constructed of any suitable material that is able to withstand the operating temperature at the openings.

Step 5. Compute sound-pressure level at worker location with mufflers at each opening on the end of the furnace. The acoustic transmission-loss of a simple expansion chamber is given by Equation (1-40), Chapter 1. One convenient method of attaching an expansion chamber is shown in Fig. 12-5.

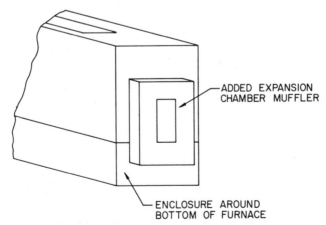

Fig. 12-5. Schematic view of muffler or expansion chamber added to entrance and exit of process furnace to reduce noise level.

Table 12-19. Sound-Pressure Levels with Bottom of Furnace Enclosed*

Row No.	Description	\\multicolumn{7}{c}{Octave-Band Center Frequency, Hz}

Row No.	Description	63	125	250	500	1000	2000	4000	8000
1–4	Same as Rows 1–4, in Table 12-18.	...	...	...	...	...	...	...	...
5	Average acoustic absorption coefficient, $\bar{\alpha}$ $\bar{\alpha} = \dfrac{1}{S_T} \sum\limits_{i=1}^{i=N} \alpha_i S_i$	.03	.03	.03	.03	.04	.04	.04	.04
6	Enclosure constant, $R' = S_T \dfrac{\bar{\alpha}}{1 - \bar{\alpha}}$	.44	.44	.44	.44	.60	.60	.60	.60
7	$\dfrac{4}{R'} + \dfrac{1}{S_T}$	9.16	9.16	9.16	9.16	6.74	6.74	6.74	6.74
8	Acoustic power inside furnace, W, watt. (From Table 12-17, Column 5)	.0998	.0908	.1017	.2944	.0748	.0125	.0013	.0011
9	Acoustic intensity inside furnace, watt/meter^2 $I = W\left(\dfrac{4}{R'} + \dfrac{1}{S_T}\right)$; row 7 times row 8.	.914	.832	.932	2.697	.504	.084	.009	.007
10	Effective acoustic power from top, watt. $W_{eff} = \eta I A = (.3)I(.064)$	.0175	.0160	.0179	.0518	.0097	.0016	.0002	.0001
11	Effective acoustic power from each end, watt. $W_{eff} = \eta I A = (.3)I(.088)$	.0241	.0220	.0246	.0712	.0133	.0022	.0002	.0002
12	Effective acoustic power from bottom, watt. $W_{eff} = \eta I A = (1)I(.176)$	.1204	.1095	.1227	.3550	.0711	.0120	.0012	.0011
13	Total effective acoustic power, W_T, watt. Row 10 plus two times row 11	.0657	.0600	.0671	.1942	.0363	.0060	.0006	.0005
14	Effective sound-power level, dB. $L_{Wi} = 10 \log_{10}(W_T/10^{-12})$	108.2	107.8	108.3	112.9	105.6	97.8	87.8	87.0
15	Sound-pressure level of worker location, dB. $L_{pi} = L_{Wi} - 20 \log_{10} r - 11$	84.5	84.1	84.6	89.2	81.9	74.1	64.1	63.3
16	A-Scale weighting factors, dB. (From Table 2-3, Chapter 2)	−26.2	−16.1	−8.6	−3.2	0.0	+1.2	+1.0	−1.1

Ch. 12 INDUSTRIAL NOISE CONTROL STUDIES

Table 12-19 (*cont.*). Sound-Pressure Levels with Bottom of of Furnace Enclosed*

Row No.	Description	Octave-Band Center Frequency, Hz							
		63	125	250	500	1000	2000	4000	8000
17	A-Weighted sound-pressure levels, dB. Row 15 plus row 16.	58.3	68.0	76.0	86.0	81.9	75.3	65.1	62.2
18	Equivalent A-Scale sound-pressure level, dBA.	88							

*The row numbers in this table correspond to the equivalent row numbers in Table 12-18. See Case Study 4.

$$\text{TL} = 10 \log_{10} \left[1 + .25 \left(\frac{S_C}{S_E} - \frac{S_E}{S_C} \right)^2 \sin^2 \left(\frac{2\pi L_C}{\lambda} \right) \right] \quad (1\text{-}40)$$

where:

TL = Transmission loss, dB
S_C = Cross-sectional area of expansion chamber, ft
S_E = Cross-sectional area of inlet to expansion chamber, ft
L_C = Length of expansion chamber, ft
λ = Wavelength of sound, ft
$\frac{2\pi L_C}{\lambda}$ = Angle in radians

The wavelength of sound is computed from,

$$\lambda = \frac{\text{Speed of sound}}{\text{frequency}} = \frac{c}{f}$$

$$c = 49.03 \sqrt{460 + °F} \text{ ft/sec}$$

a. The wavelength of sound is computed with the speed of sound at 800 °F.

$$c = 49.03 \sqrt{460 + 800} = 1740 \text{ ft/sec}$$

Then $\lambda = 1740/f$, where f is the octave-band center frequency. The values for λ are given in row 1 of Table 12-20.

b. The length of an expansion chamber is important in obtaining the maximum transmission-loss, as is shown in Fig. 1-16b, Chapter 1. In Table 12-19, the maximum sound-pressure levels at the 250, 500, and 1000 Hz octave bands at the worker location are shown by row 17. There it can be seen that the maximum transmission-loss is desired in these three bands. The maximum transmission-loss for an expansion chamber of the minimum length occurs at the frequency for which the term

$$\sin^2 \left(\frac{2\pi L_C}{\lambda} \right) = 1, \text{ in Equation (1-40)}.$$

Table 12-20. Sound-Pressure Levels for Bottom Enclosed Furnace with Muffler or Expansion Chamber at Each End (See Case Study 4)

Row	Description	Octave-Band Center Frequency, Hz							
		63	125	250	500	1000	2000	4000	8000
1	Acoustic wavelength, $\lambda = \dfrac{1740}{\text{frequency}}$ ft	27.6	13.9	7.0	3.5	1.7	.9	.4	.2
2	Transmission-loss = $10 \log_{10} \left[1 + .25 \cdot \left(\dfrac{S_C}{S_E} - \dfrac{S_E}{S_C} \right)^2 \cdot \sin^2 \left(\dfrac{.87\pi}{\lambda} \right) \right]$ (dB).	0.0	.3	1.0	2.8	4.5	.1	1.7	3.9
3	Sound-power level at top, dB. $L_W = 10 \log_{10} \left(\dfrac{W_{\text{eff}}}{10^{-12}} \right)$; W_{eff} from row 10, Table 12-19.	102.4	102.0	102.5	107.1	99.9	92.0	83.0	80.0
4	Sound-power level at each end, dB. $L_W = 10 \log_{10} \left(\dfrac{W_{\text{eff}}}{10^{-12}} \right)$; W_{eff} from Row 11, Table 12-19.	103.8	103.4	103.9	108.5	101.2	93.4	83.0	83.0
5	Sound-power level at each end with muffler, dB. Row 4 minus row 2.	102.4	103.1	102.9	105.7	96.7	93.3	81.3	79.1
6	Sound-power level from both ends combined, dB. Row 5 + 3 dB.	105.4	106.1	105.9	108.7	99.7	96.3	84.3	82.1
7	Sound-power of ends and top combined, L_W, dB. Row 6 + row 3 (logarithmic addition).	107.2	107.5	107.5	110.9	102.8	97.6	86.7	84.2
8	Sound-pressure level at worker location, dB. $L_{pi} = L_W - 20 \log_{10} r - 11$.	83.5	83.8	83.8	87.2	79.1	73.9	63.0	60.5
9	A-Weighted sound-pressure levels, dB. (From Table 2-3, Chapter 2.)	−26.2	−16.1	−8.6	−3.2	0.0	+1.2	+1.0	−1.1
10	A-Weighted sound-pressure levels, dB. Row 8 plus row 9.	57.3	67.7	75.2	84.0	79.1	75.1	64.0	59.4
11	Equivalent A-Scale sound-pressure level, dB.	86 dB							

Ch. 12   INDUSTRIAL NOISE CONTROL STUDIES   549

This will occur when $L_C = \lambda/4$. By choosing the maximum transmission-loss to occur in the 1000-Hz octave band, the length of the expansion chamber muffler should be

$$L_C = \lambda/4 = \frac{1}{4} \cdot \frac{1740}{1000} = .435 \text{ ft}$$

Then the equation for transmission loss of an expansion chamber will become

$$TL = 10 \log_{10} \left[1 + .25 \left(\frac{S_C}{S_E} - \frac{S_E}{S_C} \right)^2 \sin^2 \left(\frac{.86\pi}{\lambda} \right) \right]$$

For the above equation to be valid, the cross-sectional dimension should not exceed an area of 48 ft² and the exit area is, as before,

$S_E = .088 \text{ meter}^2 = .95 \text{ ft}^2$. Then

$$TL = 10 \log_{10} \left[1 + .25 \left(\frac{2}{.95} - \frac{.95}{2} \right)^2 \sin^2 \left(\frac{.87\pi}{\lambda} \right) \right]$$

The values of TL are computed for every octave band using values of λ from row 1 of Table 12-20. The results of the computation are listed in row 2 of the table.

b. The values for effective sound power are the same as in Table 12-19 and the values for the effective sound-power level for the top opening is listed in row 3 which is computed from

$$L_W = 10 \log_{10} \left(\frac{W_{eff}}{10^{-12}} \right) \text{ dB}$$

W_{eff} is obtained from row 6 of Table 12-19.

The effective sound-power level for each end opening is computed from the above equation with W_{eff} from row 7 of Table 12-19. The resulting sound-power levels are listed in row 4 of Table 12-20.

c. The sound-power level at each end with muffler is obtained by subtracting the values in row 2 from those in row 4. The results are listed in row 5, Table 12-20.

d. Compute the sound-power level of both ends with mufflers. Since each end has identical sound-power levels, 3 dB is added to each value in row 5 and the results are in row 6.

e. The equivalent sound-power level at the top opening and the ends is computed by combining the values in rows 3 and 6 using the following:

$$L_W = 10 \log_{10} \left[\log^{-1} \left(\frac{L_{W1}}{10} \right) + \log_{10}^{-1} \left(\frac{L_{W2}}{10} \right) \right] \text{ dB}$$

The results of this calculation are listed in row 7.

f. The sound-pressure level at the worker location is computed from,

$$L_{Wi} = L_{Wi} - 20 \log_{10} (r) - 11 \text{ dB}$$

or,

$$L_{pi} = L_{Wi} - 23.7 \text{ dB for } r = 10 \text{ ft} = 4.3 \text{ meters.}$$

The resulting sound-pressure levels are in row 8.
g. A-scale weighting factors from Table 2-3 are listed in row 9.
h. A-weighted octave-band sound-pressure levels are given in row 10 by summing values in rows 8 and 9.
i. The equivalent A-scale sound-pressure level is computed and listed in row 11, as 86 dBA.

By adding two expansion chambers to each end of the furnace, the worker-location sound level will be 86 dBA, due to the furnace only. This is acceptable to meet OSHA requirements; however, other noise sources in the area must be considered as well as the effect on the worker-location area. The above solution is easily implemented by simple modifications to the furnace with no change in the process other than, possibly, a slight change in temperature inside the furnace; this can be controlled by minor changes to fuel and air-flow rates. If the burners can be operated at a lower firing rate the sound power produced by the combustion process can be reduced.

Case Study 5. *Noise Reduction of a Cardboard-Clamping System*

Shown in Fig. 12-6 is a portion of a cardboard-box stamping machine. Corrugated cardboard is supplied to the machine from a roll. The desired shapes for shipping containers are cut to size in a shearing station beyond the section of the machine seen in the illustration. Figure 12-6 shows the clamping mechanism which firmly clamps the cardboard material for accurate positioning in the cutting section. The system is designed so that a continuous tension is applied to the stock material by the friction feed rolls as the clamp bar rapidly closes to

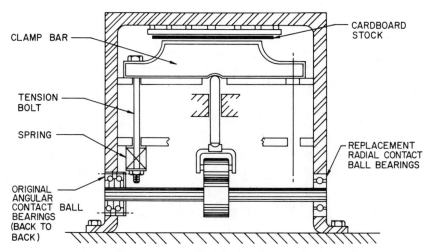

Fig. 12-6. Schematic view of cardboard clamping mechanism on cardboard box stamping machine.

Ch. 12   INDUSTRIAL NOISE CONTROL STUDIES   551

insure accurate positioning and to prevent buckling of the stock in the cutting section of the machine.

In order to achieve accurate positioning at a high production rate, the rise on the clamp-bar cam occurs in a small angular rotation. The cam rise was 0.0290 inch in 5 degrees of cam rotation. This requirement for rapid rise was determined from machine set-up studies to produce 240 strokes per minute and maintain accurate positioning.

The tensioning bolts in the figure are adjusted during set-up to provide sufficient preload on the cam and follower to maintain contact during the cycle.

Two angular-contact ball bearings were originally used to support the cam shaft at each end; these bearings were assembled back to back and loaded axially, to provide a rigid assembly. The intent of this construction was to reduce the cam shaft deflection while it is being loaded and to reduce the rotation of the cam shaft with respect to a vertical plane—not to be confused with the angular rotation of the shaft as it revolves. However, this construction actually caused the bending load on the cam shaft to be transferred to the side frames of the machines, resulting in their deflection. The periodically excited deflection of these side frames radiated acoustic energy which was later found to be the cause of the excessive noise.

In a production plant sound-pressure-level measurements were made at a location five feet above the floor and three feet away from the side frame of the clamping section of the machine. The measured A-weighted sound-pressure level was 94 dBA with all other machinery off. An oscilloscope display of the sound-pressure level as a function of time is shown in Fig. 12-7. The peaks in the response were identified as being associated with the clamping action by using a displacement transducer on the clamp bar. The repetition rate is sufficiently high to classify the repeated peaks as steady state rather than impulsive. According to OSHA, if impulses occur at intervals of less than one-

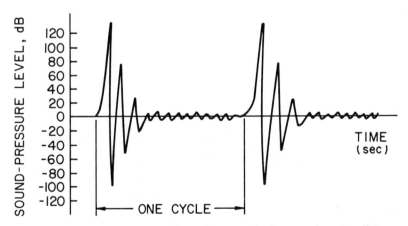

Fig. 12-7. Typical oscilloscope trace of sound-pressure level versus time at a distance of three feet from the side frame of the clamping mechanism. Peaks illustrate impact effect of clamping mechanism; lower level sound-pressure fluctuations between peaks are representative of other machine component sound; e.g., bearings, motor, drive system, etc.

half second, they shall be considered as continuous sound; at 240 rpm in this case the impulse interval is one-fourth of a second. To verify the repetition rate, peak value, rise rate, and duration of impulsive type sound, ANSI Standard S1-13-1971 recommends the oscilloscope trace method.

Accelerometers were attached to the side frame of the clamp bar system to investigate the acceleration of the frame normal to the plane of the side plates. The acceleration of the side plates, as displaced on an oscilloscope as a function of time, appeared similar to the data in Fig. 12-7 indicating that the side frame motion was the sound radiating surface. The acceleration levels of other surfaces were at least a factor of ten lower than the level on the side frames.

The analysis and noise reduction modifications for the system are described in the following paragraphs:

Since the cam rise occurs in such a small angular rotation of the cam, the system response is controlled by the masses in the system. Shown in Fig. 12-8A is a sketch of the dynamic system before the clamp bar contacts the cardboard sheet. The motion produced by the cam occurs in such a short time period that the system can be considered as shown in Fig. 12-8B. The imposed relative motion between the cam and the follower occurs in a time interval much less than the fundamental natural frequency of the system. For this type of system the springs can be eliminated in the initial motion of the masses.

The force applied to each mass is equal and opposite as given by the following relation:

$$\text{Force} = \text{mass} \times \text{acceleration}$$

$$F = ma$$

The result of the cam rise is to impose an equal and opposite force on each mass resulting in acceleration given by,

$$a = \frac{F}{m}$$

For mass one,

$$m_1 = W_1/g \quad a_1 = \frac{F}{m_1}$$

For mass two,

$$m_2 = W_2/g \quad a_2 = \frac{F}{m_2}$$

Since the forces, F, are equal (but opposite in sense), the ratio of acceleration of the two masses becomes,

$$\frac{a_2}{a_1} = \frac{F/m_2}{F/m_1} = \frac{m_1}{m_2}$$

Due to the cam rise, the acceleration applied to each mass will occur in the same

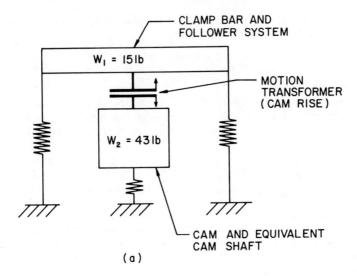

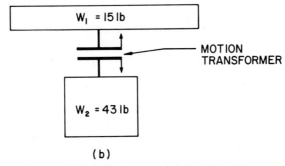

Fig. 12-8. Equivalent masses or weights of clamping mechanism for purpose of analysis: A. With equivalent springs; B. Simplified model for suddenly applied motion neglecting the spring effect.

time period; therefore, the resulting displacements will have magnitudes in the same ratio as the acceleration ratios, since,

$$\text{Velocity} = \int_0^t \text{acceleration } dt$$

$$\text{Displacement} = \int_0^t \text{velocity } dt$$

The only difference in displacement for the same functional form for acceleration will be the constant $1/m$. Therefore,

$$\frac{x_2}{x_1} = \frac{m_1}{m_2} = \frac{W_1}{W_2}$$

From the motion requirements of the system, the clamp bar travel is 0.0210 inch to clamp the cardboard sheet. Experience had indicated that the cam rise required was .0290 inch to account for the cam shaft deflection. The following calculation explains this:

W_1 = clamp bar and follower weight = 15 lb

W_2 = weight of cam plus one-half of cam shaft weight = 43 lb

One-half the shaft weight is used because the shaft is assumed to deflect in a bending manner. Not all of the shaft mass has motion for this case. The usual assumption, based on the bending mode vibration theory, is to use one-half of the total shaft weight. This is the approximation used in this example.

Using the above weights, the deflection of the cam is,

$$x_2 = \frac{W_1}{W_2} x_1 = \frac{15}{43} x_1$$

Since the follower and clamp bar motion are 0.0210 inch,

$$x_2 = \frac{15}{43} (0.0210) = 0.0073 \text{ inch}$$

This calculation illustrates that the total relative motion at the follower is,

total required motion = .0210 + .0073 = .0283 inch

The calculated cam rise of .0283 inch agrees closely with the .0290 cam rise found to be required through experience gained in the operation of the machine. The sudden motion of the cam (.0073 inch) due to the cam-shaft deflection causes bending waves to travel from the cam along the cam shaft to the bearings in the side frames. Since the angular contact bearings were rigid, with respect to rotation in the plane perpendicular to the shaft centerline, a moment was applied to the side frames resulting in motion and sound radiation.

For this type of system it is desirable to minimize the motion of the cam, which can be accomplished by reducing the mass ratio m_1/m_2. Conditions were such that only a modification of the existing machine was required rather than a completely new design, although the modification will provide an insight into the requirement for a new design. The modification of the system includes reducing the mass of the clamp-bar system by providing holes to remove all unnecessary material and making the clamp bar from magnesium instead of a gray iron casting. The weight was reduced to 2.75 lb. By bolting a disc to each side of the cam, weight was added to increase the weight of the cam and one-half the shaft weight to 131 lbs. Thus,

$$x_2 = \frac{W_1}{W_2} x_1 = \frac{2.75}{131}(.021) = 0.0004 \text{ inch}$$

Thus, the motion of the cam due to deflection was reduced by .0069 inch. With this modification the cam was reground to reduce the rise to .0214 inch.

The angular contact bearings at the side frames were found to have no influence on the cam deflection as was assumed in the original design. They were therefore replaced with radial contact ball bearings. This modification allows the shaft to rotate in the bearings with respect to a plane perpendicular to its axis as a result of bending; therefore, the bending moment applied to the side frames is reduced significantly.

Thus, based on an engineering analysis four modifications were made to the clamping system, namely: 1. the weight of the clamp was reduced; 2. the weight of the cam was increased; 3. the cam rise was reground to reflect the reduction in the cam shaft deflection; 4. the ball bearings were changed from the angular to the radial contact type in order to reduce the bending moment transferred to the side frames. It should be noted that reducing the clamp bar mass, or weight, and increasing the cam mass, or weight, resulted in a reduction of the W_1/W_2 ratio, which in turn reduced the cam-shaft deflection.

With these changes the measured sound-pressure level, at the same location and under the same conditions existing when the original measurements were made, was 85 dBA.

References

References to Chapter 1

[1] Sullivan, J. W. "Sound Waves and Acoustical Definitions." *Proceedings of Reduction of Machinery Noise Short Course.* Edited by M. J. Crocker. Purdue University, Lafayette, Indiana, 15–17 May 1974, pp. 10–20.

[2] Jones, G. R.; Hempstock, T. I.; Mulholland, K. A.; and Stott, M. A. *Teach Yourself Acoustics.* London: The English Universities Press Ltd., 1967.

[3] Beranek, L. L., ed. *Noise and Vibration Control.* New York: McGraw-Hill Book Co., 1971.

[4] Beranek, L. L., ed. *Noise Reduction.* New York: McGraw-Hill Book Co., 1960.

[5] Beranek, L. L. *Acoustics.* New York: McGraw-Hill Book Co., 1954.

[6] Kinsler, L., and Frey, A. *Fundamentals of Acoustics.* 2d ed. New York: John Wiley & Sons, 1962.

[7] Burns, W. *Noise and Man.* London: John Murray, Ltd., 1968.

[8] ANSI Standard S1.1-1971 (R-1960), *Acoustical Terminology.* American National Standards Institute, Inc., 1430 Broadway, New York, New York 10018.

[9] *Industrial Noise Manual.* 2d ed. American Industrial Hygiene Association, 1966.

[10] *Acoustics in Air Conditioning* (1967). The Trane Company, La Crosse, Wisconsin.

[11] Thomson, W. T. *The Theory of Vibration with Applications.* Englewood Cliffs, New Jersey: Prentice-Hall, 1972.

[12] Harris, C. M. *Handbook of Noise Control.* New York: McGraw-Hill Book Co., 1957.

[13] Miller, T. D. "Machine Noise Analysis and Reduction." *Sound and Vibration*, vol. 1, no. 3, March 1967, pp. 8–14.

[14] Ruzicka, J. E. "Fundamental Concepts of Vibration Control." *Sound and Vibration*, vol. 5, no. 7, July 1971, pp. 16–23.

[15] Yerges, L. F. "Noise Reduction in Metal Cutting Operations." *Proceedings of Reduction of Machinery Noise Short Course.* Edited by M. J. Crocker. Purdue University, Lafayette, Indiana. 15–17 May 1974, pp. 251–257.

[16] Davis, D. D., Jr.; Stokes, G. M.; Moore, D.; and Stevens, G. L., Jr. "Theoretical and Experimental Investigation of Mufflers with Comments on Engine-Exhaust Muffler Design." NACA Technical Report No. 1192, 1954.

[17] "Criteria for a Recommended Standard... Occupational Exposure to Noise," U.S. Department of Health, Education and Welfare, Washington, D.C. 20460.

[18] "Information on Levels of Environmental Noise Requisite to Protect Public Health and Welfare with an Adequate Margin of Safety" (March 1974). U.S. Environmental Protection Agency, Washington, D.C. 20460.

REFERENCES

References to Chapter 2

[1] Peterson, A., and Gross, E. *Handbook of Noise Measurements* (1972). General Radio Co., 300 Baker Ave., West Concord, Massachusetts 01781.
[2] Brock, J. T. *Acoustic Noise Measurements* (1967). Brüel and Kjäer Co., 5113 West 164th Street, Cleveland, Ohio 44142.
[3] *Acoustics Handbook*. Application Note 100 (November 1968). Hewlett-Packard Co., 1501 Page Mill Road, Palo Alto, California 94304.
[4] Beranek, L. L. *Acoustic Measurements*. New York: John Wiley & Sons, Third Printing, 1956.
[5] Doebelin, E. *Measurement Systems: Application and Design*. Chapter 6. New York: McGraw-Hill Book Co., 1966.
[6] Keast, D. N. *Measurements in Mechanical Dynamics*. New York: McGraw-Hill Book Co., 1967.
[7] Blake, M. P., and Mitchel, W. S. *Vibration and Acoustic Measurement Handbook*. New York: Spartan Books, 1972.
[8] Baade, P. K. "Sound Radiation of Air-Conditioning Equipment; Measurement in the Free Field Above a Reflecting Plane," *ASHRAE Transactions*, vol. 70 (1964), pp. 217-227. American Society of Heating, Refrigerating and Air-Conditioning Engineers, Inc., 345 East 47th Street, New York, N.Y. 10017.

References to Chapter 4

[1] Miller, J. D. *Effects of Noise on People*. Report No. NT1D300.7, p. 9. 31 December 1971. Prepared by the Central Institute for the Deaf for the U.S. Environmental Protection Agency, Office of Noise Control, Washington, D.C. 20460.
[2] USASI, 224-X-2. *The Relations of Hearing Loss to Noise Exposure* (1954). American National Standards Institute, Inc., 1430 Broadway, New York, New York 10018, pp. 15-18.
[3] Glorig, A.; Ward, W. Dixon; and Nixon, J. "Damage-Risk Criteria and Noise-induced Hearing Loss." *Arch. Otolaryngol.* 74, October 1961, pp. 413-423.
[4] CHABA, "Hazardous Exposure to Intermittent and Steady-State Noise." *Journal of the Acoustic Society of America*, 39 (1966), pp. 451-454.
[5] *Federal Register*, vol. 34, no. 96, May 1969, Rule 50-204.10, and 15 July 1969.
[6] *Federal Register*, vol. 36, no. 105, 29 May 1971, §1910.95, p. 10518.
[7] Jones, H. H. "Physical Agents Committee's Proposal for Limits of Noise Exposure" Presented at American Industrial Hygiene Conference (1968). See also, Botsford, J. H. "Current Trends in Hearing Damage Risk Criteria." *Sound and Vibration*, vol. 4, no. 4, April 1970, pp. 16-19.
[8] Ward, W. D., ed. "Proposed Damage Risk Criterion for Impulse Noise (Gunfire) (U)." *Report of Working Group 57, NAS-NRC, CHABA*. July 1968.
[9] Beranek, L. L., ed. *Noise and Vibration Control*, pp. 551-552. New York: McGraw-Hill Book Co., 1971.
[10] Webster, J. C. "SIL—Past, Present, and Future." *Sound and Vibration*, vol. 3, August 1969, pp. 22-26.

REFERENCES

[11] Webster, J. C. "Effects of Noise on Speech Intelligibility," *Proceedings of the Conference; Noise as a Public Health Hazard.* Washington, D.C.–13–14 June 1968, *ASHA Reports* No. 4, American Speech and Hearing Association, Washington, D.C., February 1969, pp. 49–73.

[12] ANSI, S1.4-1971, *Specification for Sound Level Meters,* American National Standards Institute, Inc., 1430 Broadway, New York, New York 10018.

[13] Harris, C. T. "Absorption of Sound in Air" *Journal of the Acoustic Society of America*, vol. 40 (1966), pp. 148-159.

[14] *Performance Data Architectural Acoustical Materials.* Acoustical and Insulating Materials Association, 205 West Touhy Avenue, Park Ridge, Illinois, 60068 (published annually).

[15] Flecher, H., and Munson, W. A. "Loudness, Its Definition, Measurement and Calculation." *Journal of the Acoustic Society of America,* vol. 5, no. 2, October 1933, pp. 82–108.

[16] Robinson, D. W., and Dadson, R. S. "A Redetermination of the Equal Loudness Relations for Pure Tones," *British Journal of Applied Physics,* vol. 7, May 1956, pp. 166-181.

[17] ANSI, S3.4–1968, *Procedure for the Computation of Loudness of Noise.* American National Standards Institute, Inc., 1430 Broadway, New York, New York 10018.

[18] Stevens, S. S. "Procedure for Calculating Loudness, Mark VI." *Journal of the Acoustic Society of America,* vol. 33, no. 11, November 1961, pp. 1577-1585.

[19] Kryter, K. D., and Pearsons, K. S. "Some Effects of Spectral Content and Duration on Perceived Noise Level. *Journal of the Acoustic Society of America,* vol. 35, no. 6, June 1963, pp. 866-883.

[20] Stevens, S. S. "Perceived Level of Noise by Mark VII and Decibels E." *Journal of the Acoustic Society of America,* vol. 51, no. 2 (part 2), February 1972, p. 597.

[21] Beranek, L. L. "Revised Criteria for Noise in Buildings." *Noise Control,* vol. 3 (1957), pp. 19–27.

[22] *ASHRAE Guide and Data Book, Systems and Equipment,* American Society of Heating, Refrigerating and Air-Conditioning Engineers, Inc. 345 East 47th Street, New York, New York 10017, Chapter 13 (1967), pp. 373–406.

[23] "The Noise Around Us, Including Technical Backup," *CTAB Panel on Noise Abatement*, Jack E. Goldman, Chairman, U.S. Department of Commerce Publications, COM 71-00147, September 1970.

References to Chapter 5

[1] ASTM C 522-73, "Standard Method of Test for Airflow Resistance of Acoustical Materials." American Society for Testing and Materials, 1916 Race Street, Philadelphia, Pa. 19103.

[2] ASTM C 384-58, "Standard Method of Test for Impedance and Absorption of Acoustical Materials by the Tube Method." American Society for Testing and Materials, 1916 Race Street, Philadelphia, Pa. 19103.

[3] Kinsler, L. E., and Frey, H. R. *Fundamentals of Acoustics.* 2d ed. New York: John Wiley & Sons, 1962.

[4] ASTM C 423-66, "Standard Method of Test for Sound Absorption of Acoustical Materials in Reverberant Rooms." American Society for Testing and Materials, 1916 Race Street, Philadelphia, Pa. 19103.

[5] ASTM E 336-71, "Standard Recommended Practice for Measurement of Airborne Sound Insulation in Buildings." American Society for Testing and Materials, 1916 Race Street, Philadelphia, Pa. 19103.
[6] ASTM E 90-70, "Standard Recommended Practice for Laboratory Measurement of Airborne Sound Transmission Loss of Building Partitions." American Society for Testing and Materials, 1916 Race Street, Philadelphia, Pa. 19103.
[7] ASTM E 413-70T, "Tentative Classification for Determination of Sound Transmission Class." American Society for Testing and Materials, 1916 Race Street, Philadelphia, Pa. 19103.
[8] Sabine, H. J. "Noise Absorbing Materials." *Symposium: The Challenge of Noise Control.* Columbus Section American Society of Mechanical Engineers, Columbus, Ohio (1970).
[9] Beranek, L. L., ed. *Noise and Vibration Control.* Chapter 7. New York: McGraw-Hill Book Co., 1971.
[10] Beranek, L. L., ed. *Noise and Vibration Control.* Chapter 11. New York: McGraw-Hill Book Co., 1971.
[11] Sullivan, J. W. Unpublished Report. Lafayette, Indiana: Purdue University, 1972.
[12] Jackson, R. S. "The Performance of Acoustic Hoods at Low Frequency." *Acoustica* 12 (1962).
[13] Fader, B. "Practical Designs for Noise Barriers Based on Lead," *American Industrial Hygiene Association Journal* 27, November–December 1966.
[14] Sabine, H. J. "The Absorption of Noise in Ventilating Ducts." *Journal of the Acoustical Society of America* 12 (1940), p. 53.
[15] Harris, C. M., ed. *Handbook of Noise Control.* Chapters 26 and 27. New York: McGraw-Hill Book Co., 1957.
[16] Beranek, L. L., ed. *Noise Reduction.* Chapter 17. New York: McGraw-Hill Book Co., 1960.
[17] Beranek, L. L., ed. *Noise and Vibration Control.* Chapter 15. New York: McGraw-Hill Book Co., 1971.
[18] *ASHRAE Guide and Data Book, 1972 Equipment.* Chapters 25 and 26. American Society of Heating, Refrigerating, and Air-Conditioning Engineers, Inc., 345 East 47th Street, New York, New York 10017.
[19] Sullivan, J. W. "Engineering Control of Industrial Noise." Paper No. MM70-613, Society of Manufacturing Engineers, 1970.
[20] Phillips, W. G., and Shadley, J. R. "Curbing Noise with Partial Enclosures." *Machine Design,* 14 April 1974.

References to Chapter 6

[1] Lazan, B. J. *Damping of Materials and Members in Structural Mechanics.* New York: Pergamon Press, 1968.
[2] Snowdon, J. C. *Vibration and Shock in Damped Mechanical Systems.* New York: John Wiley & Sons, 1968.
[3] Jones, D. I. G.; Henderson, J. P.; and Nashif, A. D. "Reduction of Vibrations in Aerospace Structures by Additive Damping." *Shock and Vibration Bulletin* 40, part 5, December 1969, pp. 1–18.
[4] Ferry, J. D. *Viscoelastic Properties of Polymers.* New York: John Wiley & Sons, 1961.
[5] Nicholas, T., and Heller, R. A. "Determination of the Complex Modulus

REFERENCES

of a Filled Elastomer from a Vibrating Sandwich Beam." *Experimental Mechanics*, vol. 7, no. 3, March 1967, pp. 110–116.

[6] Dudek, T. J. "Damping Material Effectiveness Measured by the Geiger Plate and Composite Beam Tests." *Journal of the Acoustical Society of America* 46 (1969), pp. 1384–1386.

[7] Parsons, T.; Yates, W.; and Schloss, F. *The Measurement of Dynamic Properties of Materials Using a Transfer Independence Technique*. Naval Ship Research and Development Center, Research and Development Report, no. 2981, 1969.

[8] Edwards, J. L., and Hicks, D. "Useful Range of a Mechanical Impedance Technique for Measurement of Dynamic Properties of Materials." *Journal of the Acoustical Society of America*, November 1972.

[9] Adkins, R. L. "Design Considerations and Analysis of a Complex Modulus Apparatus." *Experimental Mechanics*, 1966.

[10] Nashif, A. D. "A New Method for Determining Damping Properties of Viscoelastic Materials." *Shock and Vibration Bulletin* 3 (1967), pp. 37–47.

[11] Cannon, C. M.; Nashif, A. D.; and Jones, D. I. G. "Damping Measurements on Soft Viscoelastic Materials Using a Tuned Damper Technique." *Shock and Vibration Bulletin* 38 (1968), pp. 151–163.

[12] Birchon, D. "Hidamets, Metals to Reduce Noise and Vibration." *Engineer*, London, 5 August 1966.

[13] Bowie, G. E.; Nachman, J. F.; and Hammer, A. N. "Exploitation of Cu-Rich Damping Alloys; Part I: The Search for Alloys with High Damping at Low Stress." *American Society of Mechanical Engineers Paper* 71, Vibration (1971), p. 106.

[14] Weissmann, G. F., and Babington, W. "A High Damping Magnesium Alloy for Missile Application." *The Journal of Environmental Sciences*, October 1966.

[15] Sridharan, P., and Plunkett, R. *Damping in Porcelain Enamel Coatings*. Air Force Materials Laboratory Report AFML-TR-71-193, July 1972.

[16] Ungar, E. E. *Energy Dissipation at Structural Joints; Mechanisms and Magnitudes*. Air Force Flight Dynamics Laboratory Report FDL-TDR-64-98, July 1964.

[17] Jones, D. I. G., and Trapp, W. J. "Influence of Additive Damping on Resonance Fatigue of Structures." *Journal of Sound and Vibration*, vol. 17, no. 2 (1971), pp. 157–185.

[18] Oberst, H. "Über die Dämfung der Biegeschwingungen dünner Bleche durch festhaltende Beläge." *Acustiea 2*, leaflet 4, AB 191-194, Part I (1952), and 4, p. 433, Part II (1954).

[19] Jones, D. I. G. "Effect of Free Layer Damping on Response of Stiffened Plate Structures." *Shock and Vibration Bulletin* 41, Part 2, December 1970, pp. 105–120.

[20] Ross, D.; Ungar, E. E.; and Kerwin, E. M., Jr., "Damping of Plate Flexural Vibrations by Means of Viscoelastic Laminae." *Structural Damping*, American Society of Mechanical Engineers, 1959.

[21] Nashif, A. D., and Nicholas, T. "An Analytical and Experimental Investigation of a Two-Layer Damping Treatment." *Shock and Vibration Bulletin* 39, Part 4 (1969), p. 53.

[22] Grootenhuis, P. "Vibration Control with Viscoelastic Materials." *Environmental Engineering*, no. 38, May 1969.

[23] Nashif, A. D., and Nicholas, T. "Attenuation of Vibrational Amplitudes Through the Use of Multiple Layered Damping Treatments." *American Society of Mechanical Engineers Paper* 71, Vibration 40 (1971).

REFERENCES

[24] Nashif, A. D., and Nicholas, T. "Vibration Control by a Multiple Layered Damping Treatment." *Shock and Vibration Bulletin* 41 (1970), pp. 121-131.

[25] Mead, D. J., and Markus, S. "Loss Factors and Resonant Frequencies of Encastre Damped Sandwich Beams." *Journal of Sound and Vibration*, vol. 12, no. 1 (1970), pp. 99-112.

[26] Jones, D. I. G., and Parin, M. L. "Technique for Measuring Damping Properties of Thin Viscoelastic Layers." *Journal of Sound and Vibration*, vol. 24, no. 2 (1972).

[27] Bishop, R. E. D., and Johnson, D. C. *The Mechanics of Vibration*. New York: Cambridge University Press, 1960.

[28] Mead, D. J. *The Effect of Certain Damping Treatments on the Response of Idealized Aeroplane Structures Excited by Noise*. Air Force Materials Laboratory Report AFML-TR-65-284, August 1965.

[29] Richards, E. J., and Mead, D. J., eds. *Noise and Acoustic Fatigue in Aeronautics*. New York: John Wiley & Sons, 1968.

[30] Henderson, J. P. "Energy Dissipation in a Vibration Damper Utilizing Viscoelastic Suspension." *Shock and Vibration Bulletin* 35, Part 7, April 1968, pp. 213-229.

[31] Nashif, A. D. "Development of Practical Tuned Dampers to Operate over a Wide Temperature Range," *Shock and Vibration Bulletin* 38, Part 3 (1968), pp. 57-69.

[32] Jones, D. I. G. "Response and Damping of a Simple Beam with Tuned Dampers." *Journal of the Acoustical Society of America*, vol. 42, no. 1, July 1967, pp. 50-53.

[33] Davey, A. B., and Payne, A. R. *Rubber in Engineering Practice*, pp. 118-119. New York: Palmerton Publishing Co., 1964.

[34] Jones, D. I. G. "Effect of Isolated Tuned Dampers on Response of Multi-span Structures." *Journal of Aircraft* 4 (1967), p. 343.

[35] Jones, D. I. G., and Bruns, G. H. "Effect of Tuned Viscoelastic Dampers on Response of Multi-span Structures." *Shock and Vibration Bulletin* 36, Part 4 (1967), p. 49.

[36] Blasingame, W.; Thomas, E. V.; and DiTaranto, R. A. "Development of Damped Machinery Foundations." *Shock and Vibration Bulletin* 36, Part 4, January 1967, pp. 81-94.

[37] Ruzicka, J. E. "Vibration Control: Applications." *Electro-Technology*, January 1964.

[38] Shotwell, D. B. "Application of the Tuned and Damped Dynamic Absorber to Rubber-tired Earthmoving Machines." *American Society of Mechanical Engineers Paper* 67, Vibration 64 (1967).

[39] Grootenhuis, P. "The Attenuation of Noise and Ground Vibrations from Railways." *The Journal of Environmental Sciences*, April 1967, pp. 14-19.

[40] Baker, J. K. "Practical Examples of the Attenuation of Vibrations by Sandwich Damping." *Environmental Engineering*, no. 8, May 1969, pp. 9-11.

[41] Gayman, M. "Development of a Point Damper for the Ranger Solar Panels." *Jet Propulsion Laboratory Technical Report* No. 32-793, September 1965.

[42] Barklay, R. G., and Humphrey, P. W. "Wagging Tail Vibration Absorber." *Shock and Vibration Bulletin* 40, Part 5, December 1969, pp. 147-155.

[43] O'Leary, J. J. "Reduction in Vibration of the CH-47C Helicopter Using a Variable Tuning Vibration Absorber." *Shock and Vibration Bulletin* 40, Part 5, December 1969, pp. 191-202.

[44] Potter, J. L. "Improving Reliability and Eliminations Maintenance with Elastomeric Dampers for Rotor Systems." *Shock and Vibration Bulletin* 41, Part 2 (1970), pp. 141-149.
[45] Roberts, P. V. "Hawk Suspension System Performance on 17754 Trailed Vehicle." *Shock and Vibration Bulletin* 41, Part 2 (1970), pp. 159-169.

References to Chapter 7

[1] Crede, C. E., and Ruzicka, J. E. "Theory of Vibration Isolation." In *Shock and Vibration Handbook*. Vol. 2, Chapter 30. New York: McGraw-Hill Book Co., 1961.
[2] Muster, D., and Plunkett, R. "Isolation of Vibrations." In *Noise Reduction*. Edited by L. L. Beranek. Chapter 18. New York: McGraw-Hill Book Co., 1960.
[3] Crede, C. E. "Principles of Vibration Control." In *Handbook of Noise Control*. Edited by C. M. Harris, Chapter 12. New York: McGraw-Hill Book Co., 1957.
[4] Erhart, R., and Salerno, C. M. "Isolating Vibration—Fundamentals." *Machine Design*, 13 July 1972, pp. 112-118.
[5] Erhart, R., and Salerno, C. M. "Isolating Vibration—Isolators and Materials." *Machine Design*, 27 July 1972, pp. 92-96.
[6] Crede, C. E., and Ruzicka, J. E. "Air Suspension and Servo-Controlled Isolation Systems." In *Shock and Vibration Handbook*. Vol. 2, Chapter 33. New York: McGraw-Hill Book Co., 1961.
[7] Carson, R. W. "How to Select Vibration Isolators." *Product Engineering*, 4 March 1963, pp. 68-78.
[8] Sykes, A. "Isolation of Vibration When Machine and Foundation Are Resilient and When Wave Effects Occur in the Mount." *Noise Control*, May/June 1960, pp. 23-38.
[9] Ungar, E. E., and Dietrich, C. W. "High-Frequency Vibration Isolation." *Journal of Sound and Vibration*, vol. 4. January 1958, pp. 224-241.
[10] Muster, D., and Plunkett, R. "Isolation of Vibrations," *Noise and Vibration Control*. Edited by L. L. Beranek. Chapter 13. New York: McGraw-Hill Book Co., 1971.
[11] Fry, A. T. "Vibration Isolation." *Noise Control in Mechanical Services*. Edited by R. I. Woods, Chapter 6. Colchester, England: Anchor Press Ltd., 1972.
[12] Ruzicka, J. E. "Passive Shock Isolation, Part I." *Sound and Vibration*, vol. 4, no. 8, August 1970, pp. 14-24.
[13] Ruzicka, J. E. "Passive Shock Isolation, Part II." *Sound and Vibration*, vol. 4, no. 9, September 1970, pp. 10-22.
[14] Newton, R. E. "Theory of Shock Isolation." In *Shock and Vibration Handbook*. Vol. 2, Chapter 31. New York: McGraw-Hill Book Co., 1961.
[15] Broch, J. T. "Effect of Vibrations and Shock on Mechanical Systems and Man." *Mechanical Vibration and Shock Measurement*, Chapter 3 (1967). Brüel and Kjäer Co., 5113 West 164th Street, Cleveland, Ohio. 44142.
[16] Carson, R. W. "How to Select Vibration Isolators." *Product Engineering*, 4 March 1963, pp. 68-78.
[17] Crede, C. E. "Reduction of Excitation at the Source." In *Shock and Vi-*

REFERENCES   563

bration Concepts in Engineering Design. Chapter 3. Englewood Cliffs, N.J.: Prentice-Hall, Inc., 1965.

[18] Mitchell, L. D. "Gear Noise: The Purchaser's and the Manufacturer's Views. Noise and Vibration Control Engineering." *Proceedings of the Purdue Noise Control Conference.* Edited by M. J. Crocker. Purdue University, July 1971, pp. 95-106.

[19] Crocker, M. J., and Price, A. J. "Vibration Isolation of a Book Trimming Machine." *Applied Acoustics,* no. 3 (1970), pp. 207-216.

[20] Miller, L. N., and Dyer, I. "Printing Machine Isolation." *Noise Control* 56, July 1958, pp. 21-23.

[21] Crocker, M. J.; Hamilton, J. F.; and Price, A. J. "Vibration Isolation for Machine Noise Reduction." *Sound and Vibration,* November 1971, pp. 30-34.

[22] Seybert, A.; Crocker, M. J.; Moore, J. W.; and Jones, S. R. "Noise Reduction of a Residential Air Conditioner." *Noise Control Engineering,* vol. 1, no. 2, Autumn 1973.

[23] Mohr, J. W. "Vibration and Noise Control of Outboard Motors and Other Products." *SAE Paper No. 183 B,* Society of Automotive Engineers, June 1960.

[24] Gorenshtein, I. V.; Zaborov, V. I.; and Tyumenstseva, L. P. "Noise Reduction of Shake Tables with Eccentric Mechanical Vibrators." *Soviet Physics Acoustics,* vol. 15, no. 3, January/March 1970, pp. 311-315.

[25] Reed, D. H. "Noise Control of Aircraft Electro-magnetic Components." *Sound and Vibration,* vol. 4, 2, February 1970, pp. 16-19.

[26] Yin, T. P. "The Control of Noise and Vibration." *Scientific American,* vol. 220, no. 1, January 1969, pp. 98-106.

[27] Salerno, C. M., and Hochheiser, R. M. "How to Select Vibration Isolators for Use as Machinery Mounts." *Sound and Vibration,* vol. 7, no. 8, August 1973, pp. 22-29.

[28] Hochheiser, R. M. "How to Select Vibration Isolators for OEM Machinery and Equipment." *Sound and Vibration,* vol. 8, no. 8, August 1974, pp. 14-23.

References to Chapter 8

[1] Mitchell, L. D. "Gear Noise: The Purchaser's and the Manufacturer's Views. Noise and Vibration Control Engineering." *Proceedings of the Purdue Noise Control Conference.* Edited by M. J. Crocker. Purdue University, July 1971, pp. 95-106.

[2] Opitz, H. "Noise of Gears." Phil. Transactions, Royal Society of London, Series A, vol. 263, no. 1142, December 1968, pp. 369-380.

[3] Moeller, K. G. F. "Gear Noise," *Handbook of Noise Control.* Edited by C. M. Harris. Chapter 23. New York: McGraw-Hill Book Co., 1957.

[4] Hall, A. S., Jr.; Holowenko, A. R.; and Laughlin, H. G. *Theory and Problems of Machine Design.* Chapter 23. New York: Schaum Publishing Co., 1961.

[5] *Mark's Mechanical Engineers' Handbook,* 6th ed. New York: McGraw-Hill Book Co., 1958.

[6] Jensen, Preben W. *Cam Design and Manufacture.* New York: Industrial Press Inc., 1965.

[7] Shigley, J. E. *Kinematic Analysis of Mechanisms.* New York: McGraw-Hill Book Co., 1969.
[8] Martin, George H. *Kinematics and Design of Machines.* New York: McGraw-Hill Book Co., 1969.
[9] Ungar, E. E., and Ross, Donald. "Vibrations and Noise Due to Piston-Slap in Reciprocating Machinery." *Journal of Sound and Vibration,* vol. 2, no. 2 (1965), pp. 132–146.
[10] Binder, R. C. *Mechanics of the Roller Chain Drive.* Englewood Cliffs, New Jersey: Prentice-Hall, Inc., 1956.
[11] Shuey, S. J. *Vibration and Noise in Pumping Systems* (1972). Byron Jackson Pump Division, Borg Warner Corporation, Los Angeles, California 90054.

References to Chapter 9

[1] AMCA STANDARD 300-67, *Test Code for Sound Rating Air Moving Devices.* Air Moving and Conditioning Association, Inc., 205 West Touhy Avenue, Park Ridge, Illinois 60068.
[2] *ASHRAE Guide and Data Book, 1972, Equipment,* American Society of Heating, Refrigerating and Air-Conditioning Engineers, Inc., 345 East 47th Street, New York, New York 10017, Chapters 25 and 26.
[3] *ASHRAE Guide and Data Book, 1973, Systems,* American Society of Heating, Refrigerating and Air-Conditioning Engineers, Inc., 345 East 47th Street, New York, New York 10017, Chapter 35.
[4] Graham, J. B. "How to Estimate Fan Noise." *Sound and Vibration,* May 1972, p. 24.
[5] Peistrup, C. F., and Wesler, J. E. "Noise of Ventilating Fans." The *Journal of the Acoustical Society of America 25* (1953), p. 322.
[6] Graham, J. B. "A Method for Estimating the Sound Power Level of Fans." *ASHRAE Transactions,* vol. 72, part II (1966), p. I.1.1.
[7] Baade, P. K. "Accuracy Considerations in Fan Sound Measurement." *ASHRAE Transactions,* vol. 73, part II (1967), s. v. 1. 1.
[8] Beranek, L. L., ed. *Noise and Vibration Control.* Chapter 16. New York: McGraw-Hill Book Co., 1971.
[9] Gordon, C. G. "Spoiler-Generated Flow Noise, Part I. The Experiment," and "Part II. Results." *The Journal of the Acoustical Society of America,* vol. 43 (1969), p. 1041, and vol. 45, p. 214.
[10] Heller, H. H., and Widnall, S. E. "Sound Radiation from Rigid Flow Spoilers Correlated with Fluctuating Forces." *The Journal of the Acoustical Society of America.* vol. 47, no. 3, part 2 (1970), p. 924.
[11] Yudin, E. Y. "The Acoustic Power of the Noise Created by Airduct Elements." *Soviet Physics–Acoustics.* vol. 1 (1955), p. 383.
[12] Vér, I. L. "Prediction Scheme for the Self-Generated Noise of Silencers." *Inter-Noise 72 Proceedings,* Washington, D.C., October 1972, p. 294.
[13] Sanders, G. J. "Noise Control for Industrial Air Moving Devices." *Inter-Noise 72 Proceedings,* Washington, D.C., October 1972, p. 165.
[14] Powell, A. "On the Noise Emanating from a Two-Dimensional Jet Above the Critical Pressure." *Aeronautical Quarterly* 4 (1953), p. 103.
[15] Stucky, D. L., and Marco, S. M. "On the Noise from Jet Diffusers." *ASME Paper No. 70-WA/GT-5* (1970). American Society of Mechanical Engineers, United Engineering Center, 345 East 47th Street, New York, New York 10017.

REFERENCES

References to Chapter 10

[1] Brown, D. J. "Noise Emission and Acoustic Efficiency in Pulsating Combustion." *Combustion Science and Technology* 3 (1971), pp. 51–56.
[2] Brown, D. J., ed. *Proceedings, First International Symposium on Pulsating Combustion.* September 1971. University of Sheffield, Sheffield, England.
[3] Markstein, G. H., ed. "Nonsteady Flame Propagation." In *AGARDograph 75*. Chapters F, G, and H. Elmsford, New York: Pergamon Press, 1965.
[4] Putnam, A. A. *Combustion-Driven Oscillations in Industry.* New York: American Elsevier Publishing Co., 1971.
[5] Crocco, L., and Cheng, S-I. "Theory of Combustion Instability in Liquid-Propellant Rocket Motors." *AGARDograph 8.* London: Butterworths, 1956.
[6] Raushenbakh, B. V. "Vibratory Combustion," *Gosudarstvennoye Izdatel'stvo Fiziko-Matematkheskoy Literatury.* Moskva, 1961. Translation FTD-TT-62-942/1 + 2 (April 1963) ASTIA AD 402,909.
[7] Thring, M. W., ed. *Pulsating Combustion—The Collected Works of F. G. Reynst.* Elmsford, New York: Pergamon Press, 1961.
[8] Rayleigh, L. *The Theory of Sound.* Vol. 2. New York: Dover, 1945.
[9] Dorresteijn, W. R. "Vibrations in a Vertical Crude-Oil Heater: Causes and Remedies." *Journal of the Institute of Fuel* 41 (1968), p. 387.
[10] Giammar, R. D., and Putnam, A. A. "Noise Generation by Turbulent Flames." *Summary Report, American Gas Association*, Cat. No. M00080 (1971).
[11] Strahle, W. C. "On Combustion Generated Noise." *Journal of Fluid Mechanics* 49 (1971), pp. 399–414.
[12] Putnam, A. A. "Identification and Control of Combustion Noise." *Proceedings:* Fuels Utilization Conference, The Cleveland State University, Cleveland, Ohio, October 26, 1972.

References to Chapter 11

[1] Lighthill, M. J. "On Sound Generated Aerodynamically." *Proc. Roy. Soc. A*, London, Part 1 in Vol. 211 (1952). Part 2 in Vol. 222 (1954).
[2] Buresh, James F., and Schuder, C. B. *Development of a Universal Gas Sizing Equation for Control Valves*, TM-15. Fisher Controls Co., Marshalltown, Iowa 50158.
[3] Stiles, G. F. "Development of a Valve Sizing Relationship for Flashing and Cavitating Flow," *ISA Final Control Elements Symposium* (1970), Wilmington, Delaware.
[4] Fagerlund, A. C. "Transmission of Sound Through a Cylindrical Pipe Wall." *ASME Paper No. 73-WA/PID-4* (November 1973). American Society of Mechanical Engineers, 345 East 47th Street, New York, New York 10017.
[5] Fagerlund, A. C. *Conversion of Vibration Measurements to Sound Pressure Levels*, TM-33. Fisher Controls Co., Marshalltown, Iowa 50158.

Appendix

INTRODUCTION

The information presented here is intended for practical use in engineering and related technical fields, although other branches of industry should also find it useful. Emphasis is placed on the International System of Units (SI) which is the modernized metric system of measurement.

This emphasis is in keeping with the worldwide move to adopt SI, and the growing interest in this system in the U.S.A.

But the contents is not limited to conversions into SI units. Units from earlier metric systems are also given. The reason is that readers will meet with such units as kilogram-force, metric horsepower or bar and will need to know what the values are in SI or customary English units of measurement.

Publications that should be consulted for details of recommended metric practice and more complete information on the SI include the Standard Metric Practice Guide (designated E380-72) published by the American Society for Testing and Materials, 1916 Race St., Philadelphia, Pa. 19103 and also available from the American National Standards Institute, 1430 Broadway, New York, N. Y. 10018 as American National Standard Z210.1; ISO International Standard 1000, which covers rules for the use of SI units, multiples and submultiples, and is available from ANSI; and the National Bureau of Standards Special Publication 330, The International System of Units (SI).*

METRIC SYSTEMS OF MEASUREMENT

A metric system of measurement was first established in France in the years following the French Revolution, and various systems of metric units have been developed since that time. All metric unit systems are based, at least in part, on the International Metric Standards which are the metre and kilogram, or decimal multiples or sub-multiples of these standards.

In 1795, a metric system called the centimetre-gram-second (cgs) system was proposed, and was adopted in France in 1799. In 1873, the British Association for the Advancement of Science recommended the use of the cgs system, and since

---

* For sale by Superintendent of Documents, U. S. Government Printing Office, Washington, D. C. 20402.

APPENDIX 567

then it has been widely used in all branches of science throughout the world. From the base units in the cgs system are derived:

Unit of velocity = 1 centimetre per second.
Acceleration due to gravity (at Paris) = 981 centimetres per sec. per sec.
Unit of force = 1 dyne = 1/981 gram.
Unit of work = 1 erg = 1 dyne-centimetre.
Unit of power = 1 watt = 10,000,000 ergs per second.

Another metric system called the MKS (metre-kilogram-second) system of units was proposed by Professor G. Giorgi in 1902. In 1935, the International Electrotechnical Commission (IEC) accepted his recommendation that this system of units of mechanics should be linked with the electro-magnetic units by the adoption of a fourth base unit. In 1950, the IEC adopted the ampere, the unit of electric current, as the fourth unit, and the MKSA system thus came into being.

A gravitational system of metric units, known as the technical system, is based on the metre, the kilogram as a force, and the second. It has been widely used in engineering. Because the standard of force is defined as the weight of the mass of the standard kilogram, the fundamental unit of force varies due to the difference in gravitational pull at different locations around the earth. By international agreement, a standard value for acceleration due to gravity was chosen (9.81 metres per second squared) which for all practical measurements is approximately the same as the local value at the point of measurement.

The International System of Units (SI). — The Conference Generale des Poids et Mesures (CGPM) which is the body responsible for all international matters concerning the metric system, adopted in 1954, a rationalized and coherent system of units, based on the four MKSA units (see above), and including the *kelvin* as the unit of temperature and the *candela* as the unit of luminous intensity. In 1960, the CGPM formally named this system the Systeme International d'Unites, for which the abbreviation is SI in all languages. In 1971, the 14th CGPM adopted a seventh base unit, the *mole* which is the unit of quantity ("amount of substance").

In the period since the first metric system was established in France towards the end of the 18th century, most of the countries of the world have adopted a metric system. At the present time most of the industrially advanced metric-using countries are changing from their traditional metric system to SI. Those countries which are currently changing or considering change from the English system of measurement to metric, have the advantage that they can convert directly to the modernized system. The United Kingdom, which can be said to have led the now worldwide move to change from the English system, went straight to SI.

The SI, like the traditional metric system, is based on decimal arithmetic. For each physical quantity, units of different sizes are formed by multiplying or dividing a single base value by powers of 10. Thus, changes can be made very simply by adding zeros or shifting decimal points. For example, the metre is the basic unit of length; the kilometre is a multiple (1000 metres): and the millimetre is a submultiple (one-thousandth of a metre).

In the older metric systems, the simplicity of a series of units linked by powers of ten is an advantage for plain quantities such as length, but this simplicity is lost as soon as more complex units are encountered. For example, in different branches of science and engineering, energy may appear as the erg, the calorie, the kilogram metre, the litre atmosphere or the horsepower hour. In contrast, the SI provides only one basic unit for each physical quantity, and universality is thus achieved.

As mentioned above, there are seven base-units, which are for the basic quantities of length, mass, time, electric current, thermodynamic temperature, amount of substance and luminous intensity, expressed as the metre (m), the kilogram (kg), the second (s), the ampere (A), the kelvin (K), the mole (mol) and the candela (cd). The units are defined in the accompanying table.

The SI is a coherent system. A system is said to be coherent if the product or quotient of any two unit quantities in the system is the unit of the resultant quantity. For example, in a coherent system in which the foot is the unit of length, the square foot is the unit of area, whereas the acre is not.

Other physical quantities are derived from the base units. For example, the unit of velocity is the metre per second (m/s), which is a combination of the base units of length and time. The unit of acceleration is the metre per second squared (m/s²). By applying Newton's second law of motion — force is proportional to mass multiplied by acceleration — the unit of force is obtained which is the kilogram metre per second squared (kg·m/s²). This unit is known as the newton or N. Work, or force times distance is the kilogram metre squared per second squared (kg·m²/s²), which is the joule, (1 joule = 1 newton-metre), and energy is also expressed in these terms. The abbreviation for joule is J. Power or work per unit time is the kilogram metre squared per second cubed (kg·m²/s³), which is the watt (1 watt = 1 joule per second = 1 newton-metre per second.) The abbreviation for watt is W. The term horsepower is not used in the SI and is replaced by the watt which together with multiples and sub-multiples — kilowatt and milliwatt, for example — is the same unit as that used in electrical work.

The use of the newton as the unit of force is of particular interest to engineers. In practical work using the English or traditional metric systems of measurements, it is a common practice to apply weight units as force units. Thus, the unit of force in those systems is that force which when applied to unit mass produces an acceleration g, rather than unit acceleration. The value of gravitational acceleration g varies around the earth, and thus the weight of a given mass also varies. In an effort to account for this minor error, the kilogram-force and pound-force were introduced, which are defined as the forces due to "standard gravity" acting on bodies of one kilogram or one pound mass respectively. The standard gravitational acceleration is taken as 9.80665 metres per second squared or 32.174 feet per second squared. The newton is defined as "that force which when applied to a body having a mass of one kilogram, gives it an acceleration of one metre per second squared." It is independent of g. As a result, the factor g disappears from a wide range of formulas in dynamics. However, in some formulas in statics, where the weight of a body is important rather than its mass, g does appear where it was formerly absent (the weight of a mass of W kilograms is equal to a force of Wg newtons, where g = approximately 9.81 metres per second squared). Details concerning the use of SI units in mechanics calculations are given in the Mechanics section of *Machinery's Handbook* and the use of SI units in strength of materials calculations is covered in the section on that subject.

Decimal multiples and sub-multiples of the SI units are formed by means of the prefixes given in the following table, which represent the numerical factors shown.

Factors and Prefixes for Forming Decimal Multiples and Sub-multiples of the SI Units

Factor by which the unit is multiplied	Prefix	Symbol	Factor by which the unit is multiplied	Prefix	Symbol
10^{12}	tera	T	10^{-2}	centi	c
10^{9}	giga	G	10^{-3}	milli	m
10^{6}	mega	M	10^{-6}	micro	μ
10^{3}	kilo	k	10^{-9}	nano	n
10^{2}	hecto	h	10^{-12}	pico	p
10	deka	da	10^{-15}	femto	f
10^{-1}	deci	d	10^{-18}	atto	a

APPENDIX

International System (SI) Units

PHYSICAL QUANTITY	NAME OF UNIT	UNIT SYMBOL	DEFINITION
\multicolumn{4}{c}{Basic SI Units}			
Length	metre	m	1,650,763.73 wavelengths in vacuo of the radiation corresponding to the transition between the energy levels $2p_{10}$ and $5d_5$ of the krypton — 86 atom.
Mass	kilogram	kg	Mass of the international prototype which is in the custody of the Bureau International des Poids et Mesures (BIPM) at Sèvres, near Paris.
Time	second	s	The duration of 9,192,631,770 periods of the radiation corresponding to the transition between the two hyperfine levels of the ground state of the cesium-133 atom.
Electric Current	ampere	A	The constant current which, if maintained in two parallel rectilinear conductors of infinite length, of negligible circular cross section, and placed at a distance of one metre apart in a vacuum, would produce between these conductors a force equal to 2×10^{-7} N/m length.
Thermodynamic Temperature	degree Kelvin	K	The fraction 1/273.16 of the thermodynamic temperature of the triple point of water.
Amount of Substance	mole	mol	The amount of substance of a system which contains as many elementary entities as there are atoms in 0.012 kilogram of carbon 12.
Luminous Intensity	candela	cd	The luminous intensity, in the perpendicular direction, of a surface of 1/600,000 square metre of a black body at the temperature of freezing platinum under a pressure of 101,325 newtons per square metre.
\multicolumn{4}{c}{SI Units Having Special Names}			
Force	newton	$N = kg \cdot m/s^2$	That force which, when applied to a body having a mass of one kilogramme, gives it an acceleration of one metre per second squared.
Work, Energy, Quantity of Heat	joule	$J = N \cdot m$	The work done when the point of application of a force of one newton is displaced through a distance of one metre in the direction of the force.
Power	watt	$W = J/s$	One joule per second.
Electric Charge	coulomb	$C = A \cdot s$	The quantity of electricity transported in one second by a current of one ampere.
Electric Potential	volt	$V = W/A$	The difference of potential between two points of a conducting wire carrying a constant current of one ampere, when the power dissipated between these points is equal to one watt.
Electric Capacitance	farad	$F = C/V$	The capacitance of a capacitor between the plates of which there appears a difference of potential of one volt when it is charged by a quantity of electricity equal to one coulomb.

International System (SI) Units (*Continued*)

PHYSICAL QUANTITY	NAME OF UNIT	UNIT SYMBOL	DEFINITION
\multicolumn{4}{c}{SI Units Having Special Names}			
Electric Resistance	ohm	$\Omega = V/A$	The resistance between two points of a conductor when a constant difference of potential of one volt, applied between these two points, produces in this conductor a current of one ampere, this conductor not being the source of any electromotive force.
Magnetic Flux	weber	$Wb = V \cdot s$	The flux which, linking a circuit of one turn produces in it an electromotive force of one volt as it is reduced to zero at a uniform rate in one second.
Inductance	henry	$H = Wb/A$	The inductance of a closed circuit in which an electromotive force of one volt is produced when the electric current in the circuit varies uniformly at the rate of one ampere per second.
Luminous Flux	lumen	$lm = cd \cdot sr$	The flux emitted within a unit solid angle of one steradian by a point source having a uniform intensity of one candela.
Illumination	lux	$lx = lm/m^2$	An illumination of one lumen per square metre.

International System (SI) Units with Complex Names

PHYSICAL QUANTITY	SI UNIT	UNIT SYMBOL
\multicolumn{3}{c}{SI Units Having Complex Names}		
Area	square metre	m^2
Volume	cubic metre	m^3
Frequency	hertz*	Hz
Density (Mass Density)	kilogram per cubic metre	kg/m^3
Velocity	metre per second	m/s
Angular Velocity	radian per second	rad/s
Acceleration	metre per second squared	m/s^2
Angular Acceleration	radian per second squared	rad/s^2
Pressure	pascal ‡	Pa
Surface Tension	newton per metre	N/m
Dynamic Viscosity	newton second per metre squared	$N\ s/m^2$
Kinematic Viscosity / Diffusion Coefficient	metre squared per second	m^2/s
Thermal Conductivity	watt per metre degree Kelvin	$W/(m\ °K)$
Electric Field Strength	volt per metre	V/m
Magnetic Flux Density	tesla†	T
Magnetic Field Strength	ampere per metre	A/m
Luminance	candela per square metre	cd/m^2

* Hz = cycle/second.
† T = weber/metre².
‡ Pa = newton/metre²

APPENDIX

Metric Conversion Factors
(Symbols of SI units, multiples and submultiples are given in parentheses in the right-hand column)

Multiply	By	To Obtain
LENGTH		
centimetre	0.03280840	foot
centimetre	0.3937008	inch
fathom	1.8288*	metre (m)
foot	0.3048*	metre (m)
foot	30.48*	centimetre (cm)
foot	304.8*	millimetre (mm)
inch	0.0254*	metre (m)
inch	2.54*	centimetre (cm)
inch	25.4*	millimetre (mm)
kilometre	0.6213712	mile [U. S. statute]
metre	39.37008	inch
metre	0.5468066	fathom
metre	3.280840	foot
metre	0.1988388	rod
metre	1.093613	yard
metre	0.0006213712	mile [U. S. statute]
microinch	0.0254*	micrometre [micron] (μm)
micrometre [micron]	39.37008	microinch
mile [U. S. statute]	1609.344*	metre (m)
mile [U. S. statute]	1.609344*	kilometre (km)
millimetre	0.003280840	foot
millimetre	0.03937008	inch
rod	5.0292*	metre (m)
yard	0.9144*	metre (m)
AREA		
acre	4046.856	metre^2 (m^2)
acre	0.4046856	hectare
centimetre^2	0.1550003	inch^2
centimetre^2	0.001076391	foot^2
foot^2	0.09290304*	metre^2 (m^2)
foot^2	929.0304*	centimetre^2 (cm^2)
foot^2	92,903.04*	millimetre^2 (mm^2)
hectare	2.471054	acre
inch^2	645.16*	millimetre^2 (mm^2)
inch^2	6.4516*	centimetre^2 (cm^2)
inch^2	0.00064516*	metre^2 (m^2)
metre^2	1550.003	inch^2
metre^2	10.763910	foot^2
metre^2	1.195990	yard^2
metre^2	0.0002471054	acre
millimetre^2	0.0001076387	foot^2
millimetre^2	0.001550003	inch^2
yard^2	0.8361274	metre^2 (m^2)

* Where an asterisk is shown, the figure is exact.

Metric Conversion Factors (*Continued*)

Multiply	By	To Obtain
VOLUME (including CAPACITY)		
centimetre^3	0.06102376	inch^3
foot^3	0.02831685	metre^3 (m^3)
foot^3	28.31685	litre
gallon [U. K. liquid]	0.004546092	metre^3 (m^3)
gallon [U. K. liquid]	4.546092	litre
gallon [U. S. liquid]	0.003785412	metre^3 (m^3)
gallon [U. S. liquid]	3.785412	litre
inch^3	16,387.06	millimetre^3 (mm^3)
inch^3	16.38706	centimetre^3 (cm^3)
inch^3	0.00001638706	metre^3 (m^3)
litre	0.001*	metre^3 (m^3)
litre	0.2199692	gallon [U. K. liquid]
litre	0.2641720	gallon [U. S. liquid]
litre	0.03531466	foot^3
metre^3	219.9692	gallon [U. K. liquid]
metre^3	264.1720	gallon [U. S. liquid]
metre^3	35.31466	foot^3
metre^3	1.307951	yard^3
metre^3	1000.*	litre
metre^3	61,023.76	inch^3
millimetre^3	0.00006102376	inch^3
yard^3	0.7645549	metre^3 (m^3)
VELOCITY, ACCELERATION, and FLOW		
centimetre/second	1.968504	foot/minute
centimetre/second	0.03280840	foot/second
centimetre/minute	0.3937008	inch/minute
foot/hour	0.00008466667	metre/second (m/s)
foot/hour	0.00508*	metre/minute
foot/hour	0.3048*	metre/hour
foot/minute	0.508*	centimetre/second
foot/minute	18.288*	metre/hour
foot/minute	0.3048*	metre/minute
foot/minute	0.00508*	metre/second (m/s)
foot/second	30.48*	centimetre/second
foot/second	18.288*	metre/minute
foot/second	0.3048*	metre/second (m/s)
foot/second^2	0.3048*	metre/second^2 (m/s^2)
foot^3/minute	28.31685	litre/minute
foot^3/minute	0.0004719474	metre^3/second (m^3/s)
gallon [U. S. liquid]/min.	0.003785412	metre^3/minute
gallon [U. S. liquid]/min.	0.00006309020	metre^3/second (m^3/s)
gallon [U. S. liquid]/min.	0.06309020	litre/second
gallon [U. S. liquid]/min.	3.785412	litre/minute
gallon [U. K. liquid]/min.	0.004546092	metre^3/minute
gallon [U. K. liquid]/min.	0.00007576820	metre^3/second (m^3/s)
inch/minute	25.4*	millimetre/minute
inch/minute	2.54*	centimetre/minute
inch/minute	0.0254*	metre/minute
inch/second^2	0.0254*	metre/second^2 (m/s^2)

* Where an asterisk is shown, the figure is exact.

APPENDIX

Metric Conversion Factors (*Continued*)

Multiply	By	To Obtain
VELOCITY, ACCELERATION, and FLOW (*Continued*)		
kilometre/hour	0.6213712	mile/hour [U. S. statute]
litre/minute	0.03531466	foot^3/minute
litre/minute	0.2641720	gallon [U. S. liquid]/minute
litre/second	15.85032	gallon [U. S. liquid]/minute
mile/hour	1.609344*	kilometre/hour
millimetre/minute	0.03937008	inch/minute
metre/second	11,811.02	foot/hour
metre/second	196.8504	foot/minute
metre/second	3.280840	foot/second
metre/second^2	3.280840	foot/second^2
metre/second^2	39.37008	inch/second^2
metre/minute	3.280840	foot/minute
metre/minute	0.05468067	foot/second
metre/minute	39.37008	inch/minute
metre/hour	3.280840	foot/hour
metre/hour	0.05468067	foot/minute
metre^3/second	2118.880	foot^3/minute
metre^3/second	13,198.15	gallon [U. K. liquid]/minute
metre^3/second	15,850.32	gallon [U. S. liquid]/minute
metre^3/minute	219.9692	gallon [U. K. liquid]/minute
metre^3/minute	264.1720	gallon [U. S. liquid]/minute
MASS and DENSITY		
grain [1/7000 lb avoirdupois]	0.06479891	gram (g)
gram	15.43236	grain
gram	0.001*	kilogram (kg)
gram	0.03527397	ounce [avoirdupois]
gram	0.03215074	ounce [troy]
gram/centimetre^3	0.03612730	pound/inch^3
hundredweight [long]	50.80235	kilogram (kg)
hundredweight [short]	45.35924	kilogram (kg)
kilogram	1000.*	gram (g)
kilogram	35.27397	ounce [avoirdupois]
kilogram	32.15074	ounce [troy]
kilogram	2.204622	pound [avoirdupois]
kilogram	0.06852178	slug
kilogram	0.0009842064	ton [long]
kilogram	0.001102311	ton [short]
kilogram	0.001*	ton [metric]
kilogram	0.001*	tonne
kilogram	0.01968413	hundredweight [long]
kilogram	0.02204622	hundredweight [short]
kilogram/metre^3	0.06242797	pound/foot^3
kilogram/metre^3	0.008345406	pound/gallon [U. K. liquid]
kilogram/metre^3	0.01002242	pound/gallon [U. S. liquid]
ounce [avoirdupois]	28.34952	gram (g)
ounce [avoirdupois]	0.02834952	kilogram (kg)

* Where an asterisk is shown, the figure is exact.

Metric Conversion Factors (*Continued*)

Multiply	By	To Obtain
MASS and DENSITY (*Continued*)		
ounce [troy]	31.10348	gram (g)
ounce [troy]	0.03110348	kilogram (kg)
pound [avoirdupois]	0.4535924	kilogram (kg)
pound/foot^3	16.01846	kilogram/metre^3 (kg/m^3)
pound/inch^3	27.67990	gram/centimetre^3 (g/cm^3)
pound/gal [U. S. liquid]	119.8264	kilogram/metre^3 (kg/m^3)
pound/gal [U. K. liquid]	99.77633	kilogram/metre^3 (kg/m^3)
slug	14.59390	kilogram (kg)
ton [long 2240 lb]	1016.047	kilogram (kg)
ton [short 2000 lb]	907.1847	kilogram (kg)
ton [metric]	1000.*	kilogram (kg)
tonne	1000.*	kilogram (kg)
FORCE and FORCE/LENGTH		
dyne	0.00001*	newton (N)
kilogram-force	9.806650*	newton (N)
kilopond	9.806650*	newton (N)
newton	0.1019716	kilogram-force
newton	0.1019716	kilopond
newton	0.2248089	pound-force
newton	100,000.*	dyne
newton	7.23301	poundal
newton	3.596942	ounce-force
newton/metre	0.005710148	pound/inch
newton/metre	0.06852178	pound/foot
ounce-force	0.2780139	newton (N)
pound-force	4.448222	newton (N)
poundal	0.1382550	newton (N)
pound/inch	175.1268	newton/metre (N/m)
pound/foot	14.59390	newton/metre (N/m)
BENDING MOMENT or TORQUE		
dyne-centimetre	0.0000001*	newton-metre (N · m)
kilogram-metre	9.806650*	newton-metre (N · m)
ounce-inch	7.061552	newton-millimetre
ounce-inch	0.007061552	newton-metre (N · m)
newton-metre	0.7375621	pound-foot
newton-metre	10,000,000.*	dyne-centimetre
newton-metre	0.1019716	kilogram-metre
newton-metre	141.6119	ounce-inch
newton-millimetre	0.1416119	ounce-inch
pound-foot	1.355818	newton-metre (N · m)

* Where an asterisk is shown, the figure is exact.

APPENDIX

Metric Conversion Factors *(Continued)*

Multiply	By	To Obtain
\multicolumn{3}{c}{MOMENT OF INERTIA and SECTION MODULUS}
moment of inertia [kg · m^2]	23.73036	pound-foot^2
moment of inertia [kg · m^2]	3417.171	pound-inch^2
moment of inertia [lb · ft^2]	0.04214011	kilogram-metre^2 (kg · m^2)
moment of inertia [lb · inch^2]	0.0002926397	kilogram-metre^2 (kg · m^2)
moment of section [foot^4]	0.008630975	metre^4 (m^4)
moment of section [inch^4]	41.62314	centimetre^4
moment of section [metre^4]	115.8618	foot^4
moment of section [centimetre^4]	0.02402510	inch^4
section modulus [foot^3]	0.02831685	metre^3 (m^3)
section modulus [inch^3]	0.00001638706	metre^3 (m^3)
section modulus [metre^3]	35.31466	foot^3
section modulus [metre^3]	61,023.76	inch^3
\multicolumn{3}{c}{MOMENTUM}
kilogram-metre/second	7.233011	pound-foot/second
kilogram-metre/second	86.79614	pound-inch/second
pound-foot/second	0.1382550	kilogram-metre/second (kg · m/s)
pound-inch/second	0.01152125	kilogram-metre/second (kg · m/s)
\multicolumn{3}{c}{PRESSURE and STRESS}
atmosphere [14.6959 lb/inch^2]	101,325.	pascal (Pa)
bar	100,000.*	pascal (Pa)
bar	14.50377	pound/inch^2
bar	100,000.*	newton/metre^2 (N/m^2)
hectobar	0.6474898	ton [long]/inch^2
kilogram/centimetre^2	14.22334	pound/inch^2
kilogram/metre^2	9.806650*	newton/metre^2 (N/m^2)
kilogram/metre^2	9.806650*	pascal (Pa)
kilogram/metre^2	0.2048161	pound/foot^2
kilonewton/metre^2	0.1450377	pound/inch^2
newton/centimetre^2	1.450377	pound/inch^2
newton/metre^2	0.00001*	bar
newton/metre^2	1.0*	pascal (Pa)
newton/metre^2	0.0001450377	pound/inch^2
newton/metre^2	0.1019716	kilogram/metre^2
newton/millimetre^2	145.0377	pound/inch^2
pascal	0.00000986923	atmosphere
pascal	0.00001*	bar
pascal	0.1019716	kilogram/metre^2
pascal	1.0*	newton/metre^2 (N/m^2)
pascal	0.02088543	pound/foot^2
pascal	0.0001450377	pound/inch^2

* Where an asterisk is shown, the figure is **exact**.

APPENDIX

Metric Conversion Factors (*Continued*)

Multiply	By	To Obtain
PRESSURE and STRESS (*Continued*)		
pound/foot^2	4.882429	kilogram/metre^2
pound/foot^2	47.88026	pascal (Pa)
pound/inch^2	0.06894757	bar
pound/inch^2	0.07030697	kilogram/centimetre^2
pound/inch^2	0.6894757	newton/centimetre^2
pound/inch^2	6.894757	kilonewton/metre^2
pound/inch^2	6894.757	newton/metre^2 (N/m^2)
pound/inch^2	0.006894757	newton/millimetre^2 (N/mm^2)
pound/inch^2	6894.757	pascal (Pa)
ton [long]/inch^2	1.544426	hectobar
ENERGY and WORK		
Btu [International Table]	1055.056	joule (J)
Btu [mean]	1055.87	joule (J)
calorie [mean]	4.19002	joule (J)
foot-pound	1.355818	joule (J)
foot-poundal	0.04214011	joule (J)
joule	0.0009478170	Btu [International Table]
joule	0.0009470863	Btu [mean]
joule	0.2386623	calorie [mean]
joule	0.7375621	foot-pound
joule	23.73036	foot-poundal
joule	0.9998180	joule [International U. S.]
joule	0.9999830	joule [U. S. legal, 1948]
joule [International U. S.]	1.000182	joule (J)
joule [U. S. legal, 1948]	1.000017	joule (J)
joule	.0002777778	watt-hour
watt-hour	3600.*	joule (J)
POWER		
Btu [International Table]/hour	0.2930711	watt (W)
foot-pound/hour	0.0003766161	watt (W)
foot-pound/minute	0.02259697	watt (W)
horsepower [550 ft-lb/s]	0.7456999	kilowatt (kW)
horsepower [550 ft-lb/s]	745.6999	watt (W)
horsepower [electric]	746.*	watt (W)
horsepower [metric]	735.499	watt (W)
horsepower [U. K.]	745.70	watt (W)
kilowatt	1.341022	horsepower [550 ft-lb/s]
watt	2655.224	foot-pound/hour
watt	44.25372	foot-pound/minute
watt	0.001341022	horsepower [550 ft-lb/s]
watt	0.001340483	horsepower [electric]
watt	0.001359621	horsepower [metric]
watt	0.001341022	horsepower [U. K.]
watt	3.412141	Btu [International Table]/hour

* Where an asterisk is shown, the figure is exact.

Metric Conversion Factors (*Continued*)

Multiply	By	To Obtain
VISCOSITY		
centipoise	0.001*	pascal-second (Pa · s)
centistoke	0.000001*	metre^2/second (m^2/s)
metre^2/second	1,000,000.*	centistoke
metre^2/second	10,000.*	stoke
pascal-second	1000.*	centipoise
pascal-second	10.*	poise
poise	0.1*	pascal-second (Pa · s)
stoke	0.0001*	metre^2/second (m^2/s)
TEMPERATURE		
To Convert From	To	Use Formula
temperature Celsius, t_C	temperature Kelvin, t_K	$t_K = t_C + 273.15$
temperature Fahrenheit, t_F	temperature Kelvin, t_K	$t_K = (t_F + 459.67)/1.8$
temperature Celsius, t_C	temperature Fahrenheit, t_F	$t_F = 1.8\, t_C + 32$
temperature Fahrenheit, t_F	temperature Celsius, t_C	$t_C = (t_F - 32)/1.8$
temperature Kelvin, t_K	temperature Celsius, t_C	$t_C = t_K - 273.15$
temperature Kelvin, t_K	temperature Fahrenheit, t_F	$t_F = 1.8\, t_K - 459.67$
temperature Kelvin, t_K	temperature Rankine, t_R	$t_R = 9/5\, t_K$
temperature Rankine, t_R	temperature Kelvin, t_K	$t_K = 5/9\, t_R$

* Where an asterisk is shown, the figure is exact.

Index

Absorber, vibration, 63, 247, 287
Absorption, air, 56, 136
  air space, 167
  enclosure, 541
  factory spaces, 163
  loss, 57
  materials, 25, 146
  measurement of, 153, 154, 156
  noise level and, 117, 164, 185, 535
  normal incidence, 152, 156
  normal to random conversion, 156
  of sound in air, 56, 136
  panels, 148
  protection of materials, 167
  random incidence, 155, 156
  room, 140, 157, 162, 184, 185
  sound, 61
  table of, 148
Absorption coefficient, 16, 140
  air, 56, 136
  average of, 147
  brick, 148
  carpeting, 148
  cellulose fiber, 148
  concrete, 148
  concrete block, 148
  conversion of normal to random, 156
  definition of, 147
  effect of air space, 167
  effect of area, 148
  glass, 148
  glass fiber, 169
  in rooms, 185
  measurement of, 153, 154
  mineral fiber, 148
  normal incidence, 152
  random incidence, 155
Acceleration, 62, 63, 552
Acoustical Materials Association, 149
Acoustic calibration, 89
Acoustic impedance, 14, 153
Acoustic intensity, 23
Acoustic pressure, 57
Acoustic reflection, 13, 82, 410
Acoustic resistance, 57
Acoustic shielding, 31, 99, 150, 154, 161, 170, 177, 179
Age, effect on hearing, 39
Air, absorption, 136
  characteristic impedance, 14, 152

Air Conditioning and Refrigeration Institute, 110
Air Diffusion Council, 110
Air flow in mufflers and ducts, 404, 409, 418, 519, 528
Air-flow resistance, 149
Air jets, 429
Air mounts, 270
Air Moving and Conditioning Association, 110
Air space behind materials, 167
Air valve, 426, 525
Ambient noise, 28, 80
American Gear Manufacturers Association (AGMA), 100, 110
American National Standards Institute (ANSI), 99, 102, 110
American Society for Testing and Materials (ASTM), 99, 103, 110
American Society of Heating, Refrigeration, and Air Conditioning Engineers (ASHRAE), 99, 105, 110, 130
Amplitude, 48, 216, 225, 271, 320
Analyzer, 83, 84
Anechoic chamber, 26, 74, 117
Annoyance levels, 126
Anti-Friction Bearing Manufacturers Association, 110
Audible range, 57
Audiometer, 57
Audiometry, 57
A-weighting network, 38, 83, 87
A-weighted sound levels, calculation of, 123, 138, 514, 534
  definition of, 83, 85
  in buildings, 132
  in rooms, 132
  permissible, 121, 123, 132

Background sound levels, 28, 80
Balance quality, 296
Band pressure level, 57
Bandwidth, 86
Barriers, 31, 99, 150, 154, 161, 170, 177, 179
Base, isolation, 314
  motion, 271
Beam, 225, 228, 234, 237, 243
Bearings, 347, 353, 551
Bearing frequencies, 355

578

INDEX 579

Bearing noise, 348, 356
Beats, 57
Blade pass frequency, 395, 517
Blast furnace, 449
Blower noise, 386, 395, 514
Broad-band noise, 57
Burner, 440, 441, 458, 463, 465, 469, 470, 472, 540
B-weighted analyzer, 83

Cagi-pneurop test code, 108
Calibration, sound, 89
Cam, acceleration, 367
  follower, 367, 368
  jerk, 367
  mechanisms, 374
  motion, 366, 368
  noise, 361, 370, 550
  nomenclature, 357
  vibration, 361
Cancellation of sound, 78
Cauchy number, 489
Cavitation, 475
Center frequency, 60
CHABA, 123
Chain drive, 379
Choked flow, jets, 429
  valves, 426
Coincidence frequency, 33, 173
Combustion, amplification of noise, 443, 466
  driven oscillation, 441, 444
  noise, 439, 440, 465, 469, 470
  noise efficiency, 441
  oscillation, 440
  singing, 445, 447
  size effects, 461
  time lag, 446
Combustor, 451
Community noise, 133
Complex modulus, 213, 215, 218, 221, 241
Composite panels, 197
Continuous noise, 57
Continuous spectrum, 57
Control-room noise, 510
Critical damping, 63, 265
Critical frequency, 173
Cycle, 57, 63
Cycles per second, 5, 68

Damage risk for hearing, 120, 134
Damper, tuned, 247, 252
Damping, acoustic, 236
  air, 270
  beams, 237, 241, 243
  behavior, 224, 247
  coefficient, 265
  constrained layer, 240, 257
  coulomb, 63
  critical, 63, 265
  dry friction, 63
  elastomers, 212
  enamels, 230
  frequency effects, 231
  joint, 236
  loss factor, 216, 217, 226, 237
  materials, 207, 212
  metals, 222
  plates, 237, 241, 244
  ratio, 63, 265, 270
  stiffened panels, 239
  structures, 236
  temperature effects, 231
  unconstrained, 236, 256

Damping, vibration, 55, 56, 265
  viscoelastic, 215, 245
  viscous, 66
Dead room (see Anechoic chamber)
Decay time, 155
Decibel (dB), addition of, 26, 73
  A-weighted, 83, 85, 121, 132
  B-weighted, 83
  chart for combining, 27
  C-weighted, 83
  definition, 19, 58
  insertion loss, 175
  noise reduction, 159, 164
  reference quantities, 20, 68, 70
  scale, 19
  sound-power level, 20, 68, 73
  sound-pressure level, 19, 23
Deflection, spring, 45
  static, 45, 265, 272, 302
Degree of freedom, single, 208, 262, 270
Density, 174, 192
Diffusers, 432
Directivity pattern, 118
Direct sound, 117
Door, 192
Duct, attenuation, 194, 403, 517
  branch take-off, 402, 416
  elbows, 404, 409, 519
  end correction, 390, 410
  end reflection, 410
  lined, 191, 194, 204, 403, 409
  open end loss, 410
  outlet, 403
  sheet metal, 403
  transmission coefficient, 203
  transmission through walls, 433, 533
Dynamic magnification factor, 266, 268

Ear, 35, 36
Echo, 12, 58
Effective mass, 50, 553
Elastomer, 273
Elbows, 404, 409, 519
Electronic Industries Association, 111
Enamel, 230, 235
Enclosures, 31, 189, 191, 197, 510
End effects, 390, 410
Energy, 11, 44
Environmental Protection Agency (EPA), 100
Environment, measurement, 73
Equal loudness contours, 127
Euler number, 489
Exhaust noise, 429

Facings, 167
Fan, flow rate, 395
  inlet noise, 396, 397
  noise, 386, 395, 514
  noise attenuation, 399
  noise specification, 398, 507, 514
  pressure drop, 395
  sound-power level, 387, 389, 392, 507, 516, 540
  sound rating, 391
  specific sound power, 392
Far field, 91
Fatigue, sonic, 207
Fiber glass materials, 148
Flame, amplification of noise, 443, 466, 469
  singing, 445, 447
  size effects, 461
  time lag, 446

INDEX

Flanking, 158
Flow obstructions, acoustic power, 419
  bar, 420
  ring, 424
  rod, 420
  strip, 425
Flow resistance, 149
Fluctuating noise, 80, 122
Force, aerodynamics, 53, 300
  applied, 214
  bearings, 300
  cam, 365, 552
  electrical, 53, 300
  hydraulic, 53
  rolling elements, 53
  rotary unbalance, 297
  sliding elements, 53
  transmitted, 47
Forced oscillation, 63
Foundation flexibility, 276
Fourier coefficients, 372, 373
Fourier series, 6, 10, 371, 372
Free field, 58
Free oscillation, 63
Frequency, angular, 62
  bearing, 355
  blade pass, 395, 517
  circular, 62
  coincident, 498
  damped natural, 48, 65
  definition, 58
  double panel, 172
  narrow-band analysis, 75
  natural, 44, 47, 64, 314
  panel, 176
  pipe, 498, 501
  resonance, 64, 265, 279
  response of microphone, 83
  ring, 498
Fuel, 461
Function, cosinusoidal, 8
  harmonic, 10
  impulse, 550
  periodic, 6
  random, 6
  simple harmonic, 10
  sinusoidal, 6
  transient, 6
Furnace noise, 538

Gas sizing coefficient, 479
Gear noise, 207, 336
Gears, 329
Glass, absorption of, 148
  coincident frequency, 172
  double wall, 172
  fiber, 148, 169
  transmission loss, 170, 172
Gypsum wallboard, 171, 174, 192

Half-power bandwidth, 225
Harmonic component, 8, 75, 76, 372
Harmonic motion, simple, 6, 65, 266
Harmonic quantity, simple, 65
Harmonics, 8
Hearing, age effects, 39
  damage risk, 120, 134
  level, 58
  loss, 58
  threshold, 22, 38
  threshold level, 58
Hearing Aid Industry Conference, 111
Helmholtz resonator, 464

Hertz, 5, 68
Hydrodynamic noise, 475, 489, 490

IL (Insertion Loss), 175
ILG reference source, 94, 157, 184
Impact, 64, 299
Impedance, acoustic, 16
  air, 14, 152
  characteristic, 14
  measurement of, 153
  mechanical, 14
  mismatch, 15, 281
  specific acoustic, 14, 152
  spring, 282
  tube, 153
Impulse, gear tooth, 335
  measurement, 95
  noise, 59
Infrasound, 5
Insertion loss, 175
Institute of Electrical and Electronic Engineers (IEEE), 111
Instrumentation system, 72
Intensity, acoustic, 23
  definition, 12
  incident, 25
  level, 19, 25
  reference, 19
  reflected, 25
  related to power, 25
Interference pattern, 78
International Electrical Commission (IEC), 100, 107, 111
International Organization for Standardization (ISO), 100, 106, 111
Isolation, vibration, 49, 65, 262, 264, 301, 307, 314, 317, 321, 322
Isolators, 65, 270, 273, 274, 302, 305, 306, 310

Jerk, 64, 367
Jet engine, 207
Jet noise, 429
  efficiency, 430

Lagging of pipes, 190, 494
Lead, transmission loss of, 174, 190, 191, 192
Limp mass law, 169
Limp panels, 171
Lined ducts, 191, 194, 204, 403, 409
Lined elbows, 404
Lined enclosures, 191, 197
Liquid sizing coefficient, 479
Live room, 76, 388
Logarithmic scale, 19
Loss factor, 216, 217, 226, 237
Loudness, 37
Lubrication, 344, 349, 351, 352

Machinery foundations, 314
Mach number, 430, 478
Masonry block, 148, 170
Mass, common materials, 174
Mass law, limp, 169
Materials, absorption coefficient, 140, 147, 148
  characteristic impedance, 14, 152
  coincidence frequency, 33, 173
  critical frequency, 173
  damping, 55, 56, 66, 207, 212, 215, 224, 245, 247, 265
  density, 174
  flow resistance, 149
  loss factor, 150, 169

INDEX

Materials, porous, 167
  speed of sound, 174
  transmission loss, 150, 158, 169
Materials properties, air, 281
  alloys, 224
  Aquaplas, 227
  Blatchford, 222
  block masonry, 148, 170
  Borden, 228
  brick, 148, 174
  carpeting, 148
  cast iron, 281
  cellulose fiber, 148
  Chicago Vitreous Enamel, 235
  concrete, 174, 281
  concrete block, 148, 170
  copper, 224
  cork, 281
  Dow, 230
  drywall, 170, 171, 174, 192
  enamel, 230, 235
  foam, 191, 196
  glass, 148, 170, 172, 281
  glass fiber, 148, 169
  gypsum board, 171, 174, 192
  honeycomb, 196
  lead, 174, 190, 191, 192, 281
  Lord, 220, 221
  Lucite, 174
  manganese, 224
  mineral fiber, 148
  Mystik, 228
  neoprene, 305, 306, 307
  Paracril, 222, 223, 224, 225
  plaster, 148, 170
  plastic film, 167
  plexiglass, 174
  plywood, 148, 192, 196
  Polymer Blend, 229
  rubber, 281
  Sonoston, 234
  steel, 174, 192, 196, 281
  Stylgard, 230
  3M Company, 218, 219, 232
  timber, 174
  Uniroyal, 222, 223, 224, 225
  vinyl, 192
  vinyl floor tile, 231
  viscoelastic, 215, 245
  Vitron, 226
  wood, 170, 174, 281
Measurement, absorption coefficient, 152, 153, 157
  calibration, 89
  impedance, 153
  impulsive noise, 95
  outdoor, 79
  sound meter, 88
  sound-pressure level, 68, 73, 79, 101
  techniques, 87, 91
  transmission loss, 158
  windscreen, 81
Mechanical equipment room, 510
Mechanical impedance, 14
Mechanisms, 9, 377, 552
Membrane facings, 167
Meters, sound level, 80, 88, 122
Microbar, 59, 69
Microphone, definition, 58
  location, 91
  response, 82
  windscreen, 81
Mobility, 289
Mode, 228

Modulus, complex, 213
  Young's, 213, 217, 221, 243
Momentum flux, 475
Muffler, attenuation, 28
  burner, 464
  flow in, 418
  furnace, 545
  inline, 497
Multiple panels, 171

National Machine Builders Association, 112
NC-curves, 130, 132, 512
Near field, 91
Noise, air flow, 404, 409, 418, 519, 528
  ambient, 28, 80
  analysis, 8, 75, 76, 372
  barrier, 31, 99, 150, 154, 161, 170, 177, 179
  bearings, 398
  blast furnace, 449
  blower, 386, 395, 514
  cam, 361, 370, 550
  combustion, 439, 440, 465, 469, 470
  combustor, 451
  exposure, 121
  fan, 386, 395, 574
  flame, 445, 461, 497
  flow, 404, 409, 418, 519, 528
  furnace, 538
  gear, 207, 336
  heating and ventilating, 386
  impact, 64, 299
  jet, 429
  level, 20, 68
  mechanism, 377, 552
  meter, 80, 88, 122
  pink, 60
  pipe, 190, 482, 494, 498, 501
  radiation, 73
  random, 10, 62
  reflection, 12, 15, 80
  reverberation, 12, 93
  spectrum, 8, 75, 76, 372
  speed, 4, 5, 174, 448
  transmission, 40
  valve, 494
  white, 11, 62
Noise criteria, annoyance, 126
  cafeteria, 132
  classification, 114
  community noise, 133
  control rooms, 132
  damage-risk, 120, 134
  drafting room, 132
  electrical equipment room, 132
  laboratories, 132
  loudness, 126
  maintenance areas, 132
  manufacturing areas, 132
  mechanical equipment room, 510
  NC-curves, 130, 132, 512
  offices, 132
  OSHA, 121
  perceived noise level, 129
  PNC-curves, 133, 507, 512
  sones, 128
  speech interference, 125, 139
  tool crib, 132
  washrooms, 132
Noise effects on hearing, 120, 134
Noise reduction (NR), bearings, 347, 353, 551
  burner, 472, 540
  cam, 370, 550

INDEX

Noise reduction (NR), coefficients, 157
  definition of, 159, 164
  enclosures, 31, 189, 191, 197, 510
  furnace, 538
  isolation, 49, 65, 262, 264, 307, 314, 317, 321
  machinery, 39, 41, 42, 56, 434
  mechanical equipment room, 510
  mechanism, 552
  mufflers, 28, 497, 545
  pipes, 190, 494
  techniques, 39, 41, 42, 56, 494
  valve, 494
Noise sources, air flow, 404, 409, 418, 519, 528
  bearing, 348, 356
  blower, 386, 395, 514
  burner, 440, 441, 458, 463, 469, 470
  cam, 361, 370, 550
  combustor, 451
  directivity, 118
  enclosed, 31, 189, 191, 197, 510
  fan, 386, 395, 514
  far field of, 91
  flame, 445, 447, 461
  furnace, 538
  gear, 336
  impact, 64, 299
  in rooms, 117, 163, 164, 165, 535
  jet, 429
  mechanisms, 9, 377, 552
  near field of, 91
  pink, 60
  random, 10, 62
  reverberant field of, 164
  self noise of muffler, 418
  standard, 60, 94
  valve, 494
  white, 11, 62
Noise specification sheet, 145, 509
Nonconstrained damping layers, 236, 256
NRC (Noise reduction coefficient), 157

Octave, 60
Octave band, analyzer, 84, 86, 92
  definition, 60
  table of limits, 60
Office noise criteria, 132
Oil whirl, 352
One-third octave, 60, 84, 86
Oscillation, 60
OSHA Regulations, 67, 86, 121, 132, 551
Outdoor measurements, 79
Overall level, 83, 84, 86

Package fan attenuation units, 400
Pad-type isolators, 274, 282, 305, 306, 307
Panel absorbers, 148, 167, 168, 169
Panel damping, 239, 241, 244
Partial enclosures, 197
Particle velocity, 2, 3
Partitions (see also Walls)
  barrier, 31, 150, 158
  composite, 197, 245
  enclosures, 31, 189, 191, 197, 510
  multiple, 171
  transmission loss, 31, 158, 433, 513
Peak, level, 60
  response, 216
  sound level, 64, 299, 550
  sound pressure, 60
  valve, 60
Perceived noise level, 129
Perforated facings, 167

Period, 7, 60
Periodic, 60
Phon, 38, 127
Phon scale, 127
Pink noise, 60
Pipe, frequencies, 501
  wall attenuation, 482, 498
  wrapping, 190, 494
Pipe-wall wrapping, 190, 494
Pistonphone, 60
Pitch, 60
Plane wave, 5, 62
Plaster, 148, 170, 192
Plenum, 407, 522
Plug, valve, 476
PNC-curves, 133, 507, 512
Pneumatic isolator, 275, 308
PNL, 130
Poisson ratio, 212, 213
Power, 70 (see also Sound power)
Power Saw Manufacturers Association, 112
Prandtl number, 478
Preferred noise criteria, 133, 507, 512
Presbyacusia, 39, 60
Presbycusis, 39
Pressure, acoustic, 11, 20
  drop, 395
  level, 61
  peak, 22
  radiation, 57
  reference, 20, 68
  root mean square, 20
  sound, 20, 61
Propagation, in air, 2, 3
  in rooms, 117, 164, 535
  over barriers, 177
  plane waves, 5, 62
  speed, 4, 5, 174, 498
Protective facings, 167
PSIL, 124, 125, 139
Pure tone, 9, 20, 37

Quality, balance, 296

Radiation, sound, 73
Random, function, 6
  motion, 10
  noise, 62
  sound, 10
Rayleigh criterion, 445
Recommended practice, standard, 101
Reflection, 12, 80
Reflection coefficient, 15
Resilient materials, cork, 281
  equivalent springs, 51
  foam, 191, 196
  metal springs, 273, 282
  neoprene, 305, 306, 307
  rubber, 281
Resonance, 64, 279
Resonance amplification factor, 216
Response, harmonic, 8, 65, 75, 76, 372
  shock, 290
  sound meter, 85
  subharmonic, 65
Reverberation, 12, 93
Reverberation chamber, 76, 154, 388
Reynolds number, 489
Reynolds stress, 475
Ring frequency, 498, 501
Risk of hearing damage, 120, 134
Roller mount, 309
Room, anechoic, 26, 74, 117
  criteria for, 132

INDEX 583

Room, extended, 163, 165
  measurements in, 87, 91
  noise level, 117, 164, 535
  regular, 163
  reverberant, 76, 154, 388
  semi-anechoic, 77
  sound level and absorption, 166
  sound-pressure level, 117, 164, 535
Room constant, 118, 389
Room noise levels, 117, 132, 553
Root mean square (rms), 20, 21
Rotary unbalance, 52, 295
  grades, 298
Rotational, speed, 52
  unbalance, 52, 295
Rubber, mounts, 282, 305, 306, 307, 308

Sabin, 147, 154, 155
Sandwich panel, 197, 245
Scale, decibel, 19, 26, 58, 73
  logarithmic, 19
Semi-anechoic chamber, 77
Series, 9, 371, 373
Shear loss factor, 216, 217
Shielding, 31, 99, 150, 154, 161, 170, 177, 179, 190, 494
Shock, absorber, 65
  excitation, 289, 291
  isolation, 290
  isolator, 65, 290
  mechanical, 64
  periodic, 292
  pulse, 65
  reduction, 294
  response, 290
  spectrum, 65
  transmissibility, 290, 291
  velocity, 65
Signal, cosinusoidal, 8
  harmonic, 10
  impulse, 290, 550
  simple harmonic, 6, 266
  sinusoidal, 6
  transient, 6
Simple harmonic motion, 6, 266
Singing flame, 445, 447
Single degree of freedom, 208, 262, 270
Single wall, 31, 189, 191, 197, 510
Sinusoidal, function, 6
  motion, 9
  pressure, 18
  signal, 6
  wave, 10, 18
Snubber, 65
Society of Automotive Engineers (SAE), 100, 104, 112
Sone, 128
Sonic fatigue, 207
Sound, absorption, 126, 149
  fundamentals, 2
  level, 58
  meter, 80, 88, 122
  multiple sources, 166
  power, 11, 24, 34, 60, 70
  power level, 19, 23, 70
  pressure, 18
  pressure level, 20, 68
  propagation, 2
  radiation, 73
  radiation pattern, 75
  sources, 94, 166
Sound power, branch take-off, 402
  combustion, 442, 459

Sound power, definition, 11
  inside enclosures, 197
Sound-power level, air flow, 404, 409, 418, 519, 528
  bar in flow, 420
  blower, 387, 389, 392, 507, 516, 540
  combustion, 442
  conversion to sound-pressure level, 72, 117
  fan, 387, 389, 392, 507, 516, 540
  flame, 445, 461, 497
  furnace, 538
  jet, 429
  ring in flow, 424
  rod in flow, 420
  strip in flow, 425
Sound-pressure level, ambient (background level), 80
  combining, 73
  computation, 117, 535
  definition, 20, 68
  fluctuating, 80
  in rooms, 117, 164, 535
  typical source levels, 69
Sound source, 94 (see also Noise source)
Sound speed, 4, 5, 174, 448
Sound transmission, 99, 150, 159, 161, 170 (see also Transmission loss)
Sound transmission class (STC), 99, 159, 161, 170
Specification, new equipment, 506
Specific heat, 459, 478
Spectral analysis, 8, 75, 76, 372
Spectrum function, 411, 527
Speech communication, 125, 139
Speech interference, 125
Springs, adjustable, 304
  deflection, 302
  diameters of steel, 303
  equivalent, 49, 51
  free standing, 303
  overhead suspension, 304
  selection of, 310, 312
  stabilized, 303
  static deflection, 45, 265, 272, 302
  steel, 273, 282, 302, 303
  types, 314
Standards, definition, 99
  electrical equipment, 108
  engine powered equipment, 108
  heating, ventilating equipment, 109
  occupational safety, 67, 86, 109, 121, 551
  power transmission equipment, 109
  pneumatic equipment, 108
  selection of, 101
  table of references, 102
Standing wave, 18
Static deflection, 265, 272, 302
STC rating, 99, 159, 161, 170
Steady state, sound in rooms, 117, 163, 165, 535
  stiffness, 65
Strouhal number, 411, 467, 527
suite, transmission, 158
vibration, 47, 65
Stiffness, equivalent, 49, 51, 265
  isolator, 265
Strain, 215, 217, 229
Stress, 213, 217
Swirl burner, 467
Swirl number, 444

Third-octave band, 60, 86
  analyzer, 80

INDEX

Transmissibility, definition, 65
  efficiency, 263, 282
  force, 48, 263, 266
  motion, 263, 270
Transmission loss (TL), duct wall, 433
  mass law, 169
  materials, 150, 169
  measurement of, 151, 158
  muffler, 28, 497, 545
  panel, 31, 150, 158
  pipe wall, 190, 498
  sheet metal duct, 433
  sound transmission class, 99, 159, 161, 170
  wall, 148, 174, 192, 513
Transmission path, 40
Transmission suite, 158
Tuned absorber, 63, 247, 287
Tunnel burner, 454

Unbalance, grades, 296, 298
  quality grades, 296
  reciprocating, 298
  residual, 295
  rotary, 52, 295
  specific, 295
  translational, 52

Valve, aerodynamic noise, 477, 481
  air, 426, 525
  butterfly, 426, 525
  flow control, 426, 473, 477, 481, 525
  hydrodynamic noise, 489, 490
  liquid noise, 477
  mechanical noise, 473
  noise, 426, 473, 477, 481, 489, 490, 525
  plug, 476
  sound power, 426, 525
Velocity, 65
Velocity function, 411, 527
Vendor specification, 509
Ventilation systems, branch take-off, 402, 416
  criteria for, 105, 132
  duct attenuation, 194, 403, 517
  elbows, 404, 409, 519
  end correction, duct, 390, 410
  end reflection, 410
  fan, 386, 395, 514
  lined, 191, 194, 204, 403, 409
  obstructions in flow, 419, 420, 424, 425
  outlet, 403
  plenum, 407, 552
  sheet metal duct, 403
  wall transmission, 433, 533
Vibration, absorber, 63, 247, 287
  air mounts, 270
  amplitude, 47, 48, 216, 271, 320
  base, 314
  cam, 361

Vibration, chain drive, 380
  control, 209
  criteria, 296, 298
  gear, 335
  identification of, 318, 320, 379
  isolation, 49, 65, 262, 264, 301, 307, 314, 317, 321, 322
  isolator, 65
  meter, 66
  modes, 285, 286
  mounts, 301, 305, 307, 314
  resonant, 49
  shock excited, 289, 290, 291
  single degree of freedom, 208, 262, 270
  source, 294
  steady state, 65
  transient, 65
Viscoelastic material, 215, 245
Viscous damping, 66
Vortex, 456

Walls, absorption, 148, 152, 163, 167
  block, 148, 174, 281, 513
  brick, 148, 174
  composite, 197, 245
  concrete, 174, 281, 513
  door, 192
  double, 170
  glass, 148, 170, 172
  gypsum board, 171, 174, 192
  masonry, 174, 281, 513
  multiple, 170, 171
  plaster, 148, 170, 192
  plywood, 148, 170, 172
  steel stud, 192
  vinyl, 192
Walsh-Healey Act, 67
Wave, complex, 9
  condensative, 2
  incident, 15, 25
  in isolators, 279
  motion, 4
  plane, 5, 62
  propagation, 5, 18
  random, 10
  rarefactive, 4
  reflection, 15, 25
  simple, 9
  sound, 2
  speed, 4, 5, 174, 448
  spherical, 5, 34, 62
  standing, 17, 18
Waveform, 10
Wavelength, 4, 77, 448
Weber number, 489
White noise, 11, 62
Wind screen, microphone, 81
Wrappings, 190, 494

Young's modulus, 213, 217, 221, 243